MULTISTAGE SEPARATION PROCESSES

Fourth Edition

MULTISTAGE SEPARATION PROCESSES

Fourth Edition

FOUAD M. KHOURY

CRC Press
Taylor & Francis Group
Boca Raton London New York

CRC Press is an imprint of the
Taylor & Francis Group, an **informa** business

First published in paperback 2024

First published 2015 by CRC Press
2385 NW Executive Center Drive, Suite 320, Boca Raton FL 33431

and by CRC Press
4 Park Square, Milton Park, Abingdon, Oxon, OX14 4RN

CRC Press is an imprint of Taylor & Francis Group, LLC

© 2015, 2024 Taylor & Francis Group, LLC

Library of Congress Cataloging-in-Publication Data

Khoury, Fouad M., author.
  Multistage separation processes / Fouad M. Khoury. -- Fourth edition.
     pages cm
  Includes bibliographical references and index.
  ISBN 978-1-4822-3054-3 (hardcover : alk. paper) 1. Separation (Technology) 2.
  Vapor-liquid equilibrium. I. Title.

  TP156.S45K48 2015
  660'.2842--dc23   2014033357

ISBN: 978-1-4822-3054-3 (hbk)
ISBN: 978-1-03-291782-5 (pbk)
ISBN: 978-0-429-17251-9 (ebk)

DOI: 10.1201/b17542

Visit the Taylor & Francis Web site at
http://www.taylorandfrancis.com

and the CRC Press Web site at
http://www.crcpress.com

Contents

Preface

Multistage separation processes and other related processes constitute the building blocks for the chemical, petrochemical, and petroleum industries. These industries yield important products as common as gasoline and plastics and as specialized as medical-grade pharmaceuticals.

This book is aimed at providing the theoretical foundation and understanding the development, evaluation, design, and optimization steps of these processes. It is distinguished by its emphasis on starting with theoretical models, their role in computer simulation, followed by practical applications. The importance of relating fundamental concepts to intuitive understanding of the processes is also highlighted. A generous number of examples are provided in a wide variety of applications to relate theory to actual results and to demonstrate the performance of processes under varying conditions. Exercise problems are given at the end of chapters. These problems can be used to study the material and explore its application to various situations. The book is of value as a textbook for graduate and undergraduate students of engineering and process design and as a reference for practicing engineers in the process industry.

Improved accuracy in predicting thermodynamic and physical properties has occurred simultaneously with major advances in the development of computation techniques for solving complex process equation systems. The result has been the emergence of a variety of computer simulation programs for accurate and efficient prediction of a variety of complex processes. This has provided engineers with valuable tools that can help them make more reliable qualitative as well as quantitative decisions in plant design and operation. Frequently, however, effective use of such programs has been hampered by lack of understanding of fundamentals and incorrect selection of property prediction and simulation techniques. Improper use of simulators can be costly in time and money, which tends to defeat the purpose of computer-aided engineering. These problems are addressed in this book, and a strategy is pursued that decouples the discussion of conceptual analysis of a process and the computation techniques.

Along with rigorous mathematical methods, which are presented with a good degree of detail, attention is given throughout the book to keeping practical interpretation of the model in focus, emphasizing intuitive understanding. Graphical techniques and shortcut methods are applied wherever possible to gain a handle on evaluating performance trends, limitations, and bottlenecks. Also included are industrial practice heuristics about what ranges of operating variables will work. The student reader of this book should come away with an enhanced intuitive grasp of the material as well as a thorough understanding of the computation techniques.

The book may be used for a methodical study of the subject or as a reference for solving day-to-day problems. It follows a logical flow of ideas within each chapter and from one chapter to the next; yet each chapter is quite self-contained for quick reference. The discussion in the book starts with fundamental principles, prediction of properties, the equilibrium stage, and moves on to the different types of multistage and complex multistage and multicolumn processes, optimization, dynamics and process control, batch distillation, fluid–solid operations, and membrane separation systems.

Earlier chapters use simplified and binary models to analyze in a very informative manner some fundamentals such as the effect of reflux ratio and feed tray location, and to delineate the differences between absorption/stripping and distillation. Following chapters concentrate on specific areas such as complex distillation, with detailed analyses of various features such as pumparounds and side-strippers, and when they should be used. Also discussed are azeotropic, extractive, and three-phase distillation operations, multi-component liquid–liquid and supercritical extraction, and reactive multistage separation. The applications are clearly explained with many practical examples.

Shortcut computation methods, including modular techniques for online real-time applications, are discussed, followed by a discourse on the major rigorous algorithms in use for solving multi-component separations. The application of these methods is detailed for the various types of multistage separation processes discussed earlier. The models are also expanded to cover column dynamics.

An understanding of column hydraulics in both trayed and packed columns is essential for a complete performance analysis and design of such devices. The reader will find instructional coverage of these topics, as well as rate-based methods and tray efficiency, in subsequent chapters.

The nature of multistage separation processes presents numerous challenges to their control and optimization due to factors such as dynamic interactions and response lag. The techniques used for dealing with these problems are explored and analyzed in a chapter dedicated to this topic.

In a departure from continuous steady-state processes that characterize a greater part of the book, the subject of batch distillation is discussed. This process, which is important for the separation of pharmaceuticals and specialty chemicals, is presented, including shortcut and rigorous computation methods, along with various optimization techniques.

Another separation or purification process involves fluid–solid operations that are the basis for adsorption. Several modes of operation exist and are discussed in detail with examples.

The field of membrane separations is radically different from processes based on vapor–liquid or fluid–solid operations. This separation process is based on differences in mass transfer and permeation rates, rather than phase equilibrium conditions. Nevertheless, membrane separations share the same goal as the more traditional separation processes: the separation and purification of products. The principles of multi-component membrane separation are discussed for membrane modules in various flow patterns. Several applications are considered, including purification, dialysis, and reverse osmosis.

Many application exercise problems are included that expound on the material throughout the book and can serve both as teaching material and as an applications-oriented extension of the book. The problems cover three major aspects of the learning process: theory and derivation of model equations, engineering and problem-solving cases, and numerical and graphical exercises. The numerical problems require an algorithm definition and computations that may be done manually or with a spreadsheet program. For computer-oriented courses, these problems provide excellent material for program writing exercises.

Fouad M. Khoury

Author

Fouad M. Khoury, PhD, PE, is a specialist in multistage separation processes, their modeling, control, and optimization. He is a registered professional engineer in Texas and is a member of the American Institute of Chemical Engineers. He received his PhD in chemical engineering from Rice University and is the author of numerous articles on multistage separation processes, thermodynamics, and transport phenomena. He authored the book *Predicting the Performance of Multistage Separation Processes*, 1st edition (Gulf Publishing Company, 1995) and 2nd edition (CRC Press, 2000). He taught at Rice University and is currently teaching graduate and undergraduate courses in unit operations, advanced unit operations, and thermodynamics at the University of Houston, and is active as a consultant in the industry.

1 Thermodynamics and Phase Equilibria

The phase separation processes discussed in this book involve interactions between vapor and liquid phases, or between two liquid phases, or between a vapor phase and two liquid phases. The thermodynamic principles that govern these interactions are introduced in this chapter. Since this chapter is not intended as a full treatise on thermodynamics, only those aspects of the subject that have a direct bearing on phase separation processes are covered. To this end, theory is developed from the basic principles and carried through to the formulation of practical methods for calculating relevant thermodynamic properties, such as fugacity and enthalpy. These properties are essential for carrying out heat and material balance calculations in the separation processes described in this book.

When theoretical principles are applied to solve practical problems, various conditions of complexity are encountered. The general approach is to apply the theory to "ideal" systems and then to account for nonidealities by developing models such as equations of state and activity coefficient equations. Models are judged on their consistency with thermodynamic principles and their accuracy in representing actual data. One model may be appropriate for a given system but totally inadequate for another. Hence, the importance of properly selecting a thermodynamic model when attempting to deal with a given separation problem cannot be overemphasized.

1.1 THERMODYNAMIC FUNDAMENTALS

Thermodynamics is a science that relates properties of substances, such as their internal energy, to measurable quantities, such as their temperature, pressure, density, and composition. Thermodynamics also deals with the transformation of energy from one form to another, such as the transformation of internal energy to useful work. In this regard, thermodynamics is more general than mechanics in its formulation of the law of conservation of energy. The energy forms that mechanics is concerned with are entirely convertible to work. Thermodynamics, on the other hand, considers the conversion of thermal energy to work, recognizing that only a fraction of energy is convertible.

The principles that form the foundation of thermodynamics are embodied in several laws referred to as the laws of thermodynamics. In addition, thermodynamic functions, which interrelate the various properties of a system, are derived on the basis of these laws. A system refers to a part of space under consideration through whose boundaries energy in its different forms, as well as mass, may be transferred.

Within the context of its application to solving practical problems, thermodynamics is primarily concerned with systems at equilibrium. From an observational viewpoint, a system is at equilibrium if its properties do not change with time when it is isolated from its surroundings. The concept of equilibrium is a unifying principle that determines energy–work relationships as well as phase relationships.

The principles developed in Section 1.1 are fundamental in the sense that the system considered is not limited to any particular fluid type.

1.1.1 LAWS OF THERMODYNAMICS

The first law of thermodynamics is a formulation of the principle of conservation of energy. It states that the increase in the internal energy of a system equals the heat absorbed by the system from its surroundings minus the work done by the system on its surroundings. For infinitesimal changes, the first law is expressed mathematically by the equation

$$dU = dQ - dW \qquad (1.1)$$

where dU is the increase in internal energy, dQ is the heat absorbed, and dW is the work done by the system. The dimensional units of U, Q, and W are energy units.

The internal energy, U, is a basic property that represents the energy stored in a system. Changes in the internal energy of a system are related to work done by or on the system and to heat transferred in or out of it. In an adiabatic process, for instance, where no heat crosses the system boundaries, the work done by the system equals the change in its internal energy. This follows from the first law by setting $dQ = 0$. In a process where no work is done by or on the system, $dW = 0$, and the change in internal energy equals the heat absorbed or rejected by the system. Heat is considered positive when absorbed by the system, and work is considered positive when done by the system on its surroundings.

The second law of thermodynamics relates to the availability of energy in a system for conversion to useful work. In order for a system to perform work, it must have the capacity for spontaneous change toward equilibrium. For instance, a system comprising a hot subsystem and a colder subsystem is capable of performing work as heat passes from the hot to the cold subsystem. Part of the heat is converted to work, while the rest is rejected to the cold subsystem.

1.1.1.1 Carnot Engine

An example of a process that can deliver work by absorbing heat from a hot reservoir and rejecting heat to a cold reservoir is the Carnot engine. This is an idealized model consisting of a sequence of processes, each of which is assumed to be reversible. A reversible process is one that can be reversed by an infinitesimal change in the external conditions. For instance, in order to compress a gas reversibly, the external pressure at any moment should be $P + \Delta P$, where P is the gas pressure at that moment and ΔP is a small pressure increment. The reversible compression can be changed to a reversible expansion by changing the external pressure to $P - \Delta P$. A reversible process consists of steps in which the system is at equilibrium. In a reversible process, there are no losses due to friction or other factors.

Assume that the system used to carry out the Carnot cycle is an amount of ideal gas contained in a cylinder fitted with a frictionless piston. The concept of an ideal gas is introduced in Section 1.2. Of consequence at this point is the premise that for an ideal gas the internal energy, U, is a function of temperature only. The Carnot cycle consists of reversible isothermal and adiabatic processes. An isothermal process is one in which the system temperature is kept constant. An adiabatic process requires that no heat be transferred between the system and its surroundings. The steps are as follows:

1. Reversible adiabatic compression in which the gas temperature changes from T_1, the temperature of the cold reservoir, to T_2, the temperature of the hot reservoir. Since this is an adiabatic process, $dQ = 0$, and, from the first law, $-dW = dU$. The work done on the gas in this step is, therefore, $-W_{12} = U_2 - U_1$, where U_1 and U_2 are the values of the internal energy at temperatures T_1 and T_2, respectively.
2. Reversible isothermal expansion at temperature T_2, in which an amount of heat, Q_2, is absorbed by the gas from the hot reservoir. For an ideal gas, $dU = 0$ at constant temperature. Therefore, the work done by the gas in this step is $W_2 = Q_2$.
3. Reversible adiabatic expansion in which the gas temperature changes from T_2 to T_1. As in Step 1, the work done by the gas is $-W_{21} = U_1 - U_2$.
4. Reversible isothermal compression to the original state in which an amount of heat, Q_1, is rejected to the cold reservoir at constant temperature T_1. The work done on the system is $W_1 = -Q_1$.

The net work done by the system is the sum of the work associated with these steps:

$$W = U_2 - U_1 + U_1 - U_2 + Q_2 - Q_1$$
$$= Q_2 - Q_1$$

The Carnot efficiency is defined as

$$\eta = \frac{W}{Q_2} = \frac{Q_2 - Q_1}{Q_2}$$

The significance of the reversibility of the above processes is that at the end of the cycle the system is brought back to its starting point with no losses incurred due to friction or other causes. Thus, as the system temperature is restored to its starting point T_1, its pressure is also restored to its starting level of P_1. If the processes are not reversible, additional work would have to be done on the system to bring the pressure back to P_1. The net work would be less than $Q_2 - Q_1$, and the efficiency would be lower than the Carnot efficiency.

It can be shown that the Carnot efficiency may also be expressed as

$$\eta = \frac{T_2 - T_1}{T_2}$$

where the temperatures are on the absolute scale. This is the maximum efficiency attainable by an engine operating between a hot and a cold reservoir. Although developed for an ideal gas model, the efficiency of an engine operating with any medium between temperatures T_1 and T_2 will never exceed the above value.

1.1.1.2   Entropy

The two equivalent expressions of the efficiency imply that

$$\frac{Q_1}{T_1} = -\frac{Q_2}{T_2} \quad \text{or} \quad \frac{Q_1}{T_1} + \frac{Q_2}{T_2} = 0$$

The above equation states that the sum of the Q/T ratio along a Carnot cycle is zero. In general, any closed cycle, starting at some point and moving along reversible paths and returning to the starting point, can be represented by many small isothermal and adiabatic steps. The heat transferred in the adiabatic steps is, by definition, zero. For all the isotherms contained in the loop, the summation of the heat absorbed in each isotherm divided by its absolute temperature is zero:

$$\sum_i (Q_i/T_i) = 0$$

As the isotherms become infinitely small, the summation may be written as an integral over the closed reversible cycle:

$$\oint dQ/T = 0$$

If any two points A and B are chosen along the reversible cycle, the cycle may be broken into two reversible paths: 1 and 2. The closed cycle integral may be written as the sum of two integrals:

$$\oint dQ/T = \int_A^B \left(\frac{dQ}{T}\right)_1 + \int_B^A \left(\frac{dQ}{T}\right)_2 = 0$$

Subscripts 1 and 2 designate the integration paths. Reversing the integration limits and sign of the second integral, the above equation is rewritten as

$$\int_A^B \left(\frac{dQ}{T}\right)_1 = \int_A^B \left(\frac{dQ}{T}\right)_2$$

Thus, the value of the integral of dQ/T from point A to point B is independent of the path between A and B. A function S, the entropy, is now defined such that the change in its value from point A to point B is given by

$$S_B - S_A = \int_A^B \frac{dQ}{T}$$

Since the integral on the right-hand side is independent of the path as long as it is reversible, the change in entropy is independent of the path, and the entropy itself is a function of the thermodynamic coordinates of the system, such as its temperature and pressure. The above equation is next written for an infinitesimal change in entropy:

$$dS = dQ/T \tag{1.2}$$

where dQ is transferred reversibly. Equation 1.2 is a consequence of the second law of thermodynamics and is thus considered the mathematical formulation of this law.

1.1.2 Thermodynamic Functions

The first and second laws of thermodynamics are the basis on which thermodynamic relationships are derived. The two laws, represented by Equations 1.1 and 1.2, are combined in the following statement:

$$dU = TdS - dW$$

The work done by a system as a result of an infinitesimal volume change, dV, against a pressure, P, is $dW = PdV$. If this is the only form of work, the above equation may be written as

$$dU = TdS - PdV \tag{1.3}$$

The two expressions above for dU apply only to closed systems. A closed system is a fixed-mass body that cannot exchange matter with its surroundings, although it may exchange energy in the form of heat and work.

It was shown that the entropy, S, of a system at equilibrium is a function only of its thermodynamic coordinates such as its temperature and pressure. Such properties are said to be functions of state. The internal energy, U, is also a function of state. The internal energy and entropy, along with the temperature, pressure, and volume, are all that are needed to describe the thermodynamic state of a system. Additional functions are defined in terms of these five properties to represent other properties that might have practical significance for various applications. These properties, also the functions of state, are defined as follows:

Enthalpy $\quad\quad\quad\quad\quad$ $H = U + PV$
Helmholtz function $\quad$ $A = U - TS$

Gibbs free energy $G = U + PV - TS$
$$= H - TS$$
$$= A + PV$$

The differential of the free energy is given by

$$dG = dU + PdV + VdP - TdS - SdT$$

Combining this with Equation 1.3 gives

$$dG = -SdT + VdP \qquad (1.4)$$

It should be noted that this equation was derived for a closed system. Since such a system cannot exchange mass with its surroundings, its composition is fixed. Under these circumstances, the free energy is a function of only two variables, T and P, and can therefore be expressed as a total differential as follows:

$$dG = \left(\frac{\partial G}{\partial T}\right)_P dT + \left(\frac{\partial G}{\partial P}\right)_T dP$$

where subscripts P and T indicate that the partial derivatives are taken at constant P and T, respectively.

If the system is a homogeneous phase with a variable composition, its free energy is a function of temperature, pressure, and composition. If the system is a solution containing C components with n_1 moles of component 1, n_2 moles of component 2, and so on, then G is a function of $T, P, n_1, n_2, \ldots, n_C$, and the total differential of G is

$$dG = \left(\frac{\partial G}{\partial T}\right)_{P,n_i} dT + \left(\frac{\partial G}{\partial P}\right)_{T,n_i} dP + \sum_{i=1}^{C} \left(\frac{\partial G}{\partial n_i}\right)_{T,P,n_j(j \neq i)} dn_i \qquad (1.5)$$

The partial derivatives with respect to temperature and pressure are carried out at constant n_i for all the components; that is, at constant composition. For each term within the summation sign, the partial derivative is with respect to the number of moles of one component, keeping constant the number of moles of all the other components as well as the temperature and pressure.

Equation 1.4 is now applied to a fixed composition system to evaluate the partial derivatives of G at constant T and P:

$$\left(\frac{\partial G}{\partial T}\right)_{P,n_i} = -S$$

$$\left(\frac{\partial G}{\partial P}\right)_{T,n_i} = V$$

The partial derivative of G with respect to the number of moles of a component i is defined as μ_i, the chemical potential of that component in the phase under consideration:

$$\mu_i = \left(\frac{\partial G}{\partial n_i}\right)_{T,P,n_j(j\neq i)}$$

Equation 1.5 may thus be written in the form:

$$dG = -SdT + VdP + \sum \mu_i dn_i \qquad (1.6)$$

At constant temperature and pressure, Equation 1.6 reduces to

$$dG = \sum \mu_i dn_i$$

where dG is the differential free energy that results from mixing differential amounts of each one of the components to form a differential amount of solution. The chemical potential of each component is its contribution to the solution free energy, and is, therefore, the partial molar free energy. If many differential amounts of solution with identical temperature, pressure, and composition are mixed together, the total free energy is a simple summation of the differentials. Thus,

$$G = \sum \mu_i n_i$$

The total differential of G, resulting from the variation of either μ_i or n_i, is then

$$dG = \sum \mu_i dn_i + \sum n_i d\mu_i$$

This differential is equated to the expression given by Equation 1.6 to show that

$$-SdT + VdP = \sum n_i d\mu_i \qquad (1.7)$$

This is the Gibbs–Duhem equation, which relates the variation in temperature, pressure, and chemical potentials of the C components in the solution. Of these $C + 2$ variables, only $C + 1$ can vary independently. The Gibbs–Duhem equation has many applications, one of which is providing the basis for developing phase equilibrium relationships.

1.1.3 CONDITIONS FOR EQUILIBRIUM

A system is considered at equilibrium when it has no tendency to move away from its existing conditions. Conversely, a system not at equilibrium tends to undergo spontaneous change toward equilibrium. As a heterogeneous non-equilibrium system moves toward equilibrium, heat flows across regions within the system to equalize their temperatures, work may be generated as the pressures are equalized, and component concentrations tend to vary so as to cause the chemical potential of each component to be equal in all the regions (or phases). These processes can occur concurrently and will, in general, interact with each other.

One theoretical criterion for equilibrium is the expected change in the entropy of a system. Equation 1.2 states that

$$dS = \frac{dQ}{T}$$

for a reversible process. If the process is also adiabatic, $dQ = 0$ and, therefore, $dS = 0$. In an irreversible, or spontaneous, process, $dS > 0$ even if $dQ = 0$. In general, the second law is stated as

$$dS \geq \frac{dQ}{T} \tag{1.8}$$

where the equality sign applies to reversible processes and the inequality sign applies to irreversible processes. Thus, a system is at equilibrium if its entropy is at a maximum.

This fundamental criterion of equilibrium may be applied to practical situations such as determining phase equilibrium conditions. The relationship 1.8 is combined with the first law to arrive at the following expression:

$$dG \leq - SdT + VdP$$

It follows that for a closed system at constant T and P, $dG \leq 0$.

This statement dictates that spontaneous changes within a system at constant temperature and pressure tend to lower the free energy of the system. At equilibrium and at constant temperature and pressure, $dG = 0$.

The system may be heterogeneous, consisting of a number of homogeneous phases. Each phase is a solution containing a number of components. Assume, for simplicity, that only two phases, α and β, make up the system. Referring to Equation 1.6, a change in the free energy of the system at constant T and P is written as

$$dG = \left(\sum \mu_i dn_i\right)^\alpha + \left(\sum \mu_i dn_i\right)^\beta$$

The molecules of each component will migrate between the phases so as to minimize the total free energy. At equilibrium, $dG = 0$ and

$$\left(\sum \mu_i dn_i\right)^\alpha + \left(\sum \mu_i dn_i\right)^\beta = 0$$

Assuming that no chemical reactions take place, this relationship must apply to each component independently. Therefore,

$$(\mu_i dn_i)^\alpha + (\mu_i dn_i)^\beta = 0$$

Also, in the absence of chemical reactions, a decrease of one mole of component i in phase α causes an increase of one mole of the same component in phase β. Thus,

$$(dn_i)^\alpha = -(dn_i)^\beta$$

Hence, the condition for phase equilibrium, along with the requirement of uniformity of temperature and pressure, is the equality of the chemical potential for each component in both phases:

$$(\mu_i)^\alpha = (\mu_i)^\beta \tag{1.9}$$

EXAMPLE 1.1: ENTROPY CHANGE FOR AN IDEAL GAS

Calculate the change in entropy of 1 kmol of an ideal gas going from 100 kPa, 25°C to 200 kPa, 50°C. The heat capacity of this gas is $C_p = (7/2)R$.

SOLUTION

From the definition of enthalpy, for a differential change,

$$dH = dU + PdV + VdP$$

Substituting into Equation 1.3 and rearranging,

$$dS = dH/T - VdP/T$$

For an ideal gas,

$$dH = C_p dT, \quad V/T = R/P$$

Therefore, for an ideal gas,

$$dS = C_p dT/T - RdP/P$$
$$\Delta S = C_p \ln(T_2/T_1) - R\ln(P_2/P_1)$$
$$= \frac{7}{2}R\ln\frac{50 + 273.15}{25 + 273.15} - R\ln\frac{200}{100} = -0.411R = -(0.411)(8.314) = -3.417 \text{ kJ/kmol.K}$$

1.2   PVT BEHAVIOR OF FLUIDS

The thermodynamic principles discussed in Section 1.1 are rigorous and general. The "system" for which thermodynamic functions were derived need not be of any particular substance. The only requirement is that it satisfies the definition of a system in the thermodynamic context. In order to apply these principles to a given fluid for calculating properties such as enthalpy, free energy, and so on, the fluid behavior must be known.

The fluid behavior is commonly represented by its pressure–volume–temperature (or *PVT*) relationship, expressed in general as

$$f(P, V, T) = 0$$

This relationship might be available in the form of experimental data, or it could be represented by a model. Models are usually based on experimental data, but they also possess predictive capabilities. That is, they are expected not only to reproduce the correlated data, but also to generate data over reasonable ranges of conditions. Although many *PVT* models are semi-empirical, some are based on theoretical principles such as molecular thermodynamics and statistical thermodynamics. No single *PVT* correlation exists that can accurately predict all properties for diverse substances over wide ranges of temperature, pressure, density, and composition. Nevertheless, a number of models have demonstrated their usefulness for many applications.

Once the *PVT* relationship has been proven satisfactory for representing a fluid, it may be used, together with the thermodynamic functions, to derive expressions for other properties such as vapor pressure, vapor–liquid equilibrium (VLE) relationships, enthalpy departure from ideal gas behavior, and so on.

1.2.1 IDEAL GAS

Perhaps one of the earliest attempts at representing fluid properties centered around the concept of the ideal gas. Experiments on gases at low pressures and densities had led to the following observations: at a given temperature, the volume of a gas is inversely proportional to its absolute pressure and, at a given pressure, the volume of a gas is directly proportional to its temperature, if the latter is measured on an appropriate scale. Later work showed that this scale coincides with the absolute temperature scale associated with the Carnot engine efficiency (Section 1.1.1). The two observations were combined to form the ideal gas equation of state. An equation of state is a fluid behavior model that relates the temperature, pressure, and volume of the fluid in an equation form. The ideal gas equation of state takes the form

$$PV = nRT \qquad\qquad (1.10)$$

where V is the volume of n moles of gas, P is its absolute pressure, and T is its absolute temperature. The proportionality constant, R, is the universal gas constant. Another characteristic of an ideal gas, which was discussed with the Carnot engine (Section 1.1.1), states that the internal energy of an ideal gas is a function of temperature only and is independent of pressure. This aspect of ideal gases is discussed further in Section 1.4 on enthalpy.

Certain ideal gas characteristics follow from the definition of the ideal gas. For instance, the molar volumes of all ideal gases are the same at the same temperature and pressure because R is a universal constant. Also, it follows that the volume of a mixture of ideal gases at a given temperature and pressure is equal to the sum of the volumes of the individual gases at the same temperature and pressure. From the molecular standpoint, an ideal gas consists of molecules, each of which occupies zero volume and between which no forces of attraction or repulsion exist.

Real gases approach ideal gas behavior at low pressures. Deviations from ideality increase at higher pressures and lower temperatures, that is, as the density goes up. The ideal gas equation cannot predict the transition from gas to liquid since, according to the equation, the volume at a fixed pressure would decrease continuously and proportionately to the absolute temperature as the temperature is decreased. This, of course, is not what is observed in reality.

1.2.2 REAL FLUIDS

The ideal gas equation of state cannot describe real fluids in most situations because the fluid molecules themselves occupy a finite volume and because they exert forces of attraction and repulsion on each other. As the gas is cooled, and assuming its pressure is below the critical point, a temperature is reached where the intermolecular interactions result in a transition from the gas phase to the liquid phase. The ultimate fluid model would be one that could describe this transition as well as the fluid behavior over the entire range of temperature and pressure. Such a model would also be capable of representing mixtures as well as pure components.

Numerous attempts have been made to develop fluid models on the basis of molecular thermodynamics, taking into account the intermolecular forces. It is beyond the scope of this book to review these theories, and, in any case, the theoretical models are not necessarily the ones that are most widely used. The success of a model rests on its ability to represent real fluids. The principle of corresponding states is another approach that provides the foundation for some of these models. A number of models, or equations of state, that have proven their practical usefulness for phase equilibrium and enthalpy departure calculations are presented in this section.

1.2.2.1 Qualitative PVT Behavior of Pure Substances

The phase behavior of a pure substance may be depicted schematically on a pressure–temperature diagram as shown in Figure 1.1. The curve OC, the vapor pressure curve, separates the vapor and liquid phases. At any point on this curve, the two phases can coexist at equilibrium, both phases having the same temperature and pressure. Phase transition takes place as the curve is crossed along any path. Figure 1.1 shows two possible paths: at constant pressure (path AB) and at constant temperature (path DE). At the critical point, C, the properties of the two phases are indistinguishable and no phase transition takes place. In the entire region above the critical temperature or above the critical pressure, only one phase can exist.

The *PVT* behavior of a pure substance may also be described on a pressure–volume diagram, as shown in Figure 1.2. The variation in volume with pressure at various fixed temperatures is represented by the isotherms. If the temperature of the isotherm is above the critical, the pressure decreases continuously as the volume increases and no phase change takes place. The critical temperature isotherm is also continuous but has an inflection point at the critical pressure. On sub-critical

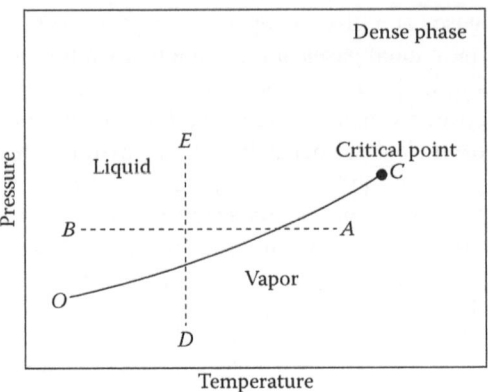

FIGURE 1.1   *P–T* diagram for a pure substance.

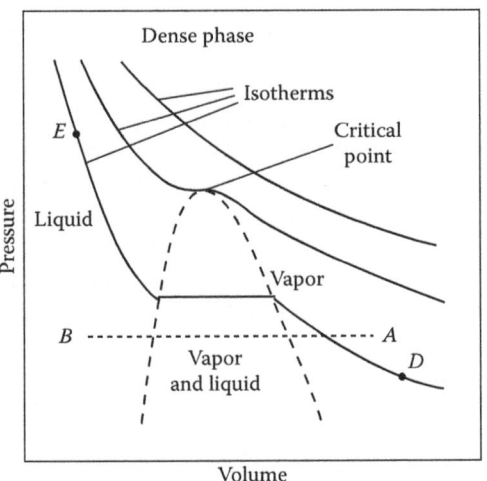

FIGURE 1.2   *P–V* diagram for a pure substance.

isotherms, the pressure of a liquid drops steeply with small increases in the volume until the liquid starts to vaporize. At this point the pressure remains constant as the total volume increases, as long as both vapor and liquid coexist at equilibrium. When all the liquid has vaporized, the pressure starts falling again as the volume is further increased. The constant pressure and constant temperature processes described by paths *AB* and *DE* in Figure 1.1 are shown again in Figure 1.2.

1.2.3   PRINCIPLE OF CORRESPONDING STATES

A qualitative observation of *PVT* behavior of pure substances indicates continuity in the isotherm at the critical point on a *PV* diagram. The existence of an inflection point on the critical isotherm at the critical pressure implies that the first and second derivatives of the pressure with respect to the volume are equal to zero at the critical point:

$$\left(\frac{\partial P}{\partial V}\right)_{T_C} = \left(\frac{\partial^2 P}{\partial V^2}\right)_{T_C} = 0 \tag{1.11}$$

These mathematical conditions at the critical point may be applied to various equations of state to determine their parameters in terms of the critical constants. The equations themselves may then be written in terms of the critical constants which are characteristic of each substance. The temperature, pressure, and volume may be replaced in the equation of state by the reduced properties, defined as

$$T_r = T/T_C; \quad P_r = P/P_C; \quad V_r = V/V_C$$

An equation of state written in terms of the reduced properties is a generalized equation that could be applied to any substance. It follows that, if two substances are at the same reduced temperature and pressure, they would have the same reduced volume. This is the concept of the principle of corresponding states, which may be used to correlate *PVT* properties of similar components.

An equation of state based on the principle of corresponding states as described above takes the form

$$f(P_r, T_r, V_r) = 0$$

Once this function is determined, it could be applied to any substance, provided its critical constants P_C, T_C, and V_C are known. One way of applying this principle is to choose a reference substance for which accurate *PVT* data are available. The properties of other substances are then related to it, based on the assumption of comparable reduced properties. This straightforward application of the principle is valid for components having similar chemical structure. In order to broaden its applicability to disparate substances, additional characterizing parameters have been introduced, such as shape factors, the acentric factor, and the critical compressibility factor. Another difficulty that must be overcome before the principle of corresponding states can successfully be applied to real fluids is the handling of mixtures. The problem concerns the definitions of P_C, T_C, and V_C for a mixture. It is evident that mixing rules of some sort need to be formulated. One method that is commonly used follows the Kay's rules (Kay, 1936), which define mixture pseudocritical constants in terms of constituent component critical constants:

$$P_{PC} = \sum_i Y_i P_{Ci}$$

$$T_{PC} = \sum_i Y_i T_{Ci}$$

Ideas derived from the principle of corresponding states have been incorporated into the development of equations of state, some of which are discussed next.

1.2.4   EQUATIONS OF STATE

New equations of state are constantly being published, and it is not the intention here to present a complete overview of these equations. The ones discussed here were selected because of their widespread use or, in some instances, because they possess some historical or theoretical significance.

1.2.4.1   van der Waals Equation

This is one of the oldest and most famous equations and one that was based on theoretical reasoning. It represents one of the earlier attempts in representing both the vapor phase and the liquid phase with the same equation although its success is mainly limited to the vapor phase. It is still used for the vapor phase portion of certain phase equilibrium calculations. The equation is given as

$$(P + a/V^2)\,(V - b) = RT \tag{1.12}$$

It is a logical requirement that equations of state approach the ideal gas equation at the limit of low pressures. As the pressure decreases, the volume increases so that at very low pressures $a/V^2 \ll P$ and $\ll V$. If these terms are dropped from the van der Waals equation, it reduces to the ideal gas form. The terms a/V^2 and b account for intermolecular forces and molecular volume. The parameters a and b are called the attraction and repulsion parameters, respectively. The parameter b is also referred to as the effective molecular volume.

Equation 1.11 may be used to evaluate a and b in terms of the critical constants, resulting in the following expressions:

$$a = 3P_c V_c^2 = 27R^2 T_c^2/64P_c$$

$$b = V_c/3 = RT_c/8P_c$$

If these expressions are substituted in the van der Waals equation, it takes the following form in terms of the reduced properties:

$$(P_r + 3/V_r^2)(3V_r - 1) = 8T_r$$

The equation in this form is applicable to any substance, provided the critical constants are known. Correlating data in this manner is one illustration of utilizing the principle of corresponding states.

1.2.4.2   Virial Equation

The ideal gas equation may be written in the form $Pv/RT = 1$, where v is the molar volume ($v = V/n$). In general, for real fluids, $Pv/RT = z$, where z is known as the compressibility factor. The compressibility factor may be expressed in terms of the pressure in a power series (or virial expansion):

$$z = 1 + B'P + C'P^2 + \cdots$$

The coefficients B', C', ... are called the second virial coefficient, the third virial coefficient, and so on. The first virial coefficient is unity. The virial coefficients (except the first) depend on the temperature and the identity of the fluid.

An alternative form of the virial equation is an expansion in a power series of $1/V$:

$$z = 1 + B/V + C/V^2 + \cdots$$

The coefficients B, C, ..., also known as second virial coefficient, third virial coefficient, and so on, are functions of the temperature and are substance-specific. The virial equation is usually truncated after the second virial term:

$$z = 1 + B'P$$

Or, substituting $B' = B/RT$,

$$z = 1 + BP/RT$$

For mixtures, the second virial coefficient is given by

$$B = \sum_i \sum_j Y_i Y_j B_{ij}$$

For $i = j$, B_{ij} is the pure component second virial coefficient. For $i \neq j$, $B_{ij} (= B_{ji})$ is the interaction second virial coefficient.

Data are readily available for pure component and binary interaction second virial coefficients for a large number of components and binaries. Binary interaction coefficients are required for extending the equation to mixtures. The simplicity of the equation, the availability of coefficient data, and its ability to represent mixtures are some of the reasons the virial equation of state is a viable option for representing gases at densities up to about 70% of the critical density. It may be used for calculating vapor phase properties at these conditions but is not applicable to dense gases or liquids.

1.2.4.3 Redlich–Kwong Equation

This equation is an improvement over the van der Waals equation in that it introduces a temperature dependency for the attraction parameter by dividing it by the term $T^{0.5}$. The equation also has a special quadratic term in the volume:

$$\left(P + \frac{a}{T^{0.5}V(V + b)} \right)(V - b) = RT \tag{1.13}$$

The parameters a and b are evaluated in terms of the critical constants by applying the critical point conditions (Equation 1.11). The results are

$$a = 0.42748R^2T_C^{2.5}/P_C$$

$$b = 0.08664RT_C/P_C$$

An alternative method for determining the parameters is by the regression of experimental data with the objective of minimizing errors between the experimental and predicted properties. For the regression, either direct *PVT* data or properties that can be derived from the equation of state, such as VLE or enthalpy departure data, may be used.

Later work on Redlich–Kwong-type equations (known as cubic equations because they are of the third degree in volume when expressed explicitly in the pressure) has shown that greater accuracy is achieved if the equation parameters are correlated as a function of temperature. The original Redlich–Kwong equation has therefore been dropped for the most part from practical applications in favor of the more recent cubic equations.

1.2.4.4   Soave Equation

In an effort to improve the equation of state representation of the effect of temperature, the Soave equation (Soave, 1972) replaces the $a/T^{0.5}$ term in the Redlich–Kwong equation with a more general temperature-dependent parameter, $a(T)$:

$$P = \frac{RT}{V - b} - \frac{a(T)}{V(V + b)}$$

The development of the equation was targeted primarily at improving the accuracy of VLE calculations. The underlying reasoning in developing the equations was that a necessary condition for an equation of state to predict mixture VLE properties was that it accurately predict pure component VLE properties, namely pure component vapor pressures.

The Soave equation uses an additional parameter, the acentric factor, to correlate vapor pressure data. The acentric factor, ω, had been defined in earlier work as a parameter that correlates the deviation of the reduced vapor pressure of a particular compound from that of simple molecules. The reduced vapor pressure is correlated with the reduced temperature as follows:

$$\log P_r^0 = C_1 - C_2/T_r$$

It was observed that the reduced vapor pressure of noble gases (simple molecules) at a reduced temperature of 0.7 is about 0.1. It is also known that, at the critical point, $T_r = 1$ and $P_r = 1$. These two conditions were applied to the above equation to evaluate C_1 and C_2. The resulting equation for simple substances is

$$\log (10P_r^0) = 10/3 - 7/(3T_r)$$

For substances in general, the reduced vapor pressure tends to deviate to varying degrees from this equation. The deviation is accounted for by including the acentric factor in the equation:

$$\log (10^{1+\omega} P_r^0) = 10/3 - 7/(3T_r) \tag{1.14}$$

The acentric factor is a characterizing parameter for each substance and is defined as

$$\omega = -\log (10P_r^0) \quad \text{at } T_r = 0.7$$

For simple compounds, $P_r^0 = 0.1$ at $T_r = 0.7$ and, therefore, $\omega = 0$.

Equation 1.14 incorporates the definition of the acentric factor and may also be used to predict the vapor pressure, once the acentric factor has been determined. Another route for calculating the vapor pressure is via an equation of state, as described below. In the Soave equation, ω is used in formulating the temperature dependency of the parameter a, which may be considered as a function of both T and ω. The function $a(T,\omega)$ was determined with the objective of fitting vapor pressures calculated by the equation of state to experimental pure component vapor pressure data.

The first step for calculating the vapor pressure from an equation of state is the familiar evaluation of parameters a and b using the critical point conditions, Equation 1.11. The results for the Soave equation are

$$a' = 0.42747R^2T_C^2/P_C$$

$$b = 0.08664RT_C/P_C$$

Here a', a constant, is the value of the parameter at the critical temperature. At other temperatures, the parameter at (T, ω) is defined as

$$a(T, \omega) = a'\alpha(T, \omega)$$

At the critical temperature, $\alpha = 1$. At other temperatures, α is determined from the vapor pressure fit, leading to the relationship

$$\alpha^{0.5} = 1 + (1 - T_r^{0.5})(0.480 + 1.574\omega - 0.176\omega^2)$$

Vapor pressures are calculated with the Soave equation by equating the pure component fugacity coefficients in the vapor and liquid. It is significant that the same equation of state is applied to both the vapor and liquid phases.

Like other cubic equations of state, the Soave equation can have three real roots for the molar volume (or density or compressibility, depending on the form in which the equation is written). When the composition, temperature, and pressure are such that three roots exist, the largest volume root is used if the system is a vapor, and the smallest volume root is used if the system is a liquid.

Equations of state are extended to mixtures by replacing pure component parameters with mixture parameters. These are expressed in terms of the mixture composition and the parameters of the pure components, using certain mixing rules. The Soave equation uses the following mixing rules:

$$a = \sum_i \sum_j X_i X_j a_i^{0.5} a_j^{0.5} (1 - k_{ij})$$ (1.15)

$$b = \sum_i X_i b_i$$

where X is the mole fraction. The summations are carried over all the components in the mixture. Parameters a_i (or a_j) and b_i are pure component parameters. The binary interaction parameters k_{ij} are equal to zero for $i = j$ and for similar components. For binaries of dissimilar molecules, such as those between carbon dioxide, hydrogen sulfide, and hydrocarbons, k_{ij} is usually in the range between 0.0 and 0.2. The parameter k_{ij} may be assumed to be independent of temperature, pressure, and composition, although better representation may be obtained with temperature-dependent interaction coefficients. The coefficients are determined from binary experimental data, particularly VLE data.

The Soave equation is widely used for hydrocarbons and related components over broad ranges of temperature and pressure. It is accurate enough for calculating enthalpy and entropy departures, vapor–liquid equilibria, and vapor density in natural gas processing and many petroleum-related operations. The equation is not very accurate in the critical region and for liquid density calculations.

1.2.4.5 Peng–Robinson Equation

This equation (Peng and Robinson, 1976) was developed with the goal of overcoming some of the deficiencies of the Soave equation, namely its inaccuracy in the critical region and in predicting liquid densities. The equation is similar to the Soave equation in that it is cubic in the volume, expresses its parameters in terms of the critical temperature, critical pressure, and acentric factor, and is based on correlating pure-component vapor pressure data. The equation is written as

$$P = \frac{RT}{V - b} - \frac{a(T)}{V(V + b) + b(V - b)}$$ (1.16)

The parameters are given by

$$a(T,\omega) = 0.45724(R^2 T_c^2 / P_c)\alpha(T,\omega)$$

$$\alpha^{0.5} = 1 + (1 - T_r^{0.5})(0.37464 + 1.5422\omega - 0.26992\omega^2)$$

$$b = 0.07780 R T_c / P_c$$

Mixtures are handled by calculating mixture parameters, using the same mixing rules used with the Soave equation (Equation 1.15).

The Peng–Robinson equation is widely used for the same applications as the Soave equation. Although it may be more accurate than the Soave equation in the critical region and for calculating liquid densities, it is not generally recommended for the latter since better methods for predicting liquid densities are available.

1.2.4.6   Benedict–Webb–Rubin (BWR) Equation

One approach toward improving the accuracy of equations of state is to produce a better fit to experimental data by including many adjustable parameters in the equation. The equation of Benedict et al. (1951) was first introduced with eight parameters. In later development, the accuracy of the equation was improved by modifying it to include 13 parameters (Starling, 1973). Since the parameters must be determined individually for each substance, the equation applicability is limited to those substances for which parameters are available. Mixtures are handled using mixing rules that require interaction coefficients for several of the parameters. Interaction coefficients are available for only a limited number of components, which further restricts the equation's usage. The equation itself and the parameter mixing rules are not presented here because of space limitation. The reader is referred to the original sources for a detailed description of the equation.

Attempts have been made to correlate the BWR parameters in terms of the critical temperature, critical pressure, and acentric factor. The equation is thus reduced essentially to a three-parameter function, not necessarily more accurate than the cubic equations.

1.2.4.7   Lee–Kesler–Plocker Equation

By applying the corresponding states principle, the deviations of the properties of a substance from those of a "simple" fluid may be correlated in terms of the acentric factor, as described above for vapor pressures (Equation 1.14). The compressibility factor has also been correlated in terms of the acentric factor in the form of a polynomial

$$z = z^{(0)} + \omega z^{(1)} + \omega^2 z^{(2)} + \cdots$$

which is usually truncated after the linear term. The compressibility factor of a fluid, z, is thus expressed in terms of the compressibility factor of a simple fluid, $z^{(0)}$, and that of a reference fluid, $z^{(1)}$. Lee and Kesler (1975) developed equations for $z^{(0)}$ and $z^{(1)}$

using data for argon, krypton, and methane to represent the simple fluid, and n-octane data to represent the reference fluid. Plocker et al. (1978) modified the parameter mixing rules and determined binary interaction parameters for many component pairs based on VLE data. With these improvements, the Lee–Kesler–Plocker equation has gained broad acceptance as a reliable generalized equation of state. Like other generalized equations, such as Soave and Peng–Robinson, the Lee–Kesler–Plocker equation is not adapted to polar compounds. The equation details, parameter values, and mixing rules are not provided here but may be found in the referenced sources.

EXAMPLE 1.2: EQUATION OF STATE CALCULATIONS

The volume of 1 kmol of methane at 300 K is 2.5 m³. Compare the methane pressure calculated by (a) the ideal gas equation, (b) the virial equation, (c) the Redlich–Kwong equation, and (d) the Soave equation. The following data are available:

Gas constant, $R = 8.314$ m³ kPa/K kmol.
Methane properties, $T_C = 190.6$ K, $P_C = 4,599$ kPa, $\omega = 0.012$.
Methane virial coefficients at 300 K, $B = -0.1$ m³/kmol, $C = 0.003$ m⁶/kmol².

Solution

a. Ideal gas equation

$$P = RT/V = (8.314)(300)/2.5 = 997.68 \text{ kPa}$$

b. Virial equation

$$PV/RT = 1 + B/V + C/V^2 = 1 - 0.1/2.5 + 0.003/(2.5)^2 = 0.96$$

$$P = (0.96)(RT/V) = (0.96)(997.68) = 957.77 \text{ kPa}$$

c. Redlich–Kwong equation

$$\left[P + \frac{a}{T^{0.5}V(V + b)}\right](V - b) = RT$$

$$a = 0.42748R^2T_C^{2.5}/P_C$$

$$= (0.42748)(8.314)^2(190.6)^{2.5}/4599 = 3222.4$$

$$b = 0.08664RT_C/P_C$$

$$= (0.08664)(8.314)(190.6)/4599 = 0.02985$$

$$\left[P + \frac{3222.4}{(300)^{0.5}(2.5)(2.5 + 0.02985)}\right](2.5 - 0.02985) = (8.314)(300)$$

$$P = 980.38 \text{ kPa}$$

d. Soave equation

$$P = \frac{RT}{V - b} - \frac{a(T)}{V(V + b)}$$

$$a(T) = a'\alpha$$

$$b = 0.08664RT_c/P_c = 0.02985$$

$$\alpha = [1 + (1 - T_r^{0.5})(0.480 + 1.574\omega - 0.176\omega^2)]^2$$

$$= [1 + (1 - (300/190.6)^{0.5})(0.480 + (1.574)(0.012) - (0.176)(0.012)^2)]^2$$

$$= 0.7617$$

$$a' = 0.42747R^2T_c^2/P_c$$

$$= (0.42747)(8.314)^2(190.6)^2/4599$$

$$= 233.40$$

$$a(T) = (233.40)(0.7617) = 177.78$$

$$P = (8.314)(300)/(2.5 - 0.02985) - 177.78/[(2.5)(2.5 + 0.02985)]$$
$$= 981.63 \text{ kPa}$$

Compared with an experimental value of 982.5 kPa, the best predictions are by the Soave equation and the Redlich–Kwong equation.

1.3 PHASE EQUILIBRIA

The conditions for equilibrium discussed in Section 1.1.3 are applied here to the problem of phase equilibria. These conditions are that, in order for two or more phases to coexist at equilibrium, they must have the same temperature and pressure and the chemical potential of each component must be equal in all the phases. The chemical potential is not a measurable quantity and is not intuitively related to observable physical properties. Applying the conditions of equilibrium to real fluids involves a transformation to more practical terms and the utilization of fluid models such as equations of state.

1.3.1 FUGACITY

The chemical potential of component i in a fluid may be derived on the basis of Equation 1.6. By holding T and n_i constant, then T, P and n_j ($j \neq i$) constant, the following relationships may be written:

$$V = \left(\frac{\partial G}{\partial P}\right)_{T,n_i}$$

$$\mu_i = \left(\frac{\partial G}{\partial n_i} \right)_{T,P,n_j}$$

If μ_i is differentiated with respect to P and the reciprocity rule applied, the following is obtained:

$$\left[\frac{\partial}{\partial P} \left(\frac{\partial G}{\partial n_i} \right)_{T,P,n_j} \right]_{T,n_i} = \left[\frac{\partial}{\partial n_i} \left(\frac{\partial G}{\partial P} \right)_{T,n_i} \right]_{T,P,n_j}$$

or

$$\left(\frac{\partial \mu_i}{\partial P} \right)_{T,n_i} = \left(\frac{\partial V}{\partial n_i} \right)_{T,P,n_j}$$

Defining the partial pressure of component i as $p_i = (n_i/n)P$, and combining this definition with the ideal gas equation to eliminate P, the ideal gas equation for component i may be written as

$$p_i V = n_i RT$$

Thus, for an ideal gas, the derivative of the volume with respect to the number of moles of component i is

$$\left(\frac{\partial V}{\partial n_i} \right)_{T,P,n_j} = \frac{RT}{p_i}$$

Substitution in the above equation gives

$$\left(\frac{\partial \mu_i}{\partial p_i} \right)_{T,n_i} = \frac{RT}{p_i}$$

In its integrated form, this equation becomes

$$\mu_i = \mu_i^0 + RT \ln p_i \tag{1.17}$$

where μ_i^0 is an integration constant, a function of temperature only.

For real fluids the partial pressure is replaced by the fugacity, a defined property, using the same form as Equation 1.17:

$$\mu_i = \mu_i^0 + RT \ln \hat{f}_i \tag{1.18}$$

where $\hat{f}_i$ is the fugacity of component i in solution. The fugacity bears the same relationship to the chemical potential for real fluids as does the partial pressure for ideal gases. Because of the direct relationship between chemical potential and fugacity, the condition for equilibrium expressed by Equation 1.9 is equivalent to the equality of component fugacities in the phases:

$$\hat{f}_i^L = \hat{f}_i^V \qquad (1.19)$$

This expression of the condition for equilibrium is used in phase equilibrium calculations more frequently than the equality of the chemical potentials because the fugacity is more closely related to observable properties than the chemical potential.

The ratio $\hat{\phi}_i = \hat{f}_i/p_i$ is called the fugacity coefficient. As the pressure approaches zero, the fluid approaches ideal gas behavior; hence,

$$\lim_{p_i \to 0} \hat{\phi}_i = 1$$

1.3.1.1 Pure Substances

To evaluate the fugacity of a pure component, Equation 1.4, which is valid for a closed system with a fixed composition, can be applied to the special case of a pure component. At constant temperature, this equation becomes

$$dG = VdP$$

For an ideal gas, $V = RT/P$ so that

$$dG = RT \, dP/P = RT \, d \ln P$$

For real fluids, the fugacity is substituted for the pressure. At constant temperature,

$$dG = VdP = RT \, d \ln f \qquad (1.20)$$

or

$$d \ln f = V \, dP/RT$$

The fugacity of a pure component is calculated by integrating this equation between zero pressure and system pressure. At $P = 0$, $f = P$ and $\phi = 1$. Also, as the pressure approaches zero, the ideal gas law applies. The integration may be carried out by first subtracting the equation

$$RT \, d \ln P = RT \, dP/P$$

from Equation 1.20:

$$RT(d \ln f - d \ln P) = V \, dP - RT \, dP/P$$

or,

$$d \ln (f/P) = d \ln \phi = (V/RT - 1/P) \, dP$$

This equation is integrated from a lower bound of $P = 0$, where $f = P$ and $\phi = 1$, to the system pressure P, where the fugacity is f and the fugacity coefficient is ϕ:

$$\ln \frac{f}{P} = \ln\phi = \frac{1}{RT} \int_0^P \left(V - \frac{RT}{P} \right) dP = \int_0^P \frac{z-1}{P} dP \qquad (1.21)$$

The compressibility factor, z, must be available as a function of pressure, either through an equation of state or directly from experimental data.

1.3.1.2 Mixtures

The fugacity of a component i in a homogeneous phase is derived in a manner analogous to pure substances. The change in the partial free energy of component i in an ideal gas mixture at constant temperature resulting from a change in its partial pressure is

$$d\hat{G}_i = RTd \ln p_i = \hat{V}_i \, dP$$

where the caret designates partial quantities, or quantities associated with a component in solution. For non-ideal solutions, vapor or liquid, the partial pressure is replaced by the fugacity:

$$RTd \ln \hat{f}_i = \hat{V}_i \, dP$$

The change in component fugacity resulting from a change in pressure from P_1 to P_2 is evaluated by integrating the partial molar volume at constant temperature:

$$\ln \left(\frac{\hat{f}_{i2}}{\hat{f}_{i1}} \right) = \frac{1}{RT} \int_{P_1}^{P_2} \hat{V}_i \, dP \qquad (1.22)$$

The fugacity coefficient of component i in a mixture is defined as

$$\hat{\phi}_i = \frac{\hat{f}_i}{p_i} = \frac{\hat{f}_i}{PX_i}$$

where p_i, the partial pressure of component i, is defined as the total pressure multiplied by X_i, the mole fraction of component i. As the pressure approaches zero, the fugacity approaches the partial pressure so that

$$\lim_{P \to 0} \hat{\phi}_i = 1$$

The lower bound of the integral is defined at $P = 0$, where $\hat{\phi}_i = 1$, resulting in the following expression for the fugacity coefficient, derived in a manner similar to Equation 1.21:

$$\ln \hat{\phi}_i = \frac{1}{RT} \int_0^P \left(\hat{V}_i - \frac{RT}{P} \right) dP = \int_0^P \left(\frac{\hat{z}_i - 1}{P} \right) dP \tag{1.23}$$

The partial molar volume or compressibility factor is evaluated from an equation of state or from experimental data.

The condition for VLE, Equation 1.19, may be stated in terms of fugacity coefficients:

$$Y_i \hat{\phi}_i^V P = X_i \hat{\phi}_i^L P \tag{1.24}$$

In this equation, X_i and Y_i are the mole fractions of component i in the liquid and vapor, respectively. The vapor–liquid distribution coefficient, or K-value, defined as $K_i = Y_i/X_i$, is therefore,

$$K_i = \frac{\hat{\phi}_i^L}{\hat{\phi}_i^V} \tag{1.25}$$

For liquids that cannot be represented by equations of state, the liquid fugacities are expressed in terms of activity coefficients, discussed in Section 1.3.3. For ideal solutions, Equation 1.24 reduces to Raoult's law, presented in Section 1.3.2.

1.3.1.3 Application to Equations of State

Vapor–liquid distribution coefficients (K-values) may be calculated from equations of state using Equations 1.21, 1.23, and 1.25. These calculations require the evaluation of partial properties of individual components, defined as the change in the total solution property resulting from the addition of a differential amount of the component in question to the solution, while holding constant the remaining component amounts and the temperature and pressure. Mathematically, the partial property $\hat{\Pi}_i$ of component i is given by

$$\hat{\Pi}_i = \left[\frac{\partial (n\Pi)}{\partial n_i} \right]_{T,P,n_j(j \neq i)} = \Pi + n \left(\frac{\partial \Pi}{\partial n_i} \right)_{T,P,n_j(j \neq i)}$$

where Π is the molar property of the solution, n is the total number of moles in the solution, and n_i is the number of moles of component i in the solution. Note that

$$\left(\frac{\partial n}{\partial n_i} \right)_{n_j(j \neq i)} = \frac{\partial (n_1 + \cdots + n_i + \cdots + n_j + \cdots n_C)}{\partial n_i} = 1$$

since all the n_js are constant except n_i (C is the number of components).

A general equation of state may be written as $PV = zRT$ or $z = PV/RT$, where V is the molar volume and z is the compressibility factor. In accordance with the partial property equation, the partial molar volume and the partial compressibility factor are expressed as follows:

$$\hat{V}_i = V + n\left(\frac{\partial V}{\partial n_i}\right)_{T,P,n_j(j \neq i)}$$

$$\hat{z}_i = z + n\left(\frac{\partial z}{\partial n_i}\right)_{T,P,n_j(j \neq i)}$$

From the equation of state,

$$\left(\frac{\partial z}{\partial n_i}\right)_{T,P,n_j(j \neq i)} = \frac{P}{RT}\left(\frac{\partial V}{\partial n_i}\right)_{T,P,n_j(j \neq i)}$$

The value of z from the general equation of state and its derivative with respect to n_i is substituted in the partial compressibility factor equation to give

$$\hat{z}_i = \frac{PV}{RT} + n\frac{P}{RT}\left(\frac{\partial V}{\partial n_i}\right)_{T,P,n_j(j \neq i)}$$

or

$$\hat{z}_i \frac{RT}{P} = V + n\left(\frac{\partial V}{\partial n_i}\right)_{T,P,n_j(j \neq i)}$$

By comparing this equation with the partial molar volume equation, a direct relationship is obtained between $\hat{V}_i$ and $\hat{z}_i$:

$$\hat{z}_i = \frac{P\hat{V}_i}{RT}$$

If equations of state are available for both the vapor and liquid phases, the above equations may be used to calculate the component fugacity coefficients in both phases by Equation 1.23, and the K-values by Equation 1.25. Alternatively, the fugacity coefficient of a component in solution may be derived from the total fugacity coefficient expression (Equation 1.21) via the definition of partial properties:

$$\ln \hat{\phi}_i = \left[\frac{\partial(n \ln \phi)}{\partial n_i}\right]_{T,P,n_j(j \neq i)} = \ln \phi + n\left(\frac{\partial \ln \phi}{\partial n_i}\right)_{T,P,n_j(j \neq i)}$$

EXAMPLE 1.3: K-VALUES FROM AN EQUATION OF STATE

Over a certain range of conditions, the vapor phase of a binary mixture may be represented by an equation of state of the form

$$PV/RT = z = 1 + BP/RT$$

The mixture coefficient is given as

$$B = Y_1B_1 + Y_2B_2$$

where B_1 and B_2 are pure component coefficients. The liquid phase may be assumed to behave as an ideal solution. Derive equations to calculate the K-values in this mixture.

SOLUTION

The K-values are calculated by Equation 1.25, with the vapor phase fugacity coefficients calculated from the equation of state and the liquid phase fugacity coefficients for an ideal solution calculated as

$$\hat{\phi}_i^L = \frac{\hat{f}_i^L}{P_i} = \frac{\hat{f}_i^L}{PX_i} = \frac{f_i^L X_i}{PX_i} = \frac{f_i^L}{P} = \frac{P_i^0}{P}$$

Thus,

$$K_i = \hat{\phi}_i^L / \hat{\phi}_i^V = P_i^0 / P\hat{\phi}_i^V$$

The following alternative routes are possible for calculating the vapor phase fugacity coefficients:

I. Use the first form of Equation 1.23, written for component 1:

$$\ln\hat{\phi}_i = \frac{1}{RT}\int_0^P \left(\hat{V}_1 - \frac{RT}{P}\right)dP$$

From the partial molar volume equation,

$$\hat{V}_i = V + n\left(\frac{\partial V}{\partial n_1}\right)_{T,P,n_2}$$

Rearrange the equation of state:

$$V = RT/P + B$$

$$B = Y_1B_1 + Y_2B_2 = (1/n)(n_1B_1 + n_2B_2)$$

$$\frac{\partial V}{\partial n_1} = \frac{\partial B}{\partial n_1} = \frac{1}{n}(B_1 - B)$$

$$\hat{V}_1 = V + B_1 - B = (RT/P) + B_1$$

$$\ln\hat{\phi}_1 = \frac{1}{RT}\int_0^P B_1\, dP = \frac{B_1 P}{RT}$$

Similarly, for component 2,

$$\ln\hat{\phi}_2 = \frac{B_2 P}{RT}$$

II. Use the second form of Equation 1.23, written for component 1:

$$\ln\hat{\phi}_1 = \int_0^P \left(\frac{\hat{z}_1 - 1}{P}\right) dP$$

$$z = 1 + BP/RT$$

$$\hat{z}_1 = z + n\left(\frac{\partial z}{\partial n_1}\right)_{T,P,n_2}$$

$$\frac{\partial z}{\partial n_1} = \frac{P}{RT}\left(\frac{\partial B}{\partial n_1}\right) = \frac{P}{RT}\frac{1}{n}(B_1 - B)$$

$$\hat{z}_1 = z + \frac{P}{RT}(B_1 - B) = 1 + \frac{B_1 P}{RT}$$

$$\frac{\hat{z}_1 - 1}{P} = \frac{B_1}{RT}$$

$$\ln\hat{\phi}_1 = \int_0^P \left(\frac{\hat{z}_1 - 1}{P}\right) dP = \int_0^P \frac{B_1}{RT}\, dP = \frac{B_1 P}{RT}$$

For component 2,

$$\ln\hat{\phi}_2 = \frac{B_2 P}{RT}$$

III. Use the partial fugacity coefficient equation derived from Equation 1.21:

$$\ln\hat{\phi}_1 = \ln\phi + n\left(\frac{\partial \ln\phi}{\partial n_1}\right)_{T,P,n_2}$$

$$\ln \phi = \int_0^P \frac{z-1}{P} dP = \int_0^P \frac{B}{RT} dP = \frac{BP}{RT}$$

$$\frac{\partial \ln \phi}{\partial n_1} = \frac{\partial \ln \phi}{\partial B} \frac{\partial B}{\partial n_1} = \frac{P}{RT} \frac{1}{n}(B_1 - B)$$

$$\ln \hat{\phi}_1 = \frac{BP}{RT} + \frac{P}{RT}(B_1 - B) = \frac{B_1 P}{RT}$$

$$\ln \hat{\phi}_2 = \frac{B_2 P}{RT}$$

Note that the results are consistent with the relationship between partial properties and total property:

$$\ln \phi = (1/n)(n_1 \ln \hat{\phi}_1 + n_2 \ln \hat{\phi}_2)$$

1.3.2 PHASE EQUILIBRIUM IN AN IDEAL SYSTEM

Mixture properties are related to constituent component properties by the general relationship

{Mixture property} = Σ{Mole fraction of i}*{Partial property of i}

This relationship was incorporated, for example, in the derivation of the equation for the free energy of a mixture. The volume of a mixture is

$$V = \sum X_i \hat{V}_i$$

where $\hat{V}_i$ is the partial volume of component i. The volume of an ideal solution is equal to the sum of the constituent component volumes at the same temperature and pressure (Section 1.2.1):

$$V = \Sigma X_i V_i$$

It follows that in an ideal solution

$$\hat{V}_i = V_i$$

It also follows from Equation 1.20 and the definition of fugacity and fugacity coefficient that for ideal solutions $\hat{\phi}_i = \phi_i$ and

$$\hat{f}_i = X_i f_i \qquad (1.26)$$

where f_i is the fugacity of the pure component. Equation 1.26 is known as the Lewis–Randall rule.

1.3.2.1   Raoult's Law

The above characterization of ideal solutions does not require the fluid to behave as an ideal gas. In fact, certain liquid mixtures behave as ideal solutions, but they obviously do not obey the ideal gas law. In a mixture forming a vapor phase and a liquid phase at equilibrium with each other, either one of the phases, or both phases, may approach ideal solution behavior. Ideal solution behavior is approached at low pressures and usually with mixtures of chemically similar components. Referring to Equation 1.24, if the vapor phase is assumed to behave as an ideal gas, $\hat{\phi}_i^V = 1$, and the left-hand side of the equation reduces to $Y_i P$, where P is the total pressure. Moreover, for ideal liquid solutions,

$$\hat{\phi}_i^L = \frac{\hat{f}_i^L}{p_i} = \frac{f_i^L X_i}{PX_i} = \frac{f_i^L}{P} = \frac{p_i^0}{P} \quad (f_i^L = f_i^V = p_i^0)$$

where p_i^0 is the vapor pressure of component i. Thus, if the vapor phase is assumed to behave as an ideal gas and the liquid phase as an ideal solution, the following relationship, known as Raoult's law, holds:

$$Y_i P = X_i p_i^0 \tag{1.27}$$

The vapor–liquid distribution coefficient in an ideal gas, ideal liquid solution system is therefore

$$K_i = Y_i/X_i = p_i^0/P$$

Since p_i^0 is a function of temperature only, the vapor–liquid distribution coefficient, or K-value, is also a function of temperature only at constant total pressure. It bears a simple relationship to the total pressure and is independent of the composition.

1.3.2.2   Binary Ideal Solutions

Raoult's law (Equation 1.27) is applied to a binary ideal system to compute the partial pressure of each component and the total pressure:

$$p_1 = Y_1 P = X_1 p_1^0$$

$$p_2 = Y_2 P = X_2 p_2^0$$

$$P = p_1 + p_2 = X_1 p_1^0 + (1 - X_1)p_2^0$$

$$= X_1(p_1^0 - p_2^0) + p_2^0$$

At constant temperature, p_1^0 and p_2^0 are constant; hence, for an ideal binary system at constant temperature, the partial pressures and the total pressure are linear functions of the molar composition (X_1 or X_2). Example 1.7 in Section 1.3.4 illustrates Raoult's law behavior and deviations from it.

1.3.2.3 Henry's Law

In deriving Raoult's law (Equation 1.27), the implied assumption was that all components are condensable; that is, their critical temperature is higher than the mixture temperature. If a component's critical temperature is below the mixture temperature, it does not condense at that temperature as a pure component. There would be no vapor pressure for that component at the system temperature and hence Raoult's law would not apply. The gas component could nonetheless exist as a solute in the liquid phase. Henry's law states that the concentration of the gas component dissolved in the liquid is directly proportional to the partial pressure of that component in the vapor phase at equilibrium with the solution:

$$p_i = PY_i = k_{Hi}X_i$$

$$K_i = Y_i/X_i = k_{Hi}/P \tag{1.28}$$

where k_{Hi} is the Henry's law constant. The constant is a function of the temperature and, to a lesser degree, of the pressure. Its value is specific to the particular solute and solvent.

Henry's law is mainly valid at low concentrations of the solute, typically below 1%, and is therefore applicable to low miscibility solutes. By extension, the solutes are not limited to non-condensables but could include condensables with low miscibility.

EXAMPLE 1.4: WATER–CO$_2$ PHASE EQUILIBRIUM

A water (1)–CO$_2$ (2) solution in a vapor–liquid separation vessel at 20°C has a CO$_2$ mole fraction of 0.01 in the liquid. Calculate the composition of the vapor at equilibrium with the liquid, the vessel pressure, and the K-values of water and CO$_2$. Henry's law constant for CO$_2$ in water at 20°C is 140,000 kPa and the vapor pressure of water is given by the Antoine equation,

$$\ln p_1^0 \text{ (kPa)} = 16.26 - 3800/[T(°C) + 226.4]$$

SOLUTION

Raoult's law is applied to the water and Henry's law to the CO$_2$. The water vapor pressure at 20°C is calculated from the Antoine equation:

$$\ln p_1^0 = 16.26 - 3800/(20 + 226.4) = 0.8379$$

$$p_1^0 = 2.312 \text{ kPa}$$

From Raoult's law for water (Equation 1.27),

$$Y_1 P = X_1 p_1^0 = (1 - 0.01)(2.312) = 2.289 \text{ kPa}$$

From Henry's law for CO_2 (Equation 1.28),

$$Y_2P = k_{H2}X_2 = (140,000)(0.01) = 1400 \text{ kPa}$$

The total pressure,

$$P = Y_1P + Y_2P = 2.289 + 1400 = 1402.289 \text{ kPa}$$

$$Y_1 = 2.289/1402.289 = 0.0016$$

$$Y_2 = 1400/1402.289 = 0.9984$$

$$K_1 = Y_1/X_1 = 0.0016/0.99 = 0.00162$$

$$K_2 = Y_2/X_2 = 0.9984/0.01 = 99.84$$

1.3.3 PHASE EQUILIBRIUM IN NON-IDEAL SYSTEMS

In many mixtures, the interactions between molecules in the liquid phase are too strong to permit adequate representation by equations of state. Such mixtures generally involve components with chemically dissimilar molecules. In the vapor phase, intermolecular forces become more significant at higher pressures, but the vapor phase fugacity coefficient could still be adequately calculated from *PVT* behavior information such as equations of state. Special vapor fugacity methods have been developed for highly non-ideal mixtures such as those where molecules associate in the vapor phase. The liquid phase deviation from ideal solution behavior is quantified by introducing the *liquid activity coefficient*.

1.3.3.1 Activity Coefficients

For a component i in the liquid, the activity coefficient γ_i is defined as

$$\gamma_i = \frac{\hat{f}_i^L}{f_i^L X_i}$$

where $\hat{f}_i^L$ is the fugacity of component i in the liquid solution and f_i^L is the standard state fugacity of component i. The standard state is defined as pure liquid component i at system temperature and pressure. In an ideal liquid solution, $\gamma_i = 1$ and

$$\hat{f}_i^L = f_i^L X_i$$

In the general case, the definitions of fugacity coefficient and activity coefficient are combined with Equation 1.24 to give

$$Y_i \hat{\phi}_i^V P = \hat{f}_i^L = X_i \gamma_i f_i^L$$

The distribution coefficient is, therefore, given by the expression

$$K_i = \frac{\gamma_i f_i^L}{\hat{\phi}_i^V P}$$ (1.29)

The vapor phase fugacity coefficient, $\hat{\phi}_i^V$, may be calculated, as before, from Equation 1.23, using, for instance, an equation of state. The pure component fugacity in the liquid state is equal to its fugacity in the vapor at equilibrium with the liquid:

$$f_i^L (\text{at } p_i^0) = f_i^V (\text{at } p_i^0)$$

Since the system pressure is not, in general, equal to the vapor pressure, the effect of pressure on the fugacity of the liquid must be taken into account in calculating f_i^L. Using the basic definition of fugacity, f_i^L is calculated by carrying out the integration over two pressure steps: from zero pressure to the vapor pressure and from the vapor pressure to the system pressure:

$$\ln f_i^L = \ln p_i^0 + \int_0^{p_i^0} \left(V_i^V - \frac{RT}{P} \right) dP + \int_{p_i^0}^{P} \frac{V_i^L}{RT} dP$$

$$= \ln p_i^0 + \ln \phi_i^0 + \int_{p_i^0}^{P} \frac{V_i^L}{RT} dP$$

The first integral is the natural logarithm of ϕ_i^0, the pure component fugacity coefficient at saturation pressure and temperature T. The integral involves the vapor volume, V_i^V, and may be evaluated using an equation of state. The second integral, known as the Poynting correction factor, takes into account the effect of pressure on the fugacity in the liquid. The volume in this integral, V_i^V, is the liquid molar volume and may be obtained from experimental data or from liquid density estimation methods (Rackett, 1970).

At low pressures, where the vapor may be assumed to behave as an ideal gas, $\hat{\phi}_i^V$ approaches unity. This follows from Equation 1.23 if $\hat{z}_i \approx 1$. Moreover, at low pressures the pure component fugacity may be assumed equal to its vapor pressure. Under these conditions, Equation 1.29 simplifies to

$$K_i = \gamma_i p_i^0 / P$$ (1.29a)

In order to calculate the distribution coefficient by Equation 1.29, the activity coefficient γ_i must be evaluated. The activity coefficients are generally determined from the experimental data and correlated on the basis of thermodynamic phase equilibrium principles. The relationship most often used for this purpose is the Gibbs–Duhem equation (Equation 1.7). At constant temperature and pressure, this equation becomes

$$\Sigma n_i d\mu_i = 0$$

or, in terms of mole fractions,

$$\Sigma X_i d\mu_i = 0$$

The chemical potential is related to the fugacity by Equation 1.18, which may be written in differential form:

$$d\mu_i = RTd\ln \hat{f}_i = RTd\ln(f_i \gamma_i X_i)$$

The last expression is based on the definition of the activity coefficient. Substituting this expression in the Gibbs–Duhem equation at constant T and P gives the following:

$$\Sigma X_i RT \, d\ln (f_i \gamma_i X_i) = 0$$

or

$$\Sigma X_i \, d\ln \gamma_i + \Sigma dX_i = 0$$

Note that $d\ln f_i = 0$ at constant temperature and pressure. Since the total number of moles is constant, $\Sigma dX_i = 0$ and hence,

$$\Sigma X_i \, d\ln \gamma_i = 0 \qquad\qquad (1.30)$$

Since Equation 1.30 was derived at constant temperature and pressure, it may be used to correlate VLE data rigorously only if such data were obtained at constant temperature and pressure. A binary VLE system can exist only at one composition at a fixed temperature and pressure since it has only two degrees of freedom. (Degrees of freedom are discussed in more detail in Chapter 2.) Nevertheless, the binary data are those that are most commonly used for correlating activity coefficient data because they are the most commonly available. Moreover, multi-component mixtures can be represented adequately in terms of binary coefficients.

If Equation 1.30 was to be applied to binary isothermal data, the term that represents the effect of composition on system pressure would be neglected. It may be shown (Walas, 1985) that the error incurred by neglecting this term is small, since it is a fraction of the change in liquid volume due to mixing, which is usually a negligible quantity.

If, on the other hand, Equation 1.30 was to be applied to isobaric binary data, the term that represents the effect of composition on the temperature would be neglected. The error incurred by neglecting this term is related to the heat of mixing, which can be significant for components with widely differing boiling points or chemical structure.

It is, therefore, more accurate to correlate isothermal than isobaric binary VLE data for predicting liquid activity coefficients on the basis of Equation 1.30.

1.3.3.2 Thermodynamic Consistency of VLE Data

Experimental VLE data, namely equilibrium temperature, pressure, and vapor and liquid compositions, can be used to calculate the activity coefficients from Equation 1.29. The K-values are calculated from the composition data, $K_i = Y_i/X_i$. The fugacities and fugacity coefficients, $f_i^L, \hat{\phi}_i^V$, are calculated from the compositions, temperature, and pressure, using, for instance, an equation of state and liquid density data as described earlier. The activity coefficients are then calculated by rearranging Equation 1.29:

$$\gamma_i = \frac{K_i \hat{\phi}_i^V P}{f_i^L}$$

An equation that satisfies the Gibbs–Duhem equation in the form of Equation 1.30 is then used to correlate, or fit, the experimentally determined activity coefficients. Normally, the binary VLE data are correlated, generating binary interaction coefficients. These coefficients may then be used to reproduce the binary VLE data and also to predict multi-component VLE behavior as described in the discussion of some of the activity coefficient equations later in this section.

Before the activity coefficients are represented with an equation, it is important to check the VLE data for thermodynamic consistency against Equation 1.30. As concluded earlier, the error introduced by applying Equation 1.30 to binary isothermal data is usually negligible. The consistency check is described for this type of data, which is the most commonly used for equation development. Equation 1.30 is written for a binary as

$$X_1 d \ln \gamma_1 + X_2 d \ln \gamma_2 = 0$$

By substituting $X_2 = 1 - X_1$ and rearranging, the following equation is obtained:

$$d \ln \gamma_2 = -X_1(d \ln \gamma_1 - d \ln \gamma_2)$$

This equation is combined with the expression for the derivative of a product,

$$d[X_1(\ln \gamma_1 - d \ln \gamma_2)] = (\ln \gamma_1 - \ln \gamma_2)dX_1 + X_1(d \ln \gamma_1 - d \ln \gamma_2)$$

to give the equation

$$\ln(\gamma_1/\gamma_2) \, dX_1 = d \ln \gamma_2 + d[X_1(\ln \gamma_1 - \ln \gamma_2)]$$

This equation is now integrated between $X_1 = 0$ and $X_1 = 1$. The limit at $X_1 = 0$ corresponds to pure component 2 with activity coefficient $\gamma_2 = 1$. Similarly, at $X_1 = 1$, $\gamma_1 = 1$. Therefore, at $X_1 = 0$, $\ln \gamma_2 = 0$, and at $X_1 = 1$, $\ln \gamma_1 = 0$. The integration leads to the following:

$$\int_{X_1=0}^{X_1=1} \ln \frac{\gamma_1}{\gamma_2} \, dX_1 = 0 \tag{1.31}$$

The consistency test is carried out by plotting $\ln (\gamma_1/\gamma_2)$ versus X_1 and graphically evaluating the integral in Equation 1.31. The curve consists of a positive part and a negative part above and below the line $\ln(\gamma_1/\gamma_2) = 0$. The data points are considered thermodynamically consistent, and the assumption of negligible effect of variable pressure is deemed valid if the areas above and below the line are equal.

The actual representation of activity coefficient data involves solving Equation 1.30. Several integrated forms of this equation for binary systems are in use, the most common of which are presented here.

1.3.3.3   Margules Equation

Different forms of this equation have been proposed (Wohl, 1946). One of these is the three-suffix or third-degree form:

$$\ln \gamma_1 = (2A_{21} - A_{12})X_2^2 + 2(A_{12} - A_{21})X_2^3 \tag{1.32}$$

$$\ln \gamma_2 = (2A_{12} - A_{21})X_1^2 + 2(A_{21} - A_{12})X_1^3$$

It can be proved that these equations are solutions to the Gibbs–Duhem equation by taking the differentials $d \ln \gamma_1$ and $d \ln \gamma_2$ and substituting them in Equation 1.30 written for a binary. It should also be recalled that $X_1 + X_2 = 1$ and that $dX_1 = -dX_2$.

The binary constants A_{12} and A_{21} are usually determined by regressing binary isothermal VLE data. If these data are accurate in the vicinities around $X_1 = 0$ and $X_1 = 1$, the binary constants may be determined from the limiting conditions:

$$\lim_{X_1 \to 0} \ln \gamma_1 = A_{12}$$

$$\lim_{X_2 \to 0} \ln \gamma_2 = A_{21}$$

The constants may, in fact, be calculated from a single equilibrium data point (vapor–liquid or liquid–liquid if intended for liquid–liquid equilibrium calculations) although the results would be heavily weighted to that point, probably resulting in less accuracy for the remaining range of compositions.

The Margules equation is generalized in its four-suffix form to multi-component mixtures as follows (Wohl, 1946):

$$\ln \gamma_i = (1 - X_i)^2 [A_i + 2X_i (B_i - A_i - D_i) + 3D_i X_i^2] \tag{1.33}$$

where

$$A_i = \sum_j X_j A_{ij} /(1 - X_i)$$

$$B_i = \sum_j X_j A_{ji} /(1 - X_i)$$

$$D_i = \sum_j X_j D_{ij} / (1 - X_i)$$

The binary interaction constants, A_{ij}, A_{ji}, and D_{ij} ($= D_{ji}$), are determined from binary phase equilibrium data.

Although it is one of the oldest activity coefficient equations, the Margules correlation is still commonly used for a wide range of applications. Its accuracy diminishes, however, as the molecules of a binary are more and more dissimilar in size or chemical structure.

1.3.3.4 van Laar Equation

The binary form of this equation is as follows:

$$\ln \gamma_1 = A_{12} \left[\frac{A_{21} X_2}{A_{12} X_1 + A_{21} X_2} \right]^2$$

$$\ln \gamma_2 = A_{21} \left[\frac{A_{12} X_1}{A_{12} X_1 + A_{21} X_2} \right]^2$$

(1.34)

This equation is also a solution to the Gibbs–Duhem equation. The constants may be calculated from infinite dilution activity coefficients or from a single VLE data point.

The multi-component van Laar equation takes the following form (Carlson and Colburn, 1942; Wohl, 1946; Prausnitz et al., 1967):

$$\ln \gamma_i = \sum_{l=1}^{C} A_{il} Z_l - \sum_{j=1}^{C} A_{ij} Z_i Z_j - \sum_{\substack{j=1 \\ j \neq i}}^{C} \sum_{\substack{k=1 \\ k \neq i}}^{C} (A_{jk} A_{ij} / A_{ji}) Z_j Z_k$$

(1.35)

where

$$Z_l = \frac{X_l A_{li} / A_{il}}{\sum_j X_j A_{ji} / A_{ij}}$$

The binary constants A_{ij} and A_{ji} are determined from binary VLE data. For a three-component system (Wohl, 1946), the liquid activity coefficients are calculated by Equation 1.35a:

$$\ln \gamma_1 = \frac{X_2^2 A_{21}^2 / A_{12} + X_3^2 A_{31}^2 / A_{13} + X_2 X_3 A_{21} A_{31} \left(A_{12} + A_{13} - A_{32} A_{13} / A_{31} \right) / \left(A_{12} A_{13} \right)}{(X_1 + X_2 A_{21} / A_{12} + X_3 A_{31} / A_{13})^2}$$

$$\ln \gamma_2 = \frac{X_3^2 A_{32}^2 / A_{23} + X_1^2 A_{12}^2 / A_{21} + X_3 X_1 A_{32} A_{12} \left(A_{23} + A_{21} - A_{13} A_{21} / A_{12} \right) / \left(A_{23} A_{21} \right)}{(X_2 + X_3 A_{32} / A_{23} + X_1 A_{12} / A_{21})^2}$$

$$\ln \gamma_3 = \frac{X_1^2 A_{13}^2/A_{31} + X_2^2 A_{23}^2/A_{32} + X_1 X_2 A_{13} A_{23}\left(A_{31} + A_{32} - A_{21} A_{32}/A_{23}\right)/\left(A_{31} A_{32}\right)}{(X_3 + X_1 A_{13}/A_{31} + X_2 A_{23}/A_{32})^2}$$

The modified van Laar equation (Black, 1959) extends the applicability of the van Laar equation to many systems, including non-symmetrical and hydrogen-bonding binaries. The equation, which is generalized to multi-component mixtures, is expressed in terms of three binary interaction parameters: A_{ij}, A_{ji}, and C_{ij} ($= C_{ji}$). These parameters must be determined from binary phase equilibrium data. The modified van Laar equation reduces to the van Laar equation if $C_{ij} = 0$. Refer to the source (Black, 1959) for detailed mathematical formulation of the equation.

Other activity coefficient equations have been proposed since the earlier equations of Margules and van Laar. It is outside the scope of this book to present the detailed development of these equations, which may be found in the indicated references. The objective here is to present the basis for each of the more commonly used equations, their features, and applicability.

1.3.3.5   Wilson Equation

This equation (Wilson, 1964) was developed on the basis of a liquid molecular model which assumes that interactions between molecules of two different components depend on the volume fraction of each component in the vicinity of a molecule of a given component. The volume fractions are determined from the probability of finding a molecule of one component or another in the vicinity of a molecule of the given component. The probabilities, and hence the volume fractions, are expressed through Boltzmann distribution functions of energy in terms of binary interaction parameters. The result for volume fractions is

$$Z_i = X_i/(X_i + A_{ij} X_j)$$

The parameters A_{ij} are temperature-dependent, given as

$$A_{ij} = \frac{V_j}{V_i} \exp\left[-\delta_{ij}/RT\right]$$

where V_i and V_j are molar volumes and $\delta_{ij} = \lambda_{ij} - \lambda_{ii}$ are energy terms and are binary constants in the Wilson equation. The activity coefficients are related to volume fractions rather than directly to mole fractions. The expression for a binary is

$$\ln \gamma_i = -\ln (X_i + A_{ij} X_j) + \beta_{ij} X_j \tag{1.36}$$

where

$$\beta_{ij} = \frac{A_{ij}}{X_i + A_{ij} X_j} - \frac{A_{ji}}{X_j + A_{ji} X_i}$$

The equation is generalized to multi-component mixtures in terms of binary parameters only, either A_{ij} and A_{ji} or $(\lambda_{ij} - \lambda_{ii})$ and $(\lambda_{ji} - \lambda_{jj})$. Either set of parameters must be determined to fit binary VLE data. Using the parameters $(\lambda_{ij} - \lambda_{ii})$ and $(\lambda_{ji} - \lambda_{jj})$ allows for a certain degree of temperature dependence through the equations that relate A_{ij} to $(\lambda_{ij} - \lambda_{ii})$.

The Wilson equation is capable of representing both polar and non-polar molecules in multi-component mixtures using only binary parameters. It cannot, however, represent liquid–liquid equilibrium systems. The activity coefficients in a multi-component mixture are calculated by the Wilson equation as follows (Prausnitz et al., 1967):

$$\ln \gamma_i = 1 - \ln \left[\sum_j X_j A_{ij} \right] - \sum_k \left[\frac{X_k A_{ki}}{\sum_j X_j A_{kj}} \right] \qquad (1.36a)$$

Note that $\lambda_{ij} = \lambda_{ji}$ and $A_{ii} = 1$. Equation 1.36a reduces to Equation 1.36 for a binary system.

1.3.3.6 Non-Random Two-Liquid (NRTL) (Renon) Equation

The model for the NRTL equation (Renon and Prausnitz, 1968) is similar to the Wilson model but includes a non-randomness constant, $\alpha_{ij} (= \alpha_{ji})$, that is characteristic of the types of components in each binary. In addition to this constant, the equation, which is generalized to multi-component mixtures, utilizes four interaction parameters for each binary: a_{ij}, a_{ji}, b_{ij}, and b_{ji}. The parameters b_{ij} and b_{ji} include a temperature dependency similar to the Wilson coefficients. The parameters a_{ij} and a_{ji} may be added to improve the ability to represent the effect of temperature. The equation may thus be used either in its three-parameter form or in its five-parameter form.

All parameters are determined by regression of binary phase equilibrium data although α_{ij} is usually fixed at some estimated value between 0.2 and 0.5.

The NRTL equation is one of the more successful equations for representing phase equilibrium data, including liquid–liquid equilibrium. It is applicable to multi-component mixtures, which may include non-symmetrical binaries. It also has built-in temperature dependency over moderate ranges.

1.3.3.7 Universal Quasi-Chemical (UNIQUAC) Equation

The UNIQUAC equation (Abrams and Prausnitz, 1975) is based on the two-liquid model in which the excess Gibbs energy is assumed to result from differences in molecular sizes and structures and from the energy of interaction between the molecules.

The equation, which is generalized to multi-component mixtures, requires pure component data for the van der Waals area and volume parameters, q_i and r_i. Additionally, the binary interaction parameters $(u_{ji} - u_{ii})$ and $(u_{ij} - u_{jj})$ are also required and are generally determined from binary phase equilibrium data. Temperature dependency is incorporated in the equation, similar to the Wilson and NRTL equations. The UNIQUAC equation is applicable to many classes of components, including mixtures containing considerably dissimilar molecules, and is also applicable to liquid–liquid equilibrium systems. It can represent temperature dependency over moderate ranges but is not necessarily more accurate than simpler equations in spite of its theoretical foundation.

The UNIQUAC equation is the basis for the development of the group contribution method, the UNIFAC equation, which predicts liquid activity coefficients from component structures on the basis of interactions between chemical functional groups.

EXAMPLE 1.5: WILSON EQUATION PARAMETERS

Parameters for activity coefficient equations can be determined from the experimental data. Describe a procedure to calculate the Wilson equation parameters for a binary solution from infinite dilution data.

SOLUTION

At $X_1 = 0$ ($X_2 = 1$), the activity coefficient of component 1 is the infinite dilution activity coefficient, γ_1^∞.

At $X_2 = 0$ ($X_1 = 1$), the activity coefficient of component 2 is the infinite dilution activity coefficient, γ_2^∞.

Write Equation 1.36 for components 1 and 2:

$$\ln\gamma_1 = -\ln(X_1 + A_{12}X_2) + X_2\left(\frac{A_{12}}{X_1 + A_{12}X_2} - \frac{A_{21}}{X_2 + A_{21}X_1}\right)$$

$$\ln\gamma_2 = -\ln(X_2 + A_{21}X_1) + X_1\left(\frac{A_{21}}{X_2 + A_{21}X_1} - \frac{A_{12}}{X_1 + A_{12}X_2}\right)$$

$$\text{At } X_1 = 0, X_2 = 1, \quad \ln\gamma_1^\infty = -\ln A_{12} + 1 - A_{21}$$

$$\text{At } X_2 = 0, X_1 = 1, \quad \ln\gamma_2^\infty = -\ln A_{21} + 1 - A_{12}$$

With the given values for γ_1^∞ and γ_2^∞, the last two equations can be solved iteratively for A_{12} and A_{21}.

EXAMPLE 1.6: ETHANOL–ISOOCTANE PHASE EQUILIBRIUM

For the binary ethanol (1)–isooctane (2) at 50°C, the Wilson coefficients are $A_{12} = 0.0995$, $A_{21} = 0.247$, and the vapor pressures are $p_1^0 = 29.5$ kPa, $p_2^0 = 19.5$ kPa. Assuming ideal gas behavior in the vapor phase, calculate the K-values and total pressure for liquid compositions at $X_1 = 0.5, 0.6, 0.7$. Does this system form an azeotrope?

SOLUTION

The activity coefficients are calculated by Equation 1.36. For $X_1 = 0.6$,

$$\beta_{12} = 0.0995/[0.6 + (0.0995)(0.4)] - 0.247/[0.4 + (0.247)(0.6)] = -0.2951$$

$$\beta_{21} = -\beta_{12} = 0.2951$$

$$\ln\gamma_1 = -\ln[0.6 + (0.0995)(0.4)] - (0.2951)(0.4) = 0.3286$$

$$\gamma_1 = 1.389$$

$$\ln\gamma_2 = -\ln[0.4 + (0.247)(0.6)] + (0.2951)(0.6) = 0.7782$$

$$\gamma_2 = 2.177$$

Because of the assumed ideal gas behavior in the vapor phase, the K-values are calculated by Equation 1.29a:

$$K_1 = Y_1/X_1 = \gamma_1 p_1^0/P$$

$$K_2 = Y_2/X_2 = \gamma_2 p_2^0/P$$

$$Y_1 + Y_2 = 1 = (X_1\gamma_1 p_1^0 + X_2\gamma_2 p_2^0)/P$$

$$P = X_1\gamma_1 p_1^0 + X_2\gamma_2 p_2^0 = (0.6)(1.389)(29.5) + (0.4)(2.177)(19.5) = 41.566 \text{ kPa}$$

$$Y_1 = X_1\gamma_1 p_1^0/P = (0.6)(1.389)(29.5)/41.566 = 0.5915$$

$$Y_2 = X_2\gamma_2 p_2^0/P = (0.4)(2.177)(19.5)/41.566 = 0.4085$$

$$K_1 = 0.5915/0.6 = 0.9858$$

$$K_2 = 0.4085/0.4 = 1.0212$$

The results for the three compositions are summarized below:

X_1	X_2	γ_1	γ_2	P (kPa)	Y_1	Y_2	K_1	K_2
0.5	0.5	1.633	1.785	41.5	0.581	0.419	1.162	0.838
0.6	0.4	1.389	2.178	41.6	0.591	0.409	0.985	1.022
0.7	0.3	1.220	2.770	41.4	0.609	0.391	0.870	1.303

A minimum boiling azeotrope is indicated by the peak in the total pressure between $X_1 = 0.5$ and $X_1 = 0.7$. Azeotropes, discussed in the following section, are equilibrium mixtures that at a certain temperature and pressure can have the same composition in the liquid and vapor. For this mixture, the azeotrope lies somewhere between $X_1 = 0.5$ and 0.7. Below the azeotrope (at $X_1 = 0.5$) $Y_1 > X_1$. Above the azeotrope (at $X_1 = 0.7$) $Y_1 < X_1$.

EXAMPLE 1.6A: ACTIVITY COEFFICIENTS IN A MULTI-COMPONENT SOLUTION

Calculate the activity coefficients of ethanol, methylcyclopentane (MCP), and benzene in solution at 63°C using the Wilson equation. The following data are given:

Component	Mole Fraction in the Liquid, X_i	Liquid Molar Volume, V_i (cm³/mol)
1. Ethanol	0.047	62
2. MCP	0.847	118
3. Benzene	0.107	94

Wilson parameters, cal/mol:

$\lambda_{12} - \lambda_{11} = 2220.6$	$\lambda_{12} - \lambda_{22} = 249.6$
$\lambda_{13} - \lambda_{11} = 1521.1$	$\lambda_{13} - \lambda_{33} = 119.9$
$\lambda_{23} - \lambda_{22} = 13.8$	$\lambda_{23} - \lambda_{33} = 248.4$

SOLUTION

Equation 1.36a is written for a three-component mixture. The Wilson coefficients at $T = 63°C = 336.15$ K with $R = 1.987$ cal/mol K are calculated as follows:

$$A_{12} = \frac{V_2}{V_1} \exp\left[-\frac{\lambda_{12} - \lambda_{11}}{RT} \right] = 0.06849$$

$$A_{21} = \frac{V_1}{V_2} \exp\left[-\frac{\lambda_{12} - \lambda_{22}}{RT} \right] = 0.36159$$

$$A_{13} = \frac{V_3}{V_1} \exp\left[-\frac{\lambda_{13} - \lambda_{11}}{RT} \right] = 0.15549$$

$$A_{31} = \frac{V_1}{V_3} \exp\left[-\frac{\lambda_{13} - \lambda_{33}}{RT} \right] = 0.55119$$

$$A_{23} = \frac{V_3}{V_2} \exp\left[-\frac{\lambda_{23} - \lambda_{22}}{RT} \right] = 0.78032$$

$$A_{32} = \frac{V_2}{V_3} \exp\left[-\frac{\lambda_{23} - \lambda_{33}}{RT} \right] = 0.86545$$

The activity coefficient calculations are tabulated below:

i	X_i	S_i	T_i	$\ln \gamma_i$	γ_i
1.	0.047	0.12158	0.77795	2.32922	10.2700
2.	0.846	0.94649	1.02735	0.02764	1.0280
3.	0.107	0.86508	0.88127	0.26367	1.3017

where

$$S_i = \sum_j X_j A_{ij}$$

$$T_i = \sum_k \frac{X_k A_{ki}}{S_k}$$

$$\ln \gamma_i = 1 - \ln S_i - T_i$$

The computation details are shown for component 1:

$$S_1 = (0.047 \times 1) + (0.846 \times 0.06849) + (0.107 \times 0.15549) = 0.12158$$

$$T_1 = \frac{0.047 \times 1}{0.12158} + \frac{0.846 \times 0.36159}{0.94649} + \frac{0.107 \times 0.55119}{0.86508} = 0.77795$$

$$\ln \gamma_1 = 1 - \ln(0.12158) - 0.77795 = 2.32922$$

$$\gamma_1 = 10.2700$$

1.3.4 VAPOR–LIQUID EQUILIBRIA: APPLICATIONS

From a practical view point, phase equilibrium behavior may be described by examining phase diagrams for binary systems. Since it is generally not feasible to represent multi-component mixtures graphically, phase diagrams of the constituent binaries may be used as an aid in understanding multi-component phase behavior. The basis for this is the fact that activity coefficients in multi-component mixtures can be related to activity coefficients in binaries. Ternary diagrams are commonly used for describing immiscible liquids (Section 1.3.5).

The binary VLE data can be represented by various types of diagrams such as pressure–composition, activity coefficient–composition, and vapor composition–liquid composition plots. Curves representing non-ideal systems can have different characteristics, as shown in Examples 1.7 through 1.11 that follow. The representation of phase behavior may be based on direct experimental data or on predictive calculations using equations such as those presented in the preceding sections. These equations have their roots in thermodynamic principles but also rely on experimental data to determine the various constants. The calculations require evaluating the different variables in Equation 1.29. These are highly complex and tedious calculations but can be performed rapidly and accurately using computer programs. The phase diagrams for the following examples are based on computer simulation. Some of the activity coefficient parameters used in the calculations are given in Table 1.1.

Assuming the vapor phase does not deviate too strongly from ideal gas behavior, the phase equilibrium characteristics are primarily controlled by the liquid activity coefficients. Under these conditions, the only term that is composition-dependent in Equation 1.29 is the activity coefficient, γ_i.

TABLE 1.1
Selected Activity Coefficient Binary Interaction Constants

Diethyl Ether (1)–Ethanol (2) at 25°C

	A_{12}	A_{21}
Margules	1.0535	0.9679
Wilson	47.5508	646.2295
UNIQUAC	362.5405	−41.9356

Ethanol (1)–Benzene (2) at 45°C

	A_{12}	A_{21}
Margules	2.1412	1.3856
Wilson	1594.1685	142.1319
UNIQUAC	−147.1802	876.8192

Methyl Acetate (1)–Chloroform (2) at 50°C

	A_{12}	A_{21}
Margules	−0.8695	−0.5955
Wilson	19.0577	−422.5855
UNIQUAC	1349.3593	−740.5781

Water (1)–3-Hydroxy-2-Butanone(2) at 750 mmHg

	A_{12}	A_{21}
Margules	−0.2977	1.5637
Wilson	642.9837	1476.1815
UNIQUAC	1018.5457	−569.7499

Note: From Gmehling, J. and U. Onken, *Vapor–Liquid Equil. Data Collection, Dechema Chemistry Data Series*, Port Washington, NY, Scholium International, Inc., 1977. With permission.

EXAMPLE 1.7: PENTANE–HEXANE AND METHANOL–WATER

The phase behavior of two binaries, pentane–hexane and methanol–water, is checked against ideal solution behavior. The partial pressure and total pressure isotherms calculated according to Raoult's law at 35°C are compared to expected actual isotherms. The component vapor pressures at 35°C are as follows:

Pentane	106 kPa	Methanol	30.2 kPa
Hexane	33 kPa	Water	6.1 kPa

The isotherms are plotted on pressure–composition diagrams (Figures 1.3a and b). The curves calculated by Raoult's law are the broken straight lines. They are obtained simply by joining the pure-component vapor pressure points to the zero pressure point on the opposite side to get the partial pressure curves and by joining the two vapor pressure points to get the total pressure curve. The solid lines are obtained by simulation and represent expected behavior of the mixtures. The

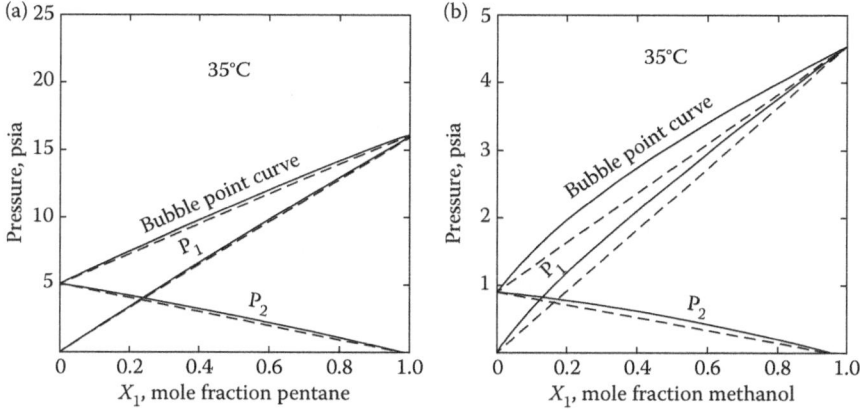

FIGURE 1.3 (a) Pressure–composition isotherms for pentane–hexane (Example 1.7), (b) pressure–composition isotherms for methanol–water (Example 1.7).

pentane–hexane binary is predicted by the Soave equation of state, and the methanol–water binary is predicted from activity coefficients calculated by the NRTL equation.

The deviation of the pentane–hexane binary from Raoult's law is negligible because the two components are non-polar and chemically similar and the system pressure is quite low. This is not the case with the methanol–water binary, where the deviations from Raoult's law are significant.

The deviation from ideality for the methanol–water system is considered positive because the actual pressure curves are higher than those calculated by Raoult's law. There are systems for which the deviation is negative; that is, actual pressures are lower than ideally calculated pressures.

EXAMPLE 1.8: DIETHYL ETHER–ETHANOL

Over the entire composition range in this binary at 25°C, the partial pressures of the individual components are greater than those predicted by Raoult's law, as is the total pressure (Figure 1.4). That is, the system exhibits positive deviation from ideality. The activity coefficients are consequently greater than unity for both components over the entire composition range (Figure 1.5). Moreover, the lower boiling component, diethyl ether, has the higher concentration in the vapor phase over the entire composition range (Figure 1.6).

EXAMPLE 1.9: ETHANOL–BENZENE

Here, too, the deviations from Raoult's law are positive for both components over the entire composition range at 45°C (Figures 1.7 and 1.8). The partial pressure deviations are large enough to result in a maximum in the total pressure (Figure 1.7). The binary thus forms a maximum pressure, or minimum temperature, azeotrope, a phenomenon discussed in more detail later in this section. In this binary, the concentration of ethanol is higher in the vapor than in the liquid below the azeotrope and is lower in the vapor than in the liquid above the azeotrope. At the azeotropic point, the concentrations in the liquid and vapor are identical (Figure 1.9).

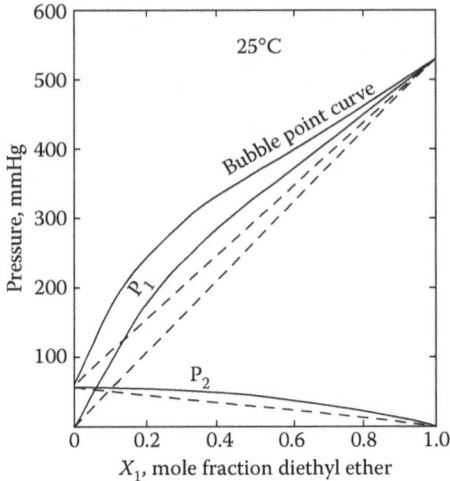

FIGURE 1.4   Pressure–composition diagram for diethyl ether–ethanol (Example 1.8).

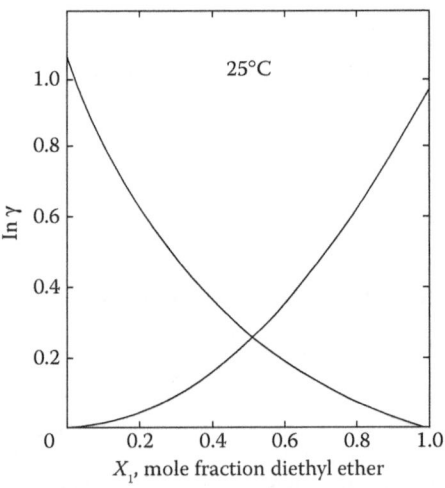

FIGURE 1.5   Activity coefficients versus composition for diethyl ether–ethanol (Example 1.8).

EXAMPLE 1.10: METHYL ACETATE–CHLOROFORM

This system is characterized by negative deviations from Raoult's law over the entire composition range at 50°C (Figures 1.10 and 1.11). The binary forms a minimum pressure, or maximum temperature, azeotrope (Figures 1.10 and 1.12).

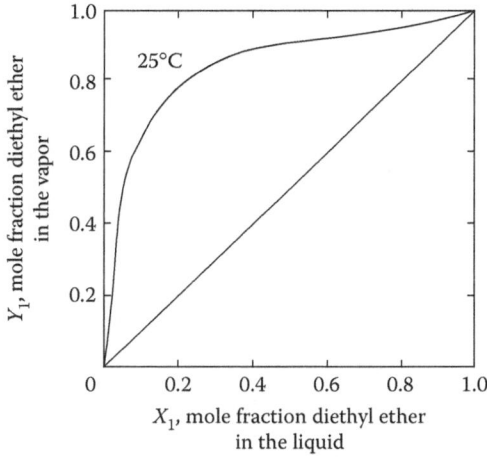

FIGURE 1.6 Y–X diagram for diethyl ether–ethanol (Example 1.8).

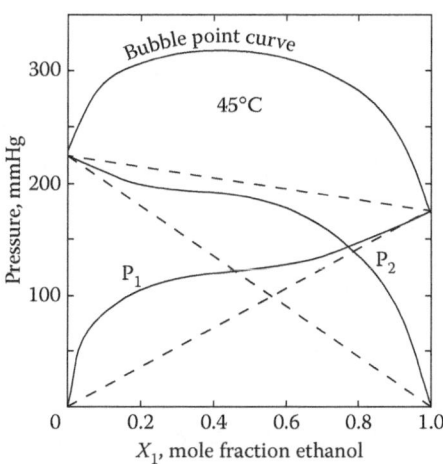

FIGURE 1.7 Pressure–composition diagram for ethanol–benzene (Example 1.9).

EXAMPLE 1.11: WATER–3-HYDROXY-2-BUTANONE

At 120°C, both components in this binary indicate positive deviation in one composition range and negative deviation in another (Figure 1.13). The system also indicates a slight maximum pressure azeotrope near a water mole fraction of 0.95 (Figures 1.13 and 1.15). Additionally, a maximum is observed for the activity coefficient of water (Figure 1.14, curve 1) and a minimum for the activity coefficient of 3-hydroxy-2-butanone (curve 2). These phenomena are best represented by the Margules equation.

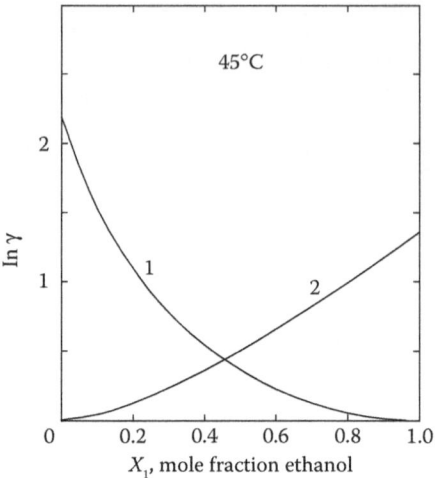

FIGURE 1.8   Activity coefficients versus composition for ethanol–benzene (Example 1.9).

1.3.4.1   Azeotropes

Example 1.9 shows that, for certain systems, deviations from Raoult's law can cause a maximum or a minimum in the vapor pressure to exist at a certain temperature and composition. At constant pressure, the boiling point or bubble point temperature curve could have a maximum or a minimum. Liquid mixtures whose vapor pressure curve or surface exhibits a maximum or a minimum are said to form azeotropes. The composition at which the azeotrope occurs is the azeotropic composition. Binaries are likely to form azeotropes if they deviate from Raoult's law and if their boiling points are not too far apart (within about 8°C). Azeotropes caused by positive deviations from Raoult's law are minimum boiling, that is, the azeotrope boils at a

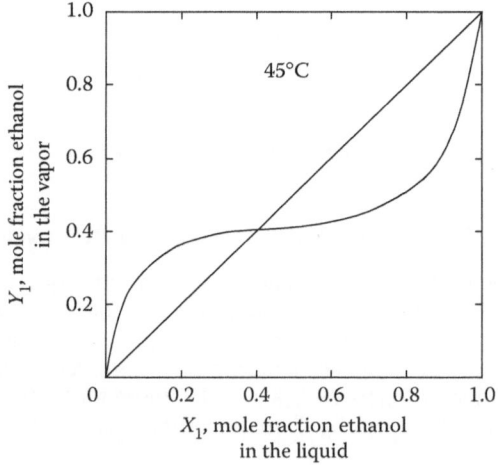

FIGURE 1.9 Y–X diagram for ethanol–benzene (Example 1.9).

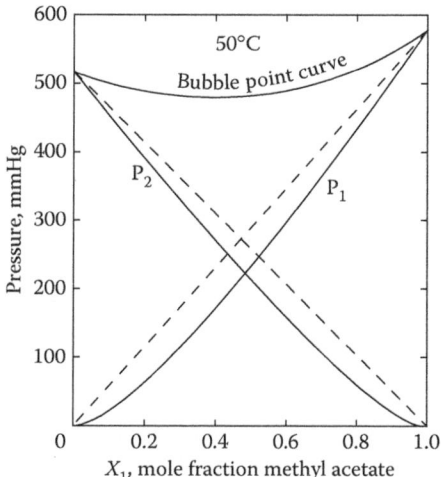

FIGURE 1.10 Pressure–composition diagram for methyl acetate–chloroform (Example 1.10).

temperature below the boiling point of either pure component. Conversely, azeotropes caused by negative deviations from Raoult's law are maximum boiling.

A binary azeotrope may be described on a vapor–liquid composition plot such as Figures 1.9, 1.12, and 1.15. On each side of the azeotropic point, a different component appears more volatile, having a higher concentration in the vapor than in the liquid. At the azeotropic point, the compositions of the two phases are identical.

The existence of an azeotrope in a given binary may be predicted on the basis of the chemical nature of the components, their boiling points, and so on. A quantitative prediction of the azeotropic composition, and the temperature and pressure at

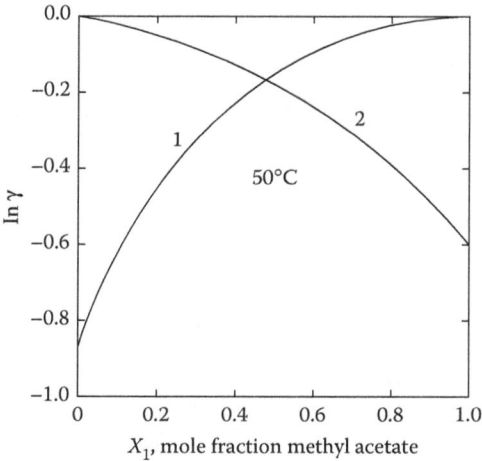

FIGURE 1.11 Activity coefficients versus composition for methyl acetate–chloroform (Example 1.10).

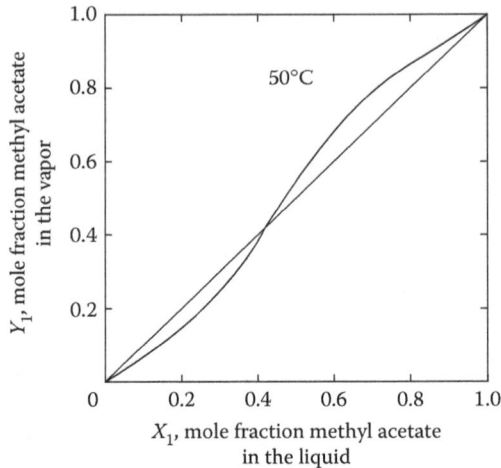

FIGURE 1.12   *Y–X* diagram for methyl acetate–chloroform (Example 1.10).

which it occurs, can be predicted with an appropriate activity coefficient equation. The interaction parameters of such an equation would have to be determined from experimental data in the first place. However, once the parameters are available, the effect on azeotropic composition of the temperature and pressure, within reasonable ranges, may be calculated from the activity coefficient equation and Equation 1.29 or 1.29a (see Example 1.6).

The separation of mixtures that form azeotropes often requires the addition of a third component called a solvent or an entrainer. These processes, described in Chapters 2

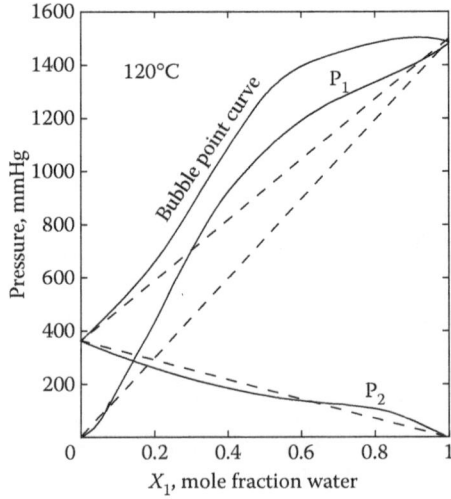

FIGURE 1.13   Pressure–composition diagram for water-3-hydroxy-2-butanone (Example 1.11).

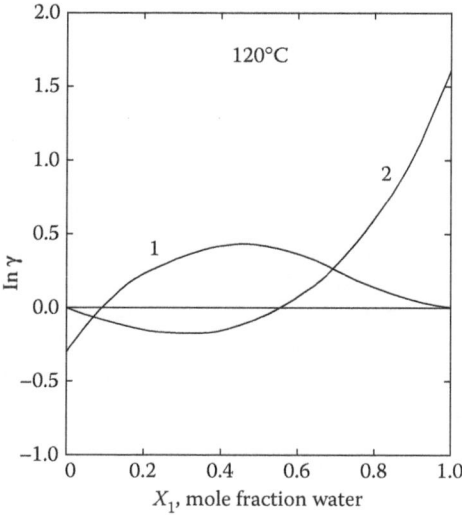

FIGURE 1.14 Activity coefficients versus composition for water-3-hydroxy-2-butanone (Example 1.11).

and 10, may result in the formation of multi-component azeotropes or heterogeneous azeotropes. Multi-component azeotropes, usually ternary, are mixtures whose equilibrium vapor and liquid phases have the same composition. An activity coefficient equation that can predict binary azeotropes could, in its generalized multi-component form, predict ternary azeotropes. Heterogeneous azeotropes are those forming two equilibrium liquid phases; they are discussed in the following section.

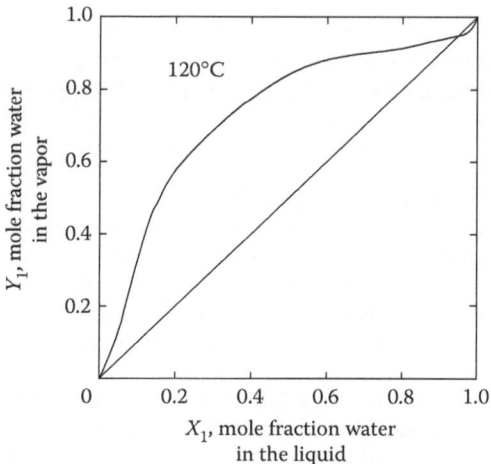

FIGURE 1.15 Y–X diagram for water-3-hydroxy-2-butanone (Example 1.11).

1.3.5 LIQUID–LIQUID AND VAPOR–LIQUID–LIQUID EQUILIBRIA

The condition for equilibrium between vapor and liquid phases, expressed by
Equation 1.19 as the equality of each component fugacity in the two phases, applies to
equilibrium between two liquid phases or any number of phases such as two liquids
and a vapor. When the deviations from ideality are large enough, mixtures can form
two immiscible liquids at equilibrium with each other. It is easy to see that an ideal
solution cannot form two liquid phases at equilibrium. In order for this to occur, the
condition $(p_i^0 X_i)^\alpha = (p_i^0 X_i)^\beta$, where α and β designate each liquid phase, must be satisfied. Since p_i^0, the vapor pressure of component i at system temperature is the same
in both phases, it follows that $(X_i)^\alpha = (X_i)^\beta$, that is, the two liquid phases are identical.

Since the liquid phases at equilibrium are non-ideal solutions, the component
fugacities in each phase are best expressed in terms of the activity coefficients.
Equation 1.19 becomes

$$\gamma_i^\alpha f_i^L X_i^\alpha = \gamma_i^\beta f_i^L X_i^\beta$$

or

$$K_i = X_i^\alpha / X_i^\beta = \gamma_i^\beta / \gamma_i^\alpha$$

1.3.5.1 Binary Systems

A non-ideal binary mixture may exist as a single liquid phase at certain compositions, temperatures, and pressures, or as two liquid phases at other conditions. Also,
depending on the conditions, a vapor phase may or may not exist at equilibrium with
the liquid. When two immiscible liquid phases coexist at equilibrium, their compositions are different, but the component fugacities are equal in both phases.

Equilibrium compositions of liquid phases at equilibrium are calculated by equating
the component fugacities, similar to vapor–liquid equilibrium calculations, described
in more detail in Chapter 2. The activity coefficients may be calculated by equations
presented in Section 1.3.3, in particular the UNIQUAC and NRTL equations. The
composition dependence of these equations is developed to the point where the same
equation with the same constants can predict activity coefficients over wide ranges of
composition, thus allowing it to predict two immiscible liquid phases at equilibrium.

EXAMPLE 1.12: N-BUTANOL–WATER

Typical non-ideal binaries forming two liquid phases is the n-butanol–water system at 1 atmospheric pressure. A solution with approximately 2 mole% n-butanol
in water exists at equilibrium with another liquid phase with approximately 38
mole% n-butanol in water. The fugacity of n-butanol in both phases is about
0.48. A phase diagram of this binary is illustrated in Figure 1.16. The curves, which
closely match the experimental data, are based on calculations using the NRTL
equation for activity coefficients.

The liquids in the immiscible region boil at a constant temperature of about
93°C. As long as two liquid phases coexist at equilibrium, the boiling point and the
equilibrium vapor composition are constant. This is a heterogeneous azeotrope:
the mixture has a minimum boiling point and the vapor is at equilibrium with two
liquid phases. The existence of a second liquid phase is a phenomenon that is

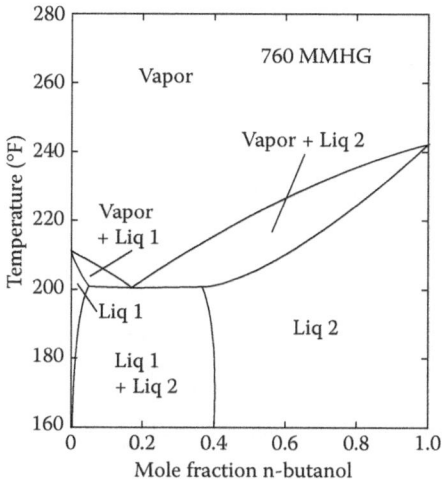

FIGURE 1.16 Temperature–composition for n-butanol–water (Example 1.12).

used to affect separation in certain processes, described in Chapter 10, by separating the phases and processing each liquid phase independently.

1.3.5.2 Ternary Systems

The phase compositions of an equilibrium binary system are determined if the temperature and pressure are given. In the n-butanol–water binary, for instance, the liquid phase compositions are given by the upper curves of the liquid 1 and liquid 2 regions in Figure 1.16. In ternaries or multi-component systems, the overall composition, as well as the temperature and pressure, must be given in order to determine the phase compositions. The difference between the two cases is a result of the phase rule, discussed in Chapter 2. A convenient means for graphically describing ternary systems forming two liquid phases is the triangular diagram such as the one described in Example 1.13.

EXAMPLE 1.13: ETHANOL–BENZENE–WATER

Figure 1.17 depicts a triangular diagram plotted for the ethanol–benzene–water ternary at a temperature of 30°C and a pressure of 150 kPa. The plots are based on calculations using the NRTL activity coefficient equation. Each of the vertices of the triangle represents a pure component, and the bases of the triangle represent binaries of the other two components. The mole percent of a component at any point in the triangle is measured by its distance from the triangle base opposite the vertex of that component.

The ethanol–benzene–water diagram has a single-liquid phase region and a two-liquid phase region, separated by the equilibrium curve ABC (Figure 1.17). If the composition of a single-liquid phase corresponding to point A is slightly altered such that the equilibrium curve is crossed to the two-liquid phase region, a second liquid phase is formed with a composition corresponding to point C. The tie line AC is the locus of points representing equilibrium liquid phases with compositions corresponding to points A and C. Thus, point D on the tie line represents two liquid phases at equilibrium with compositions given by points A and C.

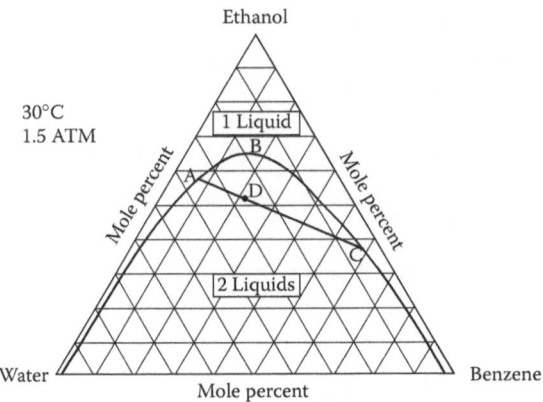

FIGURE 1.17 The ethanol–benzene–water system (Example 1.13).

The relative quantity of the phase with composition A to the phase with composition C equals the ratio of segments CD to DA, according to the lever rule.

Since ternary diagrams represent one temperature and one pressure, a set of diagrams would be required to describe the effects of temperature and pressure, as well as the effect of composition. A three-dimensional diagram could be constructed to describe the phase compositions–temperature relationship at a fixed pressure.

1.4   ENTHALPY

The calculation of multi-stage separation processes involves the solution of phase equilibrium relationships, mass balances, and energy balances. Energy balances require the computation of enthalpies of streams entering and leaving an equilibrium stage. The enthalpy is a function of state, defined in Section 1.1.2 as $H = U + PV$. It is a function of the stream composition, its temperature, and its pressure.

Energy balance calculations require evaluation of enthalpy changes rather than absolute values of the enthalpy. The variation in enthalpy with temperature and pressure is given by the equation

$$dH = \left(\frac{\partial H}{\partial T}\right)_P dT + \left(\frac{\partial H}{\partial P}\right)_T dP = C_P dT + \left(\frac{\partial H}{\partial P}\right)_T dP \qquad (1.37)$$

where C_P is the heat capacity at constant pressure. The enthalpy change resulting from a transition from T_1 and P_1 to T_2 and P_2 may be represented by three steps as shown in Figure 1.18: an isothermal process from T_1 and P_1 to T_1 and P_0, where P_0 is a low pressure such as atmospheric; an isobaric process from T_1 and P_0 to T_2 and P_0; and an isothermal process from T_2 and P_0 to T_2 and P_2.

In the low-pressure isobaric step, ideal gas behavior is assumed and ideal gas heat capacities are calculated from the experimental pure-component data. The mixture low-pressure enthalpy change is calculated as

$$H_2^0 - H_1^0 = \sum X_i (H_{i2}^0 - H_{i1}^0) = \sum X_i \int_{T_1}^{T_2} C_{Pi}^0 \, dT \qquad (1.38)$$

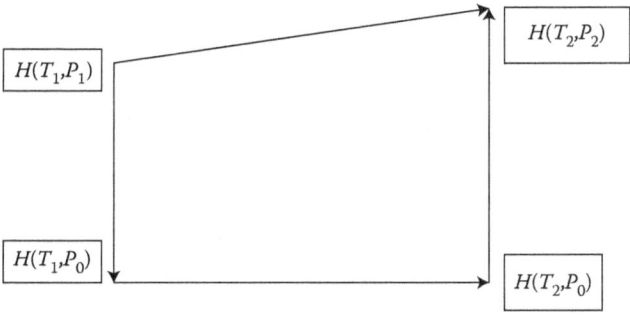

FIGURE 1.18 Enthalpy change with T and P.

where H_{i1}^0 and H_{i2}^0 are pure component ideal gas enthalpies and C_{Pi}^0 is pure component ideal gas heat capacity.

The isothermal steps represent the effect of pressure on the enthalpy. The enthalpy difference between the real state and ideal gas at the same temperature is the enthalpy departure and is generally calculated from PVT data, either experimental or predicted by an equation of state. From Equation 1.37, the enthalpy departure, or the variation in enthalpy from zero pressure to system pressure at constant temperature, is given by

$$\Delta H = \int_0^P \left(\frac{\partial H}{\partial P} \right)_T dP$$

From the definition of enthalpy (Section 1.1.2), a differential enthalpy change is given by

$$dH = dU + PdV + VdP$$

Substituting the expression for dU from Equation 1.3 gives

$$dH = TdS + VdP$$

At constant temperature,

$$\left(\frac{\partial H}{\partial P} \right)_T = T \left(\frac{\partial S}{\partial P} \right)_T + V$$

Using the Maxwell relation,

$$\left(\frac{\partial S}{\partial P} \right)_T = -\left(\frac{\partial V}{\partial T} \right)_P$$

the expression for the enthalpy departure is written as

$$\Delta H = \int_{0}^{P}\left[V - T\left(\frac{\partial V}{\partial T}\right)_{P}\right]dP = PV - RT - \int_{\infty}^{V}\left[P - T\left(\frac{\partial P}{\partial T}\right)_{V}\right]dV \quad (1.39)$$

The second form of the equation may be derived by starting with the differential enthalpy written as

$$dH = d(PV) + dU = d(PV) + TdS - PdV$$

$$\Delta H = \int_{\infty}^{V}\left(\frac{\partial H}{\partial V}\right)_{T}dV = \int_{\infty}^{V}d(PV) + \int_{\infty}^{V}\left[T\left(\frac{\partial S}{\partial V}\right)_{T} - P\right]dV$$

Substituting the Maxwell relation $(\partial S/\partial V)_{T} = (\partial P/\partial T)_{V}$ results in the second form of Equation 1.39, which would be used if the equation of state is pressure-explicit.

Equation 1.39 is general and applies to mixtures as well as pure substances in either the vapor or liquid phase. To use the equation, however, accurate PVT data for the system are required. These data may either be experimental or predicted such as by an equation of state. If the PVT method is capable of representing mixtures and the effect of composition, then the prediction of the enthalpy of mixing is inherent in Equation 1.39. Enthalpy changes accompanying phase changes can also be calculated if PVT data are available for both phases or if the equation of state is applicable to both phases. Equations of state, such as the Soave or Peng–Robinson equations, have been used for calculating enthalpy departures for both vapor and liquid phases. These equations are explicit in the pressure and may be conveniently incorporated in the second form of Equation 1.39, as detailed in the original equation of state references. The first form of Equation 1.39 may be used with equations of state that are explicit in the volume.

For liquid solutions that cannot be adequately represented by equations of state or other PVT predictive methods, the enthalpy of a mixture may be calculated by adding an excess enthalpy term, representing the enthalpy change due to mixing, to the enthalpy of an ideal solution:

$$H = \sum \bar{H}_{i} X_{i} + H^{ex}$$

In this equation, $\bar{H}_{i}$ is the enthalpy of pure component i at system temperature and pressure, and H^{ex} is the excess enthalpy. Equation 1.39 together with the equations defining the activity coefficient and the fugacity provide the basis for deriving an expression for the excess enthalpy in terms of the derivatives of the activity coefficients with respect to temperature. The result is

$$H^{ex} = -RT^{2}\sum X_{i}\left(\frac{\partial \ln\gamma_{i}}{\partial T}\right)_{P} \quad (1.40)$$

For the evaluation of the excess enthalpy by means of this equation requires the activity coefficient data as a function of temperature.

EXAMPLE 1.14: COMPUTING ENTHALPY DEPARTURE

Use the Redlich–Kwong equation of state to calculate the enthalpy departure of a mixture of acetone (1) and 1,3-butadiene (2) with mole fractions $Y_1 = 0.3$, $Y_2 = 0.7$ at 70°C and 200 kPa. Assume Kay's rules apply in calculating the equation of state mixture parameters.

$$T_{C1} = 508.2 \text{ K}, \ T_{C2} = 425.2 \text{ K}, \ P_{C1} = 4701 \text{ kPa}, \ P_{C2} = 4277 \text{ kPa}.$$

SOLUTION

By Kay's rules,

$$T_C = (0.3)(508.2) + (0.7)(425.2) = 450.1 \text{ K}$$

$$P_C = (0.3)(4701) + (0.7)(4277) = 4404.2 \text{ kPa}$$

The Redlich–Kwong equation parameters are

$$a = (0.42748)(8.314)^2(450.1)^{2.5}/4404.2 = 28836.4$$

$$b = (0.08664)(8.314)(450.1)/4404.2 = 0.0736$$

From the Redlich–Kwong equation (Equation 1.13),

$$P = \frac{RT}{V - b} - \frac{a}{T^{0.5}V(V + b)}$$

$$\left(\frac{\partial P}{\partial T}\right)_V = \frac{R}{V - b} + \frac{0.5a}{T^{1.5}V(V + b)}$$

Use the second form of Equation 1.39 to calculate the enthalpy departure:

$$\Delta H = PV - RT - \int_\infty^V \left[\frac{RT}{V - b} - \frac{a}{T^{0.5}V(V + b)} - \frac{RT}{V - b} - \frac{0.5aT}{T^{1.5}V(V + b)} \right] dV$$

$$= PV - RT + \frac{1.5a}{T^{0.5}} \int_\infty^V \frac{dV}{V(V + b)} = PV - RT + \frac{1.5a}{bT^{0.5}} \ln \frac{V}{V + b}$$

Compute V iteratively from the Redlich-Kwong equation, rearranged as follows:

$$V = \frac{RT^{1.5}V(V + b)}{PT^{0.5}V(V + b) + a} + b$$

A guess for V is used on the right-hand side of the equation to calculate V on the left-hand side. This new value is used in the next iteration on the right-hand side, and the iterations are repeated until the value of V converges. For the initial guess, use the ideal gas value:

$$V = RT/P = (8.314)(70 + 273.15)/200 = 14.26 \text{ m}^3/\text{kmol}$$

Iteration 1:

$$V = (8.314)(343.15)^{1.5}(14.26)(14.26 + 0.0736)/[(200)(343.15)^{0.5}(14.26)$$
$$\times (14.26 + 0.0736) + 28836.4] + 0.0736 = 13.82$$

Iteration 2: $V = 13.78$

Iteration 3: $V = 13.78 \text{ m}^3/\text{kmol}$ (converged)

Enthalpy departure,

$$\Delta H = (200)(13.78) - (8.314)(343.15) + (1.5)(28836.4)/[(0.0736)(343.15)^{0.5}]$$

$$\times \ln [13.78/(13.78 + 0.0736)] = -265.95 \text{ kJ/kmol}$$

EXAMPLE 1.15: EXCESS ENTHALPY FOR METHANOL–WATER MIXTURE

The enthalpy of a mixture of 0.5 kmol methanol and 0.5 kmol water at 10°C and 170 kPa is calculated as

$$H = 0.5H(\text{MeOH}) + 0.5H(\text{Water}) + H^{ex}$$

At 10°C and 170 kPa, the enthalpies of methanol and water are given as 724 and 715 kJ/kmol, respectively. The enthalpy of 1 kmol mixture at the same conditions is

$$H = (0.5)(724) + (0.5)(715) + H^{ex} = 719.5 + H^{ex}$$

The activity coefficients of methanol and water in solution at 170 kPa and two temperatures 3°C apart in the vicinity of 10°C are calculated by the van Laar equation. The derivatives are approximated by the ratio of finite differences. The results are

	γ (10°C)	γ (13°C)	$\ln \gamma$ (10°C)	$\ln \gamma$ (13°C)
Methanol	0.987	1.004	−0.0131	−0.0040
Water	0.892	0.920	−0.1143	−0.0834

The summation term in the excess enthalpy expression is approximated as

$$0.5(0.0040 + 0.0131)/3 + 0.5(-0.0834 + 0.1143)/3 = 0.008 \text{ K}^{-1}$$

The excess enthalpy is then

$$H^{ex} = -8.314(10 + 273)^2(0.008) = -5327 \text{ kJ/kmol}$$

The mixture enthalpy is

$$H = 719.5 - 5327 = -4607.5 \text{ kJ/kmol}$$

If the temperature dependence of the activity coefficients can be expressed analytically, Equation 1.40 may be used directly to calculate the excess enthalpy.

EXAMPLE 1.15A: EXCESS ENTHALPY IN A LIQUID SOLUTION OF 0.65 KMOL ETHANOL (1) AND 0.65 KMOL WATER (2) AT 2500 KPA AND 325 K

The activity coefficients are given as function of temperature (K):

$$\ln \gamma_1 = 0.4T^{0.2}X_2^2 = 0.4(325)^{0.2}(0.35)^2 = 0.1558$$
$$\ln \gamma_2 = 0.4T^{0.2}X_1^2 = 0.4(325)^{0.2}(0.65)^2 = 0.5374$$

$$\left(\frac{\partial \ln \gamma_1}{\partial T}\right)_P = (0.4)(0.2)\frac{T^{0.2}}{T}(X_2^2) = \frac{0.2 \ln \gamma_1}{T} = 0.00009588 \text{ K}^{-1}$$

$$\left(\frac{\partial \ln \gamma_2}{\partial T}\right)_P = (0.4)(0.2)\frac{T^{0.2}}{T}(X_1^2) = \frac{0.2 \ln \gamma_2}{T} = 0.00033071 \text{ K}^{-1}$$

$$H^{ex} = -(8.3145)(325)^2[(0.00009588)(0.65) + (0.00033071)(0.35)]$$
$$= -156.38 \text{ kJ/kmol}$$

1.4.1 ENTHALPY BALANCES INVOLVING PHASE CHANGE

Multi-stage separation calculations require heat balances involving transition between the phases. The enthalpy balance equations must, therefore, include heats of vaporization and condensation. Usually these heats are not calculated separately but are implied by using appropriate methods for calculating liquid enthalpies and vapor enthalpies. Consider, for instance, a system where a vapor stream and a liquid stream enter with enthalpies H_1 and h_1, respectively, and leave as a mixed phase stream with vapor enthalpy H_2 and liquid enthalpy h_2. These are total stream enthalpy rates with units such as kJ/h. If no heat is added to or removed from the system, an energy balance is written as

$$(H_2 + h_2) - (H_1 + h_1) = 0$$

Any heat of condensation or vaporization that accompanies this step is included in this energy balance.

It is possible to cause phase transition without the addition or removal of heat and without any pressure change by bringing together streams with distinct compositions. The process of mixing and attaining equilibrium may involve the transition of certain components from one phase to the other. As this takes place, temperature changes are expected due to heat of condensation or vaporization. Example 1.16 illustrates these effects.

EXAMPLE 1.16: ENTHALPY BALANCES IN THE NC4–NC6 BINARY

Normal butane and normal hexane are two chemically similar components, and a mixture of them at low pressures is expected to have characteristics approaching those of an ideal solution. Thus, for instance, the enthalpy of a liquid mixture of these two components may be approximated by the summation of the products of each pure component enthalpy at mixture conditions times its mole fraction. The excess enthalpy could be neglected.

The following are enthalpy data for a vapor stream 1 and a liquid stream 2 that mix, reach equilibrium adiabatically at the same pressure, and separate into a vapor stream 3 and a liquid stream 4. The calculations are based on the Soave equation of state.

	Stream 1 Vapor	Stream 2 Liquid	Stream 3 Vapor	Stream 4 Liquid
NC4 (kmol/h)	64.2	35.8	59.6	40.4
NC6 (kmol/h)	21.1	98.9	21.9	98.1
T (°C)	65.6	65.6	67.4	67.4
P (kPa)	276.0	276.0	276.0	276.0
Enthalpy (kJ/h)	2.73E6	1.58E6	2.66E6	1.65E6

A temperature rise from 65.6°C to 67.4°C is observed as a result of a net mass transfer from the vapor to the liquid (condensation). Since the system is assumed to behave as an ideal solution, no excess enthalpy exists and no heat of mixing contributed to the temperature rise. In nonideal mixing involving phase change, both effects, heat of mixing and phase change, can contribute to the temperature change. The two effects could either pull together or act in opposite directions.

To determine the heat associated with the net phase change, the above process could be carried out isothermally:

	Stream 1 Vapor	Stream 2 Liquid	Stream 3 Vapor	Stream 4 Liquid
NC4 (kmol/h)	64.2	35.8	53.6	46.4
NC6 (kmol/h)	21.1	98.9	17.6	102.4
T (°C)	65.6	65.6	65.6	65.6
P (kPa)	276.0	276.0	276.0	276.0
Enthalpy (kJ/h)	2.73E6	1.58E6	2.29E6	1.73E6

Maintaining the temperature at 65.6°C enhances mass transfer from the vapor to the liquid by removing the net heat of condensation, calculated as

$$\Delta H = (2.29E6 + 1.73E6) - (2.73E6 + 1.58E6) = -0.29E6 \text{ kJ/h}$$

1.5 CHARACTERIZING PETROLEUM FRACTIONS

Petroleum streams, from crude oil to products such as gasoline, kerosene, diesel fuel, fuel oil, and so on, are mixtures of large numbers of hydrocarbon components. It is impractical to analyze such mixtures and represent them compositionally on the basis of their constituent components. Petroleum streams are generally characterized in terms of small petroleum cuts, or "pseudocomponents," which are generated from their distillation curves. The pseudocomponents are identified primarily by their boiling point and specific gravity, and these two properties may be used to predict other component properties required for process calculations.

1.5.1　True Boiling Point (TBP)

In a perfect batch distillation process (Chapter 17) where a sharp separation of the components in a mixture occurs, a distillation curve can be represented by a number of steps equal to the number of components in the mixture (Figure 1.19). The boiling point at a given pressure (usually atmospheric) is plotted against the fraction distilled, usually on a volume basis. If, for instance, the mixture contains four components, the distillation curve would include four steps showing the boiling points and volumes of the components: Component 1 is of volume V_{C1} with boiling point T_{b1}, component 2 is of volume $V_{C2} - V_{C1}$ with boiling point T_{b2}, and so on.

The distillation curve of a petroleum mixture containing a large number of components would consist of many small steps. Since the separation can only approach perfect fractionation, and because of the large number of components in a complex mixture, the steps in the distillation curve or TBP curve of a petroleum mixture tend to merge into a smooth curve as shown in Figure 1.20.

1.5.2　Generating Pseudocomponents

If a petroleum mixture is represented by pseudocomponents corresponding to its TBP curve, its properties can be estimated by the same methods that apply to mixtures of chemical species. For instance, the mixture enthalpy and phase behavior can be predicted by the methods discussed in this chapter.

Generating the pseudocomponents from a TBP curve is accomplished by breaking the curve into a number of cuts, as shown in Figure 1.21. Temperatures T_{C1}, T_{C2}, T_{C3}, ... define the cut points. Thus, the first pseudocomponent boiling range is from

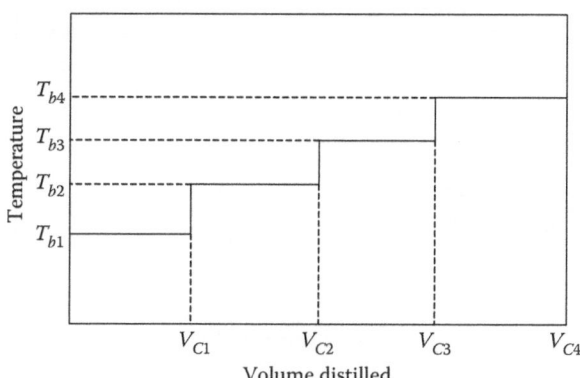

FIGURE 1.19　True boiling point curve of a mixture with a limited number of components.

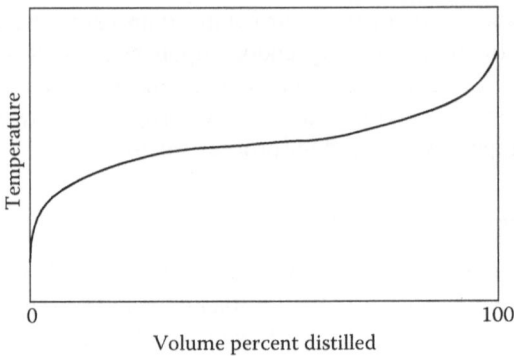

FIGURE 1.20 True boiling point curve of a petroleum mixture.

the initial boiling point of the mixture to T_{C1}, the second pseudocomponent from T_{C1} to T_{C2}, and so on. The number of cut points determines the number of pseudocomponents generated. The higher this number, the more accurately these pseudocomponents can reproduce the original TBP curve and represent the mixture. There are practical upper limits on the number of components, such as their effect on computing time in computer simulation. The spacing of the cut points does not necessarily have to be constant. It may be advantageous to define more cuts where the separation of the mixture is expected to take place, since more cuts would help produce a sharper separation.

The volume cut points correspond to the temperature cut points on the TBP curve: T_{C1} defines V_{C1}, T_{C2} defines V_{C2}, and so on (Figure 1.21). The boiling point T_{bi} of pseudocomponent i is calculated as an average:

$$T_{bi} = \frac{\int_{V_{C,i-1}}^{V_{Ci}} T dV}{V_{Ci} - V_{C,i-1}}$$

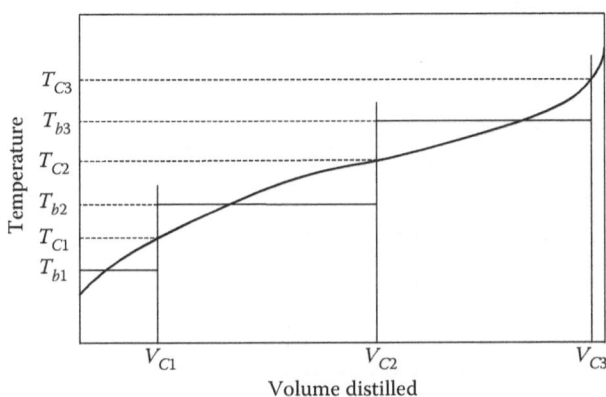

FIGURE 1.21 Generating pseudocomponents from a TBP curve.

If T, the TBP temperature, can be expressed as a mathematical function of V, the integral may be evaluated analytically. More commonly, the integration is done based on some curve fitting technique. Once the pseudocomponents are generated, their properties are estimated from correlations that are functions of the components' boiling points and specific gravities.

1.5.3 Laboratory Data

TBP curves are not usually determined directly by batch distillation since achieving complete fractionation in the laboratory is impractical. Instead, standardized batch distillation tests are conducted under closely defined conditions. Several such tests are specified by the American Society for Testing and Materials, including procedures D86, D1160, and D2887 (ASTM, 1990). These distillation curves are converted to TBP curves using the methods documented by the American Petroleum Institute (API, 1983).

The specific gravity may either be measured as an average or overall gravity of the mixture, or as discrete values at cut points obtained in the batch distillation test. In the latter case, the specific gravities at the cut points are plotted against the volume percent distilled. This gravity curve is used to determine the pseudocomponent gravities in a manner similar to the method used for determining the pseudocomponent boiling points.

If only the average specific gravity of the mixture is available, the individual pseudocomponent gravities can be estimated. One method involves the Watson characterization factor, K_W, defined as

$$K_W = (T_{ba})^{1/3}/S$$

where T_{ba} is an average boiling point of a pseudocomponent or of the mixture, in degrees Rankine, and S is the specific gravity of the pseudocomponent or the mixture. The method is based on the assumption that over reasonable ranges of boiling points, K_W is almost constant for a mixture and the cuts derived from it. The average boiling point of the mixture is calculated from its distillation curve (API, 1983). With known average boiling point and gravity, the mixture K_W can be calculated, and the pseudocomponent gravities are calculated from the above equation using their individual average boiling points. The calculated pseudocomponent gravities may have to be normalized in order to force their composite specific gravity to match the measured mixture specific gravity.

1.5.4 Pseudocomponent Properties

Stream properties required for solving material and energy balance equations and other process calculations are predicted from component properties. The properties of petroleum pseudocomponents can be estimated from their boiling points and specific gravities. The component properties include the molecular weight, critical constants, acentric factor, heat of formation, ideal gas enthalpy, latent heat, vapor pressure, and transport properties. These are predicted mainly by empirical correlations based on the experimental data. Many of these correlations are documented in the American Petroleum Institute Technical Data Book (API, 1983).

1.5.5 BLENDING STREAMS

When more than one hydrocarbon stream is involved in the same process, it is preferable to characterize all these streams using the same set of pseudocomponents. Toward this goal, the same temperature cut points for generating the pseudocomponents are used for all the streams. A common set of pseudocomponents is then generated by calculating for each temperature cut range a weighted average normal boiling point and specific gravity, and these are used to calculate a common set of pseudocomponent properties.

EXAMPLE 1.17: CHARACTERIZING A CRUDE OIL STREAM

In a laboratory analysis made on a sample of a crude oil stream, the average normal boiling point and the overall specific gravity are measured as 230.7°C and 0.80. The TBP data are also given:

Volume percent distilled	5	10	30	50	70	90	95
True boiling point (°C)	9	30	132	225	337	550	648

It is required to represent this crude oil by pseudocomponents with boiling points separated by approximately 50°C in the lower boiling range and 100°C in the upper boiling range.

SOLUTION

The TBP curve is plotted from the data provided (Figure 1.22). The cut points are set at 50, 100, 150, 200, 250, 300, 350, 400, 500, and 600°C. The TBP and volume percent of each cut are determined graphically from the plot. The Watson characterization factor is calculated from the mixture average boiling point and the overall specific gravity:

$$K_W = T^{1/3}/S = [(230.7 + 273.15)(1.8)]^{1/3}/0.80 = 12.1$$

The Watson K, assumed constant for all the cuts, is used to calculate their specific gravities:

$$S_i = (T_{bi})^{1/3}/K_W$$

For the first cut,

$$S_1 = [(14 + 273.15)(1.8)]^{1/3}/12.1 = 0.663$$

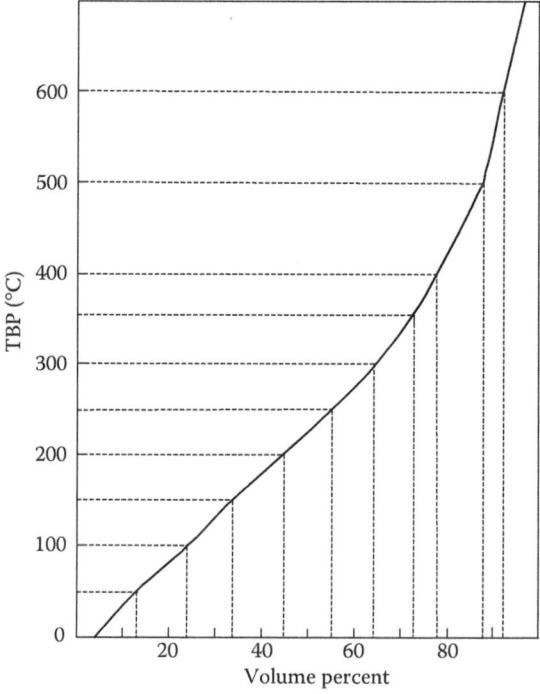

FIGURE 1.22   Crude oil TBP curve (Example 1.17).

The results are summarized in the table below. Using a volume basis of 1 m³ of sample, the masses of the pseudocomponents are also calculated from their specific gravities. The overall specific gravity calculated from the component gravities is 0.801, in close agreement with the measured value; hence no normalization of the component specific gravities is necessary.

Cut Points (°C)	Ranges (°C)	T_{bi} (°C)	S_i	Volume (m³)	Mass (kg)
50	<50	14	0.663	0.130	86.19
100	50–100	72	0.705	0.110	77.55
150	100–150	122	0.738	0.105	77.49
200	150–200	175	0.769	0.105	80.745
250	200–250	224	0.796	0.100	79.60
300	250–300	274	0.822	0.090	73.98
350	300–350	323	0.846	0.085	71.91
400	350–400	371	0.868	0.055	47.74
500	400–500	445	0.900	0.097	87.30
600	500–600	549	0.942	0.048	45.216
	600<	663	0.983	0.075	73.725
				1.000	801.446

NOMENCLATURE

A	Helmholtz function
C	Number of components
C_P	Heat capacity at constant pressure
f_i	Fugacity of pure component i
G	Gibbs free energy
H	Enthalpy
K	Vapor–liquid distribution coefficient
k_H	Henry's law constant
n	Number of moles
p	Partial pressure
P	Pressure
Q	Amount of heat
S	Entropy
T	Temperature
U	Internal energy
V	Volume
W	Work
X	Mole fraction in the liquid
Y	Mole fraction in the vapor
z	Compressibility factor
γ	Activity coefficient
ϕ	Fugacity coefficient
η	Carnot efficiency
μ	Chemical potential
ω	Acentric factor
Π	General property

SUBSCRIPTS

C	Critical conditions
i,j	Component designations
r	Reduced conditions

SUPERSCRIPTS

α, β	Phase designations
0	Ideal gas conditions or standard state conditions

PROBLEMS

1.1. An ideal gas enclosed in a cylinder is maintained at a constant temperature T. The cylinder is partitioned by a frictionless movable disk into two sections, one containing n_1 moles and the other n_2 moles. The starting pressure is P_1 in the first section and P_2 in the second. P_2 is greater than P_1.

a. Show that the system reaches equilibrium when the two pressures are equal. Hint: Calculate the entropy change. Start with Equation 1.3.

b. Calculate the entropy change if $T = 38°C$, $P_1 = 100$ kPa, $P_2 = 200$ kPa, $n_1 = 75$ kmol, and $n_2 = 25$ kmol.

1.2. An alternative form of the virial equation of state, truncated after the third virial coefficient is written as

$$z = 1 + B/V + C/V^2$$

Derive expressions for the virial coefficients in terms of the critical properties. What is the value of the critical compressibility factor, z_C?

Note: The results should be used carefully, with the understanding that the virial equation is not reliable in the critical region.

1.3. Carbon monoxide (CO) *PVT* behavior is well represented by the Soave–Redlich–Kwong (SRK) equation of state. For quick, approximate calculations over a limited range of conditions, it is proposed to use the virial equation of state of the form

$$z = PV/RT = 1 + B/V + C/V^2$$

to approximate the SRK predictions. The temperature and specific volume ranges for this application are 200 K and 0.5 m³/kmol to 250 K and 0.4 m³/kmol. Calculate virial equation coefficients to match SRK predictions at the range limits. Compare these coefficients to values obtained based on critical properties as derived in Problem 1.2. Also, compare the pressure calculated at the range limits by the two sets of coefficients.

CO properties: $T_C = 132.9$ K, $P_C = 3499$ kPa, $V_C = 0.0934$ m³/kmol, $\omega = 0.048$.

1.4. A 2 m³ tank is to be filled with 1 kmol of a mixture of propane and benzene. If the tank temperature is maintained at 200°C, calculate the final gas pressure in the tank from (a) the Redlich–Kwong equation of state using Equation 1.15 mixing rules with $k_{ij} = 0$, and (b) the virial equation of state in the form $z = 1 + B/V + C/V^2$. The coefficients in this equation may be estimated from the critical properties as derived in Problem 1.2. (See note in Problem 1.2). The mixture critical volume may be estimated from the mixing rule $V_C = (\Sigma Y_i V_{Ci}^{1/3})^3$. The gas composition and the properties of propane and benzene are as follows:

	kmol	T_C (K)	P_C (kPa)	V_C (m³/kmol)
1. Propane	0.35	369.8	4248	0.200
2. Benzene	0.65	562.2	4898	0.259

1.5. Equations of state may be applied to mixtures by determining the equation constants for the mixture from constituent pure component properties and constants using appropriate combining rules. The combining rules for the Redlich–Kwong equation constants for a binary mixture are given as

$$a = a_1 Y_1^2 + 2Y_1 Y_2 (a_1 a_2)^{0.5} + a_2 Y_2^2$$

$$b = b_1 Y_1 + b_2 Y_2$$

where Y_1 and Y_2 are the component mole fractions, and a_1, b_1, a_2, b_2 are the pure component Redlich–Kwong constants. Alternatively, mixture pseudocritical properties are calculated, and these are used to compute the mixture equation of state constants. The pseudocritical properties may be calculated using Kay's rules. Compare the compressibility factors obtained by these two methods for a 75% mol propane, 25% mole carbon dioxide mixture at 300 K, and a specific volume of 700 mL/gmol. The critical properties are given as follows:

	T_C (K)	P_C (atm)
1. Propane	369.7	41.9
2. Carbon dioxide	304.3	72.8

$$R = 8.3145 \text{ kJ/kmol K}$$

1.6. Derive equations to calculate component fugacity coefficients in a binary mixture using the virial equation of state truncated after the second virial coefficient. The mixture second virial coefficient is given as

$$B = Y_1 B_1 + Y_2 B_2 + Y_1 Y_2 \delta_{12}$$

where

$$\delta_{12} = 2B_{12} - B_1 - B_2$$

B_1 and B_2 are the second virial coefficients of components 1 and 2, and B_{12} is the interaction second virial coefficient. Express the virial equation of state in the following alternative form, truncated after the second virial coefficient:

$$PV/RT = z = 1 + BP/RT$$

Hint: The component fugacity coefficient in a mixture is a partial property, and is calculated as the partial derivative of the corresponding total property with respect to the number of moles of the component in question in the mixture.

1.7. Vapor–liquid equilibrium distribution coefficients, or K-values, may be calculated with varying degrees of accuracy. For a propylene–isobutane mixture at 50°C, where the two components are equimolar in the vapor phase, compare the K-values at 860, 1380, and 1900 kPa using each of the following methods:

a. Raoult's law

b. The virial equation of state for the vapor, truncated after the second virial coefficient. (See Problem 1.6). The following second virial coefficients are given at 50°C:

$$B_1 \text{ (propylene)} = -0.300 \text{ m}^3/\text{kmol}$$

$$B_2 \text{ (isobutane)} = -0.529 \text{ m}^3/\text{kmol}$$

The interaction second virial coefficient is given as

$$B_{12} = (1 - c_{12})(B_1 B_2)^{0.5}$$

The parameter c_{12} may be assumed close to zero for these components. What assumptions may be made to simplify the liquid fugacity calculations? Vapor pressures are obtained from the equation

$$\ln p_1^0 \text{ (kPa)} = 1.93 + a - b/[32 + c + 1.8T \text{ (°C)}$$

	a	b	c
1. Propylene	11.757	3253.6	412.6
2. Isobutane	11.592	3658.8	400.0

$$R = 8.3145 \text{ kJ/kmol K}$$

1.8. Activity coefficient data for heptane and toluene in their binary solution are given at 100 kPa:

X (heptane)	0.0	0.1	0.2	0.3	0.4	0.5	0.6	0.7	0.8	0.9	1.0
γ (heptane)	1.288	1.233	1.183	1.143	1.096	1.067	1.032	1.016	1.000	0.995	1.000
γ (toluene)	1.000	1.007	1.023	1.040	1.067	1.084	1.132	1.189	1.247	1.256	1.230

Check if these data are thermodynamically consistent based on the Gibbs–Duhem equation.

1.9. Derive expressions to calculate the van Laar constants from activity coefficient data. Such data are often available from azeotropic or infinite dilution measurements.

1.10. Hydrofluoric acid and water form an azeotrope at 120°C and atmospheric pressure, with a hydrofluoric acid mole fraction of 0.35. Calculate the activity coefficients of hydrofluoric acid and water at the azeotrope, and the van Laar constants. Assume ideal gas behavior in the vapor phase. The vapor pressures of hydrofluoric acid and water are given as follows:

$$\ln p_1^0 (\text{HF}) = 17.28 - 4495.9/(335.52 + T)$$

$$\ln p_2^0 (\text{H}_2\text{O}) = 16.65 - 4030.2/(235.00 + T)$$

where p^0 is in kPa and T is in °C.

1.11. Calculate the activity coefficients, bubble point pressure, and the vapor composition as a function of liquid composition for the hydrofluoric acid—water binary at 120°C. Use the van Laar equation with the constants calculated in Problem 1.10. Vapor pressure data may also be obtained from Problem 1.10. Assume ideal gas behavior in the vapor phase.

1.12. Derive equations to calculate the enthalpy departure using each of the following methods: (a) the ideal gas equation, (b) the virial equation of state truncated after the second virial coefficient, (c) the Soave–Redlich–Kwong equation of state.

1.13. An insulated vertical cylinder is fitted with two frictionless pistons forming compartments A and B. The upper piston is insulated and the lower piston, of negligible mass, separates the two compartments and can transfer heat. Compartment A contains n_A moles of an ideal gas at initial temperature T_{A1} and compartment B contains n_B moles of the same ideal gas at initial temperature T_{B1}. As heat is transferred across the lower piston, show that the maximum total entropy change is reached when the temperatures in both compartments are equal. Calculate the maximum total entropy change if $n_A = 60$ kmol, $n_B = 25$ kmol, $T_{A1} = 100$°C, and $T_{B1} = 150$°C. Assume C_{P0} is constant, equal to $(7/2)R$.

1.14. The virial equation as a power series of the pressure is to be used in an application for a quick estimation of the pressure–volume relationship of methane at a constant temperature of 200 K. In this application the specific volume of methane varies over a range from 2 to 4 m³/kmol. Estimate the second and third virial coefficients over this range based on data derived from the Soave equation of state at the range limits. Methane properties: $T_C = 190.6$ K, $P_C = 4599$ kPa, $\omega = 0.012$.

1.15. Compare the values of z for formaldehyde at 60°C with a specific volume of 1.0 m³/kmol using the Soave equation and the Redlich–Kwong equation. Formaldehyde properties: $T_C = 408.0$ K, $P_C = 6590$ kPa, $\omega = 0.282$.

1.16. Using the Redlich–Kwong equation, calculate the change in enthalpy per kmol of nitrogen as it is compressed from 100 to 1000 kPa at a constant temperature of 120 K. Nitrogen properties: $T_C = 126.2$ K, $P_C = 3390$ kPa.

1.17. Methane expands in an isothermal process at 200 K from 800 to 100 kPa. Calculate the enthalpy change per kmol methane using the Soave equation of state. The properties of methane are given as $T_C = 190.6$ K, $P_C = 4599$ kPa, $\omega = 0.012$.

1.18. Calculate the enthalpy departure of the mixture in Example 1.14 at the same conditions using the Redlich–Kwong equation, but with the mixing rules of Equation 1.15 instead of Kay's rules. Assume $k_{ij} = 0$.

1.19. Infinite dilution activity coefficient data are given below at two temperatures for acetone (1), and chloroform (2). Calculate the excess enthalpy of an equimolar solution of acetone and chloroform at 25°C based on the van Laar equation.

$T(°C)$	γ_1^∞	γ_2^∞
20	0.436	0.352
30	0.517	0.395

1.20. Derive an expression for the excess enthalpy based on the Wilson equation. Calculate the excess enthalpy of an equimolar solution of acetone (1)/chloroform (2) at 25°C, given the following Wilson parameters:

$$\delta_{12} = \lambda_{21} - \lambda_{11} = -304.4 \text{ J/mol}$$

$$\delta_{21} = \lambda_{12} - \lambda_{22} = -144.1 \text{ J/mol}$$

$$(\lambda_{12} = \lambda_{21})$$

The ratio of the molar volumes is $V_1/V_2 = 0.917$.

1.21. A liquid mixture of acetone(1) and methanol(2) at 80°C has 40% mole acetone and 60% mole methanol. It is required to compare the following two methods for calculating the equilibrium vapor composition and pressure of this system:

a. Use Raoult's law.

b. Assume the liquid behaves as an ideal solution and calculate the vapor phase fugacity coefficient from the virial equation,

$$z = \frac{PV}{RT} = 1 + \frac{BP}{RT}$$

Use the pressure from part (a) as an assumed value and update it in part (b), and calculate the vapor composition. The following data are given:

Pure component second virial coefficients: $B_1 = -784.97$ cm³/mol, $B_2 = -562.89$ cm³/mol.

The mixture second virial coefficient may be calculated as

$$B = B_1 Y_1 + B_2 Y_2$$

The vapor pressure is calculated from Equation 2.19, where the temperature is in K and the pressure in kPa. The equation constants are as follows:

	A_i	B_i	C_i
1. Acetone	14.392	2795.8	−43.00
2. Methanol	16.594	3644.3	−33.24

1.22. In a separation process it is required to check the severity of the effect of the liquid composition of the ethanol/MCP/benzene mixture on the activity coefficients. Using the information from Example 1.6A, calculate the activity coefficients of these components when the liquid composition is 8% mole ethanol, 18% mole MCP, and 74% mole benzene.

1.23. A liquid solution of ethanol (1) and benzene (2) at 35°C contains 30% mole ethanol. Infinite dilution activity coefficients are given: $\gamma_1^\infty = 8.9$, $\gamma_2^\infty = 4.1$. The vapor pressures of ethanol and benzene at 35°C are 16.625 and 49.875 kPa, respectively. Calculate the K-values of ethanol and benzene, the vapor phase composition, and the total pressure. The Margules liquid activity coefficient equation may be used, and the vapor phase may be assumed an ideal gas.

REFERENCES

Abrams, D. S. and J. M. Prausnitz, *A.I.Ch.E. Journal*, 21, 116, 1975.

American Society for Testing and Materials, *Annual Book of ASTM Standards*, 1990.

API Technical Data Book—Petroleum Refining, API, Washington, D.C., 1983.

Benedict, M., G.B. Webb, and L.C. Rubin, *Chem. Eng. Prog.*, 47(8), 419, 1951; 47(9), 449, 1951.

Black, C., *A.I.Ch.E. Journal*, 5(2), 249, 1959.

Carlson, H. D. and A. P. Colburn, *I & EC*, 34, 581, 1942.

Gmehling, J. and U. Onken, *Vapor–Liquid Equilibrium Data Collection, Dechema Chemistry Data Series*, Port Washington, NY, Scholium International, Inc., 1977.

Kay, W. B., *Ind. Eng. Chem.*, 28, 1014, 1936.

Lee, B. I. and M. G. Kesler, *A.I.Ch.E. Journal*, 21, 510, 1975.

Peng, D. Y. and D. B. Robinson, *Ind. Eng. Chem. Fundam.*, 15, 59, 1976.

Plocker, U., H. Knapp, and J. M. Prausnitz, *Ind. Eng. Chem. Process Des. Dev.*, 17, 324, 1978.

Prausnitz, J. M., C. K. Eckert, R. V. Orye, and J. P. O'Connell, *Computer Calculations for Multicomponent Vapor–Liquid Equilibria*, NJ, Prentice-Hall, 1967.

Rackett, J., *Chem. Eng. Data*, 15, 514, 1970.

Renon, H. and J. M. Prausnitz, *A.I.Ch.E. Journal*, 14, 135, 1968.

Soave, G., *Chemical Engineering Science*, 22, 1197, 1972.

Starling, K. E., *Fluid Thermodynamic Properties for Light Petroleum Systems*, Houston, Gulf Publishing Company, 1973.

Walas, S. M., *Phase Equilibria in Chemical Engineering*, Boston, Butterworth Publishers, 1985.

Wilson, G. W., *J. Am. Chem. Soc.*, 86, 127, 1964.

Wohl, K., *Trans. Am. Inst. Chem. Engr.*, 42, 215, 1946.

2 The Equilibrium Stage

Multistage phase separation processes operate on the principle that when binary or multi-component mixtures form two or more phases at equilibrium with each other, each phase generally has a distinct composition. If, for example, a multi-component vapor mixture is at equilibrium with a liquid phase, the vapor will, in general, be richer in the lighter components than the liquid, whereas the liquid will be richer in the heavier components. Thus, if one desires to separate a given feed stream into lighter and heavier components, a step in the right direction would be to bring the stream to such temperature and pressure that would result in the formation of two phases—a vapor and a liquid, then to separate the two phases in a flash drum. As a closer look will indicate, however, the separation power of a single equilibrium stage is rather limited.

A stage in a separation process could, for instance, be a tray in a distillation column, a section in a packed absorption column, or a partial condenser or reboiler. An equilibrium stage is a contacting device—a vessel—in which two or more phases can mix thoroughly and interact with each other for a sufficient length of time until the number of moles of any component moving from one phase to another equals the number of moles of that component moving in the opposite direction, that is, until no more net mass transfer takes place between the phases. The composition of each phase is then the equilibrium composition.

In the majority of multistage separation processes, the phases in question are vapor and liquid. Vapor–liquid–liquid equilibrium (VLLE) can exist when two partially immiscible liquids are at equilibrium with each other, as well as with a vapor phase. This is a phenomenon that can occur under certain conditions in phase separation processes. Applications in this book cover vapor–liquid, liquid–liquid, and vapor–liquid–liquid separation processes. Chapter 18 discusses membrane separation processes, where the separation is not brought about as a result of different phase compositions, but by the action of membrane barriers.

The equilibrium stage may be thought of as the building block of multistage phase separation processes. The operation of a particular process depends on the number of stages and their configuration and on the way they are designed to interact with each other. This chapter deals with the equilibrium stage, its behavior, the relationships between the parameters that determine its operation, and methods for solving the equilibrium stage equations.

Although real-life processes may not actually reach equilibrium, the assumption of an equilibrium stage is essential for a rigorous solution of a problem and for providing a sound basis for column design and performance evaluation. The application of equilibrium stage calculation results to actual performance information which is usually accomplished by utilizing tray efficiencies, a concept that is discussed in Chapter 14.

2.1   PHASE BEHAVIOR

The thermodynamic conditions of a system may be defined by a set of parameters, the independent variables. Once their values are fixed, the other parameters, the dependent variables, assume certain values that may be either measured or otherwise determined. The selection of the particular set of independent variables is generally arbitrary, although these parameters must uniquely define the system. In the case of the phase behavior of a mixture in an equilibrium stage, the independent variables could be the overall composition of the mixture, its temperature, and pressure. Fixing these parameters determines whether the system is single- or multiple-phase, the relative amounts of the phases, the composition of each phase, its enthalpy, and so on. Quantitative information about these dependent variables must rely on experimental measurements or may be predicted on the basis of thermodynamic principles, along with data representing the constituents and the interactions among them.

2.1.1   DEGREES OF FREEDOM

The number of independent variables required to define the state of a system is determined by the degrees of freedom. The degrees of freedom in an equilibrium stage are consistent with the phase rule, which may be expressed as follows:

$$F = C - \varphi + 2 \tag{2.1}$$

where F is the number of degrees of freedom, C is the number of components, and φ is the number of phases. For a single-component, two-phase system, there is one degree of freedom, the temperature or the pressure.

For a system consisting of C components, the phase rule indicates that, in the two-phase region, there are $F = C - 2 + 2 = C$ degrees of freedom. That is, it takes C independent variables to define the thermodynamic state of the system. The independent variables may be selected from a total of $2C$ intensive variables (i.e., variables that do not relate to the size of the system) that characterize the system: the temperature, pressure, $C - 1$ vapor-component mole fractions, and $C - 1$ liquid-component mole fractions. The number of degrees of freedom is the number of intensive variables minus the number of equations that relate them to each other. These are the C vapor–liquid equilibrium relations, $Y_i = K_i X_i$, $i = 1, \ldots , C$. The equilibrium distribution coefficients, K_i, are themselves functions of the temperature, pressure, and vapor and liquid compositions. The number of degrees of freedom is, thus, $2C - C = C$, which is the same as that determined by the phase rule.

If the overall composition variables in a two-phase mixture are included, the number of variables will increase by $C - 1$ overall component mole fractions. This requires $C - 1$ additional equations to relate the overall component mole fractions to the liquid and vapor component mole fractions (Section 2.3). These equations include one additional variable, the vapor mole fraction (the ratio of vapor moles to total moles). Thus, the total number of variables becomes $2C + C - 1 + 1 = 3C$, and the total number of equations becomes $C + C - 1 = 2C - 1$, bringing the number of degrees of freedom to $C + 1$. If the overall composition of a mixture is set, $C - 1$

variables are fixed, leaving two degrees of freedom. It is possible, for instance, to independently fix the temperature and the pressure, or the temperature and the vapor mole fraction, or the pressure and the vapor mole fraction, or the pressure and one component mole fraction in the vapor or the liquid, and so on. The values at which these variables may be fixed must, of course, be feasible.

2.1.2 Phase Diagrams

A graphical representation of phase behavior provides a useful tool for studying the relationships among the various interacting variables that define the system. It also helps understand many terms that are frequently used in describing multistage separation processes. The most common vapor–liquid equilibrium diagrams are the phase envelope, or P–T diagram, for binary or multi-component systems and the T–Z and Y–X diagrams for binary systems.

2.1.2.1 The Phase Envelope

The phase envelope of a mixture is analogous to the vapor pressure curve of a pure component. The vapor pressure curve defines temperature and pressure conditions at which a pure component can exist as vapor and liquid at equilibrium. In this two-phase regime only one parameter, the temperature or the pressure, may be varied independently.

A schematic of a pressure–temperature diagram for a fixed composition mixture is shown in Figure 2.1. The phase representation of a mixture on a P–T diagram is bivariant rather than univariant as in the case of a pure-component vapor pressure curve. At temperature T_1 and pressure P_1, represented by point A, the mixture is

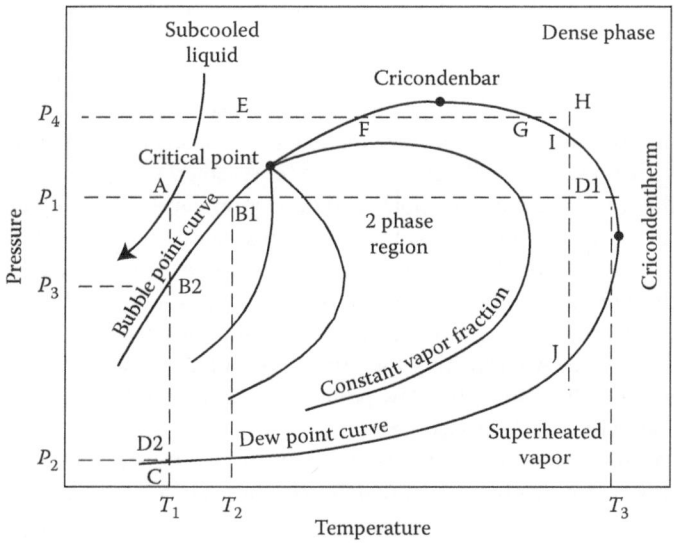

FIGURE 2.1 Schematic of a phase envelope.

a sub-cooled liquid, that is, no change of phase (vaporization) results from small changes in temperature and/or pressure. If the temperature is gradually increased at constant pressure, the vapor pressures of the constituents rise, causing the composite vapor pressure of the mixture to go up. The bubble point B1 is reached at temperature T_2 when the mixture vapor pressure equals the system pressure, P_1, and the first vapor bubble appears. Further increase in the temperature generates an increasing amount of vapor at equilibrium with the remaining liquid. The composition of the liquid changes continuously with temperature. At any given temperature in the two-phase region, the composite vapor pressure of the liquid equals the system pressure. The dew point, D1, is reached at temperature T_3 when the last liquid drop evaporates. With additional isobaric temperature rise, the vapor becomes superheated. Within the two-phase region, the vapor and liquid are saturated, that is, the vapor is at its dew point and the liquid is at its bubble point. In another example process, a superheated vapor at point C is compressed isothermally at temperature T_1 until condensation starts at pressure P_2, at the dew point (point D2). Further isothermal compression results in more condensation until all the vapor has condensed at pressure P_3, which defines bubble point B2. If the pressure is raised isothermally above the bubble point, the liquid becomes sub-cooled.

Bubble points and dew points may be generated as described above for a given mixture over ranges of temperature and pressure. The locus of bubble points is the bubble point curve and the locus of dew points is the dew point curve. The two curves together define the phase envelope. In addition to the bubble point curve (total liquid saturated) and the dew point curve (total vapor saturated), other curves may be drawn representing constant vapor mole fraction. All these curves meet at one point, the critical point, where the vapor and liquid phases lose their distinctive characteristics and merge into a single, dense phase.

The critical point of a mixture is determined by the critical points of the constituents; however, no explicit relationship is available relating the former to the latter. Mixture pseudo-critical temperatures and pressures are empirically defined averages of one kind or another of the individual components' critical temperatures and pressures. One such method follows Kay's rules, described in Section 1.2.3. Pseudo-critical properties are used in various methods for correlating thermodynamic data. The true critical point, as defined above and represented in Figure 2.1, is best determined as the limit of the bubble point and dew point curves. Moving along the bubble point curve, bubble point pressures are determined at incremental temperatures. The lowest temperature at which no phase transition can be detected by raising or lowering the pressure is the critical temperature. Similarly, the lowest pressure at which no phase transition can be detected by raising or lowering the temperature is the critical pressure. Approaching the critical point along the dew point curve is, in general, more involved since the dew point curve may include a region of retrograde condensation, which is discussed next.

Unlike a pure component, the critical point of a mixture is not necessarily the highest temperature or pressure at which two phases can coexist. This fact may be observed by studying the phase envelope shown in Figure 2.1 and in particular the dew point curve, whose shape is characteristic of a variety of mixtures. Starting at point E in the single-phase region where the pressure, P_4, is higher than the critical and the temperature is lower than the critical, the temperature is raised isobarically to point F on the phase envelope. As the mixture starts separating into two phases, it is the vapor and not the liquid that makes up most of the mixture. Thus, raising the temperature causes condensation, hence the term retrograde condensation. Point F is on the retrograde segment of the dew point curve. As the temperature is further raised isobarically, the amount of liquid increases to a maximum then starts decreasing until another dew point G is reached at the same pressure, P_4. Similar behavior occurs if the pressure is lowered isothermally from certain parts in the single-phase region, such as point H, to where retrograde condensation starts at point I. The amount of liquid then goes through a maximum, followed by complete vaporization at point J, another dew point at the same temperature. Condensation that takes place with temperature rise or with pressure drop is known as retrograde condensation, and the temperature/pressure conditions where this occurs define the retrograde region. In such systems, vapor and liquid can coexist at temperatures and pressures above the critical point. The highest temperature and the highest pressure where two-phases can coexist are called, respectively, the cricondentherm and cricondenbar, also shown in Figure 2.1.

In areas surrounding the phase envelope, the system exists as a single phase. Below the dew point curve and at higher temperatures, it is a superheated vapor, while areas above the bubble point curve and to the left of it represent a sub-cooled liquid. In areas above the phase envelope between the sub-cooled liquid and the superheated vapor, the mixture is a dense or supercritical fluid, with properties changing gradually from those typical of a liquid to those typical of a vapor.

2.1.2.2   T–Z Diagram

In this diagram, applicable mainly to binary systems, the temperature, pressure, and overall composition are the independent variables, with the pressure held constant. The diagram, shown schematically in Figure 2.2, consists of an upper curve representing dew points and a lower curve representing bubble points. The Z coordinate represents overall mole fraction of component 1, usually chosen as the more volatile component. A vertical line at $Z = 0$ corresponds to pure component 2, and point A represents its boiling point at the fixed system pressure. Similarly, pure component 1 is represented by a vertical line at $Z = 1$, and its boiling point by point B. Points above the dew point curve are in the vapor phase, and those below the bubble point curve are in the liquid phase, while the area between the two curves corresponds to the mixed phase.

The curves on a T–Z diagram are isobaric. If the pressure effect is to be included, a three-dimensional plot may be visualized, where the third coordinate would represent the pressure or, more simply, a series of curves similar to Figure 2.2 may be drawn, each representing a different pressure.

If a mixture with composition Z_1 is brought to the pressure corresponding to a given T–Z diagram and the temperature is fixed at T_0 in the two-phase region, the mixture separates into a liquid phase with composition X_1 and a vapor phase with composition Y_1. Equilibrium liquid and vapor compositions are determined by the intersection of a horizontal line through T_0 with the bubble point and dew point curves, respectively (points C and D in Figure 2.2).

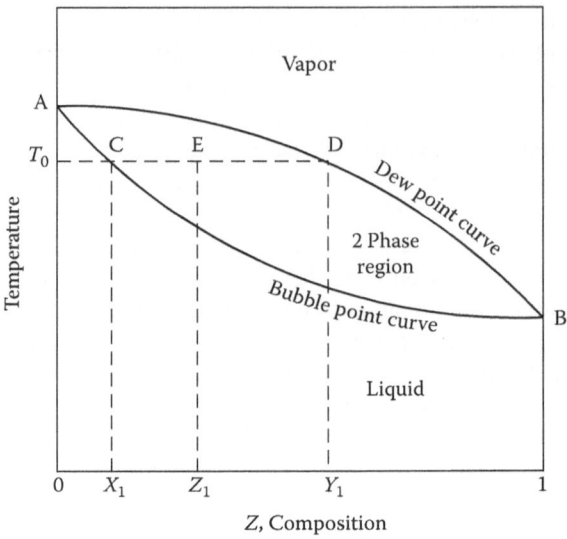

FIGURE 2.2 Schematic of T–Z diagram.

The relative amounts of vapor and liquid may also be evaluated from the T–Z diagram. A material balance on component 1 is written as

$$FZ_1 = F\psi Y_1 + F(1 - \psi)X_1 \tag{2.2}$$

where F is the total moles of vapor and liquid and ψ is the vapor mole fraction. Rearranging the equation, the vapor-to-liquid ratio is calculated as

$$\frac{\psi}{1 - \psi} = \frac{Z_1 - X_1}{Y_1 - Z_1} = \frac{\overline{CE}}{\overline{ED}} \tag{2.3}$$

where $\overline{CE}$ and $\overline{ED}$ are the line segments on the bubble point side and dew point side, respectively. This relationship is known as the lever-arm rule, which is a graphical representation of the material balance.

2.1.2.3 Y–X Diagram

As in the case of the T–Z diagram, this phase representation is applicable only to binary systems. Isobaric binary vapor–liquid equilibrium data are plotted as the mole fraction of the lower boiling component in the vapor, Y_1, versus the mole fraction of the same component in the liquid, X_1, as shown in Figure 2.3. A 45-degree diagonal representing $Y_1 = X_1$ is drawn for reference. In general, the vapor is consistently richer in the lower boiling component than the liquid, that is, $Y_1 > X_1$ over the entire composition range. Therefore, in general, the equilibrium curve lies entirely above the diagonal, as typified by curve A in Figure 2.3.

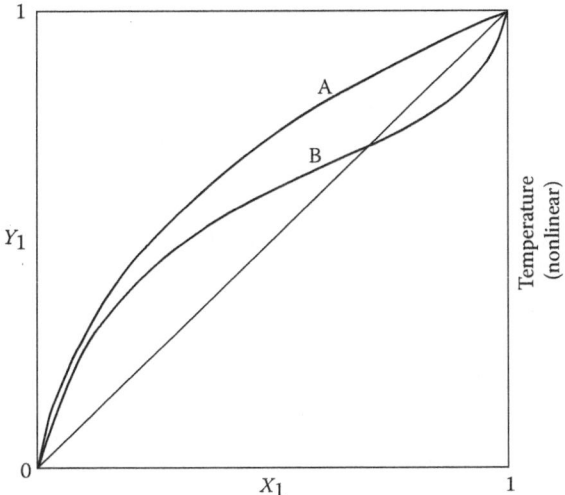

FIGURE 2.3   Schematic of Y–X diagram.

Systems that form azeotropes, however, exhibit unique characteristics as discussed in Section 1.3.4. In a typical azeotrope-forming binary, the concentration of the low boiling component in the vapor, Y_1, is greater than that in the liquid, X_1, at low concentrations of the light component. This behavior is similar to the general case. At the azeotropic point, the vapor and liquid compositions are identical, $Y_1 = X_1$, at which point the equilibrium curve crosses the diagonal. At higher concentrations of the low boiling component, the liquid becomes richer in that component, $X_1 > Y_1$, and the equilibrium curve falls below the diagonal. Curve B in Figure 2.3 represents this type of azeotropic system.

Points plotted on Y–X diagrams in general represent isobaric vapor–liquid equilibrium data. Data from a T–Z diagram may easily be plotted on a Y–X diagram. The temperature obviously changes along the equilibrium curve on a Y–X diagram. In order to show its variation on the same diagram, a nonlinear temperature coordinate may be included, parallel to the Y coordinate, as shown in Figure 2.3.

2.1.3   DISTRIBUTION COEFFICIENTS

It takes two independent variables to uniquely define a two-phase system having a fixed overall composition. Quantitative determination of the other variables requires thermodynamic data, such as enthalpy and component equilibrium distribution coefficients, or K-values. Distribution coefficients are defined as

$$K_i = \frac{Y_i}{X_i} \tag{2.4}$$

where Y_i and X_i are the mole fractions of component i in the vapor and liquid, respectively. The thermodynamic principles underlying the distribution coefficients and enthalpy, as well as methods for predicting them, are discussed in Chapter 1. Of consequence at this juncture is the fact that these properties are, in general, functions of temperature, pressure, and composition. Thus, if a mixture has fixed overall composition, its thermodynamic properties, whether measured or predicted, are uniquely determined by two independent variables. This statement applies to multiphase as well as single-phase systems. In fact, the two independent variables, plus pertinent thermodynamic data, determine whether the system exists as single phase or mixed phase.

2.1.4 Flash Operations

The main variables associated with phase relationships include the overall composition, Z_i, temperature, pressure, liquid composition, X_i, vapor composition, Y_i, vapor mole fraction, ψ, and heat transferred, Q. A process in which Z_i and two other independent variables are set, and equilibrium separation of the phases is allowed to take place, is called a flash operation. A general flash operation is shown in Figure 2.4. A feed stream initially at conditions T_1 and P_1 is controlled so that its final conditions satisfy two specifications. The feed is of fixed rate and composition, F and Z_i. A heat duty, Q, may be added to or removed from the system as required. The feed is flashed to generate a vapor product with flow rate $F\psi$ and a liquid product with flow rate $F(1 - \psi)$, where ψ is the vapor mole fraction at flash conditions T_2 and P_2. In general, ψ may be equal to zero or one or any value in between. The enthalpies of the vapor and liquid products are H_2 and h_2, respectively. The type of flash operation

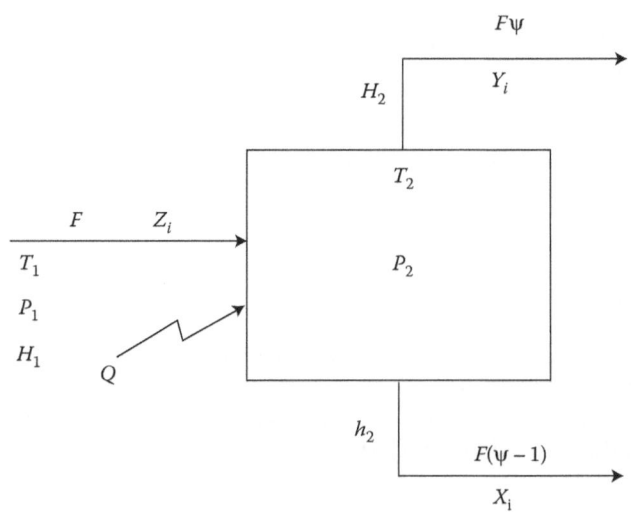

FIGURE 2.4   General flash operation.

depends on which two independent variables are specified. Following are different possible cases.

2.1.4.1 Isothermal Flash

The specified variables are the final temperature and pressure, T_2 and P_2. The dependent variables are the vapor fraction, ψ, the liquid and vapor compositions, X_i and Y_i, the total enthalpy of the two phases, $h_2 + H_2$, and the heat duty, Q. The term isothermal should not be interpreted to imply that the transition from initial conditions to final conditions is at constant temperature; T_1 is, in general, different from T_2. It simply means that within the flash drum the temperature, as well as the pressure, is fixed. The heat duty required to bring about the final conditions is equal to the enthalpy change, $Q = (H_2 + h_2) - H_1$, where H_1 is the enthalpy at T_1 and P_1. Isothermal flash conditions may be represented by a point (T_2, P_2) on the phase envelope diagram. It is clearly possible that this point may fall either within the phase envelope or outside it, in which case the system would be all vapor or all liquid (or dense phase). A flash drum operating at such conditions would have a single product and no phase separation would take place. In a single-phase situation, the dependent variables are the properties of the vapor or liquid product. The liquid or vapor composition is, of course, identical to the feed or overall composition, Z_i. Note that any set of temperature and pressure specifications is feasible.

2.1.4.2 Adiabatic Flash

In this operation, the specified variables are the temperature or pressure and the heat duty, Q. Specifying the duty is equivalent to specifying the final enthalpy, $h_2 + H_2$. The dependent variables are the pressure or temperature, the vapor fraction, and the vapor and liquid compositions. In a truly adiabatic process $Q = 0$ (or $h_2 + H_2 = H_1$), the term adiabatic flash is generally applied to a process where the heat duty is specified. The problem is to determine a temperature (or pressure if the temperature is specified) at which the total enthalpy of the products satisfies the heat duty specification. Once T_2 and P_2 are known, the problem is handled as an isothermal flash. Again, in this case, the solution could result in a single phase or a mixed phase, and any set of temperature (or pressure) and heat duty specification is feasible.

2.1.4.3 Bubble Point

If one independent variable is the temperature or pressure and the only product is a saturated liquid (i.e., the other independent variable is the vapor fraction set at the zero limit), then the operation is a bubble point flash. The dependent variables are the pressure or temperature, the heat duty, and the liquid and vapor compositions. Since the liquid is the only product, its composition is identical to the feed composition. The vapor composition refers to the composition of a vapor at equilibrium with the liquid. Although its rate is zero, its composition is determinable and is equal to the composition of the first vapor bubble resulting from infinitesimal vaporization. Bubble points are not feasible at all temperatures and pressures. As discussed in

Section 2.1.1, Figure 2.1 indicates that no bubble points can exist at temperatures or pressures above the critical point.

2.1.4.4 Dew Point

In a dew point flash, the fraction vapor is set at the unity limit, at a specified temperature or pressure. The dependent variables are the pressure or temperature, the heat duty, and the liquid and vapor compositions. The only product is a saturated vapor with composition equal to that of the feed. The equilibrium liquid composition corresponds to the first liquid drop resulting from infinitesimal condensation. As in the bubble point flash, dew points can exist only within certain ranges of temperature and pressure. Referring to Figure 2.1, no dew points can exist at temperatures above the cricondentherm or at pressures above the cricondenbar. At temperatures or pressures where retrograde condensation can occur, there can be two dew points: a normal dew point and a retrograde dew point.

2.1.4.5 General-Type Flash

In general, one may wish to set any two specifications on flash products or conditions. For example, one may specify component concentrations in one of the products, or a product rate, plus the temperature or pressure. A flash may be operated to satisfy separation specifications, such as in multistage separation processes. The following section discusses the performance of a single equilibrium stage as a separation process for different types of systems.

2.2 PERFORMANCE OF THE EQUILIBRIUM STAGE

Consider an equilibrium stage with a feed stream at fixed rate, composition, and thermal conditions (temperature and pressure). Let the two independent variables defining the system be the pressure and heat duty, making it an adiabatic flash. If now it is desired to specify another variable such as the temperature or the vapor or liquid rate or composition, then at least one of the independent variables—the pressure or heat duty—must be allowed to vary in order for the new specification to be satisfied. Certain parameters are inherently or by necessity fixed for a given system and may not be varied. In an equilibrium stage, for instance, the fact that it is a single stage is a fixed parameter. The feed rate, composition, and thermal conditions are also fixed parameters. Other parameters, such as the duty, temperature, or pressure, and, in the case of multiple feeds, some of the feed rates, may be varied to satisfy process requirements. The distinction between fixed and variable parameters reflects the actual operating practices. Thus, in order to control the vapor rate in a flash drum, one would normally vary the heat duty (or temperature) rather than the feed rate, which is usually controlled by upstream process considerations.

This section examines the effect of temperature, pressure, and, in certain situations, the feed rate on the performance of an equilibrium stage. Several examples are presented, typifying various separation processes. The different situations are mainly characterized by the nature of the feed makeup.

Although the obvious purpose of separation processes is to raise or lower the concentration of certain components to certain levels, another factor must be considered and that is the recovery of those components. Thus, the amount of a product as well as its purity is of concern to the process engineer. The concentration, commonly expressed as mole fraction of the components of interest, X_i or Y_i, is a measure of quality, while the total product rate or component recovery is a measure of quantity. The recovery of a component or group of components is defined as the fraction or percentage of these components in the feed that are recovered in a given product. If, for instance, the mole fraction of component i in the feed to a flash drum is Z_i and its mole fraction in the vapor product is Y_i, the amount of i in the feed is FZ_i and in the vapor product is $\psi F Y_i$, where F is the molar feed rate and ψ is the vapor mole fraction. The recovery of component i in the vapor product is, therefore,

$$\text{Recovery} = \frac{\psi F Y_i}{F Z_i} = \frac{\psi Y_i}{Z_i} \tag{2.5}$$

or

$$\text{Percent recovery} = 100 \frac{\psi Y_i}{Z_i} \tag{2.6}$$

The performance of an equilibrium stage may be examined in terms of components' concentration and recovery in a given product and the dependence of these quantities on the independent variables and the phase equilibrium characteristics of the system.

It should be noted that a single stage is inefficient and inflexible as a separation process. The independent variables, namely the temperature and pressure, can vary only over limited ranges, bounded by the phase envelope of the mixture. In a single stage, a component concentration can go through only a single-step change. Thus, the concentration of component i can change from X_i in the liquid to $Y_i = K_i X_i$ in the vapor. In contrast, in a multiple-stage operation, a number of steps or equilibrium stages make up the process. This allows the composition of both phases to change progressively from stage to stage, resulting in more efficient separation.

In certain systems, phase separation is enhanced by introducing an additional feed to the separation device such as in absorption or stripping and in azeotropic and extractive distillation. Although these processes are usually multistage, their characteristics are discussed in this section using the single-stage model.

2.2.1 SINGLE-FEED SYSTEMS

In the simplest form of phase separation, the stream to be separated is flashed into a vapor product and a liquid product, each having a different composition. Examples 2.1 through 2.6 of single-stage separations of certain mixtures. The distribution coefficients are predicted by the Soave–Redlich–Kwong equation of state (Soave, 1972).

EXAMPLE 2.1: BINARY MIXTURES—WIDE BOILING

Two miscible components having an appreciable difference in their boiling points form a mixture with widely separated dew points and bubble points. The mixture has a wide boiling range, and the distribution coefficients of the two components differ considerably. An example of this type of mixture is the ethane–n-hexane system, with 50–50% mole composition. The equilibrium temperature, distribution coefficients, compositions, and recovery of ethane in the vapor are calculated at a fixed pressure of 690 kPa and over a vapor mole fraction ranging from 0 to 1. Selected results are given in Table 2.1. The temperature, mole fraction ethane in the vapor, and ethane recovery in the vapor are plotted versus mole fraction vapor in Figure 2.5.

TABLE 2.1
Single-Stage Performance, 50% Mole Ethane (1)–50% Mole n-Hexane (2) at 690 kPa

T (°C)	ψ	K_1	K_2	X_1	Y_1	$\psi Y_1/Z_1$
−18.4	0.0	2.00	4.67E-3	0.500	0.998	0.000
−13.4	0.1	2.24	6.19E-3	0.445	0.997	0.199
−5.8	0.2	2.64	9.31E-3	0.376	0.994	0.398
60.7	0.5	7.22	1.38E-1	0.122	0.878	0.878
100.8	0.8	10.07	4.15E-1	0.061	0.610	0.976
111.1	1.0	10.65	5.25E-1	0.047	0.500	1.000

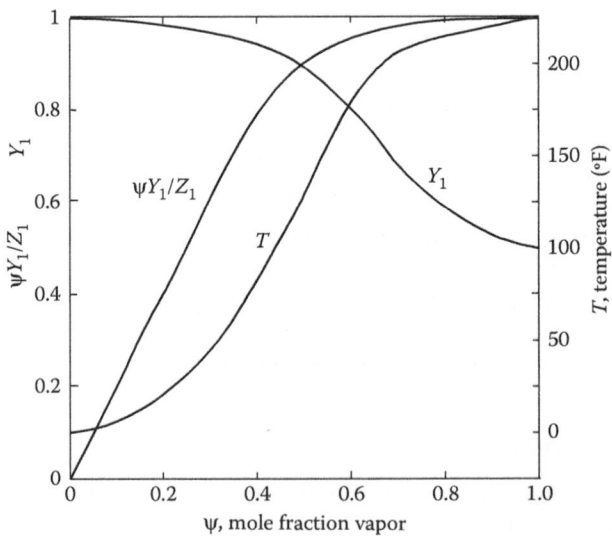

FIGURE 2.5   Single-stage performance, 50% mole ethane (1)–50% mole n-hexane (2) at 690 kPa.

Because of the large difference between the boiling points of the constituents, the temperature spans a wide range between the bubble point and the dew point, about 112°C. As a result, the separation is relatively easy, in the sense that a large recovery is attainable for a given product purity. If, for instance, a target purity is set at 80% ethane in the vapor, a recovery of about 93% can be achieved, as shown in Figure 2.5.

EXAMPLE 2.2: BINARY MIXTURES—CLOSE BOILING

The constituents of this type of mixture have relatively close boiling points. The mixture dew points and bubble points are not far apart, and the distribution coefficients are not much different for each component. An example of such a system is an equimolar ethane–propane mixture. Data representing the separation of this mixture at 690 kPa in an equilibrium stage are given in Table 2.2 and Figure 2.6.

TABLE 2.2
Single-Stage Performance, 50% Mole Ethane (1)–50% Mole Propane at 690 kPa

T (°F)	ψ	K_1	K_2	X_1	Y_1	$\psi Y_1/Z_1$
−25.0	0.0	1.66	0.342	0.500	0.830	0.000
−23.2	0.1	1.73	0.363	0.466	0.806	0.161
−21.3	0.2	1.81	0.387	0.430	0.780	0.312
−14.7	0.5	2.10	0.476	0.323	0.677	0.677
−8.8	0.8	2.38	0.570	0.238	0.566	0.906
−5.7	1.0	2.53	0.623	0.198	0.500	1.000

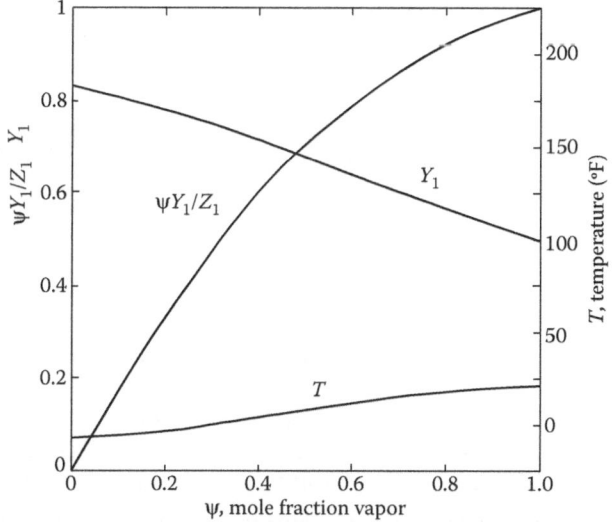

FIGURE 2.6 Single-stage performance, 50% mole ethane (1)–50% mole propane (2) at 690 kPa.

Because of the relatively small difference in the boiling points and distribution coefficients, the two components do not separate readily as shown by the low ethane recovery in the vapor of 16.2%, attainable at an ethane purity of 80%.

EXAMPLE 2.3: MULTI-COMPONENT MIXTURES

In general, mixtures occurring in chemical processes are multi-component rather than binary. It is often desired to concentrate one of the constituents, a key component, in a product in order to increase its value. To accomplish the separation, an elaborate processing scheme may be required. In a multi-component situation, components both lighter and heavier than the key component may be present. If the key component is recovered in the vapor product of a single equilibrium stage, undesired lighter components are more highly concentrated in the product than in the feed. Similarly, if the key component is recovered in the liquid product, heavier components are more highly concentrated in the product than in the feed.

An example of the performance of a single stage with a multi-component feed at 690 kPa is presented in Table 2.3 and Figure 2.7. The composition of the feed, made up of nitrogen and light hydrocarbons through hexane, is also given in Table 2.3. Consider ethane as the key component to be recovered in the vapor. Since it is not the lightest component, initial vaporization does not produce the highest ethane concentration. In fact, the initial vapor product is poorer in ethane than the feed. As more of the mixture is vaporized, the ethane concentration in the vapor increases, reaches a maximum, and then starts declining back to the feed concentration at the limit when all the feed is vaporized. Ethane recovery at maximum ethane concentration of 82.5% is about 33%. The equilibrium

---

TABLE 2.3
Single-Stage Performance, Multi-Component Mixture at 690 kPa

Component	Mole Percent
Nitrogen	1
Methane	1
Ethane	50
Propane	15
n-Butane	13
n-Pentane	10
n-Hexane	10

		Ethane		
T (°C)	ψ	X_3	Y_3	$\psi Y_3/Z_3$
−70.6	0.0	0.500	0.191	0.000
−25.5	0.1	0.468	0.784	0.157
−16.9	0.2	0.419	0.825	0.330
5.2	0.5	0.236	0.764	0.764
44.4	0.8	0.102	0.600	0.960
69.0	1.0	0.065	0.500	1.000

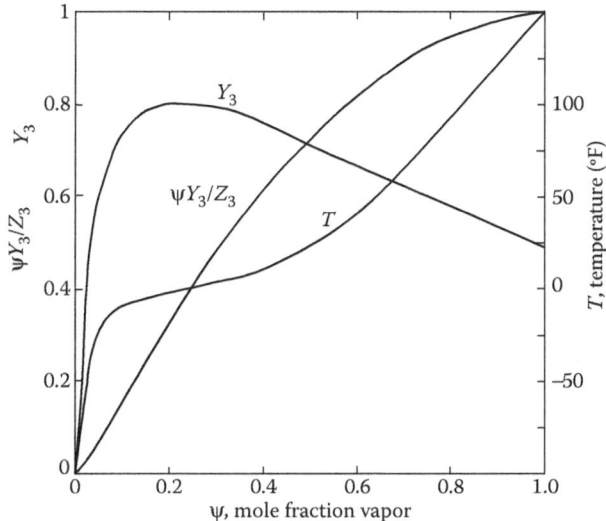

FIGURE 2.7   Single-stage performance, multi-component mixture at 690 kPa.

temperature varies as the liquid and vapor compositions change. A steep temperature rise is noticed at very low vapor fractions, where the bubble point is highly sensitive to the nitrogen concentration in the liquid.

2.2.2   SINGLE-STAGE ABSORPTION/STRIPPING

The phase separation of certain mixtures may take place at economically impractical temperatures and pressures. An example of this type of separation might be the recovery of light hydrocarbon gases dispersed in air. In these situations, a second stream may be added to the mixture to shift the vapor–liquid equilibrium region. Such a process is equivalent to a single-stage absorption or stripping device.

EXAMPLE 2.4: SINGLE-STAGE ABSORBER: NITROGEN–ETHANE

Nitrogen is to be separated out of a mixture containing 50% mole nitrogen and 50% mole ethane. Table 2.4 and Figure 2.8 represent the binary phase characteristics of this system at 690 kPa. One way of determining the level of separation is to measure the amount of nitrogen separated at a given nitrogen purity. A nitrogen recovery of about 95% is expected at a purity of 98%. In order for this to happen, however, the mixture must be brought to the two-phase region, at approximately −162°C, requiring cryogenic distillation.

As an alternative process, a liquid such as hexane may be added to the vapor mixture to form a two-phase ternary. Because of the large difference between the distribution coefficients of nitrogen and ethane, most of the nitrogen goes in the vapor phase, while most of the ethane goes in the liquid. Separating out the nitrogen in this manner is, in effect, the equivalent of a single-stage absorption process.

TABLE 2.4

Single-Stage Performance, 50% Mole Nitrogen (1)–50% Mole Ethane (2) at 690 kPa

T (°C)	ψ	X_1	Y_1	K_1	$\psi Y_1/Z_1$
−178.7	0.0	0.500	0.999	2.000	0.000
−116.7	0.5	0.027	0.973	36.037	0.973
−70.7	0.8	0.013	0.622	47.846	0.995
−63.9	1.0	0.010	0.500	50.000	1.000

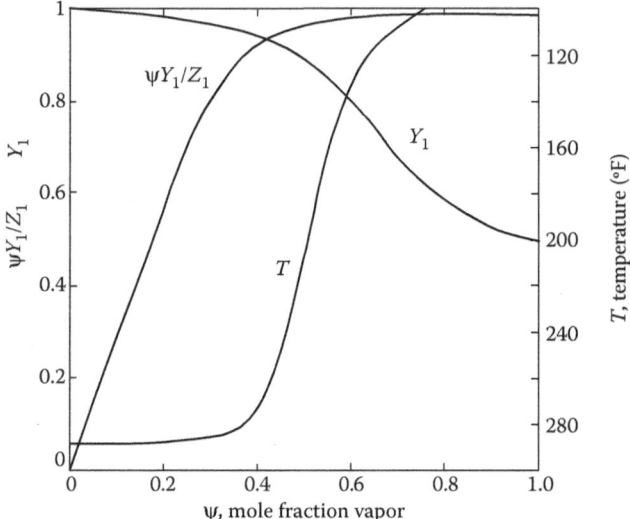

FIGURE 2.8   Single-stage performance, 50% mole nitrogen (1)–50% mole ethane (2) at 690 kPa.

Table 2.5 shows the nitrogen composition and recovery in the vapor product of an equilibrium stage at different hexane-to-vapor feed ratios. The temperature and pressure are held at −73°C and 690 kPa. A nitrogen purity of 98%, comparable to that obtained in the binary at −162°C, is attainable at −73°C by introducing hexane at a molar ratio of 7.5 to the vapor feed. The recovery of nitrogen in the vapor product is about 84%, somewhat lower than in the binary separation.

2.2.3   CLOSE BOILERS AND AZEOTROPES

Other systems that require the addition of a separating agent to enhance or effect their separation include azeotropes and mixtures of close boiling components.

The separation of close boiling components by ordinary fractionation may require too many stages or could be practically impossible because of the proximity of the distribution coefficients (*K*-values). The compositions of equilibrium vapor and liquid phases in such systems are almost identical. In general, close boiling components

TABLE 2.5
Single-Stage Performance with Multiple Feeds, 38°C, 690 kPa

		Nitrogen		
F_L/F_V	X	Y	K	Recovery
1.0	0.01	0.898	91.6	0.971
2.5	0.01	0.947	97.0	0.941
5.0	0.01	0.971	99.3	0.892
7.5	0.01	0.980	100.1	0.842
10.0	0.01	0.984	100.5	0.792

F_V— Vapor feed, 50% mole nitrogen–50% mole ethane.
F_L— Liquid feed, n-hexane.

are likely to have different chemical structures, and would, therefore, interact differently with a third component. In extractive distillation, a solvent that is less volatile than the feed components is added to the mixture for the purpose of preferentially depressing the volatility of one of the feed components. In azeotropic distillation, the added component, or entrainer, forms an azeotrope with one of the components to be separated. The azeotrope may have either a higher or lower boiling temperature than the other component and may, therefore, leave the separation device either in the bottoms or overhead product.

Azeotropes, discussed in Chapter 1, form constant boiling mixtures and are, therefore, impossible to separate unless the azeotrope is broken by the addition of an external agent, using either extractive or azeotropic distillation as in the case of close boiling mixtures.

Examples 2.5 and 2.6 demonstrate the above principles applied to a single stage.

EXAMPLE 2.5: SINGLE-STAGE AZEOTROPIC SEPARATION

Benzene and cyclohexane are difficult to separate because of their close normal boiling points (80°C and 81°C). They also form an azeotrope at a composition of about 60% mole benzene. The two components may be separated by adding ethanol, which forms azeotropes with each component. At 103.4 kPa, the cyclohexane–ethanol azeotrope boils at 67°C and the benzene–ethanol azeotrope at 69°C. The two azeotropes are separated, followed by removal of ethanol by water wash. The effect of adding ethanol to the benzene–cyclohexane mixture is shown in Figure 2.9 for a single-stage separation. Adding increasing amounts of ethanol (the entrainer) results in higher quantities of the benzene and cyclohexane azeotroping with the ethanol, thereby improving the separation. At higher ethanol concentrations, the ratio of cyclohexane to benzene in the vapor phase increases. This ratio approaches a limit after all the cyclohexane and benzene have azeotroped. In this example, the stage pressure is held at 103.4 kPa. At all ethanol feed rates, 50% of the combined benzene–cyclohexane feed is vaporized. The liquid activity coefficients were calculated with the NRTL equation.

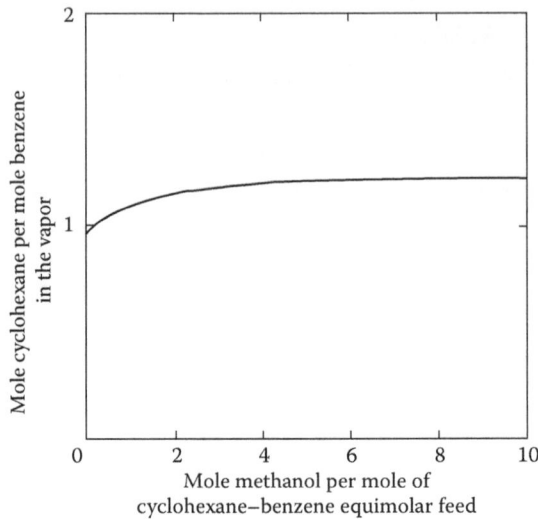

FIGURE 2.9   Single-stage azeotropic separation, benzene–cyclohexane–ethanol at 103.4 kPa.

EXAMPLE 2.6: SINGLE-STAGE EXTRACTIVE SEPARATION

Isobutane and 1-butene are close boilers and, although they do not form an azeotrope, are difficult to separate by conventional distillation. Using a single stage, a feed stream containing 40% mole isobutane and 60% mole 1-butene is flashed at 520 kPa so that 50% of this stream is vaporized. With no solvent added, the vapor and liquid product compositions are about the same. The ratio of 1-butene to isobutane in the liquid product is about 1.6 (Figure 2.10)

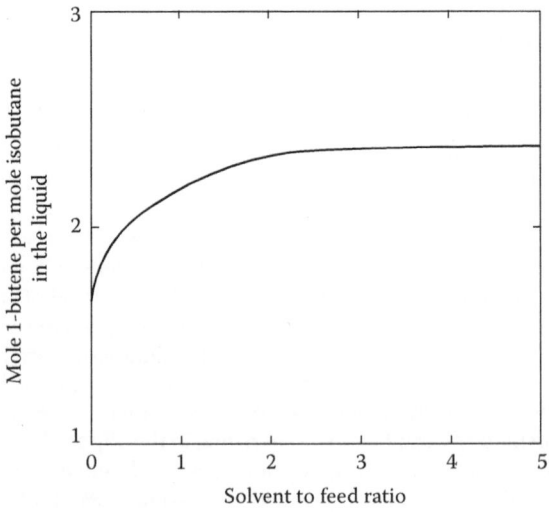

FIGURE 2.10   Single-stage extractive separation, 1-butene–isobutane–furfural at 520 kPa.

compared to 1.5 in the feed. The two components are chemically different, 1-butene being unsaturated and isobutane being saturated. Furfural, which is less volatile than both components, is added as a solvent to alter the 1-butene–isobutane relative volatilities, that is, to depress the 1-butene K-value relative to that of isobutane. This is equivalent to a statement that 1-butene is more soluble than isobutane in furfural. As a result, the ratio of 1-butene to isobutane in the liquid product is enhanced as more solvent is added while maintaining the amount of C4s vaporized at 50% mole of the feed (Figure 2.10). Furfural can be recovered from the hydrocarbons in downstream processing with relative ease because of its substantially lower volatility.

The calculations for this example are based on the Margules liquid activity coefficient equation with the following parameters:

Isobutane–furfural:	$A_{12} = 1.14$	$A_{21} = 1.31$
1-Butene–furfural:	$A_{12} = 0.84$	$A_{21} = 1.03$

2.3   SOLUTION METHODS

In order to define the quantitative relationships among the various parameters in a unit operation, a mathematical model is employed, in which the physical relationships are expressed as mathematical equations. Thus, the equilibrium stage may be simulated by a model for which the mathematical solution represents physical performance. When physical relations are translated into analytical expressions, certain assumptions must be made and the accuracy of the simulation model depends on the validity of these assumptions. For an equilibrium stage model, it is assumed that the stage is essentially at equilibrium. Additionally, it is assumed that the models used for predicting the thermodynamic properties, namely the distribution coefficients and enthalpy, are accurate. To the extent that these assumptions are met, the performance of the equilibrium stage can be accurately predicted.

The physical relations that must be satisfied in an equilibrium stage are the material balance, the energy balance, and the phase equilibrium relations. Referring to Figure 2.4, the material balance equations may be written for each component i as

$$FZ_i = F \psi Y_i + F(1 - \psi)X_i$$

or

$$Z_i = \psi Y_i + (1 - \psi)X_i \tag{2.7}$$

for $i = 1, \ldots, C - 1$, where C is the number of components. The problem definition is such that the feed conditions are set, that is, the feed composition, Z_i, is known. Equation 2.7 represents $C - 1$ independent equations, each corresponding to one of $C - 1$ components. The last component mole fraction in each phase is determined from the summation equations:

$$\sum X_i = \sum Y_i = 1$$

An energy balance on the equilibrium stage may be stated as

$$H_F + Q = h(T, P, X_i) + H(T, P, Y_i) \tag{2.8}$$

where H_F is the total feed enthalpy, Q is the heat duty added to (if positive) or removed from (if negative) the system, h is the total liquid enthalpy, and H is the total vapor enthalpy. The value of H_F is known since the feed conditions are defined, and Q is a variable parameter. The enthalpies H and h are functions of the stage temperature and pressure and the flow rate and composition of each phase.

The last set of equations defining the equilibrium stage consists of the phase equilibrium relations, which may be expressed in terms of the distribution coefficients as

$$Y_i/X_i = K_i(T,P,X_i,Y_i) \quad i = 1,\ldots,C \tag{2.9}$$

These are C equations, corresponding to the C components. The distribution coefficients are functions of the temperature, pressure, and liquid and vapor compositions. Equation 2.9 is based on the thermodynamic condition that at phase equilibrium the fugacities of each component in the liquid phase and the vapor phase are equal. Referring to Section 1.3, the fugacities can be expressed in terms of fugacity coefficients as follows:

$$\begin{aligned} f_i^V &= PY_i\phi_i^V \\ f_i^L &= PX_i\phi_i^L \end{aligned} \tag{2.10}$$

At equilibrium,

$$f_i^V = f_i^L$$

or

$$PY_i\phi_i^V = PX_i\phi_i^L$$

for $i = 1, \ldots, C$. These equations may be rearranged to give

$$K_i = Y_i/X_i = \phi_i^L/\phi_i^V \tag{2.11}$$

The fugacity coefficients may be derived from equations of state or activity coefficients (Chapter 1). In any case, the fugacities are functions of temperature, pressure, and composition. An alternative to Equation 2.9 for expressing phase equilibrium is, therefore,

$$f_i^L(T,P,X_i) = f_i^V(T,P,Y_i) \tag{2.12}$$

for $i = 1, \ldots , C$. Equations 2.7, 2.8, and 2.12 bring the total number of independent equations to $2C$. These equations involve $2C + 2$ variables, namely the temperature, T, the pressure, P, $C - 1$ component liquid mole fractions, X_i, $C - 1$ component vapor mole fractions, Y_i, the vapor mole fraction, ψ, and the heat duty, Q. With $2C$ equations and $2C + 2$ variables, the problem has two degrees of freedom, and therefore two independent variables must be fixed at certain values in order to define the equilibrium stage. These are the two degrees of freedom that were discussed in the context of the phase rule in Section 2.1. The above is a mathematical statement of the same principle.

The set of $2C$ simultaneous equations is nonlinear and fairly complex since it involves calculating fugacities and enthalpies, themselves nonlinear functions of the temperature, pressure, and composition. The equations may be solved simultaneously or by some iterative method. In general, the computational methods depend on which two variables are selected as the independent variables. Although in principle any two independent variables may be fixed, the problem complexity may vary from case to case. It is found, for instance, that a solution is more readily reached if P and T rather than P and Q are the independent variables. Since most of these calculations are carried out on computers, the solution methods should be designed for speed of convergence and reliability. Several methods have been proposed for handling the different types of flash calculations, some of which are discussed herewith.

2.3.1   ISOTHERMAL FLASH METHOD

This method starts off by fixing the temperature and pressure and iterating around the vapor fraction to calculate the equilibrium phase separation and compositions. The first step is an isothermal flash calculation. If T and P are in fact the independent variables, the solution obtained in the first step is the desired solution. If either T or P and one more variable are specified, then another, outer iterative loop is required. The outer loop iterates around P or T (whichever is not fixed) until the other specified variable is satisfied.

2.3.1.1   Basic Algorithm

The component material balance and distribution coefficient relations, Equations 2.7 and 2.9, are combined to express X_i and Y_i in terms of feed composition and vapor fraction:

$$X_i = \frac{Z_i}{1 + \psi(K_i - 1)} \tag{2.13a}$$

$$Y_i = \frac{Z_i K_i}{1 + \psi (K_i - 1)} \tag{2.13b}$$

Since

$$\sum X_i = \sum Y_i = 1$$

either of the above equations may be summed up over all the components $i = 1, \ldots ,$ C to form one equation in ψ for a given set of K_i. Alternatively, the equation

$$\sum (X_i - Y_i) = 0$$

may be solved for ψ. This equation represents a smoother function of ψ than $\sum X_i = 1$ or $\sum Y_i = 1$ and is, therefore, more amenable to solution. The function takes the form

$$f(\psi) = \sum_{i=1}^{C} \frac{Z_i(K_i - 1)}{1 + \psi(K_i - 1)} = 0 \tag{2.14}$$

If the K_i values were a function of temperature and pressure only and if a set of K_i values was available at a specified T and P, a solution of Equation 2.14 would satisfy both the material balance and equilibrium relations. This may be demonstrated by a binary system where the feed or combined mixture composition is given, along with the given fixed K-values. This implies that the temperature and pressure are fixed, and that the K-values are composition-independent.

If the overall mole fractions of the two components in the mixture are $Z_1 = 0.4$ and $Z_2 = 0.6$, and the distribution coefficients are $K_1 = 1.6$ and $K_2 = 0.7$, Equation 2.14 is written as

$$f(\psi) = \frac{0.4(1.6 - 1)}{1 + \psi(1.6 - 1)} + \frac{0.6(0.7 - 1)}{1 + \psi(0.7 - 1)} = 0$$

The solution to this equation is $\psi = 0.3333$. This is the vapor fraction of the mixture, that is, if the total mixture is 100 kmol or the feed flow rate to the phase separator is 100 kmol/h, then the products are 33.33 kmol/h vapor and 66.67 kmol/h liquid. Once ψ is determined, the liquid and vapor compositions are calculated by Equations 2.13a or 2.13b:

$$X_1 = \frac{0.4}{1 + 0.3333(1.6 - 1)} = 0.3333$$

$$Y_1 = K_1 X_1 = 1.6 \times 0.3333 = 0.5333$$

The other components may be calculated as above. For a binary, the second component may be calculated by difference:

$$X_2 = 1 - X_1 = 0.6667$$

$$Y_2 = 1 - Y_1 = 0.4667$$

In general, K_i is also a function of X_i and Y_i, themselves unknown until Equations 2.13a and 2.13b are solved. This entails the need for an iterative scheme, whereby a set of distribution coefficients is assumed and Equation 2.14 is solved. The resulting compositions are checked to determine if the equilibrium condition, Equation 1.19, is met for all the components. The distribution coefficients are updated based on calculated compositions, and the process is repeated until the equilibrium relations are met. The algorithm may be summarized in the following steps:

1. For a given feed composition, temperature, and pressure, assume initial estimates for the distribution coefficients, K_i. These estimates may be based on Raoult's law (Section 1.3).
2. Based on the current K_i values, solve Equation 2.14 for ψ. The solution to this equation is itself iterative since it is implicit in ψ and nonlinear. A method such as Newton–Raphson may be used, as described below.
3. Compute the liquid and vapor compositions using Equations 2.13a and 2.13b.
4. Compute the liquid and vapor fugacities of each component in the liquid and vapor phase, $\hat{f}_i^L$ and $\hat{f}_i^V$ (Section 1.3).
5. If $f_i^L = \hat{f}_i^V$ for all components, then the thermodynamic conditions for equilibrium are satisfied and the solution is reached at the current values of vapor fraction and vapor and liquid compositions. If the liquid and vapor fugacities are not equal within a set tolerance for any component, the distribution coefficients are updated on the basis of Equation 2.11, rewritten as

$$K_i = \frac{Y_i}{X_i} \frac{\hat{f}_i^L}{\hat{f}_i^V} \tag{2.15}$$

The iterative computation is restarted at step 2. Once a converged solution is reached, the heat duty associated with the isothermal flash may be calculated using Equation 2.8.

If a single equation of state is used in step 4 to calculate liquid and vapor densities and fugacities, situations can arise where the distinction between liquid and vapor compressibility factors may not be obvious. Using the wrong compressibility factor may result in non-convergence or convergence to a false solution. A temperature perturbation technique may be used in these situations to identify the liquid and vapor compressibility factors (Khoury, 1978). The method involves searching for multiple compressibility factor roots at temperatures around the system temperature, then selecting the smallest root as a "pseudo-vapor" compressibility factor and the largest root as a "pseudo-liquid" compressibility factor. The pseudo compressibility factors are used in intermediate iterations. In subsequent iterations, the temperature is reset to system temperature and the pseudo compressibility factors are replaced by the true compressibility factors.

Equation 2.14 may be solved by an iterative technique such as the Newton–Raphson method. The derivative of the function with respect to ψ at iteration k, $f'(\psi^{(k)})$, equals the value of the function at iteration k, $f(\psi^{(k)})$, divided by the negative of the recursive correction to the variable, $\psi^{(k)} - \psi^{(k+1)}$:

$$f'(\psi^{(k)}) = \frac{f(\psi^{(k)})}{\psi^{(k)} - \psi^{(k+1)}}$$

The value of the variable at iteration $k+1$ is updated from its value at iteration k:

$$\psi^{(k+1)} = \psi^{(k)} - f(\psi^{(k)})/f'(\psi^{(k)})$$

From Equation 2.14, the derivative of the function with respect to ψ is

$$f'(\psi^{(k)}) = -\sum_{i=1}^{C} \frac{Z_i(K_i - 1)^2}{\left[1 + \psi^{(k)}(K_i - 1)\right]^2}$$

The recursive computation usually converges in a small number of iterations due to the smooth and monotonic nature of the function $f(\psi)$.

2.3.1.2 Extension to General Flash Calculations

The isothermal flash algorithm described above may be incorporated into an iterative scheme for solving other types of flash calculations. The isothermal flash routine becomes a module in an outer computational loop in which either the temperature or the pressure is varied to satisfy a given specification.

For example, consider a fixed heat duty flash calculation at fixed pressure. A feed with a given composition and initial thermal conditions is flashed at a fixed pressure, and, in the process, a fixed heat duty, Q, is added or removed. In the special case of an adiabatic flash, $Q = 0$. The equilibrium temperature and phase compositions at the flash pressure must be determined. The temperature is determined from an energy balance, by solving Equation 2.8. The iterative solution consists of the following steps (Figure 2.11):

1. Assume a flash temperature, T.
2. Carry out an isothermal flash calculation at the given pressure, P, and the current temperature, T.
3. If Equation 2.8 is satisfied within some given tolerance, a solution to the problem has been reached. If the error is greater than the tolerance, proceed to the next step.
4. Use an interpolative–extrapolative technique to predict the temperature for the next trial and repeat the calculations beginning at step 2. In the first two iterations, the temperature is estimated independently to start the calculations.

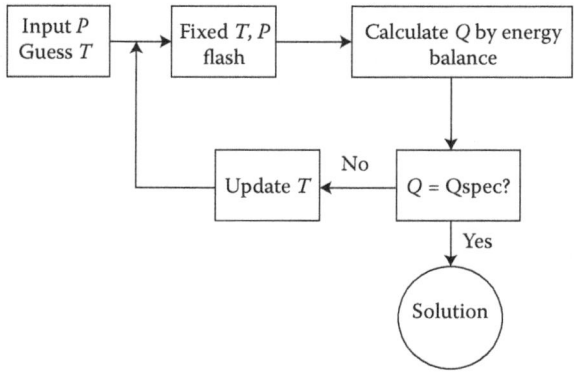

FIGURE 2.11 General flash calculations.

EXAMPLE 2.7: PHASE SEPARATION AT GIVEN TEMPERATURE AND PRESSURE

A stream containing 20% mole ethanol (1) and 80% mole isooctane (2) is flashed at 50°C and 35 kPa and separated into a vapor product and a liquid product. Calculate the fraction of the feed vaporized, and the composition of the liquid and vapor products.

The component vapor pressures at 50°C are $p_1^0 = 29.5\,\text{kPa}$ and $p_2^0 = 19.5$ kPa. The Wilson coefficients for this binary are $A_{12} = 0.0995$ and $A_{21} = 0.247$.

SOLUTION

The liquid phase is a non-ideal solution, for which liquid activity coefficients can be predicted with good accuracy by the Wilson equation (Equation 1.36). The vapor phase is assumed to behave as an ideal gas at the relatively low pressure of 35 kPa. Under these conditions, the K-values may be calculated by Equation 1.29a.

The isothermal flash algorithm described above is adapted to this system as follows:

1. For starting the iterative calculations, the liquid composition is assumed equal to the feed composition. The K-values are calculated by Equations 1.36 and 1.29a:

$$\beta_{12} = \frac{0.0995}{0.2 + (0.0995)(0.8)} - \frac{0.247}{0.8 + (0.247)(0.2)} = 0.06507$$

$$\beta_{21} = -\beta_{12} = -0.06507$$

$$\ln\gamma_1 = -\ln[0.2 + (0.0995)(0.8)] + (0.06507)(0.8) = 1.32645$$

$$\ln\gamma_2 = -\ln[0.8 + (0.247)(0.2)] - (0.06507)(0.2) = 0.15021$$

$$\gamma_1 = 3.76766$$

$$\gamma_2 = 1.16208$$

$$K_1 = (3.76766)(29.5)/35 = 3.17560$$

$$K_2 = (1.16208)(19.5)/35 = 0.64744$$

2. The current K-values are used to calculate the vapor mole fraction ψ. Equation 2.14 is solved iteratively by the Newton–Raphson method:

$$f(\psi) = \frac{0.2(3.17560 - 1)}{1 + \psi(3.17560 - 1)} + \frac{0.8(0.64744 - 1)}{1 + \psi(0.64744 - 1)}$$

$$f'(\psi) = -\frac{0.2(3.17560 - 1)^2}{\left[1 + \psi(3.17560 - 1)\right]^2} - \frac{0.8(0.64744 - 1)^2}{\left[1 + \psi(0.64744 - 1)\right]^2}$$

$$\psi^{(k+1)} = \psi^{(k)} - f(\psi^{(k)})/f'(\psi^{(k)})$$

The calculations are started with $\psi = 0.5$ as an initial guess. Convergence with the current K-values occurs at $\psi = 0.19957$.

3. The liquid and vapor compositions are calculated using Equations 2.13:

$$X_1 = \frac{0.2}{1 + (0.19957)(3.17560 - 1)} = 0.13945$$

$$X_2 = \frac{0.8}{1 + (0.19957)(0.64744 - 1)} = 0.86055$$

$$Y_1 = K_1 X_1 = (3.17560)(0.13945) = 0.44284$$

$$Y_2 = K_2 X_2 = (0.64744)(0.86055) = 0.55716$$

4. The new liquid compositions are used to calculate the activity coefficients and K-values in step 1, and the calculations are continued and the cycle is repeated until the K-values stabilize.

The converged results are as follows:

	γ	K	X	Y
Ethanol (1)	9.73764	8.20744	0.05645	0.46335
Isooctane (2)	1.02085	0.56876	0.94355	0.53665
$\psi = 0.35278$				

EXAMPLE 2.7A: AN EQUILIBRIUM STAGE WITH WILSON AND VIRIAL COEFFICIENTS

A stream flowing at the rate of 1000 kmol/h contains acetone (1), methanol (2), and water (3). The stream is sent to a flash drum maintained at 150 kPa and 80°C where it is separated into vapor and liquid products. Calculate the flow rates and compositions of the products using the Wilson equation for the activity coefficients and the truncated virial equation, $z = 1 + BP/RT$, for the fugacity coefficients. Relevant data are provided below:

i Component	Z_i	T_{ci} (K)	P_{ci} (kPa)	V_{ci} (cm³/mol)	z_{ci}	ω_i	V_i (cm³/mol) Liquid Molar Volume
1. Acetone	0.2	508.2	4701	209.0	0.233	0.307	73.52
2. Methanol	0.3	512.6	8090	117.9	0.224	0.564	40.73
3. Water	0.5	647.3	22120	57.1	0.229	0.345	18.07

Antoine constants for Equation 2.19, T in K, p_i^0 in kPa

	A_i	B_i	C_i
1.	14.392	2795.8	−43.00
2.	16.594	3644.3	−33.24
3.	16.262	3799.9	−46.65

Wilson exponential parameters, δ_{ij}, cal/mol

$\delta_{11} = 0$	$\delta_{12} = -214.95$	$\delta_{13} = 439.64$
$\delta_{21} = 664.08$	$\delta_{22} = 0$	$\delta_{23} = 205.30$
$\delta_{31} = 1405.49$	$\delta_{32} = 482.16$	$\delta_{33} = 0$

$R = 8314$ cm³ kPa/mol K.

SOLUTION

The K-values are calculated from Equation 1.29, and are modified as follows: Since f_i^L(at p_i^0) $= f_i^V$(at p_i^0), the pure-component fugacity in the liquid is calculated as

$$f_i^L = f_i^V = p_i^0 \phi_i^V$$

where ϕ_i^V is the pure-component fugacity coefficient in the vapor. Substituting in Equation 1.29,

$$K_i = \gamma_i p_i^0 \phi_i^V / (\hat{\phi}_i^V P)$$

This equation neglects the Poynting correction factor. (See discussion following Equation 1.29.) The fugacity coefficients in the vapor, $\hat{\phi}_i^V$ and ϕ_i^V, are calculated based on the virial equation, and the activity coefficients γ_i are calculated by the Wilson equation.

The pure-component fugacity coefficient is obtained by combining Equation 1.21 with the truncated virial equation:

$$\ln\phi_i^V = \int_0^P \frac{z_i - 1}{P} dP = \int_0^P \frac{B_{ii}}{RT} dP = \frac{B_{ii}P}{RT}$$

where B_{ii} is the second virial coefficient of pure component i. The partial fugacity coefficient of component i in the mixture is given by Equation 1.23:

$$\ln\hat{\phi}_i^V = \int_0^P \frac{\hat{z}_i - 1}{P} dP$$

From the definition of a partial property,

$$\hat{z}_i = 1 + \frac{P}{RT}\left[\frac{\partial(nB)}{\partial n_i}\right]_{T,P,nj\neq i}$$

Substituting in the partial fugacity coefficient equation,

$$\ln\hat{\phi}_i^V = \frac{1}{RT}\left[\frac{\partial(nB)}{\partial n_i}\right]_{T,P,nj\neq i}\int_0^P dP = \frac{P}{RT}\left[\frac{\partial(nB)}{\partial n_i}\right]_{T,P,nj\neq i}$$

The mixture second virial coefficient is given as a function of the composition:

$$B = \sum_i \sum_j Y_i Y_j B_{ij}$$

Substituting $Y_i = n_i/n$, $Y_j = n_j/n$,

$$nB = \frac{1}{n}\sum_i \sum_j n_i n_j B_{ij}$$

$$\left[\frac{\partial(nB)}{\partial n_k}\right]_{T,P,ni\neq k} = -\frac{1}{n^2}\sum_i \sum_j n_i n_j B_{ij} + \frac{2}{n}\sum_i n_i B_{ik} = -B + 2\sum_i Y_i B_{ik}$$

Thus, the partial fugacity coefficient is evaluated as

$$\ln\hat{\phi}_k^V = \frac{P}{RT}\left(2\sum_i Y_i B_{ik} - B\right)$$

The second virial coefficient may be calculated by the following correlation:

$$B_{ij} = (RT_{Cij}/P_{Cij})(B^0_{ij} + \omega_{ij} B^1_{ij})$$

where ω_{ij} is the acentric factor. This equation may be used for pure-component coefficients (B_{ii}) and for cross coefficients ($B_{ij} = B_{ji}$). For the latter, the following combining rules may be used for the pseudo-critical constants and the acentric factor:

$$T_{Cij} = (1 - k_{ij})(T_{Ci} T_{Cj})^{0.5}$$

$$V_{Cij} = \left(\frac{V_{Ci}^{\frac{1}{3}} + V_{Cj}^{\frac{1}{3}}}{2} \right)^3$$

$$z_{Cij} = 0.5(z_{Ci} + z_{Cj})$$

$$P_{Cij} = z_{Cij} R T_{Cij}/V_{Cij}$$

$$\omega_{ij} = 0.5(\omega_i + \omega_j)$$

The interaction parameter k_{ij} can be determined from the experimental data. It can be assumed zero for similar components. The parameters B^0_{ij} and B^1_{ij} are correlated in terms of the reduced temperature:

$$B^0_{ij} = 0.083 - 0.422/T^{1.6}_{rij}$$

$$B^1_{ij} = 0.139 - 0.172/T^{4.2}_{rij}$$

The activity coefficients are calculated by the multi-component Wilson equation (1.36a). For three components ($k = 1, 2, 3$),

$$\ln \gamma_i = 1 - \ln(X_1 A_{i1} + X_2 A_{i2} + X_3 A_{i3}) - X_1 A_{1i}/D_1 - X_2 A_{2i}/D_2 - X_3 A_{3i}/D_3$$

The denominators are defined as

$$D_1 = X_1 A_{11} + X_2 A_{12} + X_3 A_{13}$$

$$D_2 = X_1 A_{21} + X_2 A_{22} + X_3 A_{23}$$

$$D_3 = X_1 A_{31} + X_2 A_{32} + X_3 A_{33}$$

$$(A_{ii} = 1)$$

The binary interaction parameters A_{ij} are as defined for Equation 1.36. The vapor pressures are calculated by the Antoine Equation 2.19.

The isothermal flash algorithm (Section 2.3.1) is adapted to this example with minor modifications:

1. At the specified temperature, estimate the bubble point pressure and the equilibrium vapor composition using the K-values based on Raoult's law (Equations 2.18 through 2.20). The liquid composition is the same as the feed composition.
2. Using the current vapor and liquid compositions and the specified temperature and pressure, calculate the K-values using the modified form of Equation 1.29 as described in this example.
3. Calculate the vapor fraction ψ iteratively from Equation 2.14; then calculate the vapor and liquid compositions from Equations 2.13a and 2.13b.
4. Repeat the calculations beginning from step 2 until the compositions stabilize.

The pseudo-critical constants are calculated and tabulated along with the pure component constants:

i	j	T_{cij} (K)	V_{cij} (cm³/mol)	z_{cij} (kPa)	P_{cij}	ω_{cij}
1	1	508.2	209.0	0.233	4701	0.307
2	2	512.6	117.9	0.224	8090	0.564
3	3	647.3	57.1	0.229	22120	0.345
1	2	510.4	159.1	0.2285	6093.3	0.4355
1	3	573.5	117.1	0.231	9405.2	0.326
2	3	576.0	83.9	0.2265	12934.7	0.4545

The second virial pure component and cross coefficients are next calculated:

$$B_{11} = (8314 \times 508.2/4701)$$

$$\times \left[0.083 - 0.422 \left(\frac{508.2}{353.15} \right)^{1.6} + 0.307 \left\{ 0.139 - 0.172 \left(\frac{508.2}{353.15} \right)^{4.2} \right\} \right]$$

$$= -784.97 \text{ cm3/mol}$$
$B_{22} = -562.89$ cm³/mol
$B_{33} = -422.78$ cm³/mol
$B_{12} = -674.82$ cm³/mol
$B_{13} = -617.71$ cm³/mol
$B_{23} = -513.62$ cm³/mol.

The pure-component fugacity coefficients are

$$\phi_i^V = \exp\left[-784.97 \times 150/(8314 \times 353.15) \right] = 0.9607$$

$$\phi_2^V = 0.9717$$

$$\phi_3^V = 0.9786$$

The vapor pressures at the specified temperature of 353.15 K are

$$p_1^0 = \exp\left(14.392 - \frac{2795.8}{353.15 - 43.00}\right) = 216.51 \text{ kPa}$$

$$p_2^0 = 181.70 \text{ kPa}$$

$$p_3^0 = 47.67 \text{ kPa}$$

The flash calculations are done iteratively as described above. The sample calculations are shown here for the last (converged) pass. Liquid and vapor compositions from the previous pass are

	X_i	Y_i
Acetone (1)	0.0895	0.3761
Methanol (2)	0.2504	0.3790
Water (3)	0.6601	0.2449

Mixture second virial coefficient is given by

$$B = (0.3761)^2(-784.97) + (0.3790)^2(-562.89) + (0.2449)^2(-422.78)$$

$$+2[(0.3761)(0.3790)(-674.82) + (0.3761)(0.2449)(-617.71)$$

$$+(0.3790)(0.2449)(-513.62)] = -618.745$$

Partial fugacity coefficients are as follows:

$$\hat{\phi}_1^V = \exp\left[\left(\frac{150}{8314 \times 353.15}\right)\right.$$

$$\left. \times \left\{-2(0.3761 \times 784.97 + 0.3790 \times 674.82 + 0.2449 \times 617.71) + 618.745\right\}\right]$$

$$= 0.9607$$

$$\hat{\phi}_2^V = 0.9714$$

$$\hat{\phi}_3^V = 0.9777$$

Activity Coefficient Calculations: Wilson binary interaction parameters are

$$A_{11} = \frac{73.52}{73.52} \exp\left(\frac{-0}{1.987 \times 353.15}\right) = 1.0$$

$$A_{12} = \frac{40.73}{73.52} \exp\left(-\frac{-214.95}{1.987 \times 353.15}\right) = 0.7526$$

$$A_{13} = 0.1314$$

$$A_{21} = 0.7006$$

$$A_{22} = 1.0$$

$$A_{23} = 0.3311$$

$$A_{31} = 0.5490$$

$$A_{32} = 1.1338$$

$$A_{33} = 1.0$$

The denominators in the activity coefficient equations are defined as

$$D_1 = (0.0895)(1.0) + (0.2504)(0.7526) + (0.6601)(0.1314) = 0.3647$$

$$D_2 = 0.5317$$

$$D_3 = 0.9931$$

The activity coefficients

$$\ln\gamma_1 = 1 - \ln(0.3647) - \frac{(0.0895)(1.0)}{0.3647} - \frac{(0.2504)(0.7006)}{0.5317} - \frac{(0.6601)(0.5490)}{0.9931}$$

$$= 1.0684$$

$$\gamma_1 = 2.9106$$

$$\gamma_2 = 1.2491$$

$$\gamma_3 = 1.1666$$

K-values

$$K_1 = \frac{2.9106 \times 216.51 \times 0.9607}{0.9607 \times 150} = 4.2014$$

$$K_2 = 1.5133$$

$$K_3 = 0.3711$$

These K-values are used with the given feed composition to solve Equations 2.14, 2.13a and 2.13b for the vapor fraction and the vapor and liquid compositions. The results are as follows:

$$\psi = 0.3855$$

	X_i	Y_i
Acetone	0.0895	0.3761
Methanol	0.2504	0.3790
Water	0.6601	0.2449

The calculated compositions show no change from the previous pass, thus convergence has been reached.

EXAMPLE 2.7B: SINGLE-STAGE EXTRACTIVE SEPARATION USING THE VAN LAAR EQUATION

The effect of a solvent on the relative volatilities of close boilers discussed in Example 2.6 is examined using the van Laar multi-component equation. Using a single stage, a 100 kmol/h stream containing 50% mole isobutane and 50% mole 1-butene is flashed at 480 kPa, producing a 50 kmol/h vapor stream. It is required to determine the effect of adding furfural as a solvent on the separation. With the vapor rate maintained at 50 kmol/h, the vapor–liquid equilibrium is to be evaluated at furfural rates of 0, 100, 200, and 400 kmol/h. The multi-component van Laar equation will be used to calculate the liquid activity coefficients, and the vapor phase will be assumed to behave as an ideal gas. The vapor pressure data are calculated by the Antoine Equation 2.19 with the constants given below (p_i^0 in kPa and T in K):

	A_i	B_i	C_i
1. Isobutane	14.853	2796.78	12.470
2. 1-Butene	15.249	3071.37	22.853
3. Furfural	17.339	5705.99	13.576

van Laar constants are given for the isobutane-furfural and 1-butene-furfural binaries. The isobutane-1-butene binary constants are set to zero because these components are assumed to form an ideal solution with each other, so that

$$\gamma_1^\infty = 1 \quad \text{and} \quad \gamma_2^\infty = 1 \ (A_{12} = \ln\gamma_1^\infty \quad \text{and} \quad A_{21} = \ln\gamma_2^\infty)$$

$A_{12} = 0$	$A_{13} = 1.142$	$A_{23} = 0.842$
$A_{21} = 0$	$A_{31} = 1.310$	$A_{32} = 1.029$

Note: In order to avoid division by zero, small and equal values for A_{12} and A_{21} can be used ($A_{12} = A_{21} = 0.001$).

SOLUTION

The K-values are calculated from Equation 1.29a. The liquid activity coefficients are calculated from the three-component formulation of the van Laar equation (Equations 1.35a).

If the furfural rate is S, the fraction vapor is $\psi = V/(F + S) = 50/(100 + S)$. For each value of ψ (corresponding to each solvent rate), the flash Equation 2.14 is solved by iterating on the flash temperature. The component mole fractions in the combined feed are calculated as

$$Z_1 = 50/(100 + S), \quad Z_2 = 50/(100 + S), \quad Z_3 = S/(100 + S)$$

Since the liquid activity coefficients are functions of the liquid composition, this composition must be updated during the iterations:

1. Assume a liquid composition (use the feed composition as an initial guess).
2. Calculate the activity coefficients.
3. Assume a flash temperature.
4. Calculate the K-values.
5. Solve the flash Equation 2.14 by adjusting the temperature in Step 3.
6. Calculate the equilibrium vapor and liquid compositions.
7. If the calculated liquid composition does not match the values in Step 1, restart with the calculated liquid composition. Repeat until the liquid composition stabilizes.

The results are summarized as follows (p_i^0 in kPa, T in K, S in kmol/h):

	$S = 0$		$T = 313$			$S = 100$		$T = 337$		
	γ_i	p_i^0	K_i	X_i	Y_i	γ_i	p_i^0	K_i	X_i	Y_i
1.	1.000	523.2	1.090	0.479	0.521	1.787	944.0	3.513	0.154	0.539
2.	1.000	447.7	0.933	0.518	0.482	1.483	823.8	2.546	0.180	0.459
3.	3.216	0.875	0.006	0.0	0.0	1.108	2.89	0.007	0.665	0.004
	$S = 200$		$T = 351$			$S = 400$		$T = 372$		
	γ_i	p_i^0	K_i	X_i	Y_i	γ_i	p_i^0	K_i	X_i	Y_i
1.	2.197	1285	5.880	0.092	0.540	2.551	1956	10.40	0.052	0.535
2.	1.749	1134	4.132	0.109	0.452	1.971	1755	7.208	0.062	0.444
3.	1.036	5.408	0.012	0.798	0.009	1.011	12.68	0.027	0.886	0.023
S, kmol/h	0	100	200	400						
Y_1/Y_2	1.0809	1.1743	1.1947	1.2050						

2.3.2 PHASE BOUNDARY CALCULATIONS

Bubble points or dew points can be determined by flash calculations with ψ in Equation 2.7 or 2.13 set to either zero or one. Although the general flash calculation method described in the previous section may be applied to bubble points and dew points, phase boundary calculations are handled more efficiently by other methods. At the bubble point, the liquid composition equals the feed composition since all the feed remains in the liquid phase. The composition of a vapor bubble at equilibrium with the liquid is given by Equation 2.13b, which reduces to

$$Y_i = K_i Z_i = K_i X_i$$

At the bubble point, the following condition must, therefore, be satisfied:

$$\sum Y_i = \sum K_i X_i = 1 \qquad (2.16)$$

Similarly, at the dew point, the vapor composition equals the feed composition, and the composition of a liquid drop at equilibrium with the vapor is given by Equation 2.13a as

$$X_i = Z_i / K_i = Y_i / K_i$$

resulting in the dew point condition:

$$\sum X_i = \sum (Y_i / K_i) = 1 \qquad (2.17)$$

A bubble point or dew point calculation consists of determining the pressure at a given temperature, or the temperature at a given pressure, which would result in K_is that satisfy Equation 2.16 or 2.17. In addition, the phase equilibrium condition, Equation 2.12, must be satisfied.

2.3.2.1 Bubble Point–Dew Point Calculations for Composition-Independent K-Values

For ideal solutions, the distribution coefficients may be assumed to be functions of temperature and pressure only, and their dependence on composition may be neglected. With these assumptions, the distribution coefficients are given by Raoult's law (Section 1.3):

$$K_i = \frac{p_i^0}{P} \qquad (2.18)$$

where p_i^0 is the vapor pressure of component i, and P is the system pressure. Equation 2.18, together with a suitable expression for the vapor pressure, may be used to compute the approximate values of the bubble point or dew point pressure or temperature. These may be used as initial estimates in the iterative computations of true bubble points and dew points. Vapor pressures can be calculated from the Antoine equation:

$$p_i^0 = \exp\left(A_i - \frac{B_i}{T + C_i} \right) \qquad (2.19)$$

The constants A_i, B_i, and C_i are characteristic to each component. Combining Equations 2.18 and 2.19 with 2.16 gives an equation for the bubble point; combining with Equation 2.17 gives an equation for the dew point. Thus, at the bubble point,

$$\frac{1}{P} \sum_i X_i \exp\left(A_i - \frac{B_i}{T + C_i} \right) = 1 \tag{2.20}$$

and at the dew point,

$$P \sum_i \frac{Y_i}{\exp\left(A_i - \dfrac{B_i}{T + C_i} \right)} = 1 \tag{2.21}$$

The summations in Equations 2.20 and 2.21 are carried over all components in the system. If the temperature is fixed, the bubble point and dew point pressures may be calculated directly from Equations 2.20 and 2.21. If, however, the pressure is fixed and the bubble point or dew point temperature is required, Equation 2.20 or 2.21 must be solved by an iterative technique such as the Newton–Raphson method because the equations are implicit in the temperature.

EXAMPLE 2.8: BUBBLE POINT AND DEW POINT PRESSURES

In designing a distillation column, it is required to determine the pressure range over which the column overhead can be maintained in the two-phase region at 50°C. The overhead stream composition and the constants for the Antoine vapor pressure equation are given below, where the pressure is in kPa and the temperature is in °C.

	Z_i	A_i	B_i	C_i
n-Pentane	0.2	13.818	2477	233
Benzene	0.3	13.859	2774	220
Toluene	0.5	14.010	3103	220

SOLUTION

The required pressure range is defined by the dew point pressure and the bubble point pressure at 50°C. These are calculated by Equations 2.21 and 2.20:

	Z_i	p_i^0	Z_i/p_i^0	$Z_i p_i^0$
n-Pentane	0.2	158.4	0.00126	31.68
Benzene	0.3	36.1	0.00831	10.83
Toluene	0.5	12.4	0.04032	6.20
			0.04989	48.71

Dew point pressure

$$P_{dp} = \frac{1}{\sum Z_i/p_i^0} = \frac{1}{0.04989} = 20.04 \text{ kPa}$$

Bubble point pressure

$$P_{bp} = \sum Z_i p_i^0 = 48.71 \text{ kPa}$$

EXAMPLE 2.8A: BUBBLE POINT OF A NON-IDEAL SOLUTION

Calculate the bubble point pressure at 63°C of the liquid solution given in Example 1.6A. What is the equilibrium vapor composition? Assume ideal gas behavior in the vapor phase. The component properties are given below:

Component	Mole Fraction in the Liquid, X_i	Vapor Pressure at 63 °C, p_i^0 (atm)	Activity Coefficients from Example 1.6A
1. Ethanol	0.047	0.528	10.2700
2. MCP	0.846	0.756	1.0280
3. Benzene	0.107	0.566	1.3017

SOLUTION

Combine Equations 1.29a and 2.4:

$$PY_i = \gamma_i p_i^0 X_i$$

By summation,

$$\sum PY_i = P \sum Y_i = P = \sum \gamma_i p_i^0 X_i$$

$$= (10.2700)(0.528)(0.047) + (1.0280)(0.756)(0.846)$$
$$+ (1.3017)(0.566)(0.107)$$

$$= 0.9912 \text{ atm}$$

The equilibrium vapor composition is calculated from the equation above:

$$Y_i = \gamma_i p_i^0 X_i/P$$

The results are

Component	Mole Fraction in the Vapor, Y_i
1. Ethanol	0.2571
2. MCP	0.6634
3. Benzene	0.0795

2.3.2.2 Iterative Method for Composition-Dependent K-Values

When the distribution coefficients are composition-dependent, the above method must be modified to account for the effect of composition. A search for the unknown bubble point or dew point temperature or pressure is started on the basis of some composition-independent relationship between the K-values and the temperature and pressure, such as Equations 2.20 and 2.21. Component fugacities are then calculated for the vapor phase and the liquid phase, and the K-values are updated using Equation 2.15. The calculations are repeated until Equation 2.16 or 2.17, as well as Equation 2.12, are satisfied. The iterative scheme for the bubble point pressure calculation may proceed along the following steps:

1. Assume a bubble point pressure based, for example, on the composition-independent K-value method.
2. Using the current K-values, calculate $Y_i = K_i X_i$ for $i = 1, \ldots, C$.
3. Normalize the vapor composition by replacing Y_i with $Y_i / \Sigma Y_i$.
4. Calculate vapor and liquid fugacities, f_i^V and f_i^L, for $i = 1, \ldots, C$.
5. Update the K-values using Equation 2.15.
6. Update the pressure using an approximate relationship such as Equation 2.20.

Repeat the computations at step 2. A solution is reached when $\Sigma Y_i = 1$ and $f_i^L = f_i^V$ for all the components.

2.3.2.3 Simultaneous Method

In this method, the equations describing the system are all solved simultaneously using, for instance, the Newton–Raphson technique. Applied to phase boundary conditions, the basic relationships in Equations 2.7, 2.8, and 2.12 take a special form. Since ψ is either zero at the bubble point or one at the dew point, Equation 2.7 reduces to the trivial relationship $Z_i = X_i$ or $Z_i = Y_i$. Equation 2.8 need not be involved in the simultaneous solution of the system equations. Once the phase boundary conditions are determined, this equation may be solved independently to calculate the heat duty required to bring the feed to its bubble point or dew point. Thus, the only equation to be solved for bubble or dew point calculations is Equation 2.12.

At the bubble point, the X_is are known ($X_i = Z_i$) and the temperature or pressure is given. The unknowns are $C - 1$ vapor compositions and the pressure or temperature. At the dew point, the Y_is are known ($Y_i = Z_i$) as well as the temperature or pressure, and the unknowns are $C - 1$ liquid compositions and the pressure or temperature. In either case, C equations ($i = 1$ to C) must be solved for C unknowns.

Applying the method to bubble point pressure prediction, for example, where the liquid composition and temperature are known, Equation 2.12 may be written in the form (Peng et al., 1977)

$$F_i(P, Y_1, Y_2, \ldots, Y_{C-1}) = 0 \tag{2.22}$$

for $i = 1, 2, \ldots, C$, the number of components. The problem is to find a set of variables P, Y_1, Y_2, $\ldots$, Y_{C-1} that satisfy the set of Equation 2.22. The solution is found iteratively, beginning with estimates $P^0, Y_1^0, Y_2^0, \ldots, Y_{C-1}^0$, where the superscript refers to iteration number. The estimates can be obtained using the approximate bubble point method (Equation 2.20). To calculate the correction vector

$$\Delta P^k, \Delta Y_1^k, \Delta Y_2^k, \ldots, \Delta Y_{C-1}^k$$

the Jacobian matrix is required, which is the matrix of the partial derivatives of all the functions F_i with respect to all the variables to be evaluated:

$$\frac{\partial F_i}{\partial P} \quad \frac{\partial F_i}{\partial Y_1} \quad \frac{\partial F_i}{\partial Y_2} \quad \cdots \quad \frac{\partial F_i}{\partial Y_{C-1}}$$

for $i = 1, \ldots, C$. The correction vector is calculated from the equations

$$\frac{\partial F_i}{\partial P} \Delta P + \frac{\partial F_i}{\partial Y_1} \Delta Y_1 + \frac{\partial F_i}{\partial Y_2} \Delta Y_2 + \cdots + \frac{\partial F_i}{\partial Y_{C-1}} \Delta Y_{C-1} + F_i = 0$$

for $i = 1, \ldots, C$. The new values for the variables are then calculated as

$$P^{k+1} = P^k + \Delta P^k$$
$$Y_1^{k+1} = Y_1^k + \Delta Y_1^k$$
$$\vdots$$
$$Y_{C-1}^{k+1} = Y_{C-1}^k + \Delta Y_{C-1}^k$$

The iterations are repeated until the Equation 2.22 is satisfied within a specified tolerance, or until the corrections are within the tolerance.

EXAMPLE 2.9: BUBBLE POINT AND DEW POINT TEMPERATURES BY THE SIMULTANEOUS METHOD

Using the simultaneous method, calculate the bubble point and dew point temperatures at 35 kPa of a mixture of 60% mole benzene (1) and 40% mole toluene (2). Assume Raoult's law applies, and use the vapor pressure data in Example 2.8.

SOLUTION

Since the pressure is given and the temperature is unknown, Equations 2.20 and 2.21 cannot be solved directly for the temperature, and an iterative method is required. The simultaneous method will be used in this example. This method is more general, and may be used where the K-values can be composition-dependent.

2.3.2.4   Bubble Point Temperature

The simultaneous method is applied in this problem with one modification: the calculated variables include the vapor mole fractions of both components and the temperature, such that the total number of calculated variables is $C + 1 = 3$. This is balanced by including the vapor mole fraction summation equation along with the equilibrium relations. By Raoult's law, the equilibrium relations (Equation 2.12) for components 1 and 2 at the bubble point, and the vapor mole fraction summation equation, are written as

$$F_1 = PY_1 - Z_1 \exp\left(A_1 - \frac{B_1}{T + C_1} \right) = 0$$

$$F_2 = PY_2 - Z_2 \exp\left(A_2 - \frac{B_2}{T + C_2} \right) = 0$$

$$F_3 = Y_1 + Y_2 - 1 = 0$$

These equations will be solved for T, Y_1, and Y_2. The matrix of partial derivatives (the Jacobian) is as follows:

$$\frac{\partial F_1}{\partial T} = -Z_1 \frac{B_1}{(T + C_1)^2} \exp\left(A_1 - \frac{B_1}{T + C_1} \right); \quad \frac{\partial F_1}{\partial Y_1} = P; \quad \frac{\partial F_1}{\partial Y_2} = 0$$

$$\frac{\partial F_2}{\partial T} = -Z_2 \frac{B_2}{(T + C_2)^2} \exp\left(A_2 - \frac{B_2}{T + C_2} \right); \quad \frac{\partial F_2}{\partial Y_1} = 0; \quad \frac{\partial F_2}{\partial Y_2} = P$$

$$\frac{\partial F_3}{\partial T} = 0; \qquad\qquad\qquad\qquad\qquad \frac{\partial F_3}{\partial Y_1} = 1; \quad \frac{\partial F_3}{\partial Y_2} = 1$$

The following equations are solved for the corrections $\Delta T, \Delta Y_1, \Delta Y_2$:

$$\frac{\partial F_1}{\partial T} \Delta T + \frac{\partial F_1}{\partial Y_1} \Delta Y_1 + \frac{\partial F_1}{\partial Y_2} \Delta Y_2 + F_1 = 0$$

$$\frac{\partial F_2}{\partial T} \Delta T + \frac{\partial F_2}{\partial Y_1} \Delta Y_1 + \frac{\partial F_2}{\partial Y_2} \Delta Y_2 + F_2 = 0$$

$$\frac{\partial F_3}{\partial T} \Delta T + \frac{\partial F_3}{\partial Y_1} \Delta Y_1 + \frac{\partial F_3}{\partial Y_2} \Delta Y_2 + F_3 = 0$$

The functions F_i and their partial derivatives are calculated from the current values T, Y_1, and Y_2 (the initial estimates on the first iteration). The corrections are used to update the values of T, Y_1, Y_2 in each iteration:

$$T^{(k+1)} = T^{(k)} + \Delta T^{(k)}$$

$$Y_1^{(k+1)} = Y_1^{(k)} + \Delta Y_1^{(k)}$$

$$Y_2^{(k+1)} = Y_2^{(k)} + \Delta Y_2^{k}$$

Using the initial estimates of $T = 50°C$, $Y_1 = 0.5$, and $Y_2 = 0.5$, the computations converged in three iterations at $T = 57.25°C$, $Y_1 = 0.809$, and $Y_2 = 0.191$.

2.3.2.5   Dew Point Temperature

For the dew point calculations, two equilibrium equations and one liquid mole fraction summation equation are used:

$$F_1 = PZ_1 - X_1 \exp\left(A_1 - \frac{B_1}{T + C_1}\right) = 0$$

$$F_2 = PZ_2 - X_2 \exp\left(A_2 - \frac{B_2}{T + C_2}\right) = 0$$

$$F_3 = X_1 + X_2 - 1 = 0$$

The variables to be determined are T, X_1, and X_2. The Jacobian matrix is

$$\frac{\partial F_1}{\partial T} = -X_1 \frac{B_1}{(T + C_1)^2} \qquad \frac{\partial F_1}{\partial X_1} = -\exp\left(A_1 - \frac{B_1}{T + C_1}\right); \qquad \frac{\partial F_1}{\partial X_2} = 0$$

$$\times \exp\left(A_1 - \frac{B_1}{T + C_1}\right);$$

$$\frac{\partial F_2}{\partial T} = -X_2 \frac{B_2}{(T + C_2)^2} \qquad \frac{\partial F_2}{\partial X_1} = 0; \qquad \frac{\partial F_2}{\partial X_2} = -\exp\left(A_2 - \frac{B_2}{T + C_2}\right)$$

$$\times \exp\left(A_2 - \frac{B_2}{T + C_2}\right);$$

$$\frac{\partial F_3}{\partial T} = 0; \qquad \frac{\partial F_3}{\partial X_1} = 1; \qquad \frac{\partial F_3}{\partial X_2} = 1$$

The following equations are solved for the corrections:

$$\frac{\partial F_1}{\partial T} \Delta T + \frac{\partial F_1}{\partial X_1} \Delta X_1 + \frac{\partial F_1}{\partial X_2} \Delta X_2 + F_1 = 0$$

$$\frac{\partial F_2}{\partial T} \Delta T + \frac{\partial F_2}{\partial X_1} \Delta X_1 + \frac{\partial F_2}{\partial X_2} \Delta X_2 + F_2 = 0$$

$$\frac{\partial F_3}{\partial T} \Delta T + \frac{\partial F_3}{\partial X_1} \Delta X_1 + \frac{\partial F_3}{\partial X_2} \Delta X_2 + F_3 = 0$$

With the initial estimates of $T = 70°C$, $X_1 = 0.5$, $X_2 = 0.5$, the computations converged at $T = 63.78°C$, $X_1 = 0.3537$, and $X_2 = 0.6463$.

2.3.3 LIQUID–LIQUID AND VAPOR–LIQUID–LIQUID EQUILIBRIA

As pointed out in the introduction of this chapter, two liquid phases as well as two liquid phases and a vapor phase can exist at equilibrium. Non-ideal solution conditions can result in the co-existence at equilibrium of two liquid phases having different compositions. This phenomenon is a consequence of the high non-ideality of the mixture giving rise to a strong dependence of the activity coefficients on the composition. LLE will exist if this dependence is such that components at different concentrations in the liquid can have equal fugacities.

As an example, a mixture of benzene and water forms two liquid phases, each having one of the components at a small concentration. The activity coefficient of benzene at a certain low concentration in water and its activity coefficient at some high concentration result in equal benzene fugacities in the two phases. With a similar situation for the water component, the two phases are at equilibrium.

The liquid–liquid system crosses to the vapor–liquid–liquid equilibrium (VLLE) region if the bubble point condition of $\Sigma Y_i = 1$ is satisfied. In VLLE, the equilibrium conditions expressed in Equation 1.19 are expanded to include both liquid phases and the vapor:

$$\hat{f}_i' = \hat{f}_i'' = \hat{f}_i^V \tag{2.23}$$

where the superscripts ′ and ″ refer to the two liquid phases. Substituting the expressions for the fugacities gives

$$\gamma_i' f_i^L X_i' = \gamma_i'' f_i^L X_i'' = Y_i \hat{\phi}_i^V P \tag{2.24}$$

Assuming all the components are below their critical point at system conditions, and that the vapor behaves as an ideal gas, the above equation can be rewritten as (see Section 1.3.3)

$$\gamma_i' X_i' = \gamma_i'' X_i'' = Y_i P / p_i^0 \tag{2.25}$$

EXAMPLE 2.10: BENZENE–WATER EQUILIBRIA

A liquid mixture of 40% mole benzene (1) and 60% mole water at 450 K separates into benzene-rich and water-rich liquid phases.

a. Find the mole percent of each phase and their compositions using the van Laar equation with the following parameters:

Benzene-rich phase: $A'_{12} = 1.1736$ $A'_{21} = 4.4576$
Water-rich phase: $A''_{12} = 22.7158$ $A''_{21} = 0.3662$

b. Determine the bubble-point pressure and the equilibrium vapor composition at 450 K. The vapor pressures of benzene and water at this temperature are

$$p_1^0 = 978 \text{ kPa}$$

$$p_2^0 = 936 \text{ kPa}$$

SOLUTION

a. The equilibrium liquid–liquid distribution coefficients can be expressed as

$$K_i^L = \frac{X_i''}{X_i'} = \frac{\gamma_i'}{\gamma_i''}$$

Define the ratio of moles of water-rich liquid to total liquid moles:

$$\alpha = \frac{L''}{L' + L''}$$

By analogy to Equation 2.14,

$$f(\alpha) = \frac{Z_1(K_1^L - 1)}{1 + \alpha(K_1^L - 1)} + \frac{Z_2(K_2^L - 1)}{1 + \alpha(K_2^L - 1)} = 0$$

where Z_1 and Z_2 are the mole fractions of benzene and water in the mixture. Referring to Equations 2.13, the phase compositions are given as

$$X_i' = \frac{Z_i}{1 + \alpha(K_i^L - 1)}$$

$$X_i'' = \frac{Z_i K_i^L}{1 + \alpha(K_i^L - 1)}$$

These equations, along with the van Laar activity coefficient equations, can be solved for the compositions and the amounts of each liquid phase.

Alternatively, all the model equations may be listed singly, and then solved simultaneously using a standard nonlinear equation solver, such as a spreadsheet program. For the two-component system, the equations include the van Laar equations for both components in each liquid phase:

$$\gamma_1' = f_1(X_1', X_2', A_{12}', A_{21}')$$

$$\gamma_2' = f_2(X_1', X_2', A_{12}', A_{21}')$$

$$\gamma_1'' = f_1(X_1'', X_2'', A_{12}'', A_{21}'')$$

$$\gamma_2'' = f_2(X_1'', X_2'', A_{12}'', A_{21}'')$$

In addition, there are the summation equations:

$$X_1' + X_2' = 1$$

$$X_1'' + X_2'' = 1$$

And finally, the equilibrium conditions:

$$\gamma_1' X_1' = \gamma_1'' X_1''$$

$$\gamma_2' X_2' = \gamma_2'' X_2''$$

With the known values for the van Laar parameters, these eight equations are solved simultaneously for the four Xs and four γs. The results are listed as follows:

$$\gamma_1' = 1.02525 \qquad \gamma_1'' = 42.6697$$

$$\gamma_2' = 25.9192 \qquad \gamma_2'' = 1.13774$$

$$X_1' = 0.9570 \qquad X_1'' = 0.0230$$

$$X_2' = 0.0429 \qquad X_2'' = 0.9769$$

The relative moles of each phase are determined by a component material balance on either component. Based on a one-mole mixture, a material balance on component 1 is written as

$$Z_1 = L' X_1' + (1 - L') X_1''$$

$$L' = 0.4036$$

$$L'' = 0.5964$$

b. At the bubble point pressure,

$$Y_1 + Y_2 = 1$$

From Equation 2.25,

$$P = p_1^0 \gamma' X_1' + p_2^0 \gamma_2' X_2' = p_1^0 \gamma_1'' X_1'' + p_2^0 \gamma_2'' X_2''$$

$$= 978 \times 1.02525 \times 0.9570 + 936 \times 25.9192 \times 0.0429$$

$$= 978 \times 42.6697 \times 0.0230 + 936 \times 1.13774 \times 0.9769$$

$$= 2000.35 \text{ kPa}$$

The vapor compositions,

$$Y_1 = \gamma_1' X_1' p_1^0 / P = 1.02525 \times 0.9570 \times 978/2000.35 = 0.47970$$

$$Y_2 = \gamma_2' X_2' p_2^0 / P = 25.9192 \times 0.0429 \times 936/2000.35 = 0.52029$$

The composition of the first vapor bubble represents the azeotropic composition. If the overall liquid composition is the same as this vapor composition, then the mixture is at its azeotropic composition at the system temperature and pressure, 450 K and 2000 kPa. At these conditions, the relative amounts of the liquid phases may be expressed in terms of α as defined above. The combined liquid composition equals the vapor composition:

$$(1 - \alpha)X_1' + \alpha X_1'' = Y_1$$

$$(1 - \alpha)(0.9570) + \alpha(0.0230) = 0.4797$$

$$\alpha = 0.511$$

This is the fraction on a mole basis of the water-rich liquid phase, and the balance of 0.489 represents the benzene-rich liquid phase.

2.3.3.1 Rigorous VLLE Model

In this method, a mixed K-value is defined as the ratio of the mole fraction of a component in the vapor to its mole fraction in the mixed liquid phase (Schuil and Bool, 1985). Applied to an equilibrium stage or a flash drum, the phase equilibrium is solved using the mixed K-values instead of the usual vapor–liquid K-values to determine the flow rates and compositions of the vapor and the total liquid. The liquid phase split is then calculated on the basis of K-values for each liquid phase to determine the flow rates and compositions of the two liquid phases. An energy balance may also be included to determine the temperature or the heat transfer for the unit.

2.3.3.2 K-Value Computations

With the molar rates of the two liquid phases designated as L' and L'' and the mole fractions of a given component i in each phase as X_i' and X_i'', the mole fraction of this component in the mixed liquid phase is

$$X_{mi} = \frac{X_i' L' + X_i'' L''}{L' + L''} \tag{2.26}$$

When this equation is used in the definition of the mixed K-value, the following relationship is obtained:

$$K_{mi} = \frac{Y_i}{X_{mi}} = \frac{Y_i(L' + L'')}{X_i'L' + X_i''L''} \tag{2.27}$$

The K-values for each liquid phase with the vapor are given as

$$K_i' = \frac{Y_i}{X_i'}$$

$$K_i'' = \frac{Y_i}{X_i''}$$

When these equations are combined with Equation 2.27, the following is obtained for the mixed K-value:

$$K_{mi} = \frac{(L' + L'')K_i'K_i''}{K_i''L' + K_i'L''} \tag{2.28}$$

Defining the ratio of one liquid phase rate to the total liquid as

$$\alpha = \frac{L'}{L' + L''} \tag{2.29}$$

the mixed K-value becomes

$$K_{mi} = \frac{K_i'K_i''}{\alpha K_i'' + (1 - \alpha)K_i'} \tag{2.30}$$

Since two liquid phases coexist at equilibrium only in systems that exhibit significant departure from ideality (Chapter 1), the K-values for three-phase equilibrium must be calculated using methods that are capable of predicting this non-ideality. The method based on liquid activity coefficients, described in Chapter 1, is recommended for this purpose:

$$K_i = \frac{\gamma_i f_i^L}{\hat{\phi}_i^V P} \tag{2.31}$$

where γ_i is the liquid phase activity coefficient, f_i^L is the standard state fugacity, $\hat{\phi}_i^V$ is the vapor phase fugacity coefficient, and P is the pressure. The K-values for each phase are written as

$$K_i' = \frac{\gamma_i' f_i^L}{\hat{\phi}_i^V P} \tag{2.31a}$$

$$K_i'' = \frac{\gamma_i'' f_i^L}{\hat{\phi}_i^V P} \tag{2.31b}$$

The only quantities that are different in these equations for the two liquid phases are the activity coefficients, γ_i' and γ_i''. Equations suitable for calculating activity coefficients for liquid–liquid systems include the NRTL and UNIQUAC equations (Chapter 1). By combining Equations 2.31a and 2.31b with Equation 2.30, the following is obtained for the mixed K-value:

$$K_{mi} = \frac{\gamma_{mi} f_i^L}{\hat{\phi}_i^V P} \tag{2.32}$$

where γ_{im}, the mixed liquid phase activity coefficient, is given by

$$\gamma_{mi} = \frac{\gamma_i' \gamma_i''}{\alpha \gamma_i'' + (1 - \alpha) \gamma_i'} \tag{2.33}$$

2.3.3.3 Application to an Equilibrium Stage

The above equations are combined with the following equations to complete the VLLE model: An overall material balance relates the feed flow rate F to the liquid rates and the vapor rate V:

$$F = L' + L'' + V \tag{2.34}$$

Component material balances based on the distribution between vapor and mixed liquid are expressed as

$$X_{mi} = \frac{Z_i}{1 + \psi(K_{mi} - 1)} \tag{2.35}$$

where ψ is the vapor fraction:

$$\psi = V/F \tag{2.36}$$

The summation equations for the vapor and mixed liquid are combined in one equation:

$$\sum Y_i - \sum X_{mi} = 0 \tag{2.37}$$

where Y_i is expressed as

$$Y_i = K_{mi} X_{mi} \tag{2.38}$$

Equations 2.35 and 2.37 determine the values of X_{mi} and Y_i.

In a similar manner, X_i' and X_i'' are obtained from the following component balances and summation equations:

$$X_i'' = \frac{X_{mi}}{1 + \alpha(K_{ri} - 1)} \tag{2.39}$$

$$\sum X_i' - \sum X_i'' = 0 \tag{2.40}$$

In Equation 2.39, K_{ri} is defined as

$$K_{ri} = K_i''/K_i' \tag{2.41}$$

The mole fractions in the liquids are related by K_{ri}:

$$X_i' = K_{ri}X_i'' \tag{2.42}$$

EXAMPLE 2.11: THREE-PHASE FLASH

A 100 kmol/h stream containing ethylene glycol, furfural, and water is sent to a flash drum where the temperature and pressure are controlled such as to cause separation into two liquid streams and a vapor stream. At these conditions the K-values are determined as listed below and may be assumed constant within the range of operations. It is required to calculate the product streams flow rates and compositions.

Component	Feed Composition Mole Fraction, Z_i	K-Values Relative to L', K_i'	K-Values Relative to L'', K_i''
1. Ethylene glycol	0.2715	0.0357	0.1250
2. Furfural	0.3062	0.1878	1.0222
3. Water	0.4223	8.0720	1.1304

Solution

Since the K-values are given, the computations can be done using Equations 2.29, 2.30, and 2.34 through 2.42. For the three components, these equations total 23. The calculated variables include X_i', X_i'', Y_i, X_{mi}, K_{mi}, K_{ri}, L', L'', V, α, and ψ, also totaling 23 variables. The equations are solved simultaneously, giving the following results:

Component	X_i'	X_i''	Y_i	X_{mi}	K_{mi}	K_{ri}
1.	0.4219	0.1204	0.0149	0.3445	0.0433	3.5014
2.	0.4669	0.0856	0.0872	0.3684	0.2367	5.4430
3.	0.1112	0.7940	0.8979	0.2871	3.1275	0.1400
Flow rate, kmol/h	L' 57.98	L'' 20.02	V 22.00			

2.3.3.4 Iterative Solution

As an alternative to the simultaneous solution, an iterative approach may be used:

1. Assume α
2. Calculate K_{mi}
3. Calculate ψ from the equation

$$f(\psi) = \sum \frac{Z_i(K_{mi} - 1)}{1 + \psi(K_{mi} - 1)} = 0$$

4. Calculate X_{mi}, Y_i
5. Calculate $K_{ri} = K_i''/K_i'$
6. Calculate α from the equation

$$f(\alpha) = \sum (X_i' - X_i'') = \sum \frac{X_{mi}(K_{ri} - 1)}{1 + \alpha(K_{ri} - 1)} = 0$$

If α (calculated) $\neq \alpha$ (assumed), restart with α (calculated) until converged.

The following calculations show the converged run at a certain tolerance. Slight mismatches between the simultaneous and iterative results are due to differences between the applied convergence tolerances in the two methods.

Assume $\alpha = 0.7433$

$$K_{m1} = \frac{(0.0357)(0.1250)}{(0.7433)(0.1250)(1 - 0.7433)(0.0357)} = 0.0433$$

$$K_{m2} = 0.2367$$

$$K_{m3} = 3.1275$$

$$f(\psi) = \frac{(0.2715)(0.0433 - 1)}{1 + \psi(0.0433 - 1)} + \frac{(0.3062)(0.2367 - 1)}{1 + \psi(0.2367 - 1)}$$
$$+ \frac{(0.4223)(3.1275 - 1)}{1 + \psi(3.1275 - 1)} = 0$$

$$\psi = 0.22$$

$$X_{m1} = \frac{0.2715}{1 + (0.22)(0.0433 - 1)} = 0.3445$$

$$X_{m2} = 0.3684$$

$$X_{m3} = 0.2871$$

$$Y_1 = K_{m1}X_{m1} = (0.0433)(0.3445) = 0.0149$$

$$Y_2 = 0.0872$$

$$Y_3 = 0.8979$$

$$K_{r1} = \frac{K_1''}{K_1'} = \frac{0.1250}{0.0357} = 3.5014$$

$$K_{r2} = 5.4430$$

$$K_{r3} = 0.1400$$

$$f(\alpha) = \frac{(0.3445)(3.5014 - 1)}{1 + \alpha(3.5014 - 1)} + \frac{(0.3684)(5.4430 - 1)}{1 + \alpha(5.4430 - 1)}$$
$$+ \frac{(0.2871)(0.1400 - 1)}{1 + \alpha(0.1400 - 1)} = 0$$

$$\alpha = 0.7433$$

$X_1'' = Y_1/K_1'' = 0.0149/0.1250 = 0.1192$	$X_1' = Y_1/K_1' = 0.0149/0.0357 = 0.4174$
$X_2'' = Y_2/K_2'' = 0.0872/1.0222 = 0.0853$	$X_2' = Y_2/K_2' = 0.0872/0.1878 = 0.4643$
$X_3'' = Y_3/K_3'' = 0.8979/1.1304 = 0.7904$	$X_3' = Y_3/K_3' = 0.8979/8.0720 = 0.1112$

2.3.3.5 VLLE in Hydrocarbon–Water Systems

Hydrocarbon processing often involves water in the streams, and this can result in the coexistence of two liquid phases and a vapor phase. A valid approach in handling such VLLE conditions is to neglect the hydrocarbon solubility in water, and assume the water phase is pure water. This enables the use of a simplified mixed K-value formulation.

The solubility of water in the liquid hydrocarbon phase is given as X_s, the mole fraction of water in the water-saturated liquid hydrocarbon phase. With the hydrocarbon phase molar rate designated as L' and the water phase as L'', the overall mole fraction of water in the mixed liquid phase is calculated as

$$X_{mw} = \frac{L'X_s + L''}{L' + L''} \tag{2.43}$$

This equation is combined with Equation 2.29 to derive the following expression for α:

$$\alpha = \frac{1 - X_{mw}}{1 - X_s} \tag{2.44}$$

At equilibrium, Equation 2.25 must be satisfied for the liquid phases. The water phase is assumed pure water, therefore $X_w'' = 1$ and $\gamma_w'' = 1$, and Equation 2.25 for water in the hydrocarbon phase becomes

$$X_s \gamma_w' = 1$$

Using this expression, the K-value of water in the hydrocarbon phase is written as

$$K_w' = \frac{\lambda_w' f_w^L}{\hat{\phi}_w^V P} = \frac{f_w^L}{X_s \hat{\phi}_w^V P} \tag{2.45}$$

For other components in the hydrocarbon phase, K_i' is calculated by an equation of state or any other suitable method, as described in Section 1.3.

The K-value of water in the water phase is calculated as

$$K_w'' = \frac{\lambda_w'' f_w^L}{\hat{\phi}_w^V P} = \frac{f_w^L}{\hat{\phi}_w^V P} \tag{2.46}$$

For other components in the water phase, $X_i'' = 0$ since this phase is assumed pure water. From Equation 2.25,

$$\gamma_i'' = \frac{X_i'}{X_i''} \gamma_i'$$

It follows that for $i \neq w$, $\gamma_i'' = \infty$, and from Equation 2.31b, $K_i'' = \infty$.

The mixed K-values are now calculated from Equation 2.30. For water,

$$K_{mw} = \frac{K_w' K_w''}{\alpha K_w'' + (1-\alpha)K_w'} \tag{2.47}$$

And for all other components,

$$K_{mi} = \frac{K_i' K_i''}{\alpha K_i'' + (1-\alpha)K_i'}$$

Since $K_i'' = \infty$, this expression reduces to

$$K_{mi} = \frac{K_i'}{\alpha} \tag{2.48}$$

EXAMPLE 2.12: HYDROCARBON–WATER FLASH

A 100 kmol/h hydrocarbon–water mixture is flashed at 350 K and 101.34 kPa, producing vapor, hydrocarbon liquid, and liquid water. The feed contains 30%

mole water, 40% mole n-hexane, and 30% mole n-octane. The K-values relative to the hydrocarbon liquid at the flash conditions are 1.2483 for hexane and 0.2130 for octane. The solubility of water in the hydrocarbon liquid is 0.005288 mole fraction. The hydrocarbons are assumed insoluble in the water. The vapor pressure of water at 350 K is 41.65 kPa. It is required to determine the products rates and compositions.

SOLUTION

The vapor phase at the flash conditions is assumed to behave as an ideal gas. Hence, the K-value of water relative to the hydrocarbon liquid is written as

$$K_1' = p_1^0/PX_s = 41.65/(101.34 \times 0.005288) = 77.726$$

Relative to the liquid water,

$$K_1'' = p_1^0/P = 41.65/101.34 = 0.4110$$

The K-values of the hydrocarbons relative to the liquid water phase are infinite since hexane and octane are assumed insoluble in the water.

In this example, the problem is solved iteratively, as an alternative to the simultaneous method which requires initial estimates for the variables. The iterative solution is a natural option when workable initial estimates are not available. The iterations are started by assuming the values for α and ψ, defined as

$$\alpha = L'/(L' + L'')$$

$$\psi = V/F$$

where F is the total feed rate, and V, L', and L'' are the vapor, hydrocarbon liquid, and water flow rates. The mixed K-values are calculated by Equations 2.47 and 2.48. The following equation, similar to Equation 2.14, is then solved for ψ:

$$f(\psi) = \sum \frac{Z_i(K_{mi} - 1)}{1 + \psi(K_{mi} - 1)} = 0 \tag{2.49}$$

The mixed liquid and vapor compositions are as follows:

$$X_{mi} = \frac{Z_i}{1 + \psi(K_{mi} - 1)}$$

$$Y_i = K_{mi}X_{mi}$$

The liquid phase split is calculated by solving the following equation for α:

$$f(\alpha) = \sum \frac{X_{mi}(1 - K_{ri})}{\alpha(1 - K_{ri}) + K_{ri}} \tag{2.50}$$

where $K_{ri} = K'_i/K''_i = X''_i/X'_i$. Equation 2.49 is derived from the summation equation,

$$\sum (X'_i - X''_i) = 0$$

The liquid compositions are obtained from component balances:

$$X_{mi} = \alpha X'_i + (1-\alpha)X''_i = \alpha X'_i + (1-\alpha)K_{ri}X'_i$$

Rearranging,

$$X'_i = \frac{X_{mi}}{\alpha(1-K_{ri}) + K_{ri}}$$

$$X''_i = K_{ri}X'_i = \frac{X_{mi}K_{ri}}{\alpha(1-K_{ri}) + K_{ri}}$$

These expressions are substituted in the summation equation to generate Equation 2.50. The value obtained for α is used to update the mixed K-values, and all the computations are repeated until the convergence is achieved. The numerical calculations at convergence are shown below:

$$\alpha = 0.8714$$

$$\psi = 0.6006$$

Water:

$$K_{m1} = \frac{77.726 \times 0.411}{0.8714 \times 0.4110 + (1 - 0.8714) \times 77.726} - 3.0855$$

Hexane:

$$K_{m2} = 1.2483/0.8714 = 1.4325$$

Octane:

$$K_{m2} = 0.21301/0.8714 = 0.2444$$

$$X_{m1} = \frac{0.30}{1 + 0.6006(3.0855 - 1)} = 0.1332$$

$$X_{m2} = 0.3175$$

$$X_{m3} = 0.5492$$

$$Y_1 = 3.0855 \times 0.1332 = 0.4109$$

$$Y_2 = 0.4548$$

$$Y_3 = 0.1343$$

$$K_{r1} = K_1'/K_1'' = 77.726/0.4110 = 189.11$$

$$K_{r2} = 0$$

$$K_{r3} = 0$$

$$X_1' = \frac{0.1332}{0.8714(1 - 189.11) + 189.11} = 0.0053 \quad X_1'' = 1$$

$$X_2' = 0.3175/0.8714 = 0.3644 \quad X_2'' = 0$$

$$X_3' = 0.5492/0.8714 = 0.6303 \quad X_3'' = 0$$

$$V = F\alpha = 100 \times 0.6006 = 60.06 \text{ kmol/h}$$

$$L' + L'' = 100 - 60.06 = 39.94 \text{ kmol/h}$$

$$L' = 0.8714 \times 39.94 = 34.80 \text{ kmol/h}$$

$$L'' = 39.94 - 34.80 = 5.14 \text{ kmol/h}$$

NOMENCLATURE

A, B, C	Antoine vapor pressure constants
C	Number of components
F	Degrees of freedom
F	Feed flow rate
f	Fugacity
h	Liquid enthalpy
H	Vapor enthalpy
K	Distribution coefficient
P	Pressure
p^0	Vapor pressure
Q	Heat duty
T	Temperature
X	Mole fraction in liquid
Y	Mole fraction in vapor

Z	Mole fraction in feed
α	Fugacity coefficient
α	Number of phases
ψ	Vapor mole fraction

SUBSCRIPTS

i (or 1, 2, ...)   Component designation

SUPERSCRIPTS

k Iteration number
L Liquid phase
V Vapor phase

PROBLEMS

2.1. A hydrocarbon stream is separated into vapor and liquid streams in a flash drum maintained at 27°C and 1035 kPa. The feed composition and component K-values (assumed composition-independent) are given below. Calculate the vapor rate as a fraction of the feed, and the vapor and liquid compositions.

i	Component	Z_i	K_i
1.	Ethane	0.40	2.90
2.	Propane	0.35	0.95
3.	n-Butane	0.15	0.33
4.	n-Pentane	0.10	0.11

2.2. A stream containing ethane, butane, and pentane is fed to a temperature-controlled vessel where phase separation takes place between the vapor and the liquid. The vessel pressure is maintained at 690 kPa. The K-values, assumed composition-independent, are given by the equation

$$\ln K_i = A_i - B_i/(1.8T + C_i)$$

where T is in degrees Kelvin. The feed composition and the equation constants are given below:

		Z_i	A_i	B_i	C_i
1.	Ethane	0.27	5.75	2155	−30.3
2.	n-Butane	0.34	6.21	3386	−60.1
3.	n-Pentane	0.39	6.52	3973	−72.9

Determine the bubble point and dew point temperatures at 690 kPa, and the vapor fraction and temperature that results in a maximum n-butane mole fraction in the vapor.

2.3. The starting conditions of a given hydrocarbon stream are 38°C and 690 kPa. The stream is heated in an exchanger where 23,200 kJ/kmol are transferred to the stream. Calculate the temperature, vapor fraction, and vapor and liquid compositions at the exchanger outlet. Assume the pressure remains constant at 690 psia. Stream composition:

		z_i
1.	Ethane	0.25
2.	n-Butane	0.35
3.	n-Pentane	0.40

Use the K-values given in Problem 2.2. The enthalpies may be calculated on the assumption of ideal solution (zero excess enthalpy). The following data are given for the component enthalpies:

$$H(kJ/kmol) = 2.32[a + b(1.8T\ (°C) + 32)]$$

		Liquid		Vapor	
		a	b	a	b
1.	Ethane	2160	22.0	7200	13.5
2.	n-Butane	2668	32.5	12590	20.0
3.	n-Pentane	2880	40.0	15150	28.5

2.4. Derive equations to calculate the K-values and composition of a vapor at equilibrium with a binary liquid solution having a given composition and a given relative volatility α_{12}, or ratio of K-values. Calculate the composition of the vapor at equilibrium with a liquid having 0.40 mole fraction n-hexane and 0.60 mole fraction n-heptane. The relative volatility of n-hexane to n-heptane is 3.40.

2.5. A vapor mixture of n-hexane, n-heptane, and water is condensed at 100 kPa. The hydrocarbon K-values are given by the equation

$$\ln K_i = A_i - B_i/1.8T$$

where T is in degrees Kelvin. The mixture composition and the values of the constants are as follows:

		z_i	A_i	B_i
1.	n-Hexane	0.62	11.08	6821
2.	n-Heptane	0.05	11.29	7545
3.	Water	0.33	—	—

Assuming water and the hydrocarbons are immiscible:
  a. Calculate the water and hydrocarbon dew point temperatures and equilibrium compositions.
  b. Outline an algorithm for calculating the temperature and phase compositions of a hydrocarbon–water mixture at given total pressure P, when the fraction of the hydrocarbons condensed is β. Assume composition-independent K-values. Do the calculations for $P = 100$ kPa, $\beta = 0.5$.

2.6. A stream of hydrofluoric acid (1) in water (2) at 120°C and 200 kPa contains 12% mole hydrofluoric acid (HF). It is proposed to concentrate the HF in solution by partial vaporization in a single stage, by means of temperature and pressure control. Calculate the resulting liquid composition and the fraction vaporized at 120°C and 135 kPa. Can this process be used to concentrate the liquid for any starting composition? Use the van Laar equation for liquid activity coefficients and assume ideal gas behavior in the vapor phase. The vapor pressures of HF and water at 120°C are 1693 and 207 kPa, respectively, and the van Laar constants are $A_{12} = -6.0983$, $A_{21} = -6.9658$ (see Problems 1.8 and 1.9).

2.7. In a single equilibrium stage absorber at 25°C and 100 kPa, water is used to reduce the amount of acetone in 1000 kmol/h air from 2% mole to 0.5% mole. Assuming the flow rates of air and water leaving the absorber to be the same as the inlets, calculate the required water flow rate. The following information may be used:

Acetone vapor pressure at 25°C = 30.5 kPa.

Activity coefficient of acetone at low concentrations (infinite dilution) at 25°C = 6.5.

Water vapor pressure at 25°C = 3.2 kPa.

Assuming ideal gas behavior in the vapor, and that air is insoluble in water–acetone, use the calculated inlet water rate to determine the compositions of the outlet streams when the assumption of constant liquid and gas rates is dropped.

2.8. The separation of benzene (1) and cyclohexane (2) by distillation is complicated due to the formation of an azeotrope (Example 2.5). Vapor–liquid equilibrium data for this binary are required for the design of a workable separation process. As a first step, find the azeotropic composition and temperature at 100 kPa pressure. Use the van Laar equation for activity coefficients with parameters $A_{12} = 0.147$, $A_{21} = 0.165$. The computations can be made with assumptions consistent with low pressure conditions. The vapor pressures of benzene and cyclohexane can be represented by the Antoine Equation 2.19 with the following constants: $A_1 = 13.88$, $B_1 = 2788.5$, $C_1 = -52.36$, $A_2 = 13.74$, $B_2 = 2766.6$, $C_2 = -50.30$, where T is in K and p_i^0 is in kPa.

2.9. A three-component feed stream is flashed at 390 K to recover most of the component 3 in the liquid product, with 18% of that component leaving with the vapor product. The feed flow rate is 250 kmol/h and its composition and vapor pressures of the components at the flash temperature are as follows:

	Z_i	p_i^0 (kPa)
1.	0.24	1800
2.	0.43	800
3.	0.33	400

Using the simultaneous method, calculate the flash pressure required for this separation, and the product rates and compositions. Assume Raoult's law applies at the problem conditions.

2.10. The stream defined below is heated to 100°C to be partially vaporized in a flash drum before entering a distillation column. The fraction vaporized is controlled by the flash drum pressure. Calculate the required pressure at 100°C to have 20% mole vaporization, assuming Raoult's law applies. What are the products' flow rates and compositions? The constants for the Antoine Equation 2.19 are given for each component, with the pressure in kPa and the temperature in K.

	Feed Stream (kmol/h)	A_i	B_i	C_i
1. Benzene	210	13.86	2773.8	−53.0
2. Toluene	75	14.01	3103.0	−53.4
3. n-Heptane	30	13.86	2911.3	−56.5
	315			

2.11. A hydrocarbon vapor stream is cooled in a heat exchanger where the outlet pressure is 100 kPa. What is the required outlet temperature that would result in 75% mole condensation of the stream? The feed stream composition and the components' Antoine Equation 2.19 constants (p^0 in kPa, T in K) are given below. Raoult's law may be assumed valid at the problem conditions.

	Z_i	A_i	B_i	C_i
1. n-Pentane	0.12	13.82	2447.1	−39.9
2. n-Hexane	0.47	12.81	2154.3	−79.0
3. n-Heptane	0.41	13.86	2911.3	−56.5

2.12. The mixture given below is sent to a flash drum for a crude separation into a vapor product with 0.05 mole fraction n-pentane and a liquid product with 0.05 mole fraction ethane. Find the flash temperature and pressure required to achieve this separation. The K-values can be calculated according to Raoult's law with vapor pressures from the Antoine Equation 2.19 using the constants provided, p^0 in kPa, T in K.

	Z_i	A_i	B_i	C_i
1. Ethane	0.18	13.551	1459.7	−20.998
2. Propane	0.27	13.439	1728.4	−34.898
3. n-Butane	0.36	13.692	2140.8	−36.504
4. n-Pentane	0.19	14.246	2714.8	−27.294

2.13. Acetone and chloroform form an azeotrope and cannot be separated by conventional distillation. In extractive distillation, the separation is enhanced by adding benzene as a solvent. In a preliminary evaluation of the effectiveness of the solvent, calculations are made on a single equilibrium stage. The main feed stream is at a rate of 100 kmol/h with 50% mole acetone and 50% mole chloroform. The equilibrium stage is controlled at 70°C and 110 kPa. Determine the effect of adding 45, 50, and 60 kmol/h benzene on the separation.

Use the Wilson equation for activity coefficients and assume ideal gas behavior in the vapor phase, with component vapor pressures calculated by the Antoine Equation 2.19. The constants for this equation are consistent with pressure in kPa and temperature in Kelvin. The following data are given:

	Antoine Equation Constants			liq. mol. vol V_i
	A_i	B_i	C_i	(cm³/mol)
1. Acetone	14.636	2940.5	−35.9	77
2. Chloroform	16.055	3977.5	13.44	81
3. Benzene	13.886	2788.5	−52.4	94

The Wilson equation parameters are given below in cal/mol.

$\lambda_{12} - \lambda_{11} = -72.20$	$\lambda_{13} - \lambda_{11} = 494.92$	$\lambda_{23} - \lambda_{22} = -204.22$
$\lambda_{12} - \lambda_{22} = -332.23$	$\lambda_{13} - \lambda_{33} = -167.91$	$\lambda_{23} - \lambda_{33} = 141.62$

2.14. Benzene (1) and water (2) form a heterogeneous azeotrope at 450 K and 2000 kPa. The benzene-rich liquid phase has 95.7% mole benzene, the water-rich phase has 97.7% mole water, and the vapor phase has 48% mole benzene. Use these data to determine the van Laar parameters for the benzene-rich and the water-rich phases. An approximation of ideal gas behavior in the vapor phase may be assumed. The vapor pressures at 450 K are 936 kPa for water and 978 kPa for benzene.

2.15. The stream described below is flashed at 2026.5 kPa. It is required to determine its two-phase temperature range. Calculate the equilibrium temperature, and vapor and liquid compositions at vapor fractions of 0.001 (bubble point), 0.25, 0.50, 0.75, and 0.999 (dew point). Use the Wilson equation for activity coefficients and the truncated virial equation for the fugacity coefficients (see Example 2.7A).

Component	Feed Z_i	T_{Ci} (K)	P_{Ci} (kPa)	ω_i	Liquid Molar Volume V_i (cm³/mol)
1. Acetone	0.3	508.1	4701	0.309	73.52
2. 2-Butanone	0.3	535.6	4154	0.329	89.57
3. Ethyl acetate	0.4	523.2	3830	0.363	97.79

Pseudo-critical pressures, kPa, $P_{C12} = 4408$, $P_{C13} = 4053$, $P_{C23} = 3830$
Antoine constants (Equation 2.19), T in K, p_i^0 in kPa,

	A_i	B_i	C_i
1.	14.63615	2940.46	−35.93
2.	14.58345	3150.42	−36.65
3.	14.13645	2790.50	−57.15

Wilson exponential parameters, δ_{ij}, cal/mol,

$\delta_{11} = 0$	$\delta_{12} = 1371.31$	$\delta_{13} = -292.975$
$\delta_{21} = -650.152$	$\delta_{22} = 0$	$\delta_{23} = -405.21$
$\delta_{31} = 644.481$	$\delta_{32} = 2704.427$	$\delta_{33} = 0$

2.16. Compare the van Laar and Wilson equations by solving the problem in Example 2.7A using the van Laar equation for activity coefficients. Ideal gas behavior may be assumed for the vapor phase. Use the feed composition and vapor pressure data from Example 2.7A, and calculate the temperature and phase compositions at the pressure and vapor fraction in that example. The following are the binary van Laar constants:

$A_{12} = 0.243$, $A_{21} = 0.243$, $A_{13} = 0.89$, $A_{31} = 0.65$, $A_{23} = 0.36$, $A_{32} = 0.22$.

2.17. In an ethyl acetate (1)–ethyl alcohol (2) separation process, it is required to determine if this mixture forms an azeotrope. If it does, it is required to determine the azeotropic temperature and composition as a function of the azeotropic pressure between 70 and 120 kPa at 10 kPa intervals. Solve for the composition and temperature by doing a single computation operation at each pressure value. Additionally, for any one of the pressure points determine if the azeotrope is minimum- or maximum-boiling. Use the van Laar equation for the liquid activity coefficients and assume ideal gas behavior for the vapor phase. The following data are given:

van Laar parameters: $A_{12} = 0.855$, $A_{21} = 0.753$

Antoine constants (Equation 2.19, p_i^0 in kPa, T in °C):

$$A_1 = 14.3376, \quad B_1 = 2866.624, \quad C_1 = 217.881$$

$$A_2 = 15.4540, \quad B_2 = 2950.989, \quad C_2 = 193.768$$

2.18. One hundred kmol/h of a hydrocarbon stream is flashed at 25°C and 10 atm. The stream composition and vapor pressures at these conditions are given below. Assuming Raoult's law applies, calculate the number of kmol/h of propane in the liquid product. It is estimated that the total liquid product is between 20 and 30 kmol/h.

	Z_i	p_i^0 (atm)
Ethane	0.40	28
Propane	0.35	9
n-Butane	0.25	4

2.19. A liquid solution of isopropanol (1) and water (2) at 90°C contains 30% mole isopropanol and 70% mole water. The van Laar constants are $A_{12} = 1.042$ and $A_{21} = 0.492$. The vapor pressures of isopropanol and water at 90°C are $P_1^0 = 201.3$ kPa and $P_2^0 = 70.0$ kPa. The vapor phase may be assumed an ideal gas. Calculate the activity coefficients of isopropanol and water in this solution and its bubble point pressure at 90°C.

2.20. A stream flowing at the rate of 100 kmol/h separates in a flash drum at 80°C and 115 kPa into a vapor stream and a liquid stream at equilibrium. The overall stream composition and the component vapor pressures at 80°C are given below:

	Z_i	p_i^0 (kPa)
1. Acetone	0.45	195.75
2. Acetonitrile	0.35	97.84
3. Nitromethane	0.20	50.32

Assuming Raoult's law applies, calculate the rate of nitromethane flowing in the vapor.

2.21. A single-stage process is used to separate a binary mixture of ethyl alcohol (1) and cyclohexane (2) at 45°C. The following infinite dilution liquid activity coefficient data are given:

At $X_1 = 0$,	$\gamma_1 = 8.9$
At $X_2 = 0$,	$\gamma_2 = 3.9$

The vapor pressures at 45°C are $P_1^0 = 21.3$ kPa, $P_2^0 = 40.0$ kPa. Using the Margules liquid activity coefficient equation, calculate the activity coefficients of ethyl alcohol and cyclohexane, the total pressure, and the vapor composition if the ethyl alcohol mole fraction in the liquid is 0.30. Assume ideal gas behavior in the vapor phase.

2.22. The following vapor–liquid equilibrium data are given for benzene (1)/ cyclohexane (2) at 100 kPa:

T (°C)	X_1	Y_1	γ_1	γ_2
77.8	0.399	0.421	1.135	1.056
77.6	0.545	0.545	1.078	1.101
77.8	0.700	0.677	1.038	1.177

a. Calculate the relative volatility at each temperature and identify the azeotropic point, if one exists.

b. Calculate the van Laar constants for this binary based on the azeotropic composition.

c. For a liquid with 60% mole benzene and 40% mole cyclohexane at 80°C and assuming ideal gas behavior in the vapor phase, calculate the bubble point pressure and the equilibrium vapor composition. The component vapor pressures are given by the equation

$$\ln p_i^0 = A_i - B_i/(T + C_i),$$

where p^0 is in kPa and T is in K. The equation constants are given:

	A_i	B_i	C_i
1.	15.904	4088.766	9.043
2.	12.819	2247.980	−79.742

2.23. Streams 1 and 2, defined below, are mixed together in a vessel where the temperature is controlled at 300 K. The mixed stream leaving the vessel must be saturated vapor. What should be the pressure in the vessel? Raoult's law may be assumed, with vapor pressures calculated as in Problem 2.22 using the constants given below.

	Stream 1 (kmol/h)	Stream 2 (kmol/h)	A_i	B_i	C_i
1. Propane	30	5	13.439	1728.4	−34.989
2. n-Butane	55	25	13.692	2140.8	−36.504
3. n-Pentane	29	67	14.246	2714.8	−27.294

2.24. Phase equilibrium data are required for a process involving a binary mixture of ethanol (1) and benzene (2) at 40°C. The infinite dilution liquid activity coefficients are given:

At $X_1 = 0$,	$\gamma_1 = 9.025$
At $X_2 = 0$,	$\gamma_2 = 4.055$

The vapor pressures at 40°C are $p_1^0 = 20$ kPa, $p_2^0 = 37.3$ kPa. Using the van Laar liquid activity equation, calculate the K-values of ethanol and benzene at 40°C when the ethanol mole fraction in the liquid is 0.25, and determine the bubble point pressure. Assume ideal gas behavior in the vapor phase.

2.25. Isobutane and 1-butene are close-boilers and are difficult to separate by conventional distillation. Using a single stage, a 100 kmol/h stream containing 50% mole isobutane and 50% mole 1-butene is flashed at 480 kPa, producing a 50 kmol/h vapor stream. With no solvent added, the vapor composition is about the same as the feed. Furfural, which is less volatile than both components, is now added as a solvent to alter the isobutane-1-butene relative volatility by depressing the 1-butene K-value relative to that of isobutane.

It is required to determine the effect of the furfural rate on the separation. The multi-component van Laar equation is used to calculate the liquid activity coefficients, and the vapor phase is assumed to behave as an ideal solution. Solvent rates of 0, 100, 200, and 400 kmol/h are to be considered. The vapor pressure data are calculated using the Antoine Equation 2.19, with the constants given below (p_i^0 in kPa and T in K):

	A_i	B_i	C_i
1. Isobutane	14.853	2796.78	12.470
2. 1-Butene	15.249	3071.37	22.853
3. Furfural	17.339	5705.99	13.576

van Laar constants are given for the isobutane-furfural and 1-butene-furfural binaries. The isobutane-1-butene binary constants are set to zero because these components are assumed to form an ideal solution with each other, so that

$$\gamma_1^\infty = 1 \quad \text{and} \quad \gamma_2^\infty = 1 \quad (A_{12} = \ln \gamma_1^\infty \text{ and } A_{21} = \ln \gamma_2^\infty)$$

$A_{12} = 0$	$A_{13} = 1.142$	$A_{23} = 0.842$
$A_{21} = 0$	$A_{31} = 1.310$	$A_{32} = 1.029$

Note: In order to avoid division by zero, small and equal values for A_{12} and A_{21} may be used ($A_{12} = A_{21} = 0.001$).

For the 400 kmol/h solvent rate, in order to determine the recovery of furfural in the liquid from the extractive separation stage, the liquid is sent to an equilibrium separation stage at 480 kPa and 350°C. Find the product rates and compositions for this stage.

REFERENCES

Khoury, F. M., Hydrocarbon Processing, December 1978, pp. 155–157.
Peng, D. Y. and D. B. Robinson, A rigorous method for vapor–liquid equilibrium calculations, AIChE 70th Annual Meeting, New York, 1977.
Schuil, J.A. and Bool, K.K. Three-phase flash and distillation, *Computers and Chemical Engineering*, 9(3), 295, 1985.
Soave, G., *Chemical Engineering Science*, 27, 1197, 1972.

3 Fundamentals of Multistage Separation

In the discussion in Chapter 2 on the performance of single stage, it was noted that, for a feed stream of given composition, the range of possible product compositions obtainable with an equilibrium stage at a given pressure is bounded by the bubble point and the dew point compositions at that pressure. In order to extend this range to higher or lower component concentrations, multiple stages are used.

Multistage processes are considered to be made of a number of equilibrium stages stacked together to perform various separation functions. Using these building blocks, configurations may be designed to perform functions such as distillation, absorption, stripping, and rectification. This chapter discusses the conceptual construction of several multistage processes from equilibrium stages.

Multiple stages enhance separation in two ways: (1) temperature variation from stage to stage affects the distribution coefficients so that at lower temperatures the lighter components' concentration is driven up and vice versa at higher temperatures; and (2) the progressive enrichment or depletion of different volatility components from stage to stage resulting from changes in the compositions and flows of fluids in the process. The two factors are generally at work together to varying degrees in most multistage separation processes. A purely temperature effect process may be seen in binary distillation, while a purely compositional effect would be a constant temperature absorption process. The two effects are analyzed separately in detail in the following sections.

The coexisting liquid and vapor streams associated with phase separation may be brought about by phase creation or by phase addition. Phase creation is accomplished by heat transfer to and from the column as in distillation, where heat is added in the reboiler to vaporize some of the liquid, and heat is removed in the condenser to condense some of the vapor. Phase addition is accomplished by sending an auxiliary stream to the column: a liquid absorbent to an absorber or a stripping vapor stream to a stripper.

In contrast to phase separation processes, separation of species forming a single phase may be accomplished by several flow mechanisms through membranes. Membrane separations are discussed in Chapter 18.

The separation characteristics of a process are determined by the number of stages involved, their configuration, and the operating conditions. In general, as the process becomes more complex, the number of variables required to define it increases, making it more versatile and enabling it to handle wider separation ranges. The specification–variable relationships associated with a number of fundamental configurations are discussed next and illustrated with examples.

3.1 CASCADED STAGES

Figure 3.1 illustrates a process in which a feed stream, F, is flashed in stage 2 at temperature T_2 and pressure P_2 to produce a saturated vapor stream, V_2 at its dew point and a saturated liquid stream, L_2 at its bubble point. The vapor is sent to stage 1, where it is partially condensed at temperature T_1 and pressure P_1, and the liquid is sent to stage 3, where it is partially vaporized (or reboiled) at temperature T_3 and pressure P_3. The resulting vapor and liquid streams are V_1, L_1, V_3, and L_3 as indicated.

The pressures in the three stages are assumed equal. In order for condensation to take place in stage 1, T_1 must be below the dew point of V_2 ($T_1 < T_2$). Also, for vaporization to occur in stage 3, T_3 must be above the bubble point of L_2 ($T_3 > T_2$). The system of cascaded equilibrium stages therefore results in a temperature spread between the separated products' dew point and bubble point, a feature that is absent in a single stage. Stream V_1 is richer in lighter components than V_2, and stream L_3 is richer in heavier components than L_2, the net result being a sharper separation than that from a single stage.

Figure 3.1 shows other possible cascaded configurations for phase separation, where side products or recycled streams are considered.

3.1.1 GRAPHICAL REPRESENTATION

The operation of cascaded stages may be described for a binary mixture on a temperature-concentration diagram, Figure 3.2. Starting with a feed stream to stage 2 with mole fraction component 1 (the lighter component) equal to Z_{21}, the highest possible concentration of component 1 on that stage would be Y_{21}, which is the concentration of the first vapor bubble. (The first subscript refers to the stage and the second subscript to the component). The lowest possible concentration would be X_{21}, which corresponds to the first condensing liquid drop when the dew point curve is approached from higher temperatures.

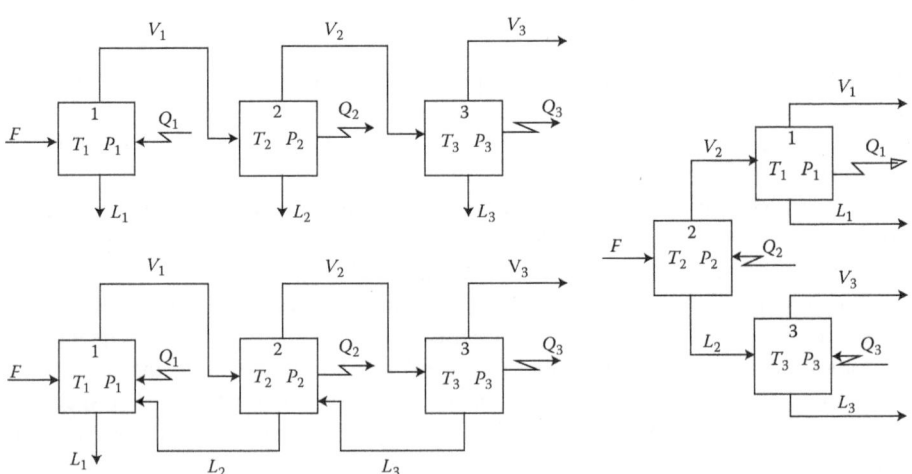

FIGURE 3.1   Cascaded stages.

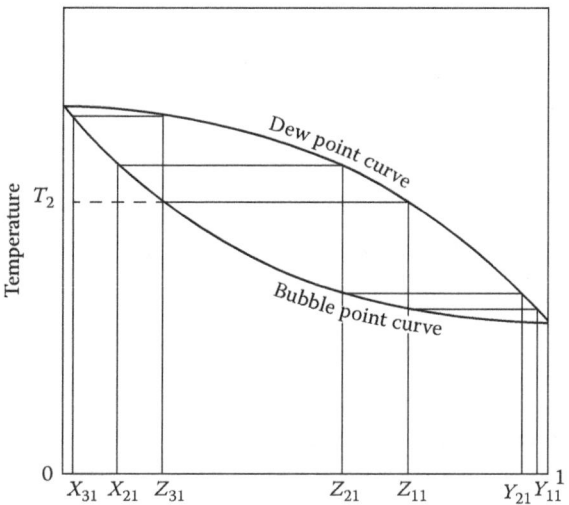

FIGURE 3.2 Graphical representation of cascaded stages.

If the feed to stage 2 is raised to some temperature T_2, intermediate between the bubble point and dew point, the vapor would have a concentration Z_{11} and the liquid Z_{31}. If the vapor with concentration Z_{11} is fed at stage 1, then the maximum possible concentration of component 1 in stage 1 would be Y_{11}. Similarly, if the liquid with concentration Z_{31} is fed at stage 3, then the minimum possible concentration of component 1 in stage 3 would be X_{31}. Clearly, Y_{11} is greater than Y_{21}, and X_{31} is smaller than X_{21}. Thus, a wider spread of product concentrations is obtainable from cascaded stages than from a single stage.

3.1.2 EQUILIBRIUM RELATIONSHIPS

The performance of cascaded stages may also be described analytically on the basis of material balances and phase equilibrium relations. Consider, for simplicity, a binary mixture of components 1 and 2 and again assume component 1 to be the lower boiling, or more volatile, component. The equilibrium equations at stage j are

$$\frac{Y_{j1}}{X_{j1}} = K_{j1} \tag{3.1}$$

$$\frac{Y_{j2}}{X_{j2}} = \frac{1 - Y_{j1}}{1 - X_{j1}} = K_{j2} \tag{3.2}$$

Equation 3.2 makes use of the fact that $X_{j1} + X_{j2} = 1$ and $Y_{j1} + Y_{j2} = 1$. Assuming the distribution coefficients K_{j1} and K_{j2} to be known quantities or estimated by some data predicting method, Equations 3.1 and 3.2 may be solved for the unknowns X_{j1} and Y_{j1}:

$$X_{j1} = \frac{1 - K_{j2}}{K_{j1} - K_{j2}} \tag{3.3}$$

$$Y_{j1} = \frac{K_{j1}(1 - K_{j2})}{K_{j1} - K_{j2}} \tag{3.4}$$

The relative volatility, defined as the ratio of distribution coefficients,

$$\alpha_{j12} = \frac{K_{j1}}{K_{j2}}$$

is now combined with Equations 3.3 and 3.4 to give

$$X_{j1} = \frac{\alpha_{j12}}{\alpha_{j12} - 1}\frac{1}{K_{j1}} - \frac{1}{\alpha_{j12} - 1} = \frac{\alpha_{j12} - K_{j1}}{K_{j1}(\alpha_{j12} - 1)} \tag{3.5}$$

$$Y_{j1} = \frac{\alpha_{j12}}{\alpha_{j12} - 1} - \frac{K_{j1}}{\alpha_{j12} - 1} = \frac{\alpha_{j12} - K_{j1}}{\alpha_{j12} - 1} \tag{3.6}$$

Or, by substituting for K_{j1} from Equation 3.1 and rearranging,

$$X_{j1} = \frac{Y_{j1}}{Y_{j1}(1 - \alpha_{j12}) + \alpha_{j12}} \tag{3.5a}$$

$$Y_{j1} = \frac{\alpha_{j12} X_{j1}}{X_{j1}(\alpha_{j12} - 1) + 1} \tag{3.6a}$$

For systems that are not highly nonideal, in which the distribution coefficients are not strong functions of the composition, it may be assumed that the relative volatility is fairly constant over reasonably wide temperature ranges.

In the cascaded flash, since $T_3 > T_2 > T_1$, and because distribution coefficients generally go up with temperature, it follows from Equations 3.5 and 3.6 that Y_{11} is greater than Y_{21}, and X_{32} is greater than X_{22}. To arrive at this, recall that component 1 is the more volatile component; hence, $K_{j1} > K_{j2}$ and $\alpha_{j12} > 1$, which implies that the ratio $\alpha_{j12}/(\alpha_{j12} - 1)$ is always greater than one and $1/(\alpha_{j12} - 1)$ is always positive. If the relative volatility, and hence these two ratios, are assumed to have the same values in stages 1 and 2, Equation 3.6 indicates that as K_{j1} decreases, Y_{j1} increases, and, since T_1 is less than T_2, K_{11} is less than K_{21}, and Y_{11} is greater than Y_{21}. Similar reasoning may be applied to Equation 3.5 to conclude that X_{32} is greater than X_{22}. This provides an analytical verification that improved separation is obtainable with cascaded flashes.

3.1.3 PARAMETER RELATIONSHIPS

The cascaded flash is also more versatile than the single flash in that more parameters may be varied to manipulate its performance. If stage 2 has a feed of fixed rate, composition, and thermal conditions, two parameters would be required to define its operation, as discussed in Chapter 2. These may be chosen as the pressure and heat duty. If we assume the pressure is fixed because of practical process considerations, one degree of freedom remains.

In a three-stage cascaded flash system, each stage could have an associated heat source or sink. If these duties are fixed and the pressure on each stage is also set, the operation of the system is defined. This implies that all the other variables, such as stage temperatures and product rates and compositions, are determinate. The three heat duties may now be varied independently to allow three of the other variables to meet certain specifications. In contrast, in a single stage only one specification is possible at fixed pressure. The three specifications apply to the system as a whole, and it is not mandatory that each specification be associated with a particular stage.

Equations 3.1 and 3.2, or their alternative forms, Equations 3.3, 3.4, and 3.5 and 3.6, represent the phase equilibrium relationships at stage j for a binary mixture. In addition to these, component material balances for each component at stage j generate the following equations:

$$F_j Z_{j1} = V_j Y_{j1} + L_j X_{j1} \tag{3.7}$$

$$F_j Z_{j2} = V_j Y_{j2} + L_j X_{j2} \tag{3.8}$$

Here F_j, V_j, and L_j are the feed, vapor product, and liquid product molar flow rates at stage j, and Z_{ji}, Y_{ji}, and X_{ji} are the feed, vapor, and liquid mole fractions of component i at stage j. Summation of Equations 3.7 and 3.8 gives the total material balance,

$$F_j = V_j + L_j$$

This equation may be substituted in Equation 3.7 or 3.8 to eliminate V_j or L_j. The result for the vapor rate is

$$V_j = \frac{F_j (Z_{j1} - X_{j1})}{Y_{j1} - X_{j1}} \tag{3.9}$$

An energy balance equation may also be written for each stage. All in all, there are five independent equations for each stage with seven variables: X_{j1}, Y_{j1}, V_j, L_j, T_j, P_j, and Q_j, where Q_j is the heat duty at stage j. Two variables per stage must be given in order to define a unique solution. Example 3.1 illustrates the solution of a three-stage cascaded flash with a binary mixture for different sets of performance specifications.

EXAMPLE 3.1: PROPANE–BUTANE CASCADED FLASH SEPARATION

A mixture of 50 kmol/h propane and 50 kmol/h butane at 38°C and 690 kPa is to be separated into propane-rich and butane-rich products using a three-stage cascaded flash arrangement similar to Figure 3.1. The pressure in each stage is fixed at 690 kPa. Calculate the product rates and compositions and the stage temperatures that correspond to each of the following operating conditions:

A. The stage temperatures are specified:

$T_1 = 32°C$
$T_2 = 38°C$
$T_3 = 44°C$

B. The following are specified:

Mole fraction propane in the vapor product from stage 1, $Y_{11} = 0.80$
Flow rate of the vapor product from stage 2, $V_2 = 60$ kmol/h
Flow rate of butane in the liquid product from stage 3, $L_3X_{32} = 25$ kmol/h

C. Same as B, with one difference: the flow rate of propane in the vapor product from stage 1, instead of the vapor product from stage 2, is specified: $V_1Y_{11} = 10$ kmol/h.

The distribution coefficients of propane and butane are given at 690 kPa and at five different temperatures in Table 3.1.

SOLUTION

The three stages involve six independent variables. Since the three-stage pressures are fixed, three other specifications are made in each of the above cases. In principle, the solution involves setting up 15 equations with 21 variables, six of which are given. The simplicity of this problem makes it more practical to group a few equations at a time and eliminate some of the variables rather than solve the whole system of equations simultaneously. This method is not particularly systematic and cannot be generalized to more complex problems but is a useful exercise for studying the relationships between variables. The specifications for cases A and B are such that the stages may be solved sequentially, one at a time. In case C, two stages must be solved simultaneously. The heat duties for each stage are not calculated, although the temperatures are. Enthalpy data would be required to carry out energy balances for duty calculations.

A. Since the stage temperatures are specified in this case, the corresponding distribution coefficients are used directly in Equations 3.3 and 3.4 to calculate the component mole fractions in the liquid and vapor in each stage. The results are given in Table 3.2.

TABLE 3.1
Distribution Coefficients of Propane and Butane at 690 kPa

T (°C)	27	32	38	43	49
K (C3)	1.370	1.540	1.710	1.900	2.090
K (C4)	0.428	0.496	0.571	0.654	0.744
a	3.20	3.10	2.99	2.91	2.81

TABLE 3.2

Calculated Mole Fractions of Propane and Butane at 690 kPa (Example 3.1)

Stage	1	2	3
T (°C)	32	38	43
K (C3)	1.540	1.710	1.900
K (C4)	0.496	0.571	0.654
X (C3)	0.483	0.377	0.278
X (C4)	0.517	0.623	0.722
Y (C3)	0.744	0.644	0.528
Y (C4)	0.256	0.356	0.472

Next, Equation 3.9 is used to calculate the vapor rate leaving stage 2:

$$V_2 = 100 \times (0.50 - 0.377)/(0.644 - 0.377) = 46.07 \text{ kmol/h}$$

The liquid rate is calculated from a material balance in stage 2:

$$L_2 = 100 - 46.07 = 53.93 \text{ kmol/h}$$

The feed at stage 1 is V_2 and the products from this stage are calculated as above:

$$V_1 = 46.07 - (0.644 - 0.483)/(0.744 - 0.483) = 28.42 \text{ kmol/h}$$

$$L_1 = 46.07 - 28.42 = 17.65 \text{ kmol/h}$$

Similarly, L_2 is the feed at stage 3 and the products are

$$V_3 = 53.93 \times (0.377 - 0.278)/(0.528 - 0.278) = 21.36 \text{ kmol/h}$$

$$L_3 = 53.93 - 21.36 = 32.57 \text{ kmol/h}$$

B. Stage 2 is solved first. Since the temperature is not known, the functional dependence of the distribution coefficients at the temperature must be determined. The data in Table 3.1 show an approximately linear relationship between the K-values and temperature. A good representation of the propane distribution coefficients within the indicated temperature range may be expressed by the equation

$$K(\text{C3}) = 1.36 + 0.03249[T°\text{C} - 26.67] \tag{3.10}$$

The butane distribution coefficients could also be expressed by a similar equation. The computations may be simplified, however, by assuming constant relative volatility, which has an average value of 3.00, based on the data in Table 3.1.

The phase equilibrium and material balance relations are applied to stage 2 using Equations 3.6 and 3.7:

$$Y_{21} = 3.00/(3.00 - 1) - K_{21}/(3.00 - 1)$$

$$(100) \times (0.50) = 60Y_{21} + 40X_{21}$$

Substituting $X_{21} = Y_{21}/K_{21}$ in the second equation and simplifying, the above two equations are rewritten as

$$Y_{21} = 1.5 - 0.5K_{21} \tag{3.11}$$

$$Y_{21}(6 + 4/K_{21}) = 5 \tag{3.12}$$

Equations 3.11 and 3.12 are combined by eliminating Y_{21}, resulting in a quadratic equation in K_{21}:

$$3K_{21}^2 - 2K_{21} - 6 = 0$$

The positive root of this equation is $K_{21} = 1.786$. The temperature is now computed directly from Equation 3.10, $T_2 = 39.8°C$, and the butane distribution coefficient is calculated from the relative volatility

$$K_{22} = K_{21}/3.00 = 1.786/3.00 = 0.595$$

The distribution coefficient of propane at stage 1 is calculated from the equivalent of Equation 3.11 written for stage 1, with $Y_{11} = 0.80$:

$$0.80 = 1.5 - 0.5K_{11}$$

$$K_{11} = 1.400$$

For butane,

$$K_{12} = 1.400/3.00 = 0.467$$

The stage 1 temperature T_1 is calculated from Equation 3.10 and rearranged as

$$T_1 = 30.78K_{11} - 15.19 = 27.9°C$$

The calculations for stage 3 are as follows: The feed at stage 3 is the liquid from stage 2:

$$L_2 = 40 \text{ kmol/h}$$

From Equation 3.11,

$$Y_{21} = 1.5 - (0.5)(1.786) = 0.607$$

$$X_{21} = Y_{21}/K_{21} = 0.607/1.786 = 0.340$$

$$X_{22} = 1 - X_{21} = 0.660$$

A butane balance is made in stage 3:

$$L_2 X_{22} = V_3 Y_{32} + L_3 X_{32}$$

The problem specifies that $L_3 X_{32} = 25$ kmol/h; hence,

$$V_3 Y_{32} = (40) \times (0.660) - 25 = 1.4 \text{ kmol/h}$$

From a total material balance in stage 3,

$$V_3 + L_3 = 40$$

Substituting for V_3, L_3, and $K_{32} = Y_{32}/X_{32}$,

$$1.4/Y_{32} + 25K_{32}/Y_{32} = 40$$

or

$$40Y_{32} = 25K_{32} + 1.4$$

Using the equivalent of Equation 3.6 for component 2 and assuming a constant relative volatility of 3.00, the following is obtained:

$$Y_{32} = \frac{1/3}{1/3 - 1} - \frac{K_{32}}{1/3 - 1}$$

or

$$Y_{32} = -0.5 + 1.5K_{32}$$

This equation is combined with the previous one to eliminate Y_{32}, yielding $K_{32} = 0.611$. The propane distribution coefficient is

$$K_{31} = (3.00) \times (0.611) = 1.833$$

The stage 3 temperature is calculated from Equation 3.10 (rearranged):

$$T_3 = 30.78K_{31} - 15.19 = 41.2°C$$

C. Two specifications in stage 1 must be satisfied by varying conditions in stages 1 and 2. Therefore, stages 1 and 2 should be solved simultaneously. The following equations apply to stage 1:

Total material balance:

$$V_2 = V_1 + L_1$$

Propane material balance:

$$V_2 Y_{21} = V_1 Y_{11} + L_1 X_{11}$$

Phase equilibrium relationship for propane:

$$K_{11} = Y_{11}/X_{11}$$

The following equations apply to stage 2:
Total material balance:

$$F_2 = V_2 + L_2$$

Propane material balance:

$$F_2 Z_{21} = V_2 Y_{21} + L_2 X_{21}$$

Phase equilibrium relationship for propane:

$$K_{21} = Y_{21}/X_{21}$$

The following variables are given or specified:

$$F_2 = 100.00 \text{ kmol/h}$$

$$Z_{21} = 0.50$$

$$Y_{11} = 0.80$$

$$V_1 Y_{11} = 10.00 \text{ kmol/h}$$

$$V_1 = 10/0.80 = 12.50 \text{ kmol/h}$$

These values are substituted in the equations in stage 1, which are combined to give the following:

$$V_2 Y_{21} = 10.00 + (V_2 - 12.50)(0.80)/K_{11}$$

Stage 2 equations are combined to give

$$50 = V_2 Y_{21} + (100.00 - V_2)Y_{21}/K_{21}$$

Additionally, Equation 3.6 is applied to stages 1 and 2, assuming the relative volatility has a constant value of 3.00:

$$Y_{11} = 1.5 - 0.5K_{11} = 0.80$$

$$Y_{21} = 1.5 - 0.5K_{21}$$

The last four equations are to be solved for Y_{21}, V_2, K_{11}, and K_{21}. The coefficient K_{11} is solved directly:

$$K_{11} = (1.5 - 0.8)/0.5 = 1.40$$

Coefficient K_{21} is eliminated, leaving two equations in V_2 and Y_{21}. An iterative solution gives $V_2 = 50.736$ mol/h and $Y_{21} = 0.632$. The other variables are readily calculated using the above system of equations. Stage 3 is solved independently as in Case B.

3.2 DISTILLATION BASICS

When the vapor product from a feed stage is condensed in another stage at a temperature lower than the feed stage temperature, a two-stage cascaded operation is formed. Vapor from the lower temperature stage is richer in lighter components and leaner in heavier components than vapor from the feed stage, as discussed in the previous section. The temperature at each stage is now held constant and the liquid from the upper stage is recycled to the feed stage (Figure 3.3). If the feed stream is a binary mixture, the compositions of streams V_1 and L_2 are essentially unchanged from the cascaded arrangement since the compositions in such systems are primarily functions of the temperature, as demonstrated in Section 3.2.2. The rates of products

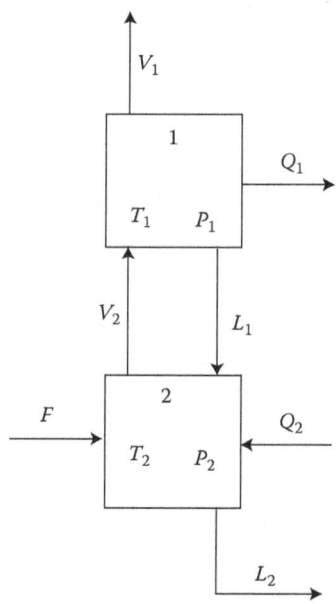

FIGURE 3.3   Two-stage distillation column.

V_1 and L_2, however, are now higher since L_1 is recycled instead of being taken as an additional product.

This arrangement represents a basic distillation process. Figure 3.3 illustrates a two-stage distillation column where heavier components of the vapor rising from the bottom stage are condensed and refluxed back to the lower stage, resulting in an overhead vapor product that has a higher concentration of lighter components. Liquid flowing down to the lower stage is reboiled and lighter components are vaporized and sent back to the upper stage so that a bottom liquid stream is produced with a higher concentration of heavier components. In order to bring about condensation on the upper stage and vaporization on the bottom stage, heat duties are applied to each stage: negative Q_1 (heat removed) and positive Q_2 (heat added).

3.2.1 TEMPERATURE EFFECT ON SEPARATION

As pointed out in the introduction to this chapter, it is helpful in understanding multistage separations to distinguish between two effects: temperature and composition. (The other factors, such as pressure, number of stages, column configuration, etc., are assumed fixed in this context.) Of the two effects, the more predominant in distillation is temperature. In the case of binary distillation where the composition effect on the distribution coefficients may be neglected, the temperature is essentially the only factor. Under these conditions, the equations derived in Section 3.1.2 are applicable.

The added separation power of a two-stage distillation column (Figure 3.3) over a single stage results from the fact that the two stages can be maintained at two different temperatures. For a binary system, the compositions are given by Equations 3.3 and 3.4 or 3.5 and 3.6. These equations indicate that separation, expressed in terms of mole fractions of the products, is a function of K-values. In this simple model, the K-values are a function of temperature only. Hence, the temperature effect is the only active one in binary distillation.

If more stages are added, the temperature spread becomes wider and therefore separation becomes sharper. Also, as discussed further along, increased vapor rate from stage 2 to 1 and liquid rate from stage 1 to 2 results in a larger temperature difference between the stages and, hence, better separation.

3.2.2 MATHEMATICAL REPRESENTATION

A mathematical formulation of the column of Figure 3.3 follows for a binary mixture. Material balances for components 1 and 2 are written for stages 1 and 2, with the first subscript referring to the stage and the second subscript to the component.

$$FZ_{21} + L_1 X_{11} = V_2 Y_{21} + L_2 X_{21} \tag{3.13}$$

$$F(1 - Z_{21}) + L_1(1 - X_{11}) = V_2(1 - Y_{21}) + L_2(1 - X_{21}) \tag{3.14}$$

$$V_2 Y_{21} = V_1 Y_{11} + L_1 X_{11} \tag{3.15}$$

$$V_2(1 - Y_{21}) = V_1(1 - Y_{11}) + L_1(1 - X_{11}) \tag{3.16}$$

Additionally, Equations 3.1 and 3.2 used for cascaded stages in Section 3.2.2 are rewritten for the present two-stage column:

$$\frac{Y_{11}}{X_{11}} = K_{11} \qquad (3.17)$$

$$\frac{Y_{21}}{X_{21}} = K_{21} \qquad (3.18)$$

$$\frac{1 - Y_{11}}{1 - X_{11}} = K_{12} \qquad (3.19)$$

$$\frac{1 - Y_{21}}{1 - X_{21}} = K_{22} \qquad (3.20)$$

Two more equations may be derived to represent energy balances in each stage. The energy balance equations provide functional relationships between stage temperatures T_1 and T_2 and heat duties Q_1 and Q_2.

All in all, 10 equations exist with 12 variables: X_{11}, X_{21}, Y_{11}, Y_{21}, L_1, L_2, V_1, V_2, T_1, T_2, Q_1, and Q_2. Hence, in order to solve the column, two of the variables must be specified. The distribution coefficients may be supplied as additional data or may be calculated from appropriate correlations. The column equations may be manipulated in different ways, depending on which of the variables are specified and which are to be solved. For instance, a chain of eliminations and substitutions may be carried out to develop an expression for L_1/V_1, the ratio of reflux to vapor from the top tray, or reflux ratio:

$$\frac{L_1}{V_1} = \frac{Y_{21} - Y_{11}}{X_{11} - Y_{21}} \qquad (3.21)$$

The application of these equations to specific cases is demonstrated in Example 3.2. The number and complexity of the equations for multistage, multi-component columns are much higher, which rules out the elimination and substitution methods. General systematic techniques for solving systems of column equations are discussed in Chapter 13.

3.2.3 PARAMETER RELATIONSHIPS

In the two-stage distillation column (Figure 3.3), the operating pressure and the feed flow rate, composition, and thermal conditions are assumed fixed. Each stage with an associated heat source or sink has one degree of freedom (Chapter 2). The column as a whole then has two degrees of freedom. The heat duties Q_1 and Q_2 may be considered as independent variables. Their magnitudes control the column operation and, if they are allowed to vary, two other performance variables may be specified. For instance, the liquid rate from the bottom stage and the reflux rate (liquid from the top stage flowing down the column) may be specified. This completely defines the column operation, and all the other variables, including the two heat duties, may be calculated.

If the process is operated at a specified liquid rate from stage 2 (L_2), the vapor rate from stage 1 (V_1) is fixed by virtue of overall material balance. A second independent variable, the liquid from stage 1, or the reflux rate L_1, is now specified. For a given bottoms rate L_2 or overhead rate V_1, the effect of reflux rate on stage temperatures and separation is examined next. As L_1 is increased, the fraction of feed vaporized on stage 2 increases because of what follows. The vapor fraction on stage 2 is

$$V_2/(F + L_1)$$

A material balance on stage 2 gives

$$V_2 = F + L_1 - L_2$$

which if substituted in the vapor fraction expression results in

$$(F + L_1 - L_2)/(F + L_1)$$

or

$$1 - L_2/(F + L_1)$$

It may be readily verified that, as L_1 increases, the vapor fraction increases since F and L_2 are fixed. A higher vapor fraction means higher temperature. Therefore, an increased reflux rate implies an elevated temperature at the vaporization stage. The reverse is true for the condensation stage. As the reflux rate increases, so does the vaporization rate, V_2, in order to maintain a constant bottoms rate, L_2. The fraction vapor on stage 1 is V_1/V_2, and, as V_2 increases, this fraction decreases since V_1 is constant. Thus, a lower temperature is obtained on the condensation stage as a result of higher reflux. The duties Q_1 and Q_2 also go up with increasing reflux rate. More cooling is required for condensation and more heating for vaporization.

Another result of a higher reflux rate or ratio is sharper separation: V_1 becomes richer in lighter components and L_2 becomes richer in heavier components. At higher reflux, the temperature gradient from one stage to the other becomes steeper, resulting in higher K-values in the lower stage and lower K-values in the upper stage. The effect of that is higher concentration of the light components in the upper stage and the opposite on the lower stage, as may be inferred from Equations 3.5 and 3.6 for a binary system.

The reflux ratio may vary between the limits of zero and infinity. At zero reflux ratio, the operation of a multistage process reduces to that of a single stage. This may be verified by setting L_1/V_1 to zero in Equation 3.21, which implies that $Y_{21} = Y_{11}$. Thus, at zero reflux, the separation is lowest, being the same as in a single stage.

As the reflux ratio is increased, the separation becomes sharper and, as L_1/V_1 approaches infinity, the maximum separation that can be achieved with a given number of stages is reached. The column is then said to operate at total reflux. In practical terms, total reflux may be approached by operating the column so that the reflux rate is very large compared to the feed and product rates. From Equation 3.21, at total reflux, $X_{11} = Y_{21}$; that is, the liquid composition from stage 1 equals the vapor composition from stage 2.

The effect of reflux ratio on the performance of a two-stage distillation column is demonstrated numerically for a binary system in Example 3.2.

EXAMPLE 3.2: PROPANE–BUTANE TWO-STAGE DISTILLATION

The propane–butane equimolar mixture of Example 3.1 at 38°C and 690 kPa is fed at a rate of 100 kmol/h at the bottom stage of a two-stage distillation column as in Figure 3.3. The liquid rate, L_2, is specified at 50 kmol/h, and it is desired to determine the effect of reflux ratio, L_1/V_1, on the products composition and stage temperatures.

SOLUTION

At the limit of zero reflux ratio, Equation 3.21 reduces to $Y_{21} = Y_{11}$. From a propane material balance on the entire column,

$$FZ_{21} = V_1Y_{11} + L_2X_{21}$$

$$(100) \times (0.50) = 50Y_{11} + 50X_{21}$$

$$1 = Y_{11} + X_{21}$$

Vapor–liquid equilibria are expressed in the form of Equation 3.6, where an average relative volatility of 3.00 is assumed, as in Example 3.1. The following equations are obtained:

$$Y_{11} = 3X_{11}/(1 + 2X_{11})$$

$$Y_{21} = 3X_{21}/(1 + 2X_{21})$$

The above equations are solved for X_{11}, X_{21}, Y_{11}, and Y_{21}. By eliminating all these variables except X_{11}, the following quadratic equation is obtained:

$$2X_{11}^2 + 2X_{11} - 1 = 0$$

One positive root exists, $X_{11} = 0.366$. The other variables are solved by substitution, giving $Y_{11} = 0.634$, $X_{21} = 0.366$, and $Y_{21} = 0.634$. The two stages have identical compositions at zero reflux ratio. The K-value of propane is calculated from its definition

$$K_{11} = K_{21} = Y_{11}/X_{11} = Y_{21}/X_{21} = 0.634/0.366 = 1.732$$

The temperature is also the same in both stages and is calculated using Equation 3.10 (rearranged):

$$T_1 = T_2 = 30.78K_{11} - 15.19 = 38.1°C$$

As the reflux ratio approaches infinity (total reflux), Equation 3.21 requires that $X_{11} = Y_{21}$. The other equations are the same as in the case of zero reflux ratio. The problem equations are solved by simple eliminations and substitutions, giving the following results: $X_{11} = 0.50$, $Y_{11} = 0.75$, $X_{21} = 0.25$, $Y_{21} = 0.50$. The K-values of propane are calculated as before:

$$K_{11} = Y_{11}/X_{11} = 0.75/0.50 = 1.5$$
$$K_{21} = Y_{21}/X_{21} = 0.50/0.25 = 2.0$$

The stage temperatures are calculated from Equation 3.10:

$$T_1 = (30.78)(1.5) - 15.19 = 31.0°C$$
$$T_2 = (30.78)(2.0) - 15.19 = 46.4°C$$

The computations for reflux ratios between zero and infinity are more complicated. Systematic solution methods of the set of generally nonlinear equations are employed. For the present two-stage binary system, the equations are solved for a reflux ratio $L_1/V_1 = 1$ by elimination.

Since $L_2 = 50$ kmol/h, it follows by overall material balance that $V_1 = 50$ mol/h. And since $L_1/V_1 = 1$, $L_1 = 50$ kmol/h. Also,

$$V_2 = F + L_1 - L_2 = 100 + 50 - 50 = 100 \text{ kmol/h}$$

Substituting the value of L_1/V_1 in Equation 3.21 gives

$$(Y_{21} - Y_{11})/(X_{11} - Y_{21}) = 1$$

or

$$2Y_{21} - Y_{11} - X_{11} = 0$$

The other equations are the same as for $L_1/V_1 = 0$ or ∞. Eliminating Y_{11} and Y_{21} results in the following two equations in X_{11} and X_{21}:

$$\frac{6X_{21}}{1 + 2X_{21}} - \frac{3X_{11}}{1 + 2X_{11}} = 0$$

$$X_{21} = \frac{1 - X_{11}}{1 + 2X_{11}}$$

By eliminating X_{21} between these two equations, the following quadratic equation in X_{11} is obtained:

$$3X_{11}^2 + X_{11} - 1 = 0$$

TABLE 3.3

Performance of Two-Stage Distillation Column (Example 3.2)

Reflux Ratio, L_1/V_1	T_1 (°C)	T_2 (°C)	Y_{11}	X_{21}
0[a]	38.1	38.1	0.634	0.366
0.2[b]	37.2	39.5	0.655	0.345
1.0[a]	34.2	42.3	0.697	0.303
20[b]	31.6	44.2	0.742	0.258
∞[a]	31.0	46.4	0.750	0.250

[a] As computed in Example 3.2.
[b] Simulation results, using Soave–Redlich–Kwong equation of state to predict K-values.
Feed rate, $F = 100$ kmol/h, 50 kmol/h propane, 50 kmol/h butane
Overhead rate, $V_1 = 50$ kmol/h
Bottoms rate, $L_2 = 50$ kmol/h
Pressure, $P = 690$ kPa

The positive root is $X_{11} = 0.434$. Working back through the other equations, the remaining compositions are calculated as $X_{21} = 0.303$, $Y_{11} = 0.697$, and $Y_{21} = 0.566$. The K-values and stage temperatures are calculated as in the first two cases:

$$K_{11} = Y_{11}/X_{11} = 0.697/0.434 = 1.606$$

$$K_{21} = Y_{21}/X_{21} = 0.566/0.303 = 1.868$$

$$T_1 = (30.78) \times (1.606) - 15.19 = 34.2°C$$

$$T_2 = (30.78) \times (1.868) - 15.19 = 42.3°C$$

The above results for reflux ratios of 0, 1, and ∞ are given in Table 3.3. Also tabulated are data points at other values of reflux ratio, calculated by computer simulation using the Soave–Redlich–Kwong (SRK) equation of state for predicting the K-values. The concentrations are also plotted in Figure 3.4, showing the effect of reflux ratio on separation. In this example, since $V_1 = L_2$, an overall material balance indicates that at any point $Y_{11} + X_{21} = 1$.

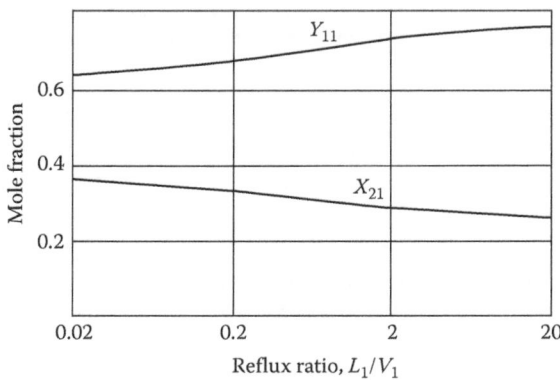

FIGURE 3.4 Performance of two-stage distillation column (Example 3.2).

3.3   ABSORPTION/STRIPPING BASICS

It was noted in Section 2.2 that the phase distribution of a mixture may be altered by introducing an additional feed to the equilibrium stage. The second feed influences the separation in different ways: by changing the values of the distribution coefficients, by shifting the azeotropic compositions, or simply by changing the vapor and liquid compositions as a result of the change in overall composition. A purely absorptive or stripping process is one in which the equilibrium phase distribution is controlled by manipulating the overall composition only and not by other factors such as temperature. In "ideal" absorption or stripping, the distribution coefficients are assumed composition-independent.

Absorption and stripping are similar in principle. The basics are discussed in this section. In some instances each process is discussed separately. In most cases the same principles apply to both, and, to avoid redundancy, both processes are simply referred to as *absorbers*.

In a binary mixture equilibrium stage, the vapor and liquid compositions are a function of only distribution coefficients and hence a function of only temperature if the pressure is held fixed (Equations 3.3 and 3.4). Thus, in a binary system, the phase compositions are independent of feed composition. If a third component is added to the mixture, the equilibrium vapor and liquid compositions are influenced by the amount and identity of the third component. This phenomenon is the basis of absorption and stripping processes. The following is a simple mathematical analysis of the effect of a third component on phase distribution.

3.3.1   TERNARY SYSTEMS

A binary mixture of components 1 and 2 is fed during the equilibrium stage held at constant temperature and pressure. The sum of mole fractions in the vapor and liquid is unity:

$$Y_1^b + Y_2^b = 1 \tag{3.22}$$

$$Y_1^b/K_1 + Y_2^b/K_2 = 1 \tag{3.23}$$

The superscript b designates binary compositions, and K_1 and K_2 are the distribution coefficients. Component 1 is assumed more volatile than component 2, that is, $K_1 > K_2$. Solving Equations 3.22 and 3.23 gives the following results, already arrived at in Equation 3.4:

$$Y_1^b = \frac{K_1(1 - K_2)}{K_1 - K_2} \tag{3.24}$$

$$Y_2^b = \frac{K_2(K_1 - 1)}{K_1 - K_2} \tag{3.25}$$

A third component is now introduced at the equilibrium stage at a given rate, keeping the stage temperature and pressure unchanged. The sums of the mole fractions in the liquid and vapor are written with superscript t, designating ternary composition:

$$Y_1^t + Y_2^t + Y_3^t = 1 \tag{3.26}$$

$$Y_1^t/K_1 + Y_2^t/K_2 + Y_3^t/K_3 = 1 \tag{3.27}$$

The distribution coefficients are assumed independent of composition; hence, $K1$ and $K2$ appearing in Equation 3.27 are equal to those in Equation 3.23. Equations 3.26 and 3.27 are solved for Y_1^t and Y_2^t in terms of K_1, K_2, K_3, and Y_3^t:

$$Y_1^t = \frac{K_1(1 - K_2)}{K_1 - K_2} + \frac{K_1}{K_3}\frac{K_2 - K_3}{K_1 - K_2}Y_3^t \tag{3.28}$$

$$Y_2^t = \frac{K_2(K_1 - 1)}{K_1 - K_2} - \frac{K_2}{K_3}\frac{K_1 - K_3}{K_1 - K_2}Y_3^t \tag{3.29}$$

The effect of the third component on phase separation is investigated by comparing Y_1^t and Y_2^t with Y_1^b and Y_2^b. From the above equations,

$$Y_1^t - Y_1^b = \left(\frac{K_1}{K_3}\frac{K_2 - K_3}{K_1 - K_2}\right)Y_3^t \tag{3.30}$$

$$Y_2^t - Y_2^b = -\left(\frac{K_2}{K_3}\frac{K_1 - K_3}{K_1 - K_2}\right)Y_3^t \tag{3.31}$$

In a typical absorption process, a stream of relatively low volatility (the absorbent) is made to interact with a gas mixture. The absorbent absorbs heavier gas components preferentially, recovering some of them and resulting in an effluent gas that is leaner in heavier components and richer in lighter components than the feed gas.

In the ternary system at hand, let component 3 represent the absorbent. It is less volatile than components 1 and 2; that is, $K_3 < K_2 < K_1$. Because of these inequalities, the coefficients of Y_3^t in parentheses in Equations 3.30 and 3.31 are always positive. This implies that $Y_1^t > Y_1^b$ and $Y_2^t < Y_2^b$. Hence, the addition of a heavier component 3 to a gas mixture of light components results in enrichment of the lighter component 1 and depletion of the heavier component 2 in the gas. Moreover, since Y_3^t increases with the amount of component 3 in the feed, the degree of enrichment or depletion is in direct relationship to the rate of heavy component added.

In a stripping process, the objective is to lower the concentration of light components in a liquid mixture by interaction with a lighter gas. The stripping gas modifies the phase equilibrium by driving more of the lighter components in the liquid to the vapor phase than the heavier components. A ternary model may be used to analyze stripping action in a manner analogous to absorption as discussed above. If the binary liquid contains components with distribution coefficients K_1 for the lighter component and K_2 for the heavier component, the stripping gas in the ternary, component 3, would be the most volatile, such that $K_3 > K_1 > K_2$. The reader may go through the exercise of deriving the mathematical relations for a ternary system that represents gas stripping.

EXAMPLE 3.3: EFFECT OF ABSORBENT ON PHASE SEPARATION

A phase separation drum charged with 100 kmol/h mixture of ethane (1) and propane (2) is maintained at 25°C and 1000 kPa. The feed composition is $Z_1 = 0.07$ and $Z_2 = 0.93$. Calculate the equilibrium vapor and liquid compositions. An absorbent, n-heptane (3), is added to the separation drum at varying rates. It is required to determine the effect of the heptane rate on the phase separation. The K-values at 25°C and 1000 kPa, assumed composition-independent, are $K_1 = 2.9$, $K_2 = 0.95$, and $K_3 = 0.009$.

SOLUTION

a. Binary Mixture
In a binary mixture the equilibrium vapor and liquid compositions are independent of the feed composition. The vapor composition is calculated by Equations 3.24 and 3.25:

$$Y_1^b = \frac{2.9(1 - 0.95)}{2.9 - 0.95} = 0.07436$$

$$Y_2^b = \frac{0.95(2.9 - 1)}{2.9 - 0.95} = 0.92564$$

The liquid compositions are as follows:

$$X_1^b = Y_1^b/K_1 = 0.07436/2.9 = 0.02564$$

$$X_2^b = Y_2^b/K_2 = 0.92564/0.95 = 0.97436$$

The fraction vapor, ψ, can be calculated by a component balance:

$$Z_i = \psi Y_i^b + (1 - \psi)X_i^b$$

Alternatively, it can be calculated from the general flash Equation 2.14:

$$f(\psi) = \frac{Z_1(K_1 - 1)}{1 + \psi(K_1 - 1)} + \frac{Z_2(K_2 - 1)}{1 + \psi(K_2 - 1)} = \frac{0.07(2.9 - 1)}{1 + \psi(2.9 - 1)} + \frac{0.93(0.95 - 1)}{1 + \psi(0.95 - 1)} = 0$$

The solution is $\psi = 0.9105$. The component flow rates are calculated as

$$V_i = F\psi Y_i$$

$$L_i = F(1 - \psi)X_i$$

The complete results are as follows:

	Z_i	F_i	Y_i	V_i	X_i	L_i
1. Ethane	0.07	7	0.07436	6.7705	0.02564	0.2295
2. Propane	0.93	93	0.92564	84.2795	0.97436	8.7205
	1.00	100	1.00000	91.0500	1.00000	8.9500

b. Ternary Mixture

Adding 1 kmol/h n-heptane to the mixture results in a new combined feed composition:

$$Z_1 = 7/101 = 0.069307$$
$$Z_2 = 93/101 = 0.920792$$
$$Z_3 = 1/101 = 0.009901$$

Equation 2.14 is solved for the vapor fraction, giving $\psi = 0.508$. The vapor composition is calculated by Equation 2.13b:

$$Y_1^t = \frac{0.069307 \times 2.9}{1 + 0.508(2.9 - 1)} = 0.102275$$

$$Y_2^t = 0.897546$$

$$Y_3^t = 0.000179$$

For a known value of Y_3^t, the other component compositions can be calculated by Equations 3.28 and 3.29:

$$Y_1^t = \frac{2.9(1 - 0.95)}{2.9 - 0.95} + \frac{2.9}{0.009} \times \frac{0.95 - 0.009}{2.9 - 0.95} \times 0.000179 = 0.102275$$

$$Y_2^t = \frac{0.95(2.9 - 1)}{2.9 - 0.95} - \frac{0.95}{0.009} \times \frac{2.9 - 0.009}{2.9 - 0.95} \times 0.000179 = 0.897546$$

The dependence of phase separation on the heptane feed rate is summarized below:

F_3 (kmol/h)	V (kmol/h)	V_1 (kmol/h)	Y_1	Y_2	Y_1/Y_2
0	91.05	6.77	0.0744	0.9256	0.0804
1	51.31	5.25	0.1023	0.8976	0.1140
2	37.33	4.38	0.1174	0.8823	0.1331
4	20.80	2.94	0.1414	0.8581	0.1648
6	10.18	1.65	0.1620	0.8375	0.1934
8.73	0	0	0.1867	0.8126	0.2298

As the heptane rate is increased, the separation is improved, as indicated by the rising ratio of ethane to propane in the vapor product. At the same time, however, the ethane recovery, expressed as the ethane flow rate in the vapor, goes down as the heptane rate is increased. This is a limitation of the single-stage separation process. At a certain heptane rate the entire combined feed is condensed, leaving a single product—a liquid at its bubble point.

3.3.2 MULTISTAGE ABSORPTION

Improved separation is achieved by employing multistage absorbers instead of a single stage. A schematic of a two-stage absorber is shown in Figure 3.5. Vapor stream, V_2, which interacts with external liquid feed L_0 in stage 1, had been enriched in lighter components and depleted in heavier components in stage 2. Separation by absorption is thus enhanced by increasing the number of stages. Better separation is also obtained by increasing the liquid feed rate (the external reflux).

Unlike distillation, an absorption or stripping process does not require the different stages to be at different temperatures. The "driving force" in absorption or stripping is a composition profile along the column induced by an external feed composition that is considerably different from the feed composition. In distillation the composition gradient is associated with a temperature profile. In absorption or stripping, the liquid and vapor streams flowing in the column originate from sources external to the column, whereas in distillation internal reflux and boil-up vapor are generated by condensing part of the overhead vapor and reboiling part of the liquid bottoms. Because of this, the difference between the vapor and liquid compositions is usually greater in an absorption or stripping stage than in a distillation stage.

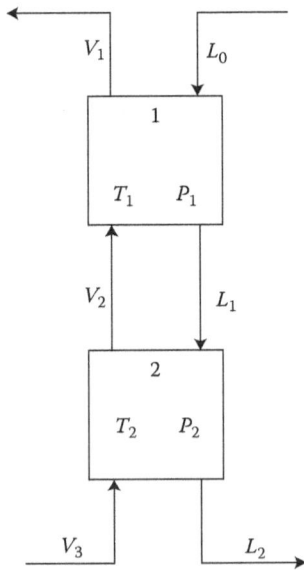

FIGURE 3.5 Two-stage absorber.

Under characteristically absorptive conditions, the transfer of a light component from one phase to the other is accompanied by a relatively small transfer of the heavy component in the opposite direction, whereas in a typical distillation stage the transfer of components occurs both ways.

In general, multistage processes are neither purely absorptive nor purely distillative but a combination of both affects to varying degrees. Predominantly distillative processes are exemplified by binary distillation, while predominantly absorptive processes are exemplified by isothermal absorption. In order to achieve isothermal absorption or stripping, heat must generally be removed or added in each stage to balance the heat associated with absorption or stripping.

3.3.3 OPERATING PARAMETERS AND MATHEMATICAL FORMULATION

The parameters required to define the operation of an absorber are discussed in relationship to a single stage. A single stage with a feed of fixed rate, composition, and thermal conditions has two degrees of freedom (Chapter 2). It is completely defined (zero degrees of freedom) if its pressure and heat duty are fixed.

The degrees of freedom for an absorber are determined in a similar manner. Each stage in an absorber receives feed streams that are either external or internal, coming from another stage. If the external streams are fixed, the internal streams at steady state are also fixed, being the products of another stage. Hence, the performance of an adiabatic absorber operating at fixed pressure is defined, with zero degrees of freedom, if the external feeds and the number of stages are defined.

The operation of an absorber may be controlled to meet performance specifications by varying one of the feed rates or by applying variable heating or cooling duties at certain stages. Consider an absorber with fixed lean oil and gas feeds. The liquid and vapor product rates, compositions, temperatures, and other properties are determined. If it is desired to meet a specified purity of certain components in one of the products such as the vapor, the lean oil rate may be used as a control variable. Its rate must be determined in order to meet the vapor purity specification.

Mathematical solution of absorbers, like other multistage separation processes, involves setting up material and energy balance equations and vapor–liquid equilibrium relations that describe the entire column. The resulting set of simultaneous equations is then solved by some suitable technique.

Referring to the model in Figure 3.5, a material balance for component i on stage j may be written as

$$L_{ji} + V_{ji} = L_{j-1,i} + V_{j+1,i} \quad i = 1, \ldots, C \quad j = 1, \ldots, N \qquad (3.32)$$

where C is the number of components and N is the number of stages. Variables L_{ji} and V_{ji} are equilibrium liquid and vapor molar rates of component i leaving stage j. By extending this nomenclature, L_{0i} represents external liquid rate of component i entering stage 1 and $V_{N+1,i}$ stands for external vapor rate of component i entering stage N.

An energy balance on stage j yields the following equation:

$$L_j h_j + V_j H_j = L_{j-1} h_{j-1} + V_{j+1} H_{j+1} \quad j = 1, \ldots, N \qquad (3.33)$$

where L_j and V_j are the total liquid and vapor molar rates leaving stage j, and h_j and H_j are the molar liquid and vapor enthalpies on stage j.

Phase equilibrium relationships are expressed in terms of V_{ji}, V_j, L_{ji}, and L_j. Since

$$Y_{ji} = V_{ji}/V_j$$
$$X_{ji} = L_{ji}/L_j$$

the distribution coefficients are written as

$$K_{ji} = \frac{V_{ji}/V_j}{L_{ji}/L_j} \quad i = 1, \ldots, C \quad j = 1, \ldots, N \qquad (3.34)$$

Each of equation sets 3.32 and 3.34 encompasses NC equations, and equation system 3.33 includes N equations. Altogether there are $N(2C + 1)$ simultaneous equations. The unknown variables are the vapor and liquid molar rates of each component in each stage (V_{ji} and L_{ji}, $j = 1, \ldots, N$ and $i = 1, \ldots, C$) and the temperatures in each stage (T_j, $j = 1, \ldots, N$). The total liquid and vapor flow rates, L_j and V_j, are determined as the sums of the component liquid and vapor rate. The stage pressures are assumed to be known, fixed values. The total number of variables is thus $2NC + N$, identical to the number of equations, with no degrees of freedom. Methods for solving the set of equations are discussed in Chapter 13.

NOMENCLATURE

C Number of components
F Feed stream designation or molar flow rate
H Vapor stream molar enthalpy
h Liquid stream molar enthalpy
K Distribution coefficient
L Liquid stream designation or molar flow rate
N Number of equilibrium stages in a column
P Pressure
Q Heat duty
T Temperature
V Vapor stream designation or molar flow rate
X Component mole fraction in the liquid
Y Component mole fraction in the vapor
Z Component mole fraction in the feed
α Relative volatility

SUBSCRIPTS

i Component designation
j Stage designation

SUPERSCRIPTS

b Binary system
t Ternary system

PROBLEMS

3.1. In the three-stage process described in Example 3.1, it is required to produce 20 kmol/h propane in the top stage vapor, 8 kmol/h propane in the top stage liquid, and 12 kmol/h propane in the bottom stage vapor. Calculate the temperature and heat duty in each stage. The vapor and liquid enthalpies of propane and butane are given by the following equations:

Propane liquid enthalpy: h (kJ/kmol) = 22,734 + 113.2T (°C)
Propane vapor enthalpy: H (kJ/kmol) = 39,427 + 38.3T (°C)
Butane liquid enthalpy: h (kJ/kmol) = 8603 + 135.7T (°C)
Butane vapor enthalpy: H (kJ/kmol) = 30,694 + 83.5T (°C)

Assume molar additive enthalpies for the mixtures.

3.2. A cascaded three-stage process similar to Figure 3.1 is used for a crude separation of a three-component mixture. Determine the number of variables required to define the process. Upstream control systems require continuous composition data for stream F. It is impractical to install an analyzer on F, but an existing analyzer is already installed on stream V_1. In addition, the following data are measured on a continuous basis: flow rate, temperature, and pressure of stream F, flow rates of L_1, L_3, and V_3, and the stage pressures P_1, P_2, and P_3. Outline a computational procedure that could be used to determine, at any given moment, the composition of stream F, as well as that of products L_1, L_3, and V_3, assuming steady-state conditions exist in the three-stage unit.

3.3. Alternative strategies are considered as backups to determine the composition of feed stream F in Problem 3.2 (Figure 3.1). Which of the following alternatives are feasible and what computational procedure could be used to determine the feed composition?
a. Replace the flow meters on L_1 and V_3 with an analyzer on L_1.
b. Replace the flow meters on L_1 and V_3 with temperature indicators in stages 1 and 3.
c. Measure the following: feed flow, temperature, and pressure; temperature and pressure in each stage; and flow rates L_1 and L_3.

3.4. A two-stage distillation column similar to the column in Example 3.2 (Figure 3.3) is to be used for separating a mixture of 60 kmol/h propane

and 40 kmol/h butane. The feed stream, at 38°C and 690 kPa, is sent to the
bottom stage, the reboiler. It is desired to recover 65% of the propane in the
distillate with a purity of 75%.

a. What is the required reflux ratio and what are the two-stage
    temperatures?
b. What reflux ratio is required to achieve a propane recovery of 90% at
    the same propane purity of 75%?
c. What is the maximum theoretical propane recovery in the distillate
    consistent with a 75% purity, and what are the stage temperatures?

3.5. A flash drum maintained at constant temperature and pressure separates a
      binary feed containing components 1 and 2 into vapor and liquid products
      for the purpose of lowering the concentration of the lighter component (2)
      in the liquid. A stripping stream consisting of a more volatile component (3)
      is added to the flash drum (a single-stage stripper) to enhance the removal
      of component 2 from the liquid.

      a. Derive a relationship between the mole fraction of component 2 in the
          liquid product and the mole fraction of component 3 in the vapor prod-
          uct. The K-values of the components are K_1, K_2, and K_3.
      b. Calculate the rate of stripping stream per 1000 mole feed stream
          required to lower component 2 mole fraction in the liquid to 0.35. The
          feed composition and K-values (assumed composition-independent) are
          given as follows:

| | Mole Fractions | | |
Component	Feed, F	Stripping Stream, S	K-Value
1.	0.54	0.00	0.72
2.	0.46	0.00	1.36
3.	0.00	1.00	3.50

3.6. A liquid absorbent A consisting of component 3 is used to selectively absorb
      component 2 from a binary gas mixture consisting of components 1 and
      2 by contacting it in a single equilibrium stage. Determine the minimum
      rate of stream A required to maintain liquid flow in the absorber. Assume
      composition-independent K-values.

      Calculate the minimum absorbent rate and the equilibrium compositions
      using the following data:

$F = 100\ \text{kmol/h}$	$X_{F1} = 0.95$	$X_{F2} = 0.05$
$K_1 = 18.0$	$K_2 = 1.5$	$K_3 = 0.1$

      Using the same data, calculate the minimum absorbent rate and the equilib-
      rium compositions for a two-stage absorber.

3.7. Compute the products of a two-stage absorber with the feeds given below,
      using successive flash calculations. In each iteration solve both stages using
      stream data from the previous iteration.

Component, i	K_i	Gas Feed, Y_{3i}	Absorbent, X_{0i}
1.	18.0	0.95	0.00
2.	1.5	0.05	0.00
3.	0.1	0.00	1.00
kmol/h		100.00	50.00

3.8. A two-stage distillation column consists of a partial condenser, stage 1, and a partial reboiler, stage 2, with no trays in between. A 100 mol/s stream consisting of a binary equimolar mixture of components 1 and 2 is fed to the reboiler. It is required to obtain online compositions of the distillate and the bottoms for control purposes. Online analyzers are not available, but the flow rates of the distillate and reflux rates can be measured in real time. The K-values of the lighter component, component 1, and the measured distillate rate, V_1, and reflux rate, L_1, are as follows:

$K_1 = 1.4$, $K_2 = 1.6$ (The subscripts denote the stage; the component subscripts are dropped: the given values are for component 1.)
$V_1 = 40$ mol/s, $L_1 = 30$ mol/s

Find the distillate and bottoms compositions.

3.9. A two-stage binary distillation column consists of stage 1, the condenser, and stage 2, the reboiler. Stream V_1 leaving stage 1 is the distillate. Stream L_1, from stage 1 to stage 2, is the reflux. Stream V_2 is the boilup from stage 2 to stage 1. Stream L_2 from stage 2 is the bottoms. The feed, going to stage 2, has a flow rate of 100 mol/s, 50 mol% propane (1) and 50 mol% butane (2). The temperature at stage 1 is $T_1 = 37°C$ and at stage 2 is $T_2 = 39°C$. The propane mole fractions are $Y_{11} = 0.655$ in the distillate, and $X_{21} = 0.345$ in the bottoms.

It is required to calculate the flow rates and compositions of L_1, L_2, V_1, and V_2. The column pressure is 690 kPa, and at this pressure the propane K-value is given as

$$K_1 = 1.36 + 0.03249[T\ (°C) - 26.7]$$

The relative volatility of propane to butane is

$$\alpha_{12} = K_1/K_2 = 3.00$$

3.10. Refer to Example 3.3 and find the n-heptane rate that will bring the propane mole fraction in the vapor product down to 0.85.

4 Material Balances in Multi-Component Separation

The separation of a multi-component mixture into products with different compositions in a multistage process is governed by phase equilibrium relations and energy and material balances. It is not uncommon in simulation studies to require certain column product rates, compositions, or component recoveries to satisfy given specifications with no concern for conditions within the column. Such would be the case when downstream processing of the products is of primary interest. In these instances, one would be concerned only with overall component balances around the column but not necessarily with heat balances or equilibrium relations. Separation would thus be arbitrarily defined, and the problem would be to calculate product rates and compositions. The actual performance of the separation process is analyzed independently in all the following chapters.

The purpose of this chapter is to set up a mathematical model for component material balances and to solve it for different sets of separation specifications.

4.1 MATHEMATICAL MODEL

The separation column is assumed to have one feed and two products: overhead (or distillate) and bottoms. The feed has a fixed flow rate and composition. The temperatures and pressures of the feed and products and of streams within the column are of no concern in the present context. Moreover, the column configuration, number of stages, and internal vapor and liquid flows are immaterial since the focus is on the net performance of a "black box" column.

Let the feed stream F, with molar flow rate F, contains C components with known mole fractions, $Z_1, Z_2, \ldots, Z_C$. The overhead and bottoms molar flow rates are D and B, respectively. As with the equilibrium stage discussed in Chapter 2, the products of the column may be completely defined in terms of the overhead mole fractions, $Y_1, Y_2, \ldots, Y_C$, the bottoms mole fractions, $X_1, X_2, \ldots, X_C$, and the distillate fraction, $\psi = D/F$.

These $2C + 1$ parameters are considered the primary variables. They are interrelated by C material balance equations:

$$Z_i = \psi Y_i + (1 - \psi)X_i \quad i = 1, \ldots, C \tag{4.1}$$

and one summation equation:

$$\sum Y_i = 1 \tag{4.2}$$

Note that the summation $\sum X_i = 1$ does not constitute an additional independent equation since it may be obtained by summing up Equation 4.1 over all the components and combining the result with Equation 4.2.

There are, therefore, $C + 1$ equations with $2C + 1$ variables, and the column has

$$(2C + 1) - (C + 1) = C \text{ degrees of freedom}$$

In order to solve the equations, C specifications, or constraints, are required, making the total number of equations equal to the number of dependent variables. The specifications define the component separation taking place in the column. In general, these specifications may be written as a function of the primary variables in the form

$$g_j(Y_1,\ldots,Y_C,X_1,\ldots,X_C,\psi) = 0 \quad j = 1,\ldots,C \tag{4.3}$$

In the rigorous treatment of columns, most of the constraints result from phase equilibrium and energy balance equations while the rest represent column performance specifications such as separation. In the present model, concerned mainly with material balances, the equilibrium and energy balance equations are replaced by specifications that completely define the separation, designed to achieve certain product rates, component purity, or recovery requirements. Equations 4.1, 4.2, and 4.3 represent a set of $2C + 1$ independent equations to be solved for the $2C + 1$ variables.

4.2 TYPES OF COLUMN SPECIFICATIONS

The g functions (Equation 4.3) that define the column performance may be general functions of the primary variables designed to meet special separation specifications, or they may directly specify the values of the primary variables. Combinations of different types of specifications are possible.

4.2.1 PRIMARY VARIABLE SPECIFICATIONS

The separation may be defined by specifying C of the primary variables $(X_1, \ldots, X_C; Y_1, \ldots, Y_C; \psi)$. The g functions representing these constraints would take the form

$$g_i = Y_i - Y_i^s = 0$$

$$g_i = X_i - X_i^s = 0$$

$$g = \psi - \psi^s = 0$$

where X_i^s, Y_i^s, and ψ^s are the specified values.

Note that specifying only X_i, $i = 1, ..., C$ or only Y_i, $i = 1, ..., C$ does not constitute a complete definition of the separation because the mole fractions are interrelated by their summation equations. Providing all the X_is or all the Y_is does not result in C independent specifications. In order to uniquely define the separation, the C specified variables could be a mix of X_is, Y_is, and ψ.

If for a given component i both X_i and Y_i are specified, then ψ is determined by Equation 4.1. Consequently, ψ may not be given as an additional independent specification. Also, there could be only one component for which both X_i and Y_i are specified. Otherwise, the column would be over specified since each pair of X_i and Y_i determines ψ. Because of these considerations, the allowable combinations of primary variable specifications are

Y_i, $i = 1, ..., C - 1$, and ψ

or X_i, $i = 1, ..., C - 1$, and ψ

or Y_i, $i = 1, ..., j$, and X_i, $i = j + 1, ..., C$

or Y_i, $i = 1, ..., j$, and X_i, $i = j, ..., C - 1$

The obvious restrictions on specification values are

$0 < \psi < 1$

$0 \leq Y_i \leq 1$

$0 \leq X_i \leq 1$

$\sum Y_i \leq 1$ $i = 1$ through j

$\sum X_i \leq 1$ $i = 1$ through j

where j is the number of components for which mole fractions in either product are specified.

Another restriction applies to situations where mole fractions of a given component are specified in both the overhead and bottoms. If both X_i and Y_i are specified, then one of the following inequalities should hold:

$$Y_i > Z_i > X_i$$

or

$$Y_i < Z_i < X_i$$

That is, if a component gets concentrated in the overhead, it must get diluted in the bottoms, or vice versa. Violation of these inequalities results in values of the distillate fraction outside the permissible range of 0 to 1, as may be inferred from Equation 4.1.

The specifications must pass another feasibility check when the distillate fraction and product component mole fractions are specified. Equation 4.1 is rearranged as

$$X_i = \frac{Z_i - \psi Y_i}{1 - \psi}$$

The calculated value of X_i must be between zero and one, and satisfy the above inequalities relative to Y_i and Z_i. An equivalent condition should be satisfied if X_i and ψ are specified.

When the column operation is defined in terms of primary specifications, Equation 4.1 is easily decoupled and each equation is solved independently.

EXAMPLE 4.1: MOLE FRACTION SPECIFICATIONS

A 100 kmol/h mixture of four components is to be separated into an overhead and bottoms products. The feed composition is as follows:

Component, i	Mole Fraction, Z_i
1.	0.1
2.	0.2
3.	0.3
4.	0.4

Calculate the product rates and compositions for the following sets of specifications:

A. $Y_1 = 0.21$, $Y_2 = 0.35$, $Y_3 = 0.20$, $\psi = 0.40$
B. $Y_1 = 0.23$, $Y_2 = 0.34$, $X_3 = 0.38$, $X_4 = 0.46$
C. $Y_1 = 0.20$, $Y_2 = 0.30$, $X_2 = 0.15$, $X_3 = 0.42$

SOLUTION

Equations 4.1 and 4.2 together with each set of specifications are solved to determine the product rates and compositions.

A. The specifications are substituted in Equation 4.1 written for components 1, 2, and 3:

$$0.10 = (0.40) \times (0.21) + (1 - 0.40)X_1$$

$$0.20 = (0.40) \times (0.35) + (1 - 0.40)X_2$$

$$0.30 = (0.40) \times (0.20) + (1 - 0.40)X_3$$

Each equation is solved separately, and the summation equations are solved for the fourth component mole fraction. The results are summarized in Table 4.1. The distillate rate is calculated from the definition of ψ:

$$D = F\psi = (100) \times (0.40) = 40 \text{ kmol/h}$$

TABLE 4.1

Results for Example 4.1, part A

i	Z_i	Y_i	X_i
1.	0.10	0.21	0.0267
2.	0.20	0.35	0.1000
3.	0.30	0.20	0.3667
4.	0.40	0.24	0.5067

The bottoms rate, by material balance,

$$B = 100 - 40 = 60 \text{ kmol/h}$$

B. Equations 4.1 and 4.2 and the specifications are listed as nine equations with nine variables:

$$Y_1 = 0.23$$
$$Y_2 = 0.34$$
$$X_3 = 0.38$$
$$X_4 = 0.46$$
$$\psi Y_1 + (1 - \psi)X_1 = 0.1$$
$$\psi Y_2 + (1 - \psi)X_2 = 0.2$$
$$\psi Y_3 + (1 - \psi)X_3 = 0.3$$
$$\psi Y_4 + (1 - \psi)X_4 = 0.4$$
$$\Sigma Y_i = 1$$

The first eight equations are combined to produce the following four equations with five variables:

$$0.23\psi + (1 - \psi)X_1 = 0.1$$
$$0.34\psi + (1 - \psi)X_2 = 0.2$$
$$\psi Y_3 + (1 - \psi)(0.38) = 0.3$$
$$\psi Y_4 + (1 - \psi)(0.46) = 0.4$$

From the last two equations,

$$Y_3 = (0.38\psi - 0.08)/\psi$$
$$Y_4 = (0.46\psi - 0.06)/\psi$$

The values of Y_i are substituted in the summation equation, which is solved for ψ:

$$0.23 + 0.34 + (0.38\psi - 0.08)/\psi + (0.46\psi - 0.06)/\psi = 1$$
$$\psi = 0.3415$$

TABLE 4.2

Results for Example 4.1, part B

i	Z_i	Y_i	X_i
1.	0.10	0.2300	0.0326
2.	0.20	0.3400	0.1274
3.	0.30	0.1457	0.3800
4.	0.40	0.2843	0.4600

TABLE 4.3

Results for Example 4.1, part C

i	Z_i	Y_i	X_i
1.	0.10	0.20	0.05
2.	0.20	0.30	0.15
3.	0.30	0.06	0.42
4.	0.40	0.44	0.38

The remaining mole fractions are calculated directly from Equation 4.1. The results are given in Table 4.2. The overhead and bottoms rates are calculated from the feed rate and the overhead mole fraction:

$$D = (100)(0.3415) = 34.15 \text{ kmol/h}$$
$$B = 100 - 34.15 = 65.85 \text{ kmol/h}$$

C. Since Y_2 and X_2 are both given for component 2, ψ may be calculated from Equation 4.1:

$$0.20 = 0.30\psi + (1 - \psi)0.15$$
$$\psi = 0.3333$$

The missing mole fractions are calculated from Equations 4.1 and 4.2 and listed in Table 4.3. The product rates are

$$D = (100) \times (0.3333) = 33.33 \text{ kmol/h}$$
$$B = 100 - 33.33 = 66.67 \text{ kmol/h}$$

4.2.2 DERIVED VARIABLE SPECIFICATIONS

A number of variables of practical importance in separation processes may be derived from the primary variables. The derived variables include component rates and recoveries, defined as follows:

Component distillate rates:

$$D_i = DY_i = F\psi Y_i \tag{4.4}$$

Component bottoms rates:

$$B_i = BX_i = F(1 - \psi)X_i \tag{4.5}$$

Component recoveries in the distillate:

$$r_i^D = \frac{DY_i}{FZ_i} = \frac{\psi Y_i}{Z_i} = \frac{D_i}{FZ_i} \tag{4.6}$$

Component recoveries in the bottoms:

$$r_i^B = \frac{BX_i}{FZ_i} = \frac{(1-\psi)X_i}{Z_i} = \frac{B_i}{FZ_i} \tag{4.7}$$

The separation specifications defined in the general form in Equation 4.3 may be written for these derived variables:

$$g_i = F\psi Y_i - D_i = 0$$

$$g_i = F(1-\psi)X_i - B_i = 0$$

$$g_i = \psi Y_i / Z_i - r_i^D = 0$$

$$g_i = (1-\psi)X_i / Z_i - r_i^B = 0$$

where D_i, B_i, r_i^D, and r_i^B are the specified values.

The C specifications required to define the separation could include a combination of g functions and primary variable specifications. It is important that the specifications be independent and feasible. Specifying both D_i and B_i, for instance, is redundant since they are not independent; they are related by material balance,

$$FZ_i = D_i + B_i$$

Possible independent derived and primary specifications include one of the variables D_i, B_i, r_i^D, r_i^B, Y_i, or X_i for each component. Specifying both D_i and Y_i, or a similar combination, is allowed for one component only. Since $D_i = F\psi Y_i$, specifying both D_i and Y_i determines ψ. Once ψ is fixed, no other D_i, Y_i pair or the equivalent may be specified.

Besides having to be independent, the specifications must be feasible. The obvious feasible limits for component product rate specifications are $D_i \le FZ_i$ and $B_i \le FZ_i$. For component recovery specifications, the feasible limits are $r_i^D \le 1$ and $r_i^B \le 1$.

EXAMPLE 4.2: COMPONENT RATES AND RECOVERY SPECIFICATIONS

The mixture of Example 4.1 is to be separated into distillate and bottoms products to meet the following sets of specifications:

A. $D_1 = 8.5$ kmol/h, $r_2^D = 0.72$, $B_3 = 23.0$ kmol/h, $r_4^B = 0.75$
B. $r_1^D = 0.82$, $r_2^D = 0.73$, $r_3^D = 0.21$, $r_4^D = 0.26$

Calculate product rates and compositions for each case.

SOLUTION

The compositions can be calculated by systematically solving Equations 4.1, 4.2, and the specifications. However, many steps may be bypassed because the problem lends itself to equation decoupling.

A. The feed component flow rates are first calculated:

$$F_i = FZ_i$$
$$F_1 = 10.0 \text{ kmol/h}$$
$$F_2 = 20.0 \text{ kmol/h}$$
$$F_3 = 30.0 \text{ kmol/h}$$
$$F_4 = 40.0 \text{ kmol/h}$$

The recovery specifications are converted to product component rate specifications:

$$D_2 = F_2 r_2^D = (20.0) \times (0.72) = 14.4 \text{ kmol/h}$$

$$B_4 = F_4 r_4^B = (40.0) \times (0.75) = 30.0 \text{ kmol/h}$$

The remaining component flow rates are calculated by simple component material balance:

$$D_3 = F_3 - B_3 = 30.0 - 23.0 = 7.0 \text{ kmol/h}$$
$$D_4 = F_4 - B_4 = 40.0 - 30.0 = 10.0 \text{ kmol/h}$$
$$B_1 = F_1 - D_1 = 10.0 - 8.5 = 1.5 \text{ kmol/h}$$
$$B_2 = F_2 - D_2 = 20.0 - 14.4 = 5.6 \text{ kmol/h}$$

Product rates:

$$D = \Sigma D_i = 8.5 + 14.4 + 7.0 + 10.0 = 39.9 \text{ kmol/h}$$
$$B = \Sigma B_i = 1.5 + 5.6 + 23.0 + 30.0 = 60.1 \text{ kmol/h}$$

Distillate fraction:

$$\psi = 39.9/100.0 = 0.399$$

Product compositions:

$$Y_1 = 8.5/39.9 = 0.2130$$
$$Y_2 = 14.4/39.9 = 0.3609$$
$$Y_3 = 7.0/39.9 = 0.1754$$
$$Y_4 = 10.0/39.9 = 0.2506$$
$$X_1 = 1.5/60.1 = 0.0250$$
$$X_2 = 5.6/60.1 = 0.0932$$
$$X_3 = 23.0/60.1 = 0.3827$$
$$X_4 = 30.0/60.1 = 0.4992$$

B. The distillate fraction is first calculated directly from the distillate recoveries. Equation 4.6 is rewritten as

$$\psi Y_i = Z_i r_i^D$$

Summation of this equation over all the components results in a convenient expression for calculating ψ:

$$\psi = \sum Z_i r_i^D = (0.10)(0.82) + (0.20)(0.73) + (0.30)(0.21) + (0.40)(0.26)$$
$$= 0.395$$

Product rates:

$$D = (100) \times (0.395) = 39.5 \text{ kmol/h}$$
$$B = 100 - 39.5 = 60.5 \text{ kmol/h}$$

The distillate mole fractions are calculated from Equation 4.6:

$$Y_i = Z_i r_i^D / \psi$$

$$Y_1 = (0.10) \times (0.82)/0.395 = 0.2076$$

$$Y_2 = (0.20) \times (0.73)/0.395 = 0.3696$$

$$Y_3 = (0.30) \times (0.21)/0.395 = 0.1595$$

$$Y_4 = (0.40) \times (0.26)/0.395 = 0.2633$$

The bottoms mole fractions are calculated from Equation 4.7 by substituting $1 - r_i^D$ for r_i^B:

$$X_i = Z_i(1 - r_i^D)/(1 - \psi)$$

$$X_1 = 0.1 \times (1 - 0.82)/(1 - 0.395) = 0.0298$$
$$X_2 = 0.2 \times (1 - 0.73)/(1 - 0.395) = 0.0893$$
$$X_3 = 0.3 \times (1 - 0.21)/(1 - 0.395) = 0.3917$$
$$X_4 = 0.4 \times (1 - 0.26)/(1 - 0.395) = 0.4892$$

4.2.3 GENERAL SPECIFICATIONS

Although column separation constraints usually consist of primary and derived variable specifications as discussed in Sections 4.2.1 and 4.2.2, in general, specifications could be any function of these variables. One could define the g functions of Equation 4.3 as, for instance, sums, differences, or ratios of component rates, recoveries, and so on. The only restrictions on the specification functions, at least from the mathematical standpoint, are that they be independent and feasible.

Depending on the functional form of the specifications, the system of equations comprising Equations 4.1, 4.2, and the specifications may be nonlinear. Nonlinear equations are solved by iterative methods such as the Newton–Raphson method. The solution method is illustrated in Example 4.3.

EXAMPLE 4.3: MIXED SPECIFICATIONS

It is required to separate a ternary mixture into two products that meet certain separation specifications. The feed consists of 25 kmol/h component 1, 33 mol/h component 2, and 42 mol/h component 3. The three specifications are as follows:

Mole fraction 2 in the bottoms: $X_2 = 0.2833$
Mole fraction 3 in the bottoms: $X_3 = 0.5333$
Recovery of 1 and 2 in the overhead: $(D_1 + D_2)/(F_1 + F_2) = 0.5172$

Calculate the product rates and compositions.

SOLUTION

The component mole fractions in the feed are calculated from the given component rates:

$$Z_1 = 0.25$$
$$Z_2 = 0.33$$
$$Z_3 = 0.42$$

From the component material balances:

$$\psi Y_1 + (1 - \psi)X_1 = 0.25$$
$$\psi Y_2 + (1 - \psi)X_2 = 0.33$$
$$\psi Y_3 + (1 - \psi)X_3 = 0.42$$

Summation equation:

$$X_1 + X_2 + X_3 = 1$$

Specifications:

$$X_2 = 0.2833$$
$$X_3 = 0.5333$$
$$(D_1 + D_2)/(F_1 + F_2) = F\psi(Y_1 + Y_2)/(F_1 + F_2) = 0.5172$$

These are seven equations with seven unknowns: X_1, X_2, X_3, Y_1, Y_2, Y_3, ψ. The last specification was converted to a function of the primary variables using Equation 4.4. The first two specifications and the summation equation define the bottoms composition:

$$X_1 = 0.1834$$
$$X_2 = 0.2833$$
$$X_3 = 0.5333$$

The values of X_1 and X_2 are substituted into the material balance equations for components 1 and 2 and the feed flows are substituted in the third specification:

$$\psi Y_1 + 0.1834(1 - \psi) = 0.25$$
$$\psi Y_2 + 0.2833(1 - \psi) = 0.33$$
$$100\psi(Y_1 + Y_2)/(25 + 33) = 0.5172$$

The material balance equation for component 3 can be solved separately once ψ is determined. The above three simultaneous equations are first solved for Y_1, Y_2, and ψ. They are nonlinear, involving products ψY_1 and ψY_2. Although they may be solved by elimination, a more general method is Newton–Raphson's multivariable iterative technique. First, rewrite the equations in the residual form:

$$g_1(Y_1, Y_2, \psi) = \psi Y_1 - 0.1834\psi - 0.0666 = 0$$
$$g_2(Y_1, Y_2, \psi) = \psi Y_2 - 0.2833\psi - 0.0467 = 0$$
$$g_3(Y_1, Y_2, \psi) = \psi Y_1 + \psi Y_2 - 0.3 = 0$$

The iteration requires starting values for the variables, Y_1^0, Y_2^0, and ψ^0, where superscript 0 refers to iteration 0, the initial estimates. In subsequent iterations, designated as k, $k + 1$, ..., the values of the variables are improved by adding corrections to current values:

$$Y_1^{k+1} = Y_1^k + \Delta Y_1^k$$
$$Y_2^{k+1} = Y_2^k + \Delta Y_2^k$$
$$\psi^{k+1} = \psi^k + \Delta \psi^k$$

Convergence is achieved when the corrections, $\Delta Y_1^k, \Delta Y_2^k$, and $\Delta \psi^k$ become smaller than a preset tolerance. The corrections are calculated from the equations

$$\frac{\partial g_1}{\partial Y_1} \Delta Y_1^k + \frac{\partial g_1}{\partial Y_2} \Delta Y_2^k + \frac{\partial g_1}{\partial \psi} \Delta \psi^k + g_1 = 0$$

$$\frac{\partial g_2}{\partial Y_1} \Delta Y_1^k + \frac{\partial g_2}{\partial Y_2} \Delta Y_2^k + \frac{\partial g_2}{\partial \psi} \Delta \psi^k + g_2 = 0$$

$$\frac{\partial g_3}{\partial Y_1} \Delta Y_1^k + \frac{\partial g_3}{\partial Y_2} \Delta Y_2^k + \frac{\partial g_3}{\partial \psi} \Delta \psi^k + g_3 = 0$$

The functions g_1, g_2, g_3, and their partial derivatives are evaluated at the current values of the variables Y_1^k, Y_2^k, and ψ^k. The following are the partial derivative expressions:

$$\frac{\partial g_1}{\partial Y_1} = \psi \qquad \frac{\partial g_1}{\partial Y_2} = 0 \qquad \frac{\partial g_1}{\partial \psi} = Y_1 - 0.1834$$

$$\frac{\partial g_2}{\partial Y_1} = 0 \qquad \frac{\partial g_2}{\partial Y_2} = \psi \qquad \frac{\partial g_2}{\partial \psi} = Y_2 - 0.2833$$

$$\frac{\partial g_3}{\partial Y_1} = \psi \qquad \frac{\partial g_3}{\partial Y_2} = \psi \qquad \frac{\partial g_3}{\partial \psi} = Y_1 - Y_2$$

The starting values of the variables are guessed or estimated. Good initial estimates help ensure convergence, although equations that are not highly non-linear are not too sensitive to the initial estimates. These values should at least be bracketed on the basis of physical realities. For instance, in this example, Y_1, Y_2, and ψ should be between 0 and 1. The estimates may further be refined by inspecting the specified values of X_2 and X_3. These specifications indicate that $X_1 < Z_1$ and $X_2 < Z_2$, which implies that the inequalities $Y_1 > Z_1$ and $Y_2 > Z_2$ should be satisfied. However, in order to check the ability of the iterative technique to recover from "bad" initial estimates, assume some arbitrary starting values: $Y_1^0 = 0.25$, $Y_2^0 = 0.25$, $\psi^0 = 0.25$. From these values, the functions g and their partial derivatives are evaluated and used in the following equations for calculating the corrections:

$$0.25\Delta Y_1 \qquad\qquad +0.0666\Delta\psi - 0.04995 = 0$$
$$0.25\Delta Y_2 - 0.0333\Delta\psi - 0.05502 = 0$$
$$0.25\Delta Y_1 + 0.25\Delta Y_2 + 0.5000\Delta\psi - 0.175 = 0$$

The solutions to these equations are

$\Delta Y_1 = 0.1598$	$\Delta Y_2 = 0.2401$	$\Delta\psi = 0.1501$

The updated values of the variables (iteration 1):

$$Y_1^1 = Y_1^0 + \Delta Y_1 = 0.25 + 0.1598 = 0.4098$$

$$Y_2^1 = Y_2^0 + \Delta Y_2 = 0.25 + 0.2401 = 0.4901$$

$$\psi^1 = \psi^0 + \Delta\psi = 0.25 + 0.1501 = 0.4001$$

Based on these values, new g functions and their partial derivatives are computed and the following equations are obtained:

$$0.4001\Delta Y_1 \qquad\qquad +0.2264\Delta\psi + 0.024 = 0$$
$$0.4001\Delta Y_2 + 0.2068\Delta\psi + 0.036 = 0$$
$$0.4001\Delta Y_1 + 0.4001\Delta Y_2 + 0.8999\Delta\psi + 0.060 = 0$$

The solutions are

$\Delta Y_1 = -0.06$	$\Delta Y_2 = -0.09$	$\Delta\psi = 0$

The updated variables (iteration 2):

$Y_1^2 = 0.3498$	$Y_2^2 = 0.4001$	$\psi^2 = 0.400$

The procedure is repeated again, generating the equations

$$0.4001\Delta Y_1 \qquad\qquad +0.1664\Delta\psi = 0$$
$$0.4001\Delta Y_2 + 0.1168\Delta\psi = 0$$
$$0.4001\Delta Y_1 + 0.4001\Delta Y_2 + 0.7499\Delta\psi = 0$$

TABLE 4.4
Results for Example 4.3

i	Z_i	Y_i	X_i	F_i	D_i	B_i
1.	0.25	0.3498	0.1834	25	14	11
2.	0.33	0.4001	0.2833	33	16	17
3.	0.42	0.2501	0.5333	42	10	32

for which the solutions are

$\Delta Y_1 = 0$	$\Delta Y_2 = 0$	$\Delta \psi = 0$

Convergence has been reached and the latest values of the variables are the solution. The other variables are easily calculated from the primary variables and are given in Table 4.4.

NOMENCLATURE

B Bottoms stream designation or molar flow rate
B_i Component molar flow rate in the bottoms
C Number of components
D Distillate or overhead designation or molar flow rate
D_i Component molar flow rate in the distillate or overhead
F Feed stream designation or molar flow rate
F_i Component molar flow rate in the feed
r Component recovery
X Component mole fraction in the bottoms
Y Component mole fraction in the distillate or overhead
Z Component mole fraction in the feed
ψ Distillate mole fraction

SUBSCRIPTS

i, j Component designation

SUPERSCRIPTS

B Bottoms
D Distillate
k Iteration number
s Specified value

PROBLEMS

4.1. A five-component stream splits into a distillate product with composition Y_i and a bottoms product with composition X_i. Which of the following sets of specifications uniquely define the separation, without taking into account vapor–liquid equilibrium or energy balance constraints?

a. X_1, X_2, X_3, X_4, X_5
b. X_1, X_2, X_3, Y_4, Y_5
c. X_1, X_2, X_3, Y_3, Y_4
d. X_1, X_2, X_3, Y_2, Y_3
e. X_1, X_2, X_3, X_4, ψ
f. X_1, X_2, X_3, Y_3, ψ.

4.2. A stream F with composition Z_i, $i = 1, ..., C$, the number of components, is separated into a distillate product D with composition Y_i and a bottoms product B with compositions X_i. It is proposed to specify the separation with X_i specifications for $i = 1, ..., j$, and Y_i specifications for $i = j + 1$, $...C$ ($j < C$). What are the feasibility conditions for these specifications? Are the following specifications feasible?

i	Z_i	Y_i	X_i
1.	0.34		0.25
2.	0.17		0.34
3.	0.22	0.45	
4.	0.27	0.19	

4.3. The feed stream of Problem 4.2 is to be separated in a distillation column to meet the following specifications:

i	Feed, Z_i	Distillate, Y_i	Bottoms, X_i
1.	0.34		0.30
2.	0.17	0.22	
3.	0.22	0.19	
4.	0.27		0.37

Downstream process design requires distillate and bottoms data. Calculate the flow rates and compositions of these streams for a fixed feed rate of 1000 kmol/h.

4.4. Based on material balance considerations only, which of the following sets of specifications uniquely define the separation of a four-component stream into a distillate and a bottoms product?

a. D_1, D_2, B_3, B_4
b. D_1, D_2, D_3, r_1^B
c. B_1, Y_1, D_2, B_3
d. B_1, X_1, D_3, Y_3

4.5. Calculate the flow rates and compositions of the distillate and bottoms of a distillation column with a feed flow rate of 1000 kmol/h and the following separation specifications (the distillate and bottoms component flow rates in kmol/h):

i	Feed, Z_i	Distillate	Bottoms
1.	0.27	$D_1 = 175.00$	$X_1 = 0.20$
2.	0.35		$B_2 = 200.00$
3.	0.21	$D_3 = 90.00$	
4.	0.17		

4.6. It is proposed to design a distillation column for coarse separation of a four-component mixture between components 2 and 3. (The components are numbered by decreasing volatility). The combined mole fraction of components 1 and 2 in the distillate should be 0.65, and the ratio of the mole fraction of components 3 and 4 in the bottoms should be 1.2. It is also required to recover 80% of component 1 in the distillate while maintaining its mole fraction in the bottoms at 0.07. Assuming the vapor–liquid equilibrium properties of this mixture are consistent with the separation specifications, calculate the distillate and bottoms compositions for the following feed composition:

Component, i	Z_i
1.	0.12
2.	0.37
3.	0.29
4.	0.22

4.7. Two binary streams each containing component 1 (the more volatile) and component 2 are to be separated in a single column equipped with a total condenser and a reboiler. Feed F1 enters the column as saturated vapor and feed F2 enters the column as saturated liquid. A vapor side draw S1 and a liquid side draw S2 are taken from the column. Using the data below, determine the correct relative locations of the feeds and products, the distillate and bottoms flow rates, and the L/V ratio in each column section. The column uses a reflux ratio of 1.8.

	F1	F2	D	B	S1	S2
Component 1 mole fraction	0.40	0.70	0.95			0.80
Component 2 mole fraction	0.60	0.30		0.90	0.70	
Stream flow rate, mol/h	150	200			75	50

4.8. A 100 kmol/h stream containing 30 mol% propane (1), 30 mol% n-butane (2), and 40 mol% n-pentane (3) is to be separated in a distillation column, such that propane is the main component in the distillate. The combined mole fraction of components 1 and 2 in the distillate should be 0.96, and the mole fraction of component 3 in the bottoms should be

0.64. It is also required to recover 94% of the propane in the distillate. A degrees-of-freedom analysis indicates that it should be possible to meet all these specifications in one column. However, before a detailed design is carried out, it is recommended to do material balance calculations. Determine the expected product flow rates and compositions.

4.9. It is required to design a distillation column to separate the stream defined below into distillate and bottoms products, according to the indicated specifications. Preliminary calculations based on material balances alone are recommended in order to define the expected products' flow rates and compositions. Calculate the distillate and bottoms flow rates and compositions.

	Feed (kmol/h)	Specifications
1. Propane	54	Concentration of pentane in bottoms, $X_{B3} = 0.39$
2. n-Butane	28	Recovery of pentane in bottoms, 95%
3. n-Pentane	18	Ratio of mole fractions C3 to C4 in distillate, $Y_{D1}/Y_{D2} = 19.5$

5 Binary Distillation
Principles

Binary distillation is a multistage process for separating a mixture of two components using the principles of vapor–liquid equilibria. Although binary distillation is not a very common occurrence in practice (impurities are normally present along with the two primary components to be separated), developing methods for solving it is beneficial in two ways. In the first place, binary distillation is helpful in representing the principles of distillation in general by using a simplified model. Second, many real-life distillation problems can be approximated by a binary system. The simplifying assumptions and the graphical techniques described in this chapter were necessary before the advent of computers and the fast and rigorous programs for solving columns. Why, then, bother with learning the simplified outdated methods? The graphical methods have the advantage of providing a qualitative, as well as quantitative, albeit perhaps approximate, representation of the problem. They allow the engineer to visualize the effects of the different variables and their feasible ranges before proceeding to the rigorous numerical solution. The simplified methods should not be used as a substitute but in conjunction with rigorous methods. Computer programs can, in fact, enhance the use of graphical techniques for evaluating column performance trends because the very same data needed to construct the diagrams are usually readily available from the simulation output.

This chapter starts with a discussion of a column section, defined as a group of adjacent equilibrium stages with no side feeds or draws and with no heat gains or losses. The column section serves as a model for formulating the necessary assumptions and for developing analytical and graphical methods for representing binary multistage equilibrium separation.

The concepts developed for a column section are then applied to the solution of a total column, and a graphical technique is implemented on the Y–X diagram.

The simplifying assumptions that are necessary for the Y–X solution method are then relaxed to include enthalpy balances. A graphical solution of this more rigorous model, which requires enthalpy–composition data, is presented.

Solution steps are described for cases with different sets of column specifications. A detailed analysis of column performance viewed through example applications will be covered in the next chapter.

5.1 COLUMN SECTION

For the purpose of discussion in this chapter, a column section is defined as any part of a distillation or other multistage separation process that includes no side feeds or draws and no heaters or coolers. A column section does necessarily have a liquid feed at the top and a vapor feed at the bottom. It also has a vapor product at the top and a liquid product at the bottom. It operates adiabatically, that is, no heat is transferred across the system boundary to or from the surroundings, although enthalpy is transferred with the streams themselves.

A simple absorber is itself a column section. However, in the case of binary systems, it is not practical to have a column section as a unit operation on its own; it must be part of a total column. The reason for this is that, in order to operate such a column section, one would need binary liquid and vapor feeds with disparate compositions. Generating such streams would have required some separation process in the first place.

5.1.1 Development of the Model

A column section model is shown in Figure 5.1. It consists of N equilibrium stages designated by subscript j. Liquid and vapor molar flow rates leaving stage j are designated as L_j and V_j, respectively. Mole fractions X_j and Y_j are of the more volatile component in L_j and V_j, respectively.

The column section can be solved by simultaneous solution of the component mass balance and energy balance equations and the vapor–liquid equilibrium relations. Additional equations include the temperature, pressure, and composition dependence of the equilibrium coefficients and enthalpies. The equations for stage j are as follows:

Component balance on the lighter component:

$$L_{j-1}X_{j-1} + V_{j+1}Y_{j+1} = L_jX_j + V_jY_j \tag{5.1}$$

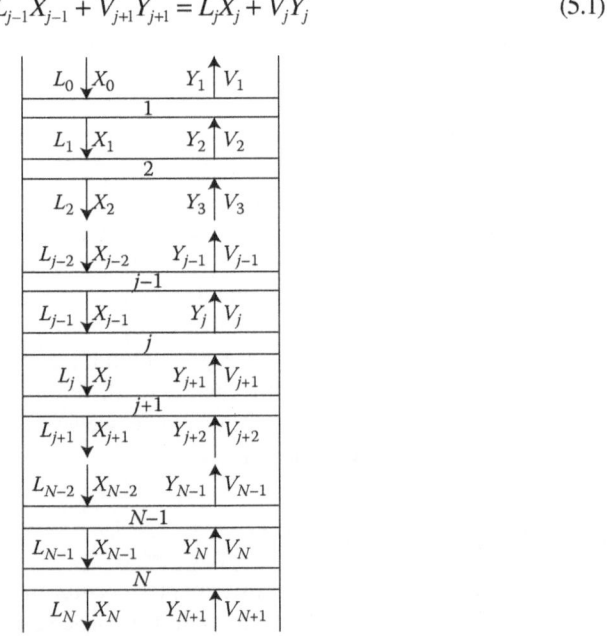

FIGURE 5.1   Schematic of a column section.

A similar equation can be written for the less volatile component. Instead, an overall material balance is written:

$$L_{j-1} + V_{j+1} = L_j + V_j \tag{5.2}$$

Equilibrium relations,

$$Y_j/X_j = K_{j1} \tag{5.3}$$

$$(1 - Y_j)/(1 - X_j) = K_{j2} \tag{5.4}$$

The second subscript of the equilibrium coefficients refers to components 1 and 2, the more volatile and the less volatile components.

Enthalpy balance, where H_j is the vapor molar enthalpy and h_j is the liquid molar enthalpy at tray j,

$$L_{j-1}h_{j-1} + V_{j+1}H_{j+1} = L_jh_j + V_jH_j \tag{5.5}$$

Data or correlations for predicting equilibrium coefficients and enthalpies can be represented as functions of the temperature, pressure, and composition:

$$K_{j1} = f(P_j, T_j, X_j, Y_j) \tag{5.6}$$

$$K_{j2} = f(P_j, T_j, X_j, Y_j) \tag{5.7}$$

$$H_j = f(P_j, T_j, Y_j) \tag{5.8}$$

$$h_j = f(P_j, T_j, X_j) \tag{5.9}$$

Also, finally, the stage pressure equals the sum of the partial pressures:

$$P_j = P_{j1} + P_{j2} \tag{5.10}$$

For each stage j, the inlet streams are assumed to be of known rates, compositions, and thermal conditions, while outlet stream parameters are unknown. Thus, L_{j-1}, V_{j+1}, X_{j-1}, Y_{j+1}, h_{j-1}, and H_{j+1} are known quantities. The column or stage pressures are usually fixed and are therefore known, which renders Equation 5.10 redundant. This leaves nine equations for each stage with nine unknowns: L_j, V_j, X_j, Y_j, h_j, H_j, T_j, K_{j1}, and K_{j2}.

By extending this analysis to an N-stage column section, the parameters L_0, V_{N+1}, X_0, Y_{N+1}, h_0, and H_{N+1}, as well as the column section pressure, are assumed to be of known quantities. That is, the feed rates, compositions, thermal conditions, and the column pressure are known. The column section is described by $9N$ equations with $9N$ variables. Since the number of equations equals the number of variables, the column section is completely specified if the external feeds, number of stages, and stage pressures are fixed.

5.1.1.1 Assumptions and Simplifications

At this point, we are not concerned with the developing methods for rigorous solution of the above system of equations. Discussion on computational methods for the general case of multi-component systems is covered in Chapter 13. This chapter considers a simplified model that lends itself to graphical solution and provides a tool for qualitative understanding of the operation of a distillation column. The model has a relatively low level of complexity because of its binary nature and also because of other simplifying assumptions.

It will be shown that if certain assumptions hold, the vapor and liquid flows V_j and L_j may be considered constant throughout the column section. If the sensible heat of the vapor and liquid is neglected when compared to the latent heat, that is, if $h_{j-1} \approx h_j$ and $H_{j+1} \approx H_j$, Equation 5.5 becomes

$$L_{j-1}h_j + V_{j+1}H_j = L_jh_j + V_jH_j$$

Also, if the stream enthalpy equals the sum of component enthalpies, that is, it exhibits ideal solution behavior, the above equation becomes

$$\sum_{i=1}^{2}(L_{j-1,i}h_{ji} + V_{j+1,i}H_{ji}) = \sum_{i=1}^{2}(L_{ji}h_{ji} + V_{ji}H_{ji})$$

Variables L_{ji} and V_{ji} refer to component molar flow rates in the liquid and vapor phases. The second subscript identifies the component. The component liquid and vapor enthalpies, both at saturated conditions, are related by the equation

$$H_{ji} = h_{ji} + \lambda_{ji}$$

where λ_{ji} is the heat vaporization of component i at stage j. A key assumption at this point is that the two components have equal molar heats of vaporization:

$$\lambda_{j1} = \lambda_{j2} = \lambda_j$$

The enthalpy balance equation becomes

$$\sum_{i=1}^{2}(L_{j-1,i}h_{ji} + V_{j+1,i}h_{ji} + V_{j+1,i}\lambda_j) = \sum_{i=1}^{2}(L_{ji}h_{ji} + V_{ji}h_{ji} + V_{ji}\lambda_j)$$

The material balance equation (Equation 5.1) is now written for each component in terms of component flow rates, and then multiplied by the component liquid enthalpy:

$$L_{j-1,1}h_{j1} + V_{j+1,1}h_{j1} = L_{j1}h_{j1} + V_{j1}h_{j1}$$

$$L_{j-1,2}h_{j2} + V_{j+1,2}h_{j2} = L_{j2}h_{j2} + V_{j2}h_{j2}$$

These two equations are subtracted from the enthalpy balance equation. The resulting equation, after canceling λ_j, is

$$V_{j+1,1} + V_{j+1,2} = V_{j1} + V_{j2}$$

or

$$V_{j+1} = V_j$$

Also, combining with Equation 5.2,

$$L_{j-1} = L_j$$

Thus, the assumptions of enthalpy additivity and equality of heats of vaporization of the two components in a binary lead to the approximation of constant molar overflow of liquid and vapor in a column section. The implication is that for each mole of component 1 vaporizing at a given stage, one mole of component 2 condenses. Substituting $V = V_j = V_{j+1}$ and $L = L_{j-1} = L_j$ in Equation 5.1 yields the following:

$$Y_{j+1} = (L/V)X_j + Y_j - (L/V)X_{j-1} \tag{5.11}$$

This equation is the end result of combining Equations 5.1, 5.2, and 5.5, together with the simplifying enthalpy assumptions.

Next, Equations 5.3 and 5.4 are combined into one equation, and the system of Equations 5.1 through 5.5 is reduced to two equations. Each of Equations 5.3 or 5.4, aided by a vapor–liquid equilibrium coefficient prediction method such as Equations 5.6 or 5.7, could be used to establish the vapor–liquid equilibrium relationships. However, further simplification is possible if Equations 5.3 and 5.4 are combined with each other:

$$K_{j1}/K_{j2} = (Y_j/X_j) \times (1 - X_j)/(1 - Y_j)$$

Rearranging the equation gives

$$Y_j = \frac{X_j(K_{j1}/K_{j2})}{1 + X_j(K_{j1}/K_{j2} - 1)} \tag{5.12}$$

If the relative volatility, $\alpha_{12} = K_1/K_2$, is assumed constant and its value is known, Equation 5.12 can be used to generate the entire Y–X equilibrium curve.

Equations 5.11 and 5.12 can be used to obtain either an analytical or a graphical solution for the column section. Equation 5.12 relates the vapor and liquid compositions on the same tray, and Equation 5.11 relates the composition of liquid descending from a given tray to the composition of vapor rising from the tray below.

5.1.2 ANALYTICAL SOLUTION

As stated earlier in Section 5.1.1, if the external feeds to a column stage or section and its pressure are defined, the stage or section performance is defined. Equations 5.11 and 5.12 are now examined from this perspective. Consider first a single stage. If the external feeds are defined, then L, V, X_{j-1}, and Y_{j+1} are known. This leaves two unknowns, X_j and Y_j, between Equations 5.11 and 5.12; hence the problem is completely specified.

If the external feeds in an N-stage column section are defined, L, V, X_0, and Y_{N+1} are known. There are $2N$ unknowns (X_j and Y_j, $j = 1, ..., N$) and $2N$ equations (5.11 and 5.12, $j = 1, ..., N$) and the problem is completely defined. It is implied that N is also known, as well as the column pressure, which is assumed fixed. Thus, the performance of a column section is defined if N, L, V, X_0, and Y_{N+1} and the pressure are known.

With a fixed number of stages N, the parameters L, V, X_0, and Y_{N+1} can be substituted by other parameters as specified variables. For instance, one may specify, N, L, X_0, Y_{N+1}, and Y_1 instead of V, and the column section would still be completely specified. The vapor feed rate is relaxed to satisfy the vapor product composition specification, Y_1.

If the number of stages is relaxed, another parameter must be specified in order to define the column section. If, for example, L, V, X_0, and Y_{N+1} are specified, but N is not, the column section is not defined. If X_N or Y_1 is specified, the column section would be defined and N would have to be calculated to meet the specification. Table 5.1 lists the various possible ways of specifying a column section, together with the corresponding unknowns or dependent variables for each case.

Note that the system of simultaneous Equations 5.11 and 5.12 can easily be decoupled and solved independently and sequentially for $j = 1, ..., N$. The computational steps for an analytical solution of a column section are as follows for the case where N, L, V, X_0, and Y_1 are specified:

1. Calculate X_1 from Equation 5.12 ($j = 1$).
2. Calculate Y_2 from Equation 5.11 ($j = 1$).

Repeat steps 1 and 2 for $j = 2, ..., N$. Cases where other sets of variables are specified may require a trial and error procedure.

5.1.3 GRAPHICAL REPRESENTATION ON THE Y–X DIAGRAM

Equations 5.11 and 5.12 are plotted on a Y–X diagram in Figure 5.2. Curve A is the equilibrium curve representing Equation 5.12 or any other set of vapor–liquid equilibrium data. Curve B is the operating line, or material balance relationship, expressed by Equation 5.11. The binary column section can be solved on this diagram by graphically duplicating the steps in the analytical procedure.

Any point on curve A represents compositions of vapor and liquid leaving an equilibrium stage. The points on curve B relate the compositions of passing vapor and liquid streams between stages. It can be shown that curve A must lie above curve B by comparing Y_A and Y_B, which are the values of Y on curves A and B corresponding to a given X. Mole fraction Y_A is in the vapor at some stage, while Y_B is the mole

TABLE 5.1
Parameters in a Fixed-Pressure Binary Column Section

Specifications	Unknowns
Class 1	
N, L, V, X_0, X_N	$X_j(j = 1, ..., N-1), Y_j(j = 1, ..., N+1)$
N, L, V, X_0, Y_{N+1}	$X_j(j = 1, ..., N), Y_j(j = 1, ..., N)$
N, L, V, Y_1, Y_{N+1}	$X_j(j = 0, ..., N), Y_j(j = 2, ..., N)$
N, L, V, Y_1, X_N	$X_j(j = 0, ..., N-1), Y_j(j = 2, ..., N+1)$
Class 2	
N, L, V, X_0, Y_1	$X_j(j = 1, ..., N), Y_j(j = 2, ..., N+1)$
N, L, V, X_N, Y_{N+1}	$X_j(j = 0, ..., N-1), Y_j(j = 1, ..., N)$
Class 3	
N, L (or V), X_0, Y_1, X_N	V (or L), $X_j(j = 1, ..., N-1), Y_j(j = 2, ..., N+1)$
N, L (or V), X_0, Y_1, Y_{N+1}	V (or L), $X_j(j = 1, ..., N), Y_j(j = 2, ..., N)$
N, L (or V), X_N, Y_{N+1}, X_0	V (or L), $X_j(j = 1, ..., N-1), Y_j(j = 1, ..., N)$
N, L (or V), X_N, Y_{N+1}, Y_1	V (or L), $X_j(j = 0, ..., N-1), Y_j(j = 2, ..., N)$
Class 4	
L, V, X_0, Y_1, X_N	$N, X_j(j = 1, ..., N-1), Y_j(j = 2, ..., N+1)$
L, V, X_0, Y_1, Y_{N+1}	$N, X_j(j = 1, ..., N), Y_j(j = 2, ..., N)$
L, V, X_N, Y_{N+1}, X_0	$N, X_j(j = 1, ..., N-1), Y_j(j = 1, ..., N)$
L, V, X_N, Y_{N+1}, Y_1	$N, X_j(j = 0, ..., N-1), Y_j(j = 2, ..., N)$
Class 5	
L (or V), X_0, X_N, Y_1, Y_{N+1}	N, V (or L), $X_j(j = 1, ..., N-1), Y_j(j = 2, ..., N)$

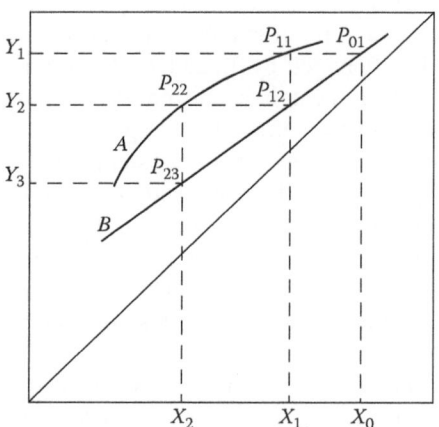

FIGURE 5.2 Equilibrium stages in a binary column section.

fraction in the vapor on the tray below. Since Y is the mole fraction of the more volatile component, Y is larger on trays that are higher up in the column, therefore Y_A is greater than Y_B.

As indicated by Equation 5.11, the operating line is a straight line if L/V is constant. Plotting it requires knowledge of the slope L/V and a point on the line, or two points on the line. Note that a point on the operating line has coordinates X_j and Y_{j+1}, which represent the inlet and outlet compositions at either end of the column section: X_0 and Y_1 or X_N and Y_{N+1}.

If only one point on the operating line or only its slope is known, the number of equilibrium stages must be known in order for the column section to be defined. The operating line would then be determined by trial and error to match the specified number of stages.

5.1.3.1 Constructing Equilibrium Stages

The equilibrium curve and the operating line drawn on a Y–X diagram (Figure 5.2) together describe the performance of a column section. The horizontal line $P_{01}P_{11}$ represents the transition from the region above stage 1 to stage 1 proper, and the vertical line $P_{11}P_{12}$ represents the transition from stage 1 to the region between stages 1 and 2. Thus, the point P_{01} corresponds to liquid composition entering stage 1 and vapor composition leaving stage 1, P_{11} corresponds to liquid and vapor compositions leaving stage 1, P_{12} represents liquid composition from stage 1 to stage 2 and vapor composition from stage 2 to stage 1, and so on. The points along connecting lines $P_{01}P_{11}$ and $P_{11}P_{12}$ do not represent actual compositions since the vapor and liquid compositions are assumed to undergo discrete changes from stage to stage rather than continuous variation. The graphical construction of stages is applied next to the different classes of column section specifications listed in Table 5.1.

Class 1: The operating line slope L/V is known and the line is bounded by one coordinate on each end. Held at a fixed slope, the operating line is moved to a position where the number of steps between it and the equilibrium curve equals the given number of stages. In Figure 5.2, assume X_0 and X_2 are given as well as the slope and the number of stages, $N = 2$. Line B is moved to a position that generates two steps between curves A and B, bounded by X_0 and X_2. Once line B is determined, other compositions such as Y_1 and Y_3 can be read from the diagram.

Class 2: The operating line is defined by the specified values of the slope and one end of the line. The other end is determined by stepping off the specified number of stages. Referring to Figure 5.2, if the compositions of the liquid entering and vapor leaving the top of the column section are given, point $P_{01}(X_0, Y_1)$ is determined. Since the slope L/V is also given, the operating line B can be drawn. If the number of stages is given as 2, two steps are drawn between curves A and B down to point $P_{23}(X_2, Y_3)$, which determines the compositions of liquid leaving and vapor entering the bottom of the column section.

Class 3: Here, one end of the operating line is fixed and the other end is bounded by one coordinate. The operating line is free to rotate around the fixed point since the slope is not given. The correct position is determined by finding the slope which results in a number of steps equal to the given number of stages.

Class 4: This class of specifications fixes one end of the operating line and its slope. The other end is bounded by one coordinate. The required number of stages is found by constructing steps starting at the fixed point and ending at the intersection of the specified coordinate with the operating line. If the point of intersection does not coincide with the completion of a step, the last step should extend past the intersection.

Class 5: Finally, if two points on the operating line are specified, the line is fixed. The number of stages required for the specified separation is determined by stepping stages between the operating line and the equilibrium curve, starting at one of the specified points and ending at the other. Again, if the last step does not coincide with the given point, it is extended past it.

In Classes 3 and 4, the unknown flow rate, L or V, is calculated from the slope L/V as determined from the graphical construction, and the specified flow rate V or L.

5.2   TOTAL COLUMN

A total column differs from a column section in that the former may have side feeds and products and coolers and heaters. Specifically, a conventional distillation column has one feed stream, an overhead (distillate) product and a bottoms product, a condenser, and a reboiler, as shown schematically in Figure 5.3.

Vapor rising from the top of the column is partially condensed; the vapor and liquid are separated; the vapor is taken out as a distillate or overhead product at its

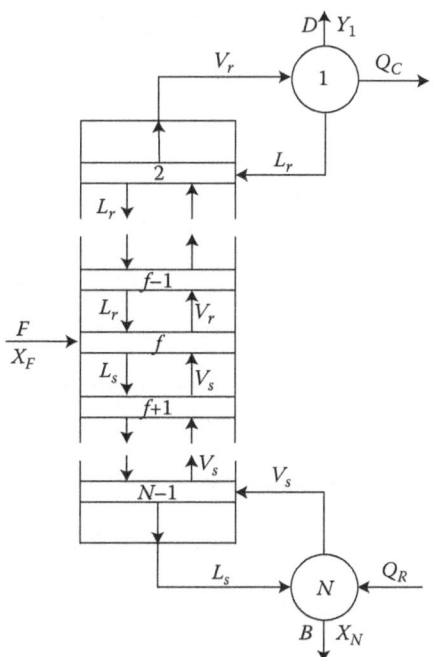

FIGURE 5.3   Schematic of a distillation column.

dew point; and the liquid is returned to the column as reflux. This partial condenser acts as an additional equilibrium stage since the vapor and liquid interact on it and are then separated. It is counted as stage 1, followed by the column top tray as stage 2, and so on. The vapor from the column could be totally condensed; part of the condensate taken as liquid distillate and the remaining part refluxed to the column. With this mode of operation, the condenser is a total condenser and does not act as an equilibrium stage. No further change in the composition of the vapor leaving the column top tray is achieved with a total condenser, since no additional phase separation takes place. The column model discussed in this section is configured with a partial condenser.

Liquid from the column bottom tray is sent to the reboiler, stage N, where it is partially vaporized. The vapor is sent back to the column bottom tray, stage $N-1$, and the liquid is taken out as a bottoms product at its bubble point. In this configuration, the reboiler is a partial reboiler, acting as an additional equilibrium stage.

The feed to the column could be at any thermal condition: superheated or saturated vapor, mixed phase, or saturated or subcooled liquid. The feed tray could be any tray between (and sometimes including) the condenser and reboiler. Regardless of its thermal condition, the feed is assumed to mix and equilibrate with the liquid and vapor on the feed tray.

The top section of the column, or the rectifying section, is that part of the column above the feed and including the condenser. In the rectifying section, liquid flowing down selectively absorbs the heavier component, thereby raising the purity of the lighter component in the distillate and the recovery of the heavier component at the bottoms.

The bottom section, or the stripping section of the column, is the part that includes the reboiler and all stages below the feed. In the stripping section, vapor flowing up selectively strips the lighter component, thereby raising the purity of the heavier component at the bottoms and the recovery of the lighter component in the distillate.

The techniques developed earlier for solving column sections are now extended to include total columns. A mathematical model is set up which is used to derive graphical methods of solution on a Y–X diagram. Solution methods are described for different sets of specifications.

5.2.1 MATHEMATICAL MODEL

The assumptions and simplifications made for a column section equally apply to a total column. They are summarized as follows:

1. The two components have equal and constant enthalpies of vaporization (latent heats).
2. The sensible enthalpy changes of the vapor and liquid are negligible compared to the latent heats.
3. The binary mixture behaves as an ideal solution.
4. The stages are adiabatic except at designated locations (condenser and reboiler).
5. The pressure is constant throughout the column.

The implication of these assumptions is that for each mole of liquid vaporized, one mole of vapor is condensed, so that liquid and vapor molar flows from tray to tray are constant within each section of the column. As in the case of the column section, the component and enthalpy balances reduce to operating line equations, which must be solved in conjunction with the vapor–liquid equilibrium equation or data. Each section within a column is represented by a different operating line.

5.2.1.1   Rectifying Section Operating Line

Referring to Figure 5.3, a material balance on component 1 on any part of the rectifying section that includes stages 1 through j (for j less than f, the feed tray) gives

$$V_r Y_{j+1} = L_r X_j + DY_1$$

or

$$Y_{j+1} = (L_r/V_r)X_j + (D/V_r)Y_1 \qquad (5.13)$$

for $j = 1, \ldots, f-1$. The distillate is of molar flow rate D, with Y_1 mole fraction component 1. (The mole fraction subscripts refer to the stage; the component subscript is dropped since the equations are written only for component 1, the more volatile component in the binary). The liquid and vapor molar flows, L_r and V_r, are assumed constant throughout the rectifying section.

5.2.1.2   Stripping Section Operating Line

A component material balance on component 1 on any part of the stripping section that includes stages j through N (for j greater than f, the feed stage) gives

$$L_s X_{j-1} = V_s Y_j + BX_N$$

where B and X_N are the bottoms molar rate and component 1 mole fraction, respectively. The equation is rearranged as follows:

$$Y_j = (L_s/V_s)X_{j-1} - (B/V_s)X_N \qquad (5.14)$$

where $j = f+1, \ldots, N$. The liquid and vapor molar flows, L_s and V_s, are assumed constant throughout the stripping section.

5.2.1.3   Feed Stage

The feed stream, F, with composition X_F enters the column on stage f, mixes with the vapor and liquid entering that stage from adjacent stages, and reaches equilibrium. A material balance on component 1 on the stage f gives

$$FX_F + L_r X_{f-1} - L_s X_f = V_r Y_f - V_s Y_{f+1} \qquad (5.15)$$

An energy balance on the feed stage determines the relative flow rates of liquid and vapor above and below the feed stage, namely, L_r, L_s, V_r, and V_s. As the feed stream enters the column, it changes from its initial thermal conditions (T_F, P_F) to the feed tray conditions (T_f, P_f). The energy balance on the feed stage is written as

$$FH_F + L_r h + V_s H = L_s h + V_r H$$

where H_F, h, and H are the molar enthalpies of the feed, the liquid, and the vapor, respectively. Set an arbitrary enthalpy reference point of zero for saturated vapor at P_f and define the amount of heat required for the conversion of one mole of feed at T_F and P_F to saturated vapor at P_f as ΔH_F. Neglecting sensible enthalpies relative to latent heat, leads to the following equalities, expressing enthalpies relative to saturated vapor enthalpy:

$$H = 0$$

$$h = -\lambda$$

$$H_F = -\Delta H_F$$

where λ is the molar heat of vaporization. Substituting these values in the energy balance equation gives

$$(L_s - L_r)\lambda = F\Delta H_F$$

Define the quantity $q = \Delta H_F / \lambda$, and the following equation is obtained:

$$L_s = L_r + qF \qquad (5.16)$$

Equation 5.16 is combined with an overall material balance on the feed stage,

$$F + L_r + V_s = L_s + V_r$$

to give

$$V_r = V_s + (1 - q)F \qquad (5.17)$$

If the feed is a saturated vapor at T_f and P_f, then $\Delta H_F = 0$ and $q = 0$. The vapor rate leaving the feed tray is simply the sum of vapor from the bottom column section and the feed, $V_r = V_s + F$. Since no external liquid is involved, the liquid rate entering the tray from the upper section equals the liquid rate leaving it: $L_r = L_s$.

If the feed is a saturated liquid at T_f and P_f, then $\Delta H_F = \lambda$ and $q = 1$. The liquid rate leaving the feed stage is the sum of liquid from the upper column section and the feed, $L_s = L_r + F$. Since the external feed contains no vapor, the vapor rate leaving the tray equals the vapor rate entering it from the bottom column section: $V_r = V_s$.

For a mixed phase feed, that is part vapor and part liquid, at T_f and P_f, the liquid portion joins the liquid leaving the feed tray and the vapor portion joins the vapor leaving it. The amount of liquid is qF and that of the vapor is $(1 - q)F$. For a partially vaporized feed, q is equal to the liquid fraction $(0 < q < 1)$.

Note in the above cases that V_r is greater than or equal to V_s, and L_s is greater than or equal to L_r. That is, the feed has no tendency to "dry up" either the liquid or the vapor.

A feed that is superheated at T_f and P_f has a negative ΔH_F and, therefore, a negative q. A superheated feed will increase the vapor rate above the feed tray ($V_r > V_s$) and decrease the liquid rate below the feed tray ($L_s < L_r$). If the feed is sufficiently superheated, it may dry up the liquid in the lower column section, making L_s approach zero.

If the feed is a subcooled liquid at T_f and P_f, $\Delta H_F > 1$ and $q > 1$. In this case, L_s is greater than L_r, and V_r is less than V_s. Thus, if the feed is sufficiently subcooled, it may dry up the vapor in the upper section (V_r approaches zero).

5.2.1.4 Analytical Solution

The equations describing total column operation include vapor–liquid equilibrium relations, Equation 5.12; component balances in the rectifying and stripping sections, Equations 5.13 and 5.14; feed stage component balance, Equation 5.15; feed stage energy balance and overall material balance, expressed as Equations 5.16 and 5.17; and overall column component balance, Equation 5.18:

$$FX_F = DY_1 + BX_N \tag{5.18}$$

The problem definition is such that the feed is of fixed rate, composition, and thermal conditions. Also, the column pressure and the number of stages are known. The condenser and reboiler duties are not presently of concern; calculating them would require energy balance equations for the condenser and reboiler. Aside from these, the system of equations includes the following:

Equation 5.12	N	Equations
Equation 5.13	$f-1$	Equations
Equation 5.14	$N-f$	Equations
Equation 5.15	1	Equation
Equation 5.16	1	Equation
Equation 5.17	1	Equation
Equation 5.18	1	Equation
Total	$2N+3$	Equations

The variables are $X_j(j = 1, \ldots, N)$, $Y_j(j = 1, \ldots, N)$, L_r, L_s, V_r, V_s, and B (or D), totaling $2N + 5$ variables. Thus, in order to define the column operation, two variables must be specified. If the number of stages is not fixed, three parameters would have to be specified.

The method for solving the above set of equations depends on which parameters are specified. Although the simplified binary distillation model discussed in this chapter is mainly intended for developing a graphical solution, an analytical solution is first presented as an exercise for better understanding the interrelations among the different variables. An algorithm is provided only for one set of specifications: the distillate and bottoms compositions, Y_1 and X_N, for a given number of stages and a given feed stage. Graphical solutions for these and other specifications are presented in the following section.

Since the analytical method requires given number of stages and feed stage, it is more suited for evaluating the performance of an existing column than it is for column design, although it could be used for design by doing multiple runs or case studies. The availability of fast and easy to use computer programs, such as spreadsheet programs, offers a viable option in the numerical approach that can supplement the graphical method whose accuracy may be limited by the graphics.

The computational steps of the analytical method are as follows (see Example (6.7):

1. Given Y_1, X_N, F, X_F, T_F, P_F, N, and f.
2. A column overall material balance, $F = D + B$, combined with Equation 5.18 gives the following:

$$FX_F = (F - B)Y_1 + BX_N$$

All the parameters in this equation are known except B. Solve for B, then compute $D = F - B$.
3. Assume a reflux rate, L_r, and calculate $V_r = D + L_r$.
4. Calculate X_1 from Equation 5.12 with $j = 1$.
5. Calculate Y_2 from Equation 5.13 with $j = 1$.
6. Repeat steps 4 and 5 with $j = 2, ..., f - 1$. The last compositions calculated are X_f and Y_f.
7. From vapor–liquid equilibrium versus temperature data, calculate T_f that corresponds to $K_f = Y_f/X_f$.
8. Calculate ΔH_F, the difference between the molar feed enthalpy at initial conditions T_F and P_F and the enthalpy of a mole of saturated vapor at T_f and P_f.
9. Calculate q from its definition, $q = \Delta H_F/\lambda$.
10. Calculate L_s and V_s from Equations 5.16 and 5.17.
11. Calculate Y_{f+1} from Equation 5.15.
12. Calculate X_{f+1} from Equation 5.12 with $j = f + 1$.
13. Calculate Y_{f+2} from Equation 5.14 with $j = f + 2$.
14. Repeat steps 12 and 13 for $j = f + 2, ..., N$. The last compositions calculated are Y_N and X_N.
15. If X_N calculated matches X_N specified, the problem is solved. Otherwise, start over at step 3 with a new value for the reflux rate L_r.

5.2.1.5 The Description Rule

By taking an inventory of equations and variables for the above column, it was determined that two or three variables (depending on whether or not the number of stages is fixed) must be specified in order to define the column performance. An alternative to this mathematical analysis is to apply the description rule (Hanson et al., 1962), which can be a less tedious method for determining the number of independent variables, based on a practical evaluation of the process.

The description rule states that the number of independent variables that are required to define a process uniquely is equal to the number of variables set by construction of the process equipment plus the number of variables that can be directly and independently controlled during operation of the process. The rule is simply a statement of the nature of the actual process which the mathematical model must represent.

Considering again the column described above, the engineer has the freedom during the design phase to decide on the number of trays as well as the feed location. These 2 degrees of freedom are used up at construction time. With an existing column, the operator starts up the process by pumping in the feed stream to the column. The process devices may include, in addition to the pump, a valve at the pump discharge and a feed preheater. If these pieces of equipment are considered part of the system, they provide the operator with the means to control the feed rate, temperature, and pressure by independently manipulating the pump speed, the valve setting, and the steam rate to the preheater. This particular subsystem therefore requires three independent variables to specify it. The setup does not include a device for controlling the feed composition and, hence, this parameter is not included in the list of independent variables.

If the system boundary includes just the feed inlet and the column, the feed conditions are set and no devices are available to manipulate them. With a fixed feed and column configuration, the operator has the task of operating the column to achieve the desired separation. The variables available for controlling the separation are the operating devices that can be physically and independently manipulated. One variable that must be set is the column pressure, which is usually controlled by manipulating the vapor or vent flow out of the column. The column pressure is normally dictated by the temperature of the condenser cooling medium and is not, in general, used to control the separation.

Two more operating variables exist: the reboiler heating medium rate (steam) and the condenser cooling medium rate (water). Therefore, two more variables must be set in order to specify the column operation. The steam and water rates themselves may be set or, alternatively, they may be controlled by two other variables such as the desired separation specifications.

Note that the column may have additional controllers such as condenser and reboiler level controls. The levels, however, are not independent variables since they must be maintained within defined bounds. The liquid levels, which have no effect on the separation, may be controlled by manipulating the distillate and bottoms product rates.

In conclusion, for an existing column with a fixed feed, two variables are required to define the column performance, which is consistent with the result reached by the mathematical analysis.

5.2.2 Graphical Solution on the Y–X Diagram

The material balance equations, or operating lines (Equations 5.13 and 5.14) together with the feed stage heat balances (Equations 5.16 and 5.17), and the equilibrium relation, (Equation 5.12) form the basis for the McCabe–Thiele (McCabe and Thiele, 1925) graphical solution of binary distillation on a Y–X diagram.

The phase equilibrium data, or Equation 5.12, are first plotted on a Y–X diagram (Figure 5.4). The operating lines, assumed straight lines, are then plotted on the basis of known points, slopes, and/or number of stages, depending on which variables are specified. Together, these curves completely define the column operation, and a graphical solution readily follows.

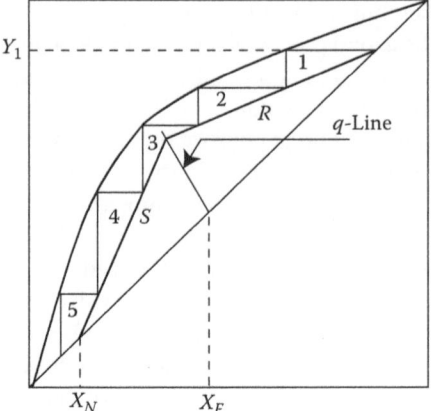

FIGURE 5.4 Performance of a binary distillation column.

Each operating line is the locus of discrete points, (X_1, Y_2), (X_2, Y_3), and so on, relating the compositions of liquid and vapor counter-flowing between stages. For the purpose of plotting them, the operating lines, Equations 5.13 and 5.14, are considered continuous in Y and X, with slopes L_r/V_r and L_s/V_s. Extrapolating them toward the diagonal determines one fixed point for each line. At the intersection of the rectifying section operating line with diagonal, since $Y = X$, Equation 5.13 becomes

$$V_r X = L_r X + D Y_1$$

It is also known that $D = V_r - L_r$, and when this expression for D is substituted in the above equation, the result is $Y = X = Y_1$. That is, the rectifying section operating line intersects the diagonal at the distillate composition. Similarly, it may be shown that the stripping section operating line intersects the diagonal at $Y = X = X_N$, the bottoms composition.

Another point that determines the operating lines is their common intersection point. Equations 5.13 and 5.14 are both extrapolated toward the feed stage. Equation 5.13 is valid above that stage, and Equation 5.14 is valid below it. The operating line of the total column is made up of two straight lines, R and S (Figure 5.4). Since R corresponds to stages above the feed and S corresponds to stages below the feed, the transition from R to S, or their intersection, occurs at the feed stage. At the intersection point, Y_{j+1} and X_j in Equation 5.13 are identical to Y_j and X_{j-1} in Equation 5.14. Subtracting these equations gives

$$Y(V_r - V_s) = X(L_r - L_s) + D Y_1 + B X_N$$

which, when combined with Equations 5.16, 5.17, and 5.18, gives

$$Y = X q/(q - 1) - X_F/(q - 1) \qquad (5.19)$$

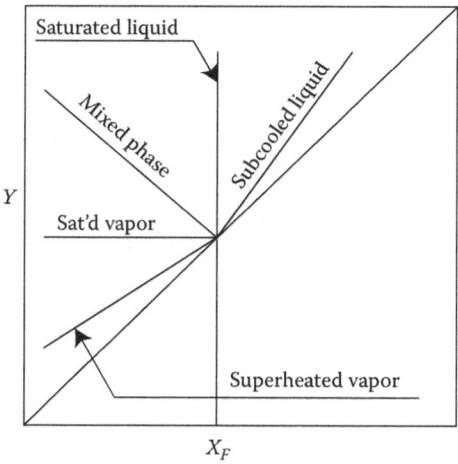

FIGURE 5.5   Plots of q-lines for different feed thermal conditions.

Equation 5.19 is known as the q-line equation. The q-line represents the feed thermal conditions and is the locus of intersection points of the operating lines above and below the feed tray. It is a straight line with a slope of $q/(q-1)$ and an intersection point with the diagonal at $Y = X = X_F$. This may be shown by setting $Y = X$ in Equation 5.19 and solving for X.

The values of q for different feed thermal conditions at feed tray temperature and pressure, T_f and P_f, were discussed in Section 5.2.1. The corresponding values of the q-line slope are summarized in the table below. Example of q-lines are shown in Figure 5.5.

Feed thermal condition at	T_f, P_f	q-Line slope
Superheated vapor	$q < 0$	$0 <$ Slope < 1
Saturated vapor	$q = 0$	0
Mixed phase	$0 < q < 1$	$-\infty <$ Slope < 0
Saturated liquid	$q = 1$	∞
Subcooled liquid	$q > 1$	$\infty >$ Slope > 1

5.2.2.1   Representing a Total Column

Once the operating lines are plotted, the equilibrium stages are stepped off between the operating line and equilibrium curve as described in Section 5.1.3. The optimum feed tray location is at the step that crosses from one operating line to the other over their intersection point. If the number of stages is fixed, the operating lines are determined by trial and error to match the specified number of stages.

The mathematical model described in Section 5.2.1 indicates that two variables must be specified in order to define the performance of a column with fixed pressure, number of stages, feed tray location, and feed rate, composition, and thermal conditions. Three specifications are required if the number of stages is unknown. Graphical methods for solving distillation columns depend on the column specifications. A partial condenser with vapor distillate and a partial reboiler with liquid bottoms are assumed. As such, the condenser and reboiler are both counted as equilibrium stages. The column solution is described for the following sets of specifications.

5.2.2.2   Separation and Reflux Ratio Specified

The separation is specified in terms of the composition of one component in the distillate and bottoms, Y_1 and X_N. (The subscripts refer to stages. When component subscripts are dropped, implied reference is to the more volatile component). The separation may alternatively be specified in terms of variables that are functions of Y_1 and X_N. The reflux ratio expresses relative flow rates of liquid and vapor within the column. It may be specified in terms of L_r, V_r, L_s, or V_s, or a function thereof such as L_r/V_r.

The number of stages is unknown; therefore three independent specifications are required to define the column operation. Consider the case where the specifications are Y_1, X_N, and L_r/V_r. The first step in a graphical solution is to plot the equilibrium curve corresponding to the given column pressure on a Y–X diagram based on vapor–liquid equilibrium data (Figure 5.4). Next, the q-line is plotted on the basis of the feed composition and its thermal conditions at feed tray temperature and pressure. The feed tray pressure is known, but the feed tray temperature must be assumed for the purpose of calculating the q-line slope. The rectifying section operating line, R, has one point fixed at the intersection of a horizontal line through Y_1 and the diagonal. The slope of R is L_r/V_r, as indicated by Equation 5.13. With one point fixed and the slope defined, the operating line, R, is next drawn on the diagram. The stripping section operating line, S, is drawn by joining X_N on the diagonal to the intersection of R with the q-line.

This completes the operating line for the entire column. The equilibrium stages are next stepped off between the operating line and the equilibrium curve beginning at Y_1, as shown in Figure 5.4. (Construction of the stages could just as well be started at X_N.) The diagram shows that four full equilibrium stages plus a fraction of a stage are required to achieve the specified separation at the given reflux ratio. A fraction of a theoretical stage could translate into one or more actual trays, depending on their efficiency. Fractional stages are generally rounded up, bringing the total in this case to five. The resulting separation would be slightly sharper than specified, that is, the bottoms composition would be somewhat less than X_N. The diagram also shows that the feed should be introduced at the third stage from the top. The top stage is the condenser and the bottom stage is the reboiler.

The assumed feed tray temperature may now be checked and, if necessary, the q-line is redrawn using the updated slope. The above procedure is then repeated for the new q-line slope. To check the feed tray temperature, it is necessary to make use of the vapor–liquid equilibrium data from which the equilibrium curve was drawn.

To complete the column solution, the product rates are computed from a total material balance and one component balance around the column:

$$F = D + B$$

$$FX_F = DY_1 + BX_N$$

The values of L_r and V_r are determined from their given ratio and a material balance on the condenser, $D = V_r - L_r$. The stripping section flows, V_s and L_s, can be calculated from Equations 5.16 and 5.17.

5.2.2.3 Distillate Composition, Reflux Ratio, and Number of Stages Specified

The equilibrium curve and the q-line are plotted as above. Since Y_1 and L_r/V_r are given, the rectifying section operating line can be drawn. The point of intersection of this line with the q-line fixes one point on the stripping section operating line. The slope of this line must be determined by trial and error to meet the specified number of stages. The stripping section operating line is rotated around its intersection with the q-line and, for each position, equilibrium stages are stepped off between the operating lines and the equilibrium curve. The correct position is found when the stepped number of stages matches the specified number. The bottoms composition, X_N, is then determined by dropping a vertical line through the intersection of the lower operating line with the diagonal. The solution also determines the feed location.

A similar technique may be used if X_N and L_s/V_s are specified instead of Y_1 and L_r/V_r.

5.2.2.4 Separation and Number of Stages Specified

The separation is specified in terms of Y_1 and X_N or quantities that are functions thereof, such as recoveries. The intersections of a horizontal line through Y_1 and the diagonal and a vertical line through X_N and the diagonal determine one point on each operating line. The q-line is drawn on the basis of feed data as before. This line is the locus of intersection points of the two operating lines. The correct intersection point is determined by trial and error until the stepped number of stages equals the specified number. The closer the intersection point to the equilibrium curve, the larger the number of stages. The resulting operating lines may be used to determine the feed location, L_r/V_r, L_s/V_s (or reflux ratio), tray compositions, and so on.

5.2.2.5 Reflux Ratio, Product Rates, and Number of Stages Specified

If the reflux ratio, or L_r/D, and the distillate rate, D, are specified, the rectifying section operating line slope, L_r/V_r can be calculated. The feed composition and thermal conditions are also known, which allows the q-line to be plotted and L_s and V_s to be calculated from Equations 5.16 and 5.17. Thus, the slopes of the two operating lines are known as well as the locus of their intersection points.

The column is solved by trial and error by moving the operating lines with their intersection point along the q-line, keeping their slopes fixed (Figure 5.6). The number of stages for each position is determined by stepping off stages as usual. The position that results in the specified number of stages is the required solution. Information obtained from the solution includes the separation achievable with the given reflux ratio and the number of stages and the optimum feed location.

5.2.2.6 Columns with Multiple Feeds, Side Draws, and Side Heaters/Coolers

The methods developed for graphical construction of equilibrium stages on the Y–X diagram for single-feed, two-product distillation columns can be generalized to include columns with multiple feeds, side draws, and side heaters or coolers. When

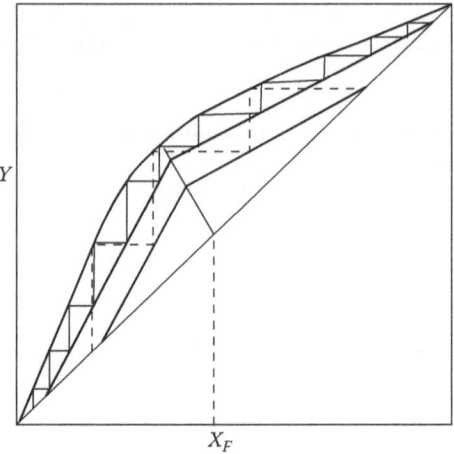

FIGURE 5.6 Binary distillation: given number of stages, reflux ratio, and product rate.

these discontinuities exist, each column section operating line has, in general, a different L/V ratio or slope. A column section operating line is defined if its L/V ratio and one point on the line are known or if two points on it are known.

The way the operating lines are actually constructed depends on how the particular column is specified. Figure 5.7 shows a demonstration column with two feeds, a liquid draw, a vapor draw, a partial condenser and a reboiler, a side cooler, and a side reboiler. Also shown is the Y–X diagram and operating lines for this column. If we assume the distillate rate and composition and the reflux ratio are known, the calculations can be started at the top. A vertical line drawn through the distillate composition intersects the diagonal at a point that defines the upper end of the top operating

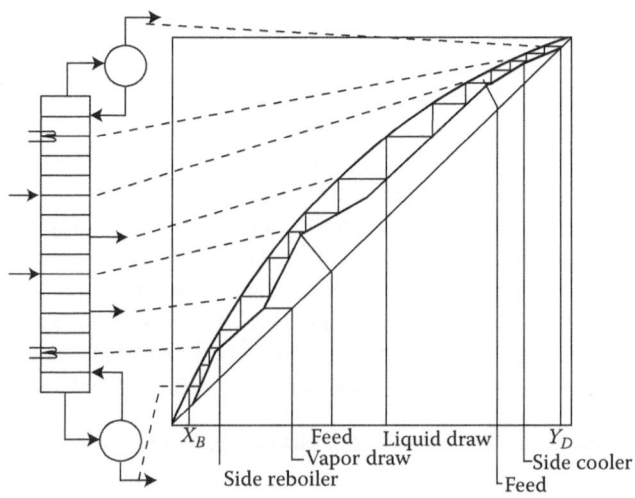

FIGURE 5.7 Column with multiple feeds, side draws, and side heaters/coolers.

line. The L/V ratio, calculated from the distillate rate and reflux ratio, determines the slope of the line. Equilibrium stages are then stepped off between this operating line and the equilibrium curve. This is continued until the side cooler stage is reached.

Side Heaters/Coolers: The effect of a side heater or cooler on the L/V ratio is determined from an energy balance at stage j, the stage where heat duty per unit time, q_j, is transferred. If the side heater/cooler is placed at stage j, the streams directly affected by the heat transfer are L_j and V_j (compositions X_j and Y_j). The role of the side heater/cooler in modifying the vapor and liquid profiles above it and below it is analyzed based on energy and material balances. The energy balance is written as

$$q_j + L_{j-1}h_{j-1} + V_{j+1}H_{j+1} = L_jh_j + V_jH_j$$

When the McCabe–Thiele assumptions are applied, the sensible heat is neglected in comparison with the latent heat, which is assumed identical for both components in the binary. With these assumptions, the energy balance is simplified, resulting in the equation

$$-q_j = H(V_{j+1} - V_j) + h(L_{j-1} - L_j)$$

where H and h are average vapor and liquid molar enthalpies, respectively. The material balance equation at tray j,

$$L_{j-1} - L_j = V_j - V_{j+1}$$

is combined with the above equation to give

$$-q_j = (H - h)(V_{j+1} - V_j) = (H - h)(L_j - L_{j-1})$$
$$= \lambda(V_{j+1} - V_j) = \lambda(L_j - L_{j-1})$$

The operating line slope below the heater or cooler stage is, therefore,

$$\frac{L_j}{V_{j+1}} = \frac{L_{j-1} - q_j/\lambda}{V_j - q_j/\lambda}$$

This equation relates the liquid and vapor profiles above and below the heater/cooler tray to the heat duty q_j. Referring to Figure 5.7, if the slope of the operating line below the side cooler stage is known or can be calculated, the line is constructed directly. If the number of stages between the side cooler and the upper feed is known, the slope of the operating line must be varied to match the number of stages.

The upper feed is represented graphically by its q-line drawn with a slope corresponding to the feed thermal conditions. The q-line starts at the intersection with the diagonal of a vertical line drawn through the feed composition. The intersection of the q-line with the previous operating line determines the upper end of the next operating line. With a known slope, or L/V ratio, this line is now drawn, and the stages between the upper feed and the next discontinuity, the liquid draw, are stepped off.

Side Draws: The q-line of a side draw passes through the intersection of the operating lines above and below the draw tray. A liquid side draw from stage j is a saturated liquid with composition X_j. The liquid draw composition is the same as the draw tray liquid composition, and therefore the liquid draw q-line is a vertical line passing through X_j. A material balance on a tray with liquid draw S_j^L is written as

$$L_j = L_{j-1} - S_j^L$$
$$V_j = V_{j+1}$$

A vapor draw from stage j is a saturated vapor with composition Y_j. The vapor draw composition is the same as the draw tray vapor composition, and therefore the vapor draw q-line is a horizontal line passing through Y_j. For a tray with vapor draw S_j^V,

$$L_j = L_{j-1}$$
$$V_j = V_{j+1} - S_j^V$$

The operating line above the liquid draw ends at its intersection with the q-line of the liquid draw. A vertical line constructed at the draw tray liquid composition represents this side stream and determines its composition.

The liquid flow below the liquid draw tray is reduced by an amount equal to the side draw rate, and if the side draw is taken at tray j, $L_j = L_{j-1} - S_j^L$. Under the present simplifying assumptions, the vapor flow does not change at the liquid draw tray. Therefore, the L/V ratio below the liquid draw is less than the ratio above it, and the slope of the operating line below the draw is reduced.

The next discontinuity occurs at the lower feed, whose q-line is constructed in the usual manner. The slope of the operating line between the liquid draw and the lower feed is adjusted to match the number of stages in that section.

The slope of the operating line between the lower feed and the vapor draw is again determined such that the stepped stages equal the known number of stages in that section. The q-line of the vapor draw is horizontal since it is a saturated vapor. Its composition is read on the Y-axis, also determined by dropping a vertical line through the intersection of the q-line with the diagonal.

The next section operating line has a smaller slope since the vapor flow is reduced above the draw tray j by the vapor draw rate S_j^V and the liquid flow is assumed constant. The operating line slope is determined such that the number of stepped off stages matches the number of stages between the vapor draw and the side reboiler.

The effect of the side reboiler on the operating line slope is determined as described above for the side cooler, taking into account that the heat duty here is positive.

If the number of stages between the side reboiler and bottom reboiler is known, the slope of the last section operating line is adjusted to match the number of stages. If the bottoms composition is known, the operating line is determined and the equilibrium stages are stepped off.

5.2.2.7 Columns with Stripping Vapor Feed

Multistage columns can be designed and operated such that the stripping action is achieved by injecting a vapor stream below the bottom stage. The stripping vapor, usually steam, replaces the reboiler. Open steam (meaning steam fed directly to the column rather than through a reboiler) is commonly used in stripper columns, discussed in Chapter 8. In binary distillation the less volatile component (steam) may be injected as the stripping agent, and no reboiler would be necessary.

A material balance on the lighter component around a column section between any stage j in the stripping section and the bottom stage is written as

$$L_{j-1}X_{j-1} + V_{N+1}Y_{N+1} = BX_N + V_jY_j$$

where V_{N+1} and Y_{N+1} are the flow rate and composition of the stripping vapor. If the stripping vapor contains only the less volatile component, then $Y_{N+1} = 0$. The flows L_{j-1} and V_j are constant from tray to tray within the section and are designated as L_s and V_s. The above equation, rearranged, is the stripping section operating line:

$$Y_j = \frac{L_s}{V_s}X_{j-1} - \frac{B}{V_s}X_N$$

Applied to the bottom stage, $j-1 = N$, $j = N + 1$, $B = L_s$, $Y = Y_{N+1} = 0$, and $X = X_B$. That is, the operating line intersects the X-axis at $X = X_B$. This intersection point represents passing streams below the column. With open steam, the operating line crosses the diagonal and ends at the X-axis. Equilibrium stage steps are constructed in the usual manner, between the operating line and the equilibrium curve.

5.2.3 TRAY EFFICIENCY

Most of the column tray-by-tray calculations, including the graphical methods described in this chapter, are based on the concept of the equilibrium stage. Since actual column trays can generally only approach equilibrium stage performance to varying degrees, it is common practice to use tray efficiencies to relate actual performance to equilibrium conditions. There are different types of tray efficiencies; a detailed discussion of which is deferred to Chapter 14. The Murphree tray efficiency (Murphree, 1925) is introduced at this point because it can be readily incorporated in the Y–X graphical construction of stages.

Figure 5.8 shows a segment of an operating line and the equilibrium curve. Tray j, with a liquid composition X_j, receives vapor with composition Y_{j+1} from the tray below. If tray j were an equilibrium stage, the vapor leaving it would have a composition Y_j^*, determined by the equilibrium curve. The change in the vapor composition as it passes through tray j would be $Y_j^* - Y_{j+1}$. In an actual tray, the vapor leaving tray j has a composition Y_j, resulting in a change in vapor composition of $Y_j - Y_{j+1}$. The value of Y_j depends on the tray efficiency. The Murphree tray vapor efficiency is defined as

$$E_V = \frac{Y_j - Y_{j+1}}{Y_j^* - Y_{j+1}}$$

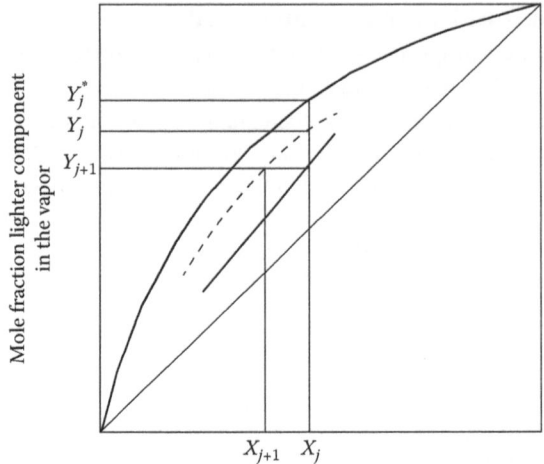

FIGURE 5.8   Implementing Murphree tray efficiency on the Y–X diagram.

The tray efficiency may generally be estimated on the basis of past experience with similar columns or from operating data of an existing column. Given the tray efficiency, a pseudo-equilibrium curve can be drawn, such as the dotted line in Figure 5.8. This curve is constructed based on the definition of the tray vapor efficiency by marking off vertical segments between the operating line and the equilibrium curve such that

$$Y_j - Y_{j+1} = E_V(Y_j^* - Y_{j+1})$$

Once this curve is in place, the graphical construction of stages proceeds as before.

5.3   COLUMN SOLUTION WITH MATERIAL AND ENTHALPY BALANCES

The graphical binary distillation solution technique described on the Y–X diagram in the previous section is based on a number of simplifying assumptions. Simplifications in the energy balances were made, which lead to the approximation that the vapor and liquid flows are constant in each column section and that the operating lines can be represented by straight lines. A more rigorous solution must include accurate energy balances. In order to accomplish this, enthalpy–composition data for the binary mixture are needed, generally presented in the form of an enthalpy–composition diagram.

The method described here was originally presented by Ponchon (1921) and Savarit (1922). Besides the enthalpy–composition diagram, the method makes use of the lever rule for relating the rates of vapor and liquid products of a binary equilibrium stage to their compositions and enthalpies, as described below.

5.3.1 Single-Stage Mass and Energy Balances

If a feed stream with mole rate F separates in an equilibrium stage into a vapor product with mole rate V and a liquid product with mole rate L, then by overall material balance,

$$F = V + L \qquad (5.20)$$

A component material balance on either component is written as

$$FZ = VY + LX \qquad (5.21)$$

where Z, Y, and X are the mole fractions of one of the components in the feed, vapor product, and liquid product, respectively. The mole fractions are not subscripted since the material balance applies to each component, and no distinction between them is necessary at this point.

A total enthalpy balance on the equilibrium stage is written as

$$FH_F = VH + Lh \qquad (5.22)$$

where H_F, H, and h are the enthalpy per mole of the feed, vapor product, and liquid product, respectively.

By combining Equations 5.20 and 5.21, and 5.20 and 5.22, the following equations are obtained:

$$V/F = (Z - X)/(Y - X) \qquad (5.23)$$

$$L/F = (Y - Z)/(Y - X) \qquad (5.24)$$

$$V/F = (H_F - h)/(H - h) \qquad (5.25)$$

$$L/F = (H - H_F)/(H - h) \qquad (5.26)$$

Equation pairs 5.23 and 5.25, and 5.24 and 5.26 are now combined to give

$$\frac{H_F - h}{Z - X} = \frac{H - h}{Y - X} \qquad (5.27)$$

$$\frac{H - H_F}{Y - Z} = \frac{H - h}{Y - X} \qquad (5.28)$$

The enthalpies and compositions of the feed, vapor product, and liquid product are now plotted on an enthalpy–composition diagram, Figure 5.9. The point F represents the feed with coordinates (Z, H_F); the point V represents the vapor product with coordinates (Y, H); and the point L represents the liquid product with coordinates (X, h). It can be seen from Equations 5.27 and 5.28 that the line segments LF and FV are actually the segments on the same straight line since they

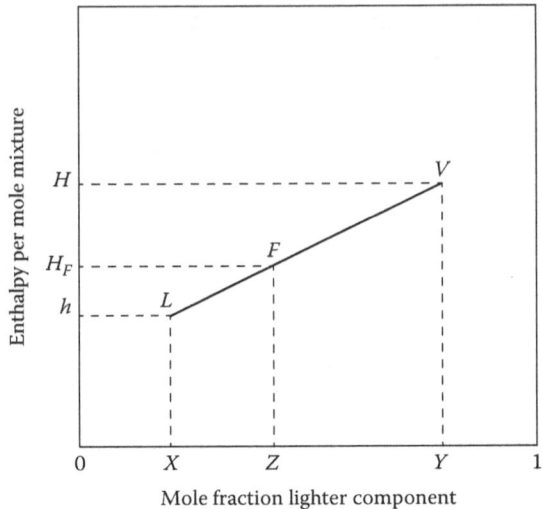

FIGURE 5.9   Mass and enthalpy balances on the *H–X* diagram.

have identical slopes and a common point, *F*. The fact that points *F*, *V*, and *L* lie on a straight line is a result of material and enthalpy balances and can be used to determine unknown variables graphically. If, for instance, the points *L* and *F* and composition *Y* are known, *H* can be determined independently of enthalpy–composition data and based solely on material and enthalpy balances. Graphically, *H* is determined by joining *L* and *F* with a straight line, then extending it until it intersects the vertical line through *Y*. The intersection point corresponds to point *V* with coordinates (*Y*,*H*).

Looking back at Equations 5.23, 5.24, 5.25, and 5.26, it can be seen that the rates of *F*, *L*, and *V* obey the lever rule with respect to compositions and enthalpies. If, for instance, *F* is considered a pivot point, the product of vapor rate *V* times compositional line segment $\underline{ZY}$ equals the product of liquid rate *L* times compositional line segment $\underline{XZ}$. Similar relations apply to enthalpy line segments $\underline{H_FH}$ and $\underline{hH_F}$ and also to line segments $\underline{FV}$ and $\underline{LF}$. Also, points *V* and *L* could be selected as the pivot points instead of point *F*.

The above concepts amount to a graphical representation of the overall material, component material, and enthalpy balances around an equilibrium stage. A column solution may be developed by applying these graphical rules on an enthalpy–composition diagram.

5.3.2   BINARY *H–X* DIAGRAMS

As explained in Section 1.4, the enthalpy of a mixture is a function of temperature, pressure, and composition. These parameters determine the phase, so that in vapor–liquid equilibrium calculations, the enthalpy is also implicitly a function of the phase. For a binary mixture at constant pressure, the equilibrium vapor and liquid

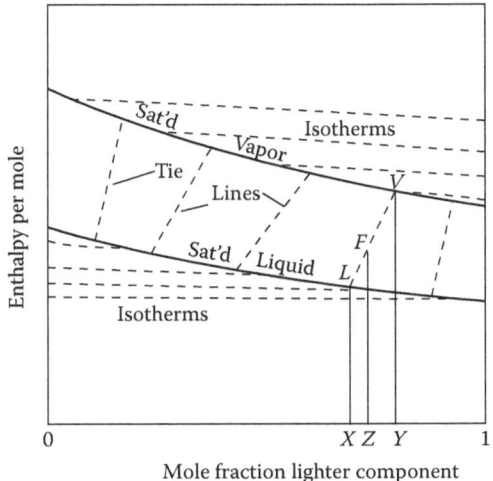

Mole fraction lighter component

FIGURE 5.10 Binary enthalpy–composition diagram.

temperatures vary with composition as represented by the dew point and bubble point curves (Figure 2.2). The enthalpy at each point may be plotted as a function of the composition, resulting in a saturated vapor enthalpy curve and a saturated liquid enthalpy curve, as shown in Figure 5.10. The composition is plotted as the mole fraction of the lighter component.

At temperatures above the dew point curve and below the bubble point curve, superheated vapor and subcooled liquid enthalpies are represented by isotherms as indicated.

Each point on the saturated liquid curve is associated with a point on the saturated vapor curve at equilibrium with it. The equilibrium vapor and liquid compositions may be obtained from Y–X or T–X diagrams. Saturated vapor and liquid points on the H–X diagram at equilibrium with each other are joined by straight lines called tie lines. The single-stage graphical representation described in Section 5.3.1 is an illustration of a tie line. If, for instance, X is known, L can be determined as a point on the saturated liquid curve with composition coordinate X. Point V must lie on the other end of the tie line on the saturated vapor curve. Point F can then be determined either from information on the relative rates of feed, liquid, and vapor or from its composition or enthalpy.

The H–X diagram may be constructed on the basis of either experimental data or computations using any of the prediction methods such as equations of state (Chapter 1).

5.3.3 Solving Distillation Columns on the H–X Diagram

Binary distillation columns are solved on the H–X diagram by graphically implementing the appropriate material and enthalpy balances on an H–X diagram prepared for the given binary at a given pressure. The detailed solution steps somewhat depend on the way the column is specified.

A column having a completely defined single feed, an overhead product, a bottoms product, a condenser, a reboiler, and a fixed column pressure has three degrees of freedom if the number of trays is variable (Section 5.2). The column model depicted in Figure 5.3 meets these definitions and is used here to describe the method. Note that the condenser in this model is a partial condenser, so that the distillate is a saturated vapor. With three degrees of freedom, the column requires three specifications to define its operation. Let the condenser duty, q_C, the distillate composition, Y_D, and the bottoms composition, X_B, be specified.

Starting the calculations at the top of the column, an energy balance around the condenser is written as

$$V_2 H_2 = L_1 h_1 + D H_1 + q_C$$
$$= L_1 h_1 + D H_1 + D Q_C = L_1 h_1 + D(H_1 + Q_C)$$
$$= L_1 h_1 + D'$$

where D' is the difference point for streams V_2 and L_1, q_C is the condenser duty per unit time, and Q_C is the condenser duty per mole distillate: $Q_C = q_C/D$. A difference point on the enthalpy–composition diagram represents the enthalpy difference between two streams. It lies on the straight line joining the points representing these streams and extended to one side of them. Point D' represents the distillate with its enthalpy incremented by Q_C. Its coordinates are Y_D and $H_1 + Q_C$. The distillate itself is represented by point D, with coordinates Y_D and H_1. Since the distillate composition Y_D (or Y_1) and the condenser duty Q_C are known, D' can

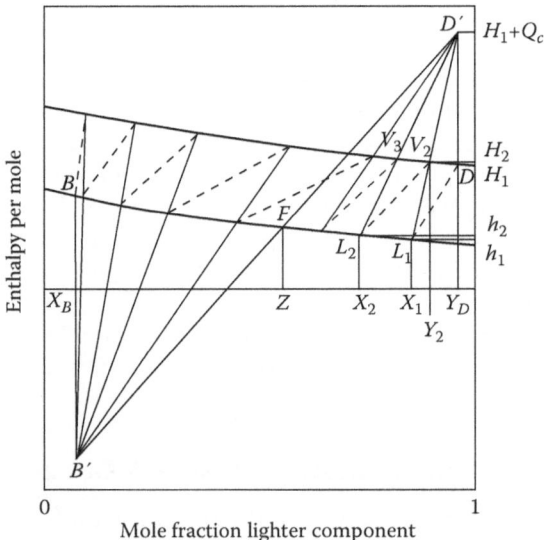

FIGURE 5.11 Solving a distillation column on the *H–X* diagram.

be located on the H–X diagram (Figure 5.11). The intersection of the vertical line through Y_D with the saturated vapor curve provides a handy way for determining H_1 at point D. A tie line drawn through this point intersects the saturated liquid curve at X_1 and h_1, the coordinates of the reflux L_1. A straight line joining D' and L_1 is the operating line corresponding to the mass and enthalpy balances on the condenser (stage 1). It represents the addition of streams D' and L_1 to generate stream V_2, the vapor leaving stage 2. By mass and enthalpy balances alone, V_2 could be represented by any point on the operating line. However, since V_2 must also lie on the saturated vapor curve, its coordinates are set by the intersection point at Y_2 and H_2.

Next, an energy balance is written around the top two stages:

$$V_3 H_3 = L_2 h_2 + D(H_1 + Q_C)$$

Stream L_2 coordinates (X_2, h_2) are determined by the intersection of the tie line through (Y_2, H_2) and the saturated liquid curve. Stream V_3 coordinates (Y_3, H_3) lie at the intersection of the operating line through D' and L_2 and the saturated vapor curve.

This procedure is repeated for all trays above the feed. For operating lines relating passing liquid and vapor streams, L_j and V_{j+1}, below the feed tray, the energy balance is performed around the top j trays that include the feed:

$$Fh_F + V_{j+1} H_{j+1} = L_j h_j + DH_D + DQ_C$$
$$= L_j h_j + D'$$

An energy balance around the entire column relates the condenser and reboiler duties:

$$Fh_F = B(h_B - Q_B) + D(H_D + Q_C)$$
$$= B' + D'$$

These two equations are combined to derive an operating line equation in terms of B':

$$V_{j+1} H_{j+1} = L_j h_j + B'$$

As this equation indicates, the operating lines for the trays below the feed are constructed using B' as a pivot (Figure 5.11). This point has coordinates X_B and $h_B - Q_B$, and can be located graphically from the overall enthalpy balance, represented by a straight line through D' and F. Point B' is the intersection of this line with the vertical line through X_B, one of the specified variables. The feed composition Z and enthalpy H_F determine the coordinates of F. For a feed at its bubble point, F falls on the saturated liquid curve.

Continuing with the procedure of defining equilibrium stages down the column, the pivot point is switched from D' to B' when the feed tray is crossed. Alternatively, the procedure could be started at the reboiler, defining stages up the column and switching from B' to D' as the feed tray is crossed.

For an existing column, or if the feed tray is given, the graphical construction of equilibrium stages on the H–X diagram proceeds as described above, using D' as a pivot point above the feed and B' below it. In a design situation or in a case where the feed tray is to be determined, the optimum feed tray would be the object. For a given number of trays, the optimum feed tray is that which minimizes the reflux required to achieve a specified separation or maximizes the separation obtainable with a fixed reflux. When designing a new column, the optimum feed tray is that which minimizes the number of trays required for a specified separation, assuming other conditions such as reflux ratio and column pressure are fixed.

The optimum feed tray location is determined on the H–X diagram as the stage or tie line that crosses the overall column enthalpy balance line joining D' and B'. This fact, which is one of the interesting features of the H–X diagram method, can be verified graphically by constructing a few stages around the presumed optimum feed tray, as shown in Figure 5.12. The dotted lines represent tie lines, and the stage just above the feed is designated as stage m. The solid lines indicate the operating lines originating at D', the pivot point used above the feed. The overall column energy balance line $\underline{D'B'}$ crosses tie line n. If this is determined to be the feed stage, the next stage, k, should be constructed with B' as the pivot, using the indicated dashed line. This results in a liquid at stage k with composition X_k. If the stage below n is used as the feed tray, then this stage, k', is constructed with D' as the pivot. The resulting liquid is of composition X_k'. Thus, with the same number of stages, moving the feed

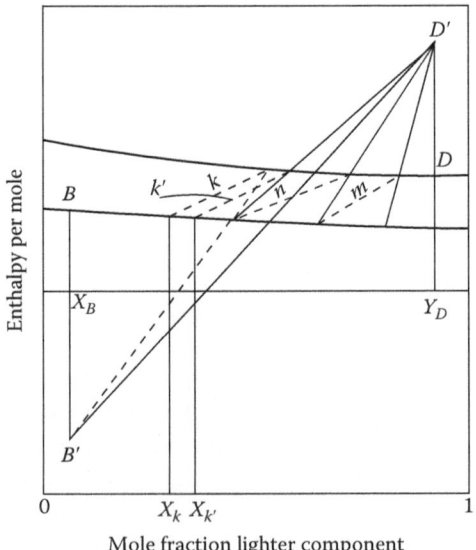

FIGURE 5.12   Effect of feed tray location on the separation.

from tray n to k' raises the liquid composition from X_k to X_k'. Since these are mole fractions of the lighter component, it can be seen that a poorer separation results from lowering the feed. A similar conclusion is reached by raising the feed; hence, the presumed optimum feed tray is indeed optimal.

Finally, the distillate and bottoms rates are calculated from the lengths of line segments $\underline{D'B'}$, $\underline{D'F}$, and $\underline{B'F}$, using the lever rule. Other tray liquid or vapor rates can be calculated from segment lengths on the appropriate operating line. For instance, the reflux rate can be calculated from segments $\underline{D'V_2}$ and $\underline{V_2L_1}$. The reboiler duty is calculated as the difference between the enthalpy coordinates of B' and B.

5.3.4 OTHER COLUMN FEATURES REPRESENTED ON THE H–X DIAGRAM

The H–X diagram offers a convenient and versatile, yet rigorous, means for preliminary evaluation of the performance of multistage separation prior to proceeding to numerical solutions. This graphical representation of columns is not hampered by the assumptions made in the Y–X method for the purpose of generating straight operating lines. In view of this, the method is applicable to a wider range of columns, including non-ideal distillation. It is still, however, essentially for binary systems and is contingent on the availability of H–X data. It lends itself to modeling a number of column features, some of which are described next.

5.3.4.1 Condenser Types

The column discussed in Section 5.3.3 is equipped with a partial condenser where part of the overhead vapor from the top tray (stage 2) is condensed and returned to the top tray as reflux. The vapor leaving the condenser is the distillate. Part of the liquid in a partial condenser can be drawn out as an additional product—a liquid distillate—alongside the vapor distillate. These two products, coming from the same stage, are at equilibrium with each other. The energy balance around a partial condenser with vapor and liquid distillates is stated as

$$V_2 H_2 = L_1 h_1 + D^V H_1 + D^L h_1 + q_C$$

where V_2 is the vapor rate from the top of the column, L_1 is the reflux rate, D^V is the vapor distillate, and D^L is the liquid distillate. Since D^V and D^L are at equilibrium, the points representing them on the H–X diagram fall on a tie line (Figure 5.13a). This line also represents a mass balance comprising the two distillate products. The total distillate D must, therefore, lie on this line. Its position on the line is determined by the relative rates of D^V and D^L. In terms of the total distillate, the energy balance around the condenser is written as

$$\begin{aligned}
V_2 H_2 &= L_1 h_1 + D H_D + q_C \\
&= L_1 h_1 + D(H_D + Q_C) \\
&= L_1 h_1 + D'
\end{aligned}$$

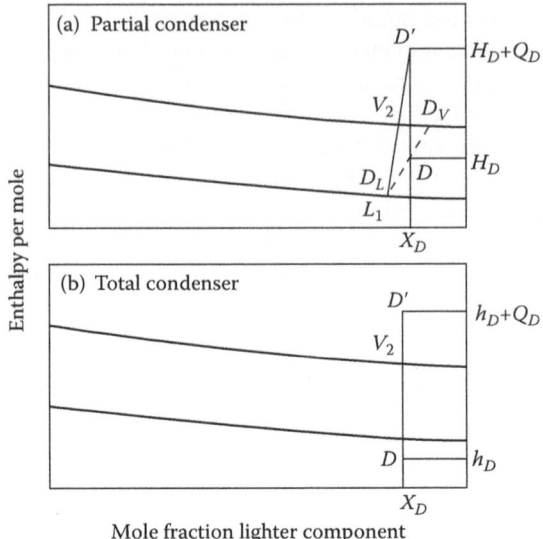

FIGURE 5.13 Modeling condensers on the H–X diagram: (a) partial condenser, (b) total condenser.

Point D' has coordinates X_D and $H_D + Q_C$. The reflux, L_1, has the same composition and enthalpy per mole as D^L and is, therefore, represented by the same point. The operating line representing the condenser is obtained by joining D' and L_1.

In a total condenser, the entire overhead vapor is condensed; part of it is refluxed, while the remainder is taken as liquid distillate. The distillate point, D, could either be on the saturated liquid curve or below it, depending on whether the overhead vapor is condensed to the bubble point or is subcooled (Figure 5.13b). From an energy balance around the condenser,

$$
\begin{aligned}
V_2 H_2 &= L_1 h_1 + D^L h_D + q_C \\
&= L_1 h_1 + D^L (h_D + Q_C) \\
&= L_1 h_1 + D'
\end{aligned}
$$

Note that L_1 and D' have the same composition, X_D, and, therefore, the operating line is vertical.

5.3.4.2 Multiple Feeds, Side Draws

The versatility of the H–X method is also demonstrated in the analysis of the performance of multiple-feed and multiple-product columns. The basic approach is unchanged: relate the passing vapor and liquid streams at the point of interest in the column by performing an energy balance, then implement the relationship graphically on the H–X diagram.

The situation may arise, for instance, where it is necessary to study the effect of a liquid side draw rate on the performance of the column below the draw tray.

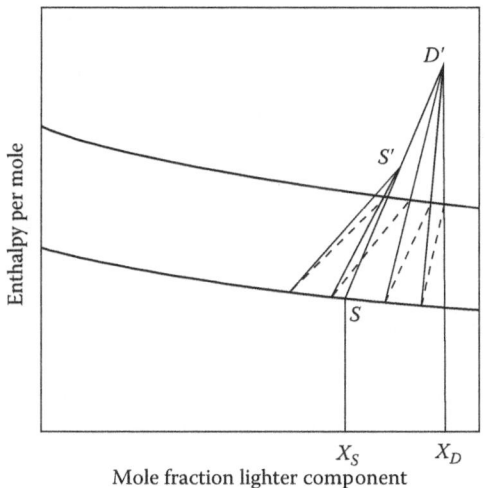

FIGURE 5.14   Modeling side draws on the H–X diagram.

Consider any tray j above which the only heat or material transfer across the column boundary occurs at the condenser, the distillate, and the side draw. An energy balance around the top part of the column down to and including tray j is written as

$$V_{j+1}H_{j+1} = L_j h_j + DH_D + q_c + Sh_S$$
$$= L_j h_j + D(H_D + Q_C) + Sh_S$$

where S is the side draw molar rate and h_S is its molar enthalpy. Graphical construction of the trays above the draw proceeds as usual, using point D' with coordinates X_D and $H_D + Q_C$ as the pivot (Figure 5.14). Starting at the tray just below the draw, the pivot point is switched to a point S' representing the combined effect of D' and S. The latter corresponds to the draw composition, X_S, and its enthalpy, h_S. Since the draw is a saturated liquid, S lies on the saturated liquid curve. Point S' lies on a straight line joining D' and S. Its position is determined by the relative rates of D' and S.

With a small draw rate, S' is not far from D' and its effect on fractionation below the draw tray is small. As the draw rate is increased, S' moves farther away from D' toward S. Comparing operating lines using S' as a pivot vs. those using D' indicates that fractionation below the draw tray is compromised as the draw rate is increased. If S is large enough, S' can reach a point where a tie line coincides with the operating line for the passing vapor and liquid streams below the draw tray. This implies that no additional fractionation can take place by adding more trays below that tray.

Analyzing the effect of a feed above a given tray is analogous to the case with a side draw. The energy balance that includes the upper part of the column through any tray j below the feed takes the form

$$V_{j+1}H_{j+1} = L_j h_j + D(H_D + Q_C) - Fh_F$$
$$= L_j h_j + D' - Fh_F$$

The point D' is the usual pivot that corresponds to the distillate and the condenser. The point F has the coordinates of the feed composition and enthalpy, (X_F, h_F). The combined effect is represented by a new pivot point F', which corresponds to the difference $D'-Fh_F$. To locate F', a straight line is drawn from F to D' and extended past D' to F' so that, by the lever rule,

$$F \times \underline{FF'} = D \times \underline{D'F'}$$

where $\underline{FF'}$ and $\underline{D'F'}$ are the line segment lengths.

5.3.4.3   Side Coolers, Heaters

Heat sources or sinks other than the condenser and reboiler may exist at different points in the column such as in the case of pumparounds. As an example, a column with a partial condenser may have a side cooler, a few trays down the column. For trays above the side cooler, the energy balance around passing vapor and liquid streams and the overhead is unchanged. Therefore, the graphical construction of trays starting at the top is not affected by the side cooler. When the side cooler is crossed, the energy balance becomes

$$
\begin{aligned}
V_{j+1}H_{j+1} &= L_jh_j + D(H_D + Q_C + Q_S) \\
&= L_jh_j + D''
\end{aligned}
$$

where Q_S is defined as q_S/D and q_S is the side cooler duty. When the side cooler tray is crossed, the pivot point is switched to D'' with coordinates X_D and $H_D + Q_C + Q_S$.

5.3.4.4   Tray Efficiency

Modifying the graphical construction of stages to account for tray efficiency can easily be incorporated in the H–X diagram method. Using the column model described in Section 5.3.3 with a partial condenser as an example, the graphical construction is started at the condenser by marking the distillate composition at Y_1 (Figure 5.15). Assuming the condenser performs as an equilibrium stage, the liquid reflux composition is X_1, as determined by the tie line. Next, Y_2, the vapor composition in stage 2, is set by the operating line. If stage 2 is at an equilibrium stage, its liquid composition, X_2^*, would be determined by the tie line passing through Y_2. In an actual tray, the liquid composition, X_2, falls short of the equilibrium composition and the difference between X_1 and X_2 is some fraction of the difference between X_1 and X_2^*. The Murphree tray liquid efficiency is defined simply as the ratio of these two differences:

$$E_L = \frac{X_1 - X_2}{X_1 - X_2^*}$$

or, in general,

$$E_L = \frac{X_j - X_{j+1}}{X_j - X_{j+1}^*}$$

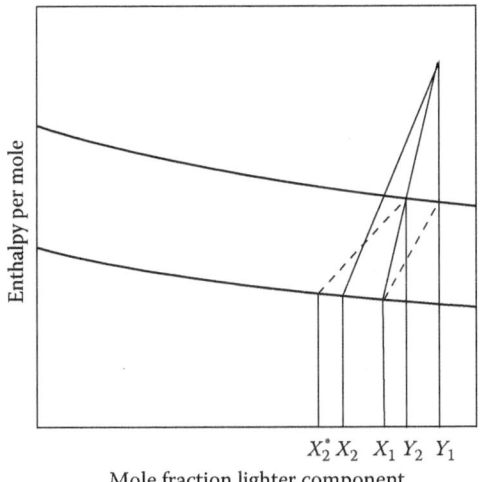

Mole fraction lighter component

FIGURE 5.15 Implementing Murphree tray efficiency on the H–X diagram.

This definition of the Murphree tray efficiency should be compared with the Murphree tray vapor efficiency, defined in Section 5.2.3. The tray efficiency may generally be estimated on the basis of past experience with similar columns or from operating data of an existing column. The tray efficiency may then be used in the graphical construction of actual trays on the H–X diagram. The procedure follows the usual steps, except that the operating lines are drawn from the pivot point to the actual tray compositions, X_{j+1}, rather than the equilibrium compositions, X_{j+1}^*. The length of each incremental segment, $X_j - X_{j+1}$, is calculated from the tray efficiency:

$$X_j - X_{j+1} = E_L(X_j - X_{j+1}^*)$$

An analogous procedure is followed if the Murphree tray vapor efficiency is known. In this case, it is more convenient to start the graphical construction at the reboiler.

NOMENCLATURE

B Bottoms stream designation or molar flow rate
D Distillate or overhead designation or molar flow rate
E Murphree tray efficiency
F Feed stream designation or molar flow rate
H Vapor molar enthalpy
h Liquid molar enthalpy
K Vapor–liquid equilibrium distribution coefficient
L_j Liquid molar flow rate leaving tray j
L_{ji} Component i molar flow rate in liquid leaving tray j
N Number of equilibrium stages in a column

P	Pressure
Q	Heat duty per mole
q	Heat duty per unit time
q	Feed or draw thermal condition parameter
S	Side draw designation or molar flow rate
T	Temperature
V_j	Vapor molar flow rate leaving tray j
V_{ji}	Component i molar flow rate in vapor leaving tray j
X	Mole fraction in the liquid
Y	Mole fraction in the vapor
α	Relative volatility
λ	Molar heat of vaporization

SUBSCRIPTS

B	Bottoms or reboiler designation
C	Condenser designation
F	Feed conditions designation
f	Feed tray designation
i	Component designation
j	Stage designation
r	Rectifying section designation
s	Stripping section designation

SUPERSCRIPTS

L	Liquid phase
V	Vapor phase

REFERENCES

Hanson, D. N., J. H. Duffin, and G. F. Somerville, *Computation of Multistage Separation Processes*, New York, Reinhold Publishing Corporation, 1962.

McCabe, W. L. and E. W. Thiele, *Ind. Eng. Chem.*, 17, 605, 1925.

Murphree, E. V., *Ind. Eng. Chem.*, 17, 747, 1925.

Ponchon, M., *Tech. Moderne*, 13, pp. 20, 55, 1921.

Savarit, R., *Arts et Metiers*, 65, 1922.

6 Binary Distillation
Applications

The principles of binary distillation presented in Chapter 5 are applied in this chapter to study column performance under different operating conditions. The objective is to identify relevant column parameters and determine performance trends for a variety of applications.

As a learning tool, the binary model is useful for qualitatively studying the characteristics of multistage separation. The model is used in this chapter to answer such questions as what effect the reflux ratio or product rate or number of trays has on separation or what is the minimum number of trays or minimum reflux ratio required to achieve a given separation, or over what ranges column performance specifications are feasible.

Although rigorous computer programs can provide design data for a particular separation process, understanding the process characteristics is necessary in order to evaluate its performance under varying operating conditions. Reams of multiple cases of computer output may cover all the expected operating ranges, but a visual representation of the process can be invaluable. Thus, the graphical methods can be applied to get the most out of computer simulation results.

Moreover, the availability of computer programs provides an added impetus to the use of graphical methods, because these programs generally provide the data required for graphical representation.

Multi-component separations can also be approximated with the graphical methods by selecting appropriate key components to define the separation.

Using the graphical techniques developed in Chapter 5, Section 6.1 examines the effect on column performance of each parameter separately, and Section 6.2 considers the interactions of different parameters. Section 6.3 presents example applications where graphical methods are used in conjunction with simulation results to check the design data.

6.1 PARAMETERS AFFECTING COLUMN PERFORMANCE

The binary model, conveniently represented on the $Y–X$ or $H–X$ diagrams, demonstrates the effect on column performance of the reflux ratio, product rate, number of trays, feed location, and thermal conditions. In the cases that follow in this section, the feed is assumed of fixed flow rate and composition. The effect of column pressure, which influences column performance through its effect on the equilibrium curve, is not considered here and is held constant.

6.1.1 EFFECT OF REFLUX RATIO AND PRODUCT RATES

The reflux ratio is a measure of the relative flows of liquid and vapor counter-flowing in the column. It is commonly defined as $R = L_1/D$, where L_1 is the reflux rate or rate of liquid flowing from the condenser back to the column, and D is the distillate rate. The operating line of the column rectifying section given by Equation 5.13 has a slope of L_r/V_r, where L_r and V_r are the liquid and vapor flow rates in the rectifying section. Since the molar flow is assumed constant in each column section, $L_1 = L_r$ and $R = L_r/D$. Also, writing a material balance around the condenser, $V_r = D + L_r$, the operating line slope may be expressed in terms of the reflux ratio:

$$\frac{L_r}{V_r} = \frac{L_r}{D + L_r} = \frac{R}{1 + R} \tag{6.1}$$

As R increases, so does L_r/V_r, and as R goes from zero to infinity, L_r/V_r goes from 0 to 1.

The effect of reflux ratio on the separation may be examined by looking at the McCabe–Thiele Y–X diagram. If the total number of stages and feed location are fixed, the number of stages in the rectifying section is also fixed. Line r_1 in Figure 6.1 represents the rectifying section operating line at some reflux ratio R_1. The corresponding number of stages in the rectifying section is four. If the reflux ratio is increased, the slope of r_1 increases. In order to maintain the same number of stages in the rectifying section, the new operating line, r_2, must intersect the diagonal at a higher concentration of the light component. The overhead product is therefore richer in the lighter component, and a better separation is achieved at a higher reflux ratio when all other parameters are held constant.

The number of stages in the stripping section is also constant. Lines s_1 and s_2 in Figure 6.1 show that, as the reflux ratio increases, the slope of the stripping section operating line must decrease in order to maintain a fixed number of stages. With a lower L_s/V_s slope, the stripping section operating line intersects the diagonal at a lower concentration of the lighter component. Consequently, the bottoms is richer in the heavier component as the reflux ratio is increased, resulting in better separation.

As the reflux ratio approaches zero, L_r/V_r also approaches zero and the rectifying section operating line becomes a horizontal line intersecting the equilibrium curve at its intersection with the q-line (Figure 6.1). The stripping section operating line becomes vertical since any other slope would imply an infinite number of stripping stages. At zero reflux, the number of stages is indeterminate and has no effect on the separation. A single stage would result in the same separation as any number of stages. This conclusion was arrived at in Section 3.2.3 on the basis of a two-stage model. The overhead and bottoms compositions at zero reflux correspond to the Y and X coordinates of the intersection of the q-line with the equilibrium curve.

At total reflux, R goes to infinity and $L_r/V_r = 1$ so that both rectifying and stripping section operating lines coincide with the diagonal. The distance between the equilibrium curve and the operating lines is largest at total reflux, resulting in the largest

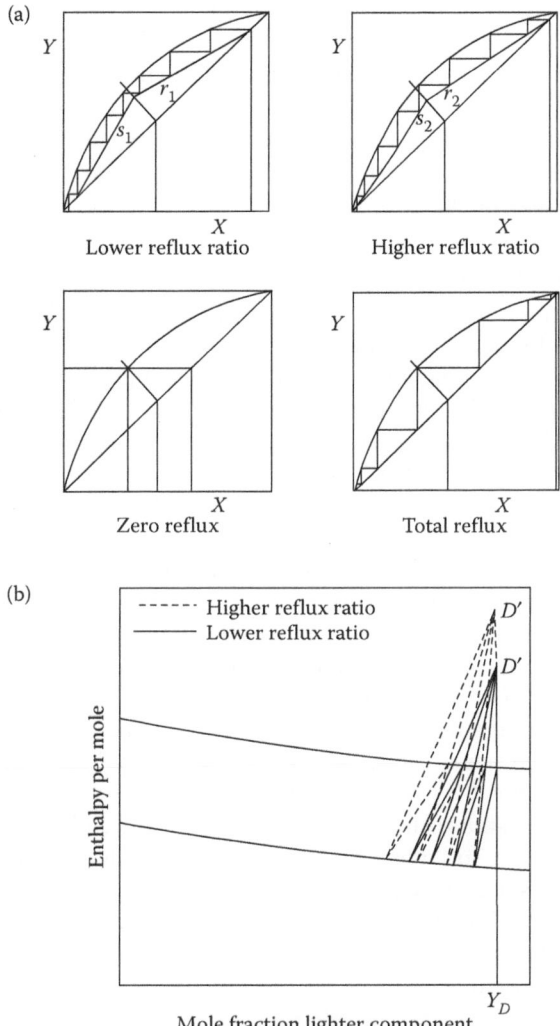

FIGURE 6.1 (a) Effect of reflux ratio on column performance—*Y–X* diagram. (b) Effect of reflux ratio on column performance—*H–X* diagram.

size steps between equilibrium stages. Hence, maximum separation is obtained at total reflux for a given number of stages. This is another conclusion arrived at earlier in Section 3.2.3 using the simplified two-stage model. Graphically, the separation obtainable at total reflux with a given number of stages may be determined on the *Y–X* diagram by stepping off the number of stages in each section, starting at the intersection of the *q*-line with the equilibrium curve (Figure 6.1).

The effect of reflux ratio on column performance may also be inspected using the *H–X* diagram. Given a fixed distillate rate, a higher reflux ratio implies larger vapor and liquid traffic in the rectifying section, requiring a higher vapor condensation rate in the condenser. A higher condenser duty, q_C (or $Q_C = q_C/D$), results in a higher pivot point D' on the H–X diagram and greater slopes for the operating lines. This causes the stepping off of stages to take larger strides, thereby enhancing the separation (Figure 6.1a).

At total reflux L/D and Q_C go to infinity, causing the operating lines to become vertical. These conditions determine the minimum number of stages required to achieve a given separation.

6.1.1.1 Product Rates

The complete definition of column performance when the number of stages and feed location are fixed requires two specifications (Section 5.2.2). In studying the effect of reflux ratio on separation, other parameters, including product rates, are assumed constant. In Y–X graphical representation, a fixed product rate is implied by fixing the q-line slope. This slope depends on the feed thermal conditions at feed tray temperature and pressure (Section 5.2.2). The column pressure is fixed, but its temperature depends on the product rates. Therefore, in order to maintain a fixed product rate, the q-line slope must be held constant. The actual product rates corresponding to a given q-line slope are calculated by material balance once the product compositions have been determined.

The qualitative effect of product rates on product purities may be examined graphically by comparing Y–X diagrams with a vertical q-line and a horizontal q-line (Figure 6.2). For a feed with given thermal conditions, the q-line is vertical if the feed tray temperature is such that the feed is saturated liquid at this temperature. To convert the same feed to saturated vapor with a horizontal q-line requires a higher temperature feed tray. A vertical q-line column thus corresponds to a generally colder column with a lower overhead rate and a higher bottoms rate. The converse is true for a horizontal q-line column. In the former case, the rectifying and stripping steps are shifted to the right of the Y–X diagram, resulting in higher purity overhead and lower purity bottoms. The shift is to the left for a hotter column with high overhead and low bottoms rates, resulting in lower purity overhead and higher purity bottoms.

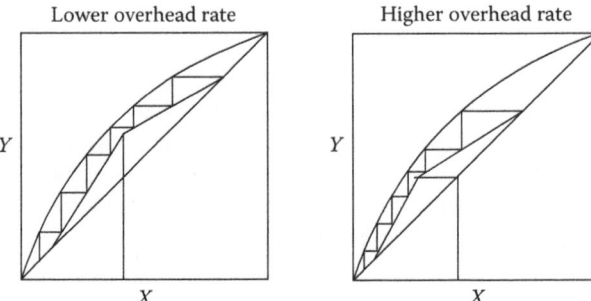

FIGURE 6.2   Effect of product rates on column performance.

EXAMPLE 6.1: BENZENE–TOLUENE COLUMN—REFLUX RATIO SPECIFIED

A feed stream made up of 40% mole benzene and 60% mole toluene is to be separated into benzene-rich and toluene-rich products using a distillation column. The column has ten equilibrium stages including a partial condenser and a partial reboiler and is operated at 172 kPa. The feed stream, with a flow rate of 100 kmol/h, is at its bubble point at 172 kPa and is placed in the fourth stage from the top. It is required to determine the compositions of the two products at different reflux ratios. Vapor–liquid equilibrium data for the benzene–toluene system are provided in Table 6.1 at 172 kPa.

Solution

The McCabe–Thiele graphical method is applied using the Y–X diagram. The equilibrium curve is first plotted from the given data (Figure 6.3). A temperature coordinate is also included in the diagram (nonlinear) to determine the condenser, reboiler, and tray temperatures.

The next step is to plot the q-line, which intersects the diagonal at the feed composition. The slope of the q-line is computed on the basis of the feed thermal conditions at the feed tray temperature, which is dependent on the product rates. It is assumed that the product rate (distillate or bottoms) is such that the feed is almost a saturated liquid at feed tray conditions so that the q-line is close to vertical.

The operating lines are drawn next. Assuming a reflux ratio $R = 1$, the slope of the rectifying section operating line is $L_r/V_r = 0.5$, Equation (6.1). The rectifying section operating line is determined by trial and error by drawing lines with a slope of 0.5 between the q-line and the diagonal. The correct line is that which results in four rectifying stages, which is the specified number of theoretical stages above the feed tray (Figure 6.3).

The stripping section operating line must pass through the intersection of the q-line with the rectifying section operating line. The stripping section operating line slope is varied until the number of stages in the stripping section is six.

TABLE 6.1
Vapor–Liquid Equilibrium Data for the Benzene–Toluene Binary at 172 kPa

X (Benzene)	Y (Benzene)	T (°C)
0.00	0.00	131
0.10	0.19	126
0.20	0.35	122
0.30	0.49	118
0.40	0.60	115
0.50	0.70	112
0.60	0.77	108
0.70	0.84	106
0.80	0.90	103
0.90	0.95	101
1.00	1.00	98

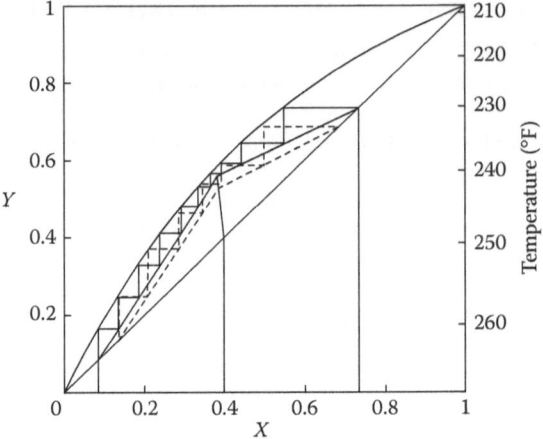

FIGURE 6.3 Y–X diagram for benzene–toluene.

TABLE 6.2
Results of Example 6.1

Reflux Ratio	0	1	10	∞
Distillate composition Y (benzene)	0.540	0.735	0.796	0.950
Bottoms composition X (benzene)	0.340	0.070	0.004	0.002
Distillate rate (kmol/h)	30	50	50	40
Condenser temperature (°C)	117	110	108	101
Reboiler temperature (°C)	117	127	130	131

The compositions of the products are read from the intersections of the operating lines with the diagonal. The results are tabulated in Table 6.2 for different reflux ratios. Also shown are the condenser and reboiler temperatures as read from Figure 6.3, and the product flow rates, calculated from a component material balance:

$$FX_F = DY_D + (F-D)X_B$$

The product rates at high and low reflux ratios are sensitive to the q-line slope, which varies slightly depending on feed tray conditions. As a result, certain variations in the product rates are observed.

6.1.2   Effect of Number of Stages and Feed Location

The effect of the number of stages and feed location on separation may be readily inspected using the binary Y–X diagram. Other parameters, namely the reflux ratio and the product rates, are assumed constant. The condition of constant product rate is implied by fixing the q-line slope.

The equilibrium curve and q-line are first drawn as usual on a Y–X diagram. The slope of the rectifying section operating line is known from the specified reflux ratio. For a given number of stages in the rectifying section, the correct operating line is found by trial and error aimed at matching the given number of stages while maintaining a constant slope. As the operating line gets closer to the equilibrium line, it intersects the diagonal at a higher concentration of the light component, and the number of rectifying stages increases.

The stripping section operating line is drawn by joining the q-line intersection with the rectifying section operating line to the diagonal. The correct stripping section operating line slope is one that produces the given number of stripping stages. Stripping section operating lines closer to the equilibrium curve imply more stripping stages and lower concentrations of the light component in the bottoms.

In conclusion, with other parameters held constant, more stages in a given section result in higher purity of the corresponding product, and vice versa. Consequently, if the total number of stages is held constant, but the feed location is raised, higher purity bottoms and lower purity distillate products are expected. The reverse is true if the feed location is lowered.

Using the H–X diagram, if the feed tray is not fixed, its optimum location can be determined as discussed in Section 5.3.3.

EXAMPLE 6.2: BENZENE–TOLUENE COLUMN—NUMBER OF STAGES AND FEED LOCATION SPECIFIED

The benzene–toluene mixture of Example 6.1 is to be separated into approximately 50 kmol/h distillate and 50 kmol/h bottoms products. The column is to be operated at a reflux ratio of one. Determine the effect of the number of stages and feed location on the purity of the products.

SOLUTION

Based on the results of Example 6.1, a distillate rate of about 50 kmol/h corresponds to a near vertical q-line, that is, a bubble point feed. The equilibrium curve is drawn from data given in Table 6.1.

The operating lines are drawn as described above. Figure 6.3 illustrates the effect of varying the number of stages in each section, and Table 6.3 summarizes the results. The first two cases in Table 6.3 represent columns with the same number of stripping stages (four) but with different numbers of rectifying stages (three and four). The results show that the bottoms concentrations are about the same for the two cases, but the distillate has a higher purity when more rectifying

TABLE 6.3
Results of Example 6.2: Effect of Number of Stages and Feed Location

Number of Stages	7	8	10	12
Feed location (from top)	3	4	4	4
Distillate composition Y (benzene)	0.690	0.759	0.761	0.761
Bottoms composition X (benzene)	0.0681	0.0704	0.0394	0.0185

stages are available. The columns corresponding to the last three cases have the same number of rectifying stages (four) but increasing numbers of stripping stages (four, six, and eight). With more stripping stages, the distillate compositions do not change much, but the bottoms purity goes up.

6.1.3 NUMBER OF STAGES VERSUS REFLUX RATIO

If the separation is specified, one could calculate the number of stages required with a given reflux ratio or the reflux ratio required with a given number of stages. Higher reflux ratios are required with fewer trays, and lower reflux ratios are required with columns having more trays. Using multiple calculations, a curve could be generated of the required reflux ratio vs. the number of trays available for a specified separation. Each point on this curve would correspond to a combination of the number of trays and reflux ratio that satisfies the specified separation. In a design situation, the selection of a particular combination is based on, among other factors, economic considerations of capital expenditure vs. operating cost.

The separation is specified by fixing the distillate and bottoms compositions, Y_D and X_B. The product compositions determine the product rates, which are calculated by material balance. The product compositions are used to locate one end point of each operating line on the Y–X diagram. The feed conditions determine the q-line (Figure 6.4).

If the reflux ratio is specified, then the slope of the rectifying section operating line is known. Its intersection with the q-line defines another point on the stripping section operating line. The equilibrium stages are then stepped off as usual between the operating lines and the equilibrium curve. The optimum feed location is the stage that lies on both operating lines, that is, the stage through which the q-line passes.

If the reflux ratio is increased, the slope of the rectifying section operating line increases, the operating lines move farther away from the equilibrium curve, and

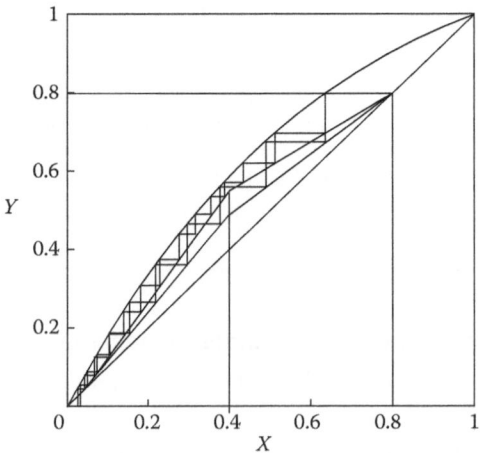

FIGURE 6.4 Product purities specified.

the required number of stages decreases. At total reflux $L_r/V_r = 1$, and the operating lines coincide with the diagonal. Total reflux corresponds to the minimum number of stages required to achieve a specified separation. With fewer stages this separation is impossible regardless of the reflux ratio.

As the reflux ratio is lowered, the L_r/V_r slope decreases and the operating lines move closer to the equilibrium curve, thereby requiring more stages for the same separation. If the reflux ratio is lowered to a point where one of the operating lines touches the equilibrium curve, the number of stages becomes infinite. This corresponds to the minimum reflux ratio, below which the specified separation cannot be met with any number of stages. The point where the operating line touches the equilibrium curve, or the pinch point, can occur in the vicinity of the feed stage at the intersection of the operating lines, or anywhere else, depending on the shape of the equilibrium curve.

The H–X diagram (Figure 6.1b) may also be used to see these effects. As the reflux ratio is reduced, the pivot point D' moves down toward the saturated vapor enthalpy curve. A point is reached at which the operating line originating at D' coincides with one of the H–X diagram tie lines. When this happens, it becomes impossible to change the composition from stage to stage, theoretically requiring an infinite number of stages to achieve the specified separation. This position of D' defines the minimum reflux ratio.

If the number of stages is fixed, the problem is to calculate the reflux ratio required to meet the specified separation. In this situation, two cases may arise—the feed location is fixed or is to be determined.

If the feed location is fixed, the number of rectifying stages is known and the reflux ratio is calculated by adjusting the L_r/V_r slope until the number of stages is matched. The intersection of the rectifying section operating line with the q-line is then joined to the bottoms composition projection on the diagonal. The resulting number of stripping stages may or may not match the existing number, meaning that the feed location may or may not be optimum. If the required number of stripping stages exceeds the actual, the bottoms specified composition cannot be met. If the required number is less than the actual, then too many unnecessary stages exist below the feed.

In the case where the feed location is free to move up or down the column, the problem becomes to determine the optimum feed location as well as the reflux ratio. The calculation proceeds as above, starting with an assumed number of rectifying stages. If the resulting number of stripping stages brings the total stages to match the existing number, a solution has been reached. Otherwise, a new trial is started with an altered number of rectifying stages. The procedure is repeated until a match is obtained for the number of stages.

The H–X diagram technique for determining the number of stages and optimum feed location for a specified separation was described in Section 5.3.3. It was implied that the reflux ratio was also specified since the condenser and reboiler duties were given. If the number of stages is specified, but the condenser and reboiler duties are allowed to vary, it becomes necessary to repeat the above procedure with different pivot points (corresponding to different condenser and reboiler duties) until the number of stages is matched.

EXAMPLE 6.3: BENZENE–TOLUENE COLUMN—SEPARATION SPECIFIED

It is required to separate the benzene–toluene mixture of Example 6.1 into a benzene-rich distillate with 0.80 mole fraction benzene and a toluene-rich bottoms
with 0.05 mole fraction benzene. The separation is to be made using a distillation
column with 15 theoretical stages that include a partial condenser and a partial
reboiler. Calculate the reflux ratio required to achieve the specified separation and
determine the optimum feed location. What effect would lowering the number of
stages to ten have on the reflux ratio and the optimum feed location?

SOLUTION

The product rates consistent with the specified separation are calculated by a
material balance on the benzene:

$$FX_F = DY_D + BX_B$$

$$(100)(0.40) = D(0.80) + (100 - D)(0.05)$$

$$D = 46.7 \text{ kmol/h}$$

Using the Y–X diagram, the q-line is assumed vertical, corresponding to a saturated liquid feed. The procedure outlined in the above section amounts to sliding
the intersection of the operating lines up and down the q-line until the number
of stepped stages equals 15. The reflux ratio is then calculated from the slope
of the rectifying section operating line. From Figure 6.4, $L_r/V_r = 0.55$, and from
Equation 6.1,

$$R = \frac{L_r/V_r}{1 - L_r/V_r} = 1.22$$

According to the diagram, the feed location is between the fourth and fifth stages
from the top. This feed location is optimum because it requires the least amount
of reflux for the given total number of stages. If the feed has to be introduced at a
higher stage, a smaller number of rectifying stages would be available, thus requiring a higher reflux ratio to achieve the same distillate purity. The stripping section
would have more stages than necessary. If the feed is sent to a lower stage, fewer
stripping stages would be available and a higher boilup rate would be required to
achieve the same bottoms purity. The higher boilup rate results in a higher reflux
ratio and more fractionation than necessary in the rectifying section.

Figure 6.4 shows that with ten stages the optimum feed location is the third
stage and the L_r/V_r ratio is 0.66, corresponding to a reflux ratio of 1.94.

6.2   PARAMETER INTERACTIONS IN FIXED
CONFIGURATION COLUMNS

A column with a fixed number of stages and feed location may be operated over
certain ranges of conditions. It may be operated to satisfy given separation requirements or product rates or component recoveries. The column tray size and type,

and condenser and reboiler capacities, set practical limits on throughput, reflux ratio, and condenser and reboiler duties and temperatures. Thermodynamics and material and energy balances also place constraints on the operable ranges. From a rating standpoint for studying the performance of existing columns or from the standpoint of evaluating column design options to meet preset process specifications, the engineer must consider the ranges of different column variables and their interactions. Each variable has certain ranges of allowable values and, when one variable assumes a certain value, the others become restricted to varying degrees in their allowable ranges.

6.2.1 Column Operable Ranges

A column operator can control the column performance by manipulating the reboiler and condenser duties. Consider starting up a column with a mixed-phase feed introduced at some intermediate tray between the condenser and reboiler. With no condenser or reboiler duties, the liquid flows down the column and out as bottoms, and the vapor flows up the column and out as overhead. The column thus acts as a flash drum.

The reboiler is now started, the bottoms is partially reboiled, and the vapor is sent back to the bottom of the column. The ascending vapor interacts with the descending liquid and, if sufficient vapor is generated at the reboiler, a net amount of vapor reaches the feed stage and joins the vapor portion of the feed. The total vapor flows out the top of the column as overhead product. If the condenser is not put in service, the column operates as a stripper, that is, fractionation takes place on the trays between the feed and the bottoms but not above the feed. By increasing the reboiler duty, one can strip the bottoms of more of the lighter components, thus generating a bottoms product with a higher concentration of the heavy components. The overhead purity cannot be improved as long as the condenser is not in operation.

If the reboiler is turned off and the condenser is activated, the column operates as a rectifier. The condensed liquid is returned to the top of the column as reflux and fractionation takes place on the trays between the overhead and the feed, with no fractionation occurring below the feed. In this mode of operation, higher reflux results in higher concentration of the lighter components in the distillate, but the bottoms cannot be enriched in the heavier components as long as the reboiler is inactive.

The stripping and rectifying modes of operating the column represent limiting conditions of the operable range. Between these limits, the column has both a stripping section and a rectifying section. At reflux ratios approaching zero with finite boilup ratios, the column acts primarily as a stripper, and at boilup ratios approaching zero with finite reflux ratios, it performs mostly as a rectifier.

The column operable range is determined in part by the requirement that no tray be allowed to "dry up," that is, liquid and vapor must exist on each tray to maintain phase equilibrium. This range may be defined by the limits over which the condenser and reboiler duties may vary. As such, the condenser and reboiler duties are considered the two independent variables required to define the column performance (Sections 3.2.3 and 5.2.1). Alternatively, other pairs of variables may be chosen as the independent variables defining the column performance and each set can vary within certain feasible limits. The following sections look at different possible pairs of independent column variables.

6.2.2 FEASIBLE RANGES OF PRODUCT RATES AND REFLUX RATIOS

Consider the bottoms product rate and the reflux ratio as the independent variables. The feed to the column is assumed of mixed-phase vapor and liquid. The bottoms product rate could vary from zero to the feed flow rate. At a bottoms rate approaching zero, the reboiler duty must be sufficient to vaporize all the liquid flowing down the column, leaving an infinitesimal amount of bottoms. The vaporized liquid joins the feed vapor, and the combined vapor flows up the column to the condenser. A minimum amount of reflux is required to maintain liquid and vapor phases on all the trays. The conditions of zero bottoms rate and minimum reflux ratio are represented for the benzene–toluene system of Example 6.1 by point A in Figure 6.5. The reboiler and condenser duties may be increased simultaneously, raising the reflux ratio while keeping a fixed bottoms rate. Thus, the column is operable at any point above A. The overhead composition does not change with reflux ratio since practically all the feed leaves as overhead product. The bottoms product, although approaching zero flow rate, becomes more concentrated in the heavy components as the reflux ratio is increased.

With higher bottoms flow rates, lower reboiler duties are required at minimum reflux since less liquid must be vaporized. For any given bottoms rate, a minimum amount of reflux is required to maintain both liquid and vapor phases on all the trays. Curve AB in Figure 6.5 is a plot of the minimum reflux ratio vs. bottoms rate. This minimum is not to be confused with the minimum reflux ratio required to bring about a specified separation with a given number of stages. The minimum reflux curve in Figure 6.5 represents reflux ratios below which either the liquid or the vapor dries up on some of the trays. The curve characteristics depend mainly on the feed

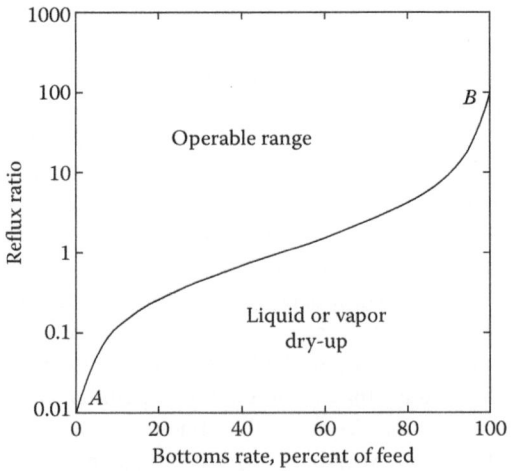

FIGURE 6.5 Column operable range.

thermal conditions. The upper limit of the reflux ratio is determined by practical considerations such as duty limitations and column diameter.

Figure 6.5 shows that a column can operate over wide ranges of product rate and reflux ratio specifications. Expressed mathematically, this fact results from a high degree of independence between the two variables. This independence may also be verified by inspecting the Y–X diagram. The effect of changing a product rate is equivalent to varying the q-line slope (Section 6.1.1). Changing the reflux ratio is expressed graphically by varying the operating lines' slopes. The q-line and the operating lines may be moved quite independently of each other, subject only to maintaining a fixed number of stages.

6.2.3 FEASIBLE RANGES OF DISTILLATE AND BOTTOMS COMPOSITIONS

If the overhead and bottoms compositions (Y_D and X_B) are specified, the product rates are determined by material balance. The specified separation also determines the reflux ratio. (The number of stages and feed location are assumed fixed.)

The product compositions can vary only over portions of the entire composition range. For instance, the following inequalities must be observed: $Y_D > X_F$ and $X_B < X_F$, where X_F is the feed composition. Also, the upper limit of Y_D and the lower limit of X_B are determined by the existing number of stages operating at or close to total reflux. At the other extreme, the reflux ratio must be high enough to prevent column dry-up. Figure 6.6 charts the feasible ranges of Y_D and X_B combinations for the benzene–toluene binary.

The relatively small operable range of composition specifications results from the lesser degree of independence between Y_D and X_B. The higher the dependence between the specified variables, the smaller the operable range. Examined on a Y–X equilibrium diagram, the dependence between Y_D and X_B is underscored by the

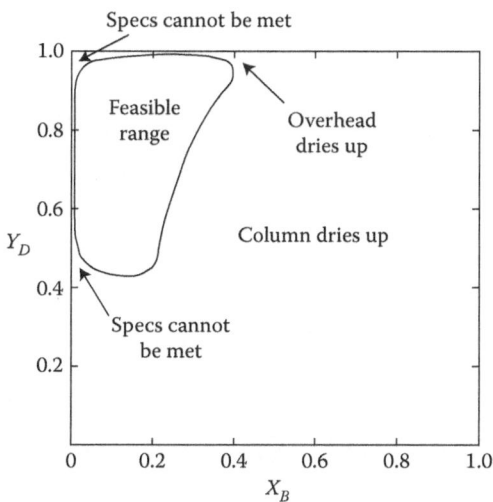

FIGURE 6.6   Ranges of feasible compositions.

limited flexibility in finding a solution if Y_D and X_B, as well as the number of stages and feed location, are fixed. In this situation, the q-line slope offers the only flexibility in solving the problem.

6.2.4 FEASIBLE RANGES OF DISTILLATE COMPOSITION AND REFLUX RATIO

With a fixed number of stages and feed location, if the reflux ratio is specified, a distillate composition specification may be satisfied within a limited range by varying the product rates. On a Y–X equilibrium diagram, this is equivalent to fixing the rectifying section operating line slope and intersection with the diagonal and varying the q-line slope. The specified variables (Y_D and reflux ratio) are, therefore, not highly independent; they define a relatively small feasible region. This is illustrated for the benzene–toluene system in Figure 6.7.

6.2.5 FEASIBLE RANGES OF DISTILLATE COMPOSITION AND BOTTOMS RATE

If the distillation bottoms product rate is fixed (resulting also in a fixed distillate rate), a distillate composition specification may be satisfied within a certain range by varying the reflux ratio. Higher distillate purities may be obtained by increasing the reflux ratio, with maximum Y_D obtainable at total reflux. At the lowest possible reflux required to keep all column trays from drying up, a low Y_D is expected, which at the limit would approach the vapor composition obtainable from a single-stage flash.

Operating the column at a specified product rate and composition is equivalent to fixing the q-line slope and the Y_D-diagonal intersection of the rectifying section operating line on a Y–X equilibrium diagram. With a given number of rectifying trays, there is a limited degree of independence between the product rate and the distillate composition and, therefore, a limited feasible range can be defined in terms of these two variables. This is shown in Figure 6.8 for the benzene–toluene binary.

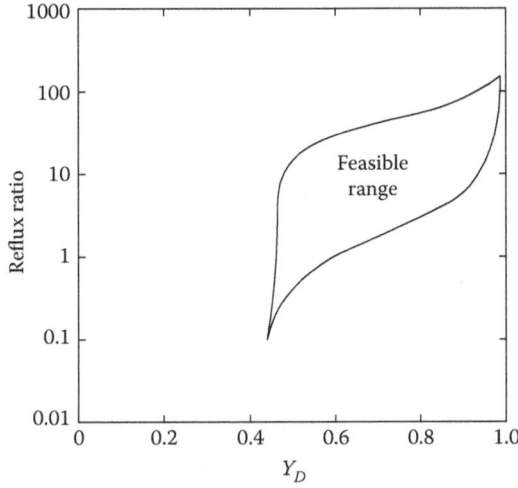

FIGURE 6.7 Feasible compositions and reflux ratios.

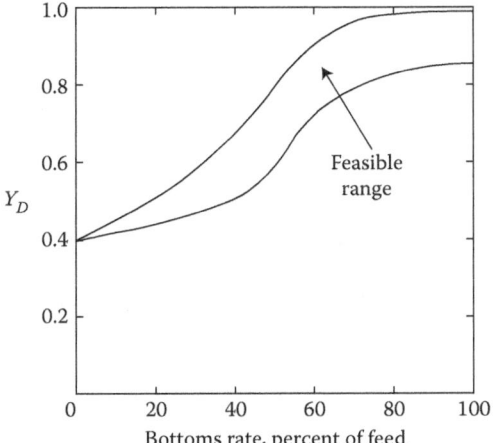

FIGURE 6.8 Feasible compositions and product rates.

6.3 DESIGN STRATEGIES GUIDED BY GRAPHICAL REPRESENTATION

The design of a separation process must take into account anticipated variations in operating conditions. The process is usually expected to operate not only at design conditions and, therefore, flexibility must be built into the design to ensure stable performance over the expected parameter ranges. The basic design typically relies on a rigorous computer simulation of the process. In conjunction with this, the graphical methods can help understand the functioning of the process and provide a visual backup to check reasonableness and flexibility of the design.

The following applications illustrate some of the factors considered in the validation of a column design using graphical representation.

EXAMPLE 6.4: METHANOL DEHYDRATION

A distillation column is designed for separating methanol from a methanol–water mixture containing 60% mole methanol. The overhead methanol product must have a purity of 98% mole, and the methanol in the bottoms should not exceed a concentration of 4% mole. The column operating pressure is 100 kPa, with the feed entering the column at its bubble point. It is proposed to design the column at a reflux ratio of 1.0, or an L_r/V_r ratio of 0.5.

Using the Y–X Diagram: For this application, a preliminary design is done graphically on a Y–X diagram. The equilibrium properties are provided by the curve in Figure 6.9. The graphical representation of the column is done in the usual manner for the known values of X_D, X_B, and feed definition, and for a given column pressure. The q-line is vertical since the feed is a saturated liquid. The rectifying section operating line is drawn with a slope of $L_r/V_r = 0.5$. The graphical design indicates that 13 theoretical stages are required to meet the desired specifications at a reflux ratio of 1 and that the optimum feed location is the tenth stage from the top.

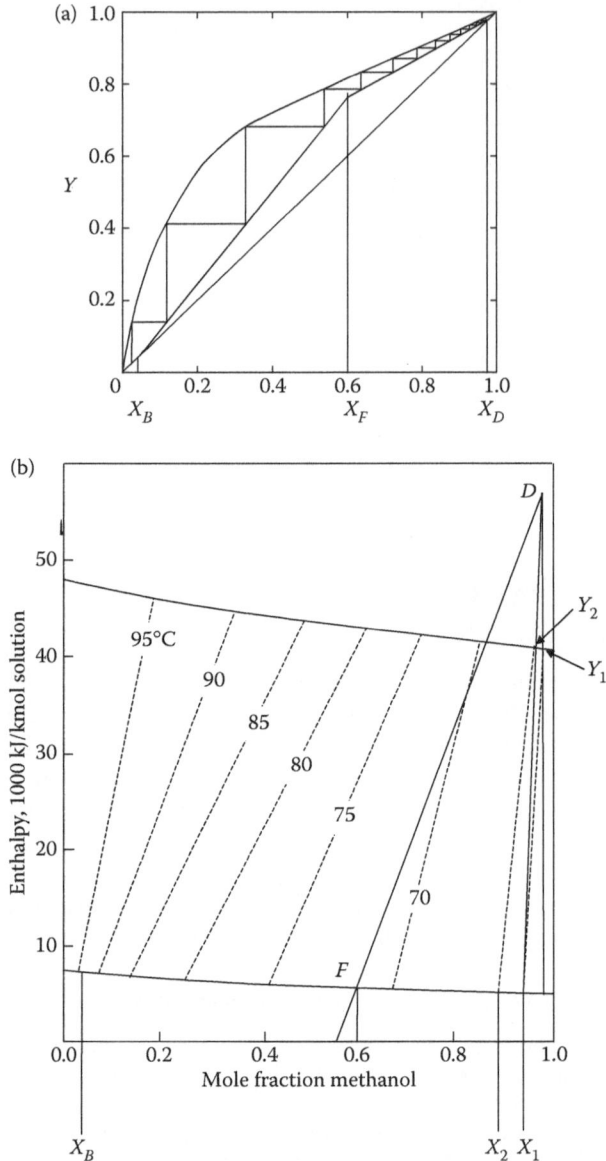

FIGURE 6.9 (a) Methanol dehydration. (b) Solving the methanol–water column on the H–X diagram.

The next step in the column design is to decide on a column diameter that will handle the expected reflux. The reflux, in turn, depends on the number of trays. With an assumed overall tray efficiency of 65%, the actual number of trays would be $13/0.65 = 20$, including condenser and reboiler. A column could then be designed based on this reflux ratio and number of trays, although economic considerations might suggest alternative trays vs. reflux combinations. Along with economic considerations, an evaluation of the design from a process perspective is in order before the actual construction is begun. An examination of the Y–X diagram (Figure 6.9) can reveal potential operating problems that might require re-evaluating the design.

Figure 6.9 shows that the rectifying section operating line is close to the equilibrium curve, while the stripping section operating line is considerably farther away. It is harder to produce a given purity methanol in the overhead than a comparable purity water in the bottoms, and more rectifying trays are needed than stripping trays.

The proximity of the rectifying section operating line to the equilibrium curve gives the column the tendency to "pinch out" in that region. That is, any slight change in some of the operating conditions could cause the operating line to move even closer to the equilibrium curve, requiring many more trays in that region. Since the number of trays is fixed, the changes in operating conditions could prevent the column from making the required separation at the design reflux. Increasing the reflux ratio may not be possible with the existing column diameter. A design that is pinched or that is close to the minimum reflux ratio is sensitive to small changes in the reflux. The column is not stable under these conditions because a slight drop in the reflux can cause a large reduction in the separation. Because of this, it is good practice to avoid designing columns close to the minimum reflux ratio. The methanol column should, therefore, be redesigned with a higher reflux ratio, such as 1.5.

The changes in operating conditions that can cause problems may be identified by referring again to Figure 6.9. If, for instance, the feed is partially vaporized, the q-line tilts to the left, requiring either more rectifying trays or a higher reflux ratio. A similar situation would occur if the methanol concentration in the feed drops slightly or the methanol product purity specification is made a bit more stringent.

In contrast, the stripping section of the column is quite stable since the operating line in that section is a good distance from the equilibrium curve. Lowering or raising reflux ratio by reasonably small amounts causes minor changes in the bottoms composition.

Using the H–X Diagram. Enthalpy and phase equilibrium data for the methanol (1)/water (2) mixture are given below:

			Enthalpy of Saturated Solution	
T (°C)	X_1	Y_1	h (kJ/kmol)	H (kJ/kmol)
100	0.000	0.000	7536	48,195
95	0.029	0.185	7269	46,543
90	0.072	0.342	7001	45,357
85	0.139	0.491	6687	44,171
80	0.243	0.618	6269	43,240
75	0.404	0.733	5885	42,333
70	0.664	0.852	5548	41,449
64.5	1.000	1.000	5234	40,449

The design specifications are the same as those used in the Y–X diagram method: $Y_D = 0.98$, $X_B = 0.04$, with a reflux ratio of 1 (or $L_r/V_r = 0.5$). If a partial condenser is used, designated as stage 1, then $Y_1 = Y_D = 0.98$. The equilibrium liquid composition at stage 1 is obtained from the $Y-X$ diagram (Figure 6.9), $X_1 = 0.94$. The vapor composition at stage 2 is calculated by component 1 material balance in stage 1:

$$V_2Y_2 = (L_1 + D)Y_2 = L_1X_1 + DY_1$$

$$Y_2 = \frac{Y_1 + X_1(L_1/D)}{1 + (L_1/D)} = \frac{0.98 + (0.94)(1.0)}{1.0 + 1.0} = 0.96$$

The thermodynamic data are plotted on an $H-X$ diagram, Figure 6.9b. Following the graphical construction method described in Section 5.3, a vertical line is drawn through $Y_1 = 0.98$. The other end of the tie line defines $X_1 = 0.94$. Next, X_1 is joined to Y_2 and the line is extended to intersect the vertical line through Y_1. The intersection defines the upper pivot point D'. The next tie line joins Y_2 to X_2. The next operating line is a line connecting D' to X_2, and more stages are constructed in a similar manner (not shown in Figure 6.9b). The feed point F lies on the saturated liquid curve since the feed is assumed to be at its bubble point at the feed tray conditions. The lower pivot point B' (not shown) is defined by the intersection of a line through D' and F and the vertical line through $X_B = 0.04$, the bottoms composition. The stages below the feed are constructed using B' as the pivot.

Similar conclusions are drawn as in the $Y-X$ method. The proximity of the operating lines in the rectifying section indicates low separation power on each stage in this section, which suggests using a higher reflux ratio. This would move D' further up on the diagram, resulting in operating lines that are farther apart, so that fewer stages are required.

EXAMPLE 6.5: DEMETHANIZER COLUMN

In this application, the column is designed with a computer simulation program and then the computer output is used for plotting the distillation diagram to check the design. This example, which is based on two articles by Johnson and Morgan (1985, 1986), also shows how the principles of binary distillation can be applied to multi-component mixtures.

In a typical computer-aided column design, the number of trays and feed location are fixed based on past experience, shortcut distillation, or a graphical method. The column is solved rigorously on a simulator using the pre-determined configuration. Following this, the design is checked and perhaps optimized using additional computer runs. A graphical check of the initial simulation run prior to the optimization runs can help define the direction of these simulations and thereby avoid unnecessary runs.

The column in this application is a demethanizer in an expander gas plant that recovers ethane and heavier components from natural gas. The feed to the column contains 58% mole methane and lighter components. The product specifications are 98% mole methane and lighter components in the overhead and 3% mole methane and lighter components in the bottoms.

Using the Y–X Diagram: In the initial simulation, ten theoretical stages are assumed, with the feed going onto the fourth stage from the top. The results from this run are used to construct a McCabe–Thiele diagram, shown schematically in Figure 6.10. The multi-component mixture is represented on a binary $Y-X$

diagram by lumping the light key component, methane, and lighter components into one pseudocomponent. The equilibrium curve is then plotted as Y vs. X for the pseudocomponent. These data are usually readily available from the simulation run. The stage compositions are then plotted from the simulation results, and the equilibrium stages are stepped off. Note that the graphical construction of equilibrium stages from simulation results does not require an operating line since the tray vapor and liquid compositions are known. In fact, the operating lines are generated after the stages have been stepped off by drawing the best straight line through the corners of the steps in each column section. The points obtained from simulation may not, in general, lie exactly on one straight line since the simplifying assumptions that are the basis for the straight line derivation are not used in the rigorous methods.

To complete the construction of the Y–X diagram from simulation results, the feed line must be drawn. The intersection with the diagonal of a straight line drawn through the feed composition determines one point on the q-line. One other point is determined by the feed equilibrium vapor and liquid compositions at the feed tray conditions. If the feed is a saturated liquid, the equilibrium liquid composition is the same as the feed composition, and the equilibrium vapor composition is the bubble point composition on the equilibrium curve. In this case the q-line is vertical. For a saturated vapor feed, the equilibrium vapor composition is the same as the feed composition, the equilibrium liquid composition is the dew point composition, and the q-line is horizontal. For a mixed-phase feed, the q-line slope is determined by the feed thermal condition (Section 5.2.2). Note that, for a multicomponent mixture, the feed equilibrium vapor and liquid compositions from the simulation output may not lie exactly on the equilibrium curve because of the discrepancies resulting from lumping the light components in one pseudocomponent.

A Y–X distillation diagram has thus been completed based on the same data used in the rigorous solution of the column. The Y–X diagram can now be used for a graphical check of the design. The following observations may be made after examining Figure 6.10.

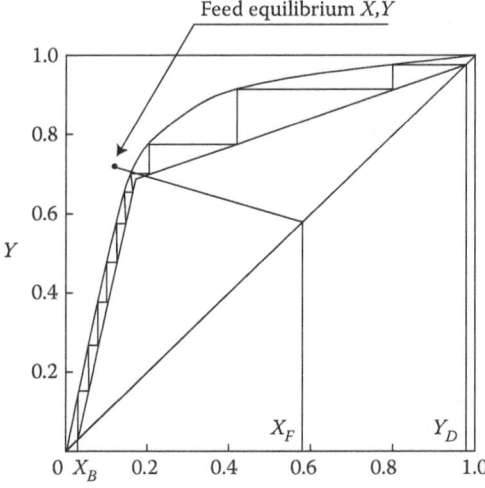

FIGURE 6.10 Demethanizer column—10-tray design.

The q-line passes through the same stage that straddles both operating lines. This verifies that the initial selection of the feed stage was correct. Also, the operating lines are adequately distanced from the equilibrium curve, indicating that the column is stable and should have no tendency to pinch out as a result of small changes in operating conditions. The separation progresses steadily from stage to stage, showing no overtraying anywhere in the column. The reflux ratio required to achieve the specified separation, although readily available from the computer output, can be calculated from the slope of the rectifying section operating line. All in all, the graphical representation indicates that the initial design in this case is basically sound, requiring no significant modifications.

Figure 6.11 represents schematically another case where different design characteristics may be observed. The Y–X diagram is drawn based on the results of a simulation run using 13 stages, with the feed going onto the fourth stage from the top. The stripping section operating line is distinctly closer to the equilibrium curve and, consequently, about two or three trays below the feed are achieving little separation. It is also noted that the intersection of the rectifying section operating line and the stripping section operating line is a few number of stages above the q-line, meaning that the feed location is a few trays above optimum.

The effect of redistributing the reboiler duty between the bottom reboiler and a side reboiler a few trays up the column can also be studied on the Y–X diagram. As discussed in Section 5.2.2, a side reboiler in this case has the effect of increasing the slope of the operating line below the side reboiler. Given the other problem constraints, the side reboiler tends to move the operating line closer to the equilibrium curve. One result of this is the redistribution of the separation power per tray above and below the side reboiler. Moreover, the proximity of the operating line to the equilibrium curve in the vicinity of the side reboiler may have the potential to cause a pinch point in that area.

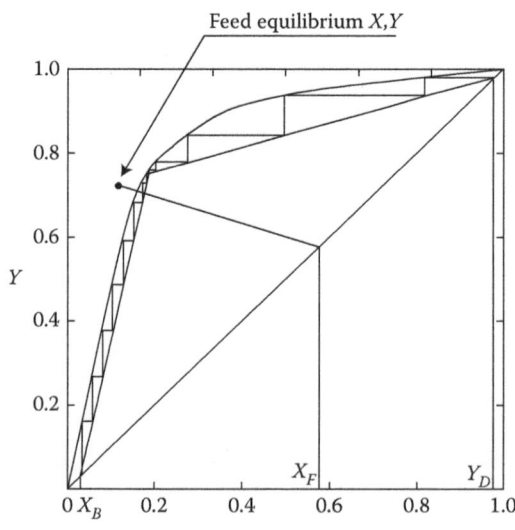

FIGURE 6.11 Demethanizer column—13-tray design.

Using the H–X Diagram: The construction of the Ponchon–Savarit *H–X* diagram for binary distillation calculations was described in Section 5.3. A multi-component mixture can be represented on the binary diagram as in the McCabe–Thiele case, by lumping the light key and lighter components into one light pseudocomponent. Representing the column on an *H–X* diagram when the tray compositions are not known proceeds by stepping off stages starting at the distillate or bottoms (Section 5.3). To use the *H–X* diagram for evaluating simulation results, the procedure is modified to take advantage of enthalpy vs. composition data, tray compositions, and other data from the simulation (Johnson and Morgan, 1985, 1986).

The tray vapor and liquid compositions and enthalpies, generally available from the computer output, are first used to construct the saturated vapor and liquid enthalpy–composition curves on a light pseudocomponent basis. The *H–X* method is especially suited for nonideal systems where the McCabe–Thiele assumptions may be in gross error. The *H–X* method takes into account the variation of enthalpy with composition (Section 5.3), which is reflected in the shape of the enthalpy curves. Ideal mixtures are characterized by straight, horizontal lines while nonideal mixture lines exhibit varying slopes and curvature.

The tie lines are drawn next by joining corresponding dew points and bubble points on the saturated vapor and liquid curves. These points also correspond to individual tray vapor and liquid compositions. The pivot points are plotted next, based on the distillate enthalpy and condenser duty and the bottoms enthalpy and reboiler duty. The rectifying section operating lines are drawn by joining the distillate pivot point to the compositions on the saturated liquid curve corresponding to the trays above the feed. The stripping section operating lines are drawn by joining the bottoms pivot point to the compositions on the saturated vapor curve corresponding to the trays below the feed. Since a consistent set of data is used both for constructing the enthalpy–composition curves and for determining the tray compositions, each rectifying section operating line should intersect the saturated vapor curve at the vapor composition of the tray below. Likewise, each stripping section operating line intersects the saturated liquid curve at the liquid composition of the tray above.

The completed *H–X* diagram can now be used for a visual evaluation of the proposed column design. The feed location is optimum if its tie line crosses the line joining the distillate and bottoms pivot points (Section 5.3). If this condition is not satisfied, consideration should be given to moving the feed tray up or down as indicated. The spacing or density of the operating lines gives a visual measure of the extent of separation achieved per tray at various points in the column. The number of trays in the different column sections may then be considered for modification. Finally, the effect of varying the condenser and reboiler duties can be visualized on the *H–X* diagram by moving the pivot points up or down and observing the positions of the operating lines, and so on.

EXAMPLE 6.6: SEPARATION OF ACETONE/ISOPROPANOL SOLUTION

A distillation column is to be designed for the separation of an acetone(1)/isopropanol(2) solution coming from two sources: Feed $F1$ is 60 kmol/h with 56% mole acetone, and feed $F2$ is 100 kmol/h with 40% mole acetone. Feed $F1$ may be assumed saturated liquid and feed $F2$ saturated vapor at feed tray conditions. The column, operating at atmospheric pressure, will be equipped with a bubble point total condenser and a partial reboiler.

The distillate will be produced at a rate of 60 kmol/h with a purity of 94% mole acetone. Additionally, a small stream S of 20 kmol/h with 74% mole acetone, required for further downstream processing, will be produced as a liquid side draw from the column. The column will be operated at a reflux ratio of 2.0.

Determine the number of theoretical trays, the optimum feed trays, and the side draw tray. Phase equilibrium data at 100 kPa are given below:

X_1	Y_1	X_1	Y_1
0.00	0.000	0.40	0.654
0.05	0.165	0.45	0.698
0.10	0.273	0.50	0.732
0.15	0.362	0.60	0.796
0.20	0.440	0.70	0.850
0.25	0.500	0.80	0.893
0.30	0.558	0.90	0.945
0.35	0.610	1.00	1.000

SOLUTION

The McCabe–Thiele method will be used. The side draw is more concentrated in acetone than either feed and should therefore be located above the feeds. Calculate the internal liquid and vapor flow rates and the slope of the operating line between the condenser and the side draw:

$$L_r = RD = (2)(60) = 120 \text{ kmol/h}$$

$$V_r = D + L_r = 60 + 120 = 180 \text{ kmol/h}$$

$$L_r/V_r = 120/180 = 0.667$$

Because the side draw is a saturated liquid, the liquid and vapor flows below the side draw S are

$$L_r' = L_r - S = 120 - 20 = 100 \text{ kmol/h}$$

$$V_r' = V_r = 180 \text{ kmol/h}$$

The slope of the operating line between the side draw and the upper feed is

$$L_r'/V_r' = 100/180 = 0.556$$

Feed $F1$ is the one more concentrated in acetone and should be the upper feed. This stream is saturated liquid at feed tray conditions, hence the flows and operating line slope between the two feeds are

$$L_s = L_r' + F1 = 100 + 60 = 160 \text{ kmol/h}$$

$$V_s = L'_r = 180 \text{ kmol/h}$$

$$L_s/V_s = 160/180 = 0.889$$

The lower feed is saturated vapor, therefore, between the lower feed and the reboiler,

$$L'_s = L_s = 160 \text{ kmol/h}$$

$$V'_s = V_s - F2 = 180 - 100 = 80 \text{ kmol/h}$$

$$L'_s/V'_s = 160/80 = 2$$

The bottoms rate,

$$B = L'_s - V'_s = 160 - 80 = 80 \text{ kmol/h}$$

or, by overall material balance,

$$B = F1 + F2 - D - S = 60 + 100 - 60 - 20 = 80 \text{ kmol/h}$$

The graphical solution (Figure 6.12) is started by drawing vertical lines through $X_D = 0.94$, the distillate composition, $X_s = 0.74$, the side draw composition, $X_{F1} = 0.56$, feed F1 composition, and $X_{F2} = 0.40$, feed F2 composition. Since a total condenser is used, the vapor from the top tray has the same composition

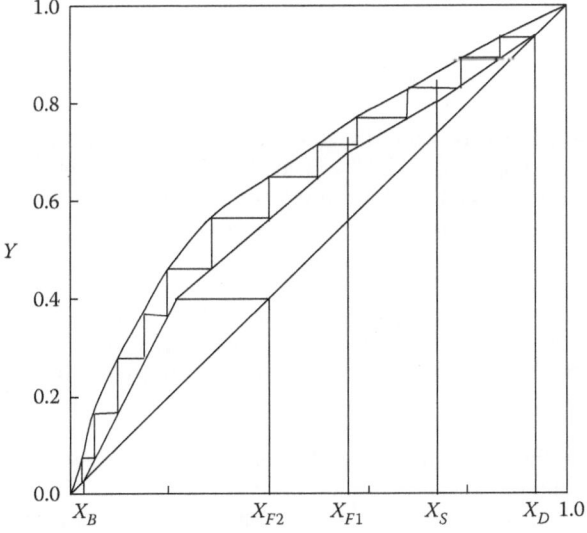

FIGURE 6.12   Acetone–isopropanol column solution.

as the distillate product. The intersection of the vertical line through X_D with the diagonal defines the vapor composition from the top tray. The q-lines are drawn from the intersection points of the vertical lines with the diagonal. Their orientations are vertical for S and $F1$ because they are saturated liquid, and horizontal for $F2$ because it is saturated vapor.

The operating lines are then drawn for each column section with the slopes calculated above, and with their intersections defined by the intersections with each q-line. The slopes are 0.667 between D and S, 0.556 between S and $F1$, 0.889 between $F1$ and $F2$, and 2.0 between $F2$ and B. The bottoms composition is calculated by an acetone over all material balance:

$$F1X_{F1} + F2X_{F2} = DX_D + SX_S + BX_B$$

$$X_B = \frac{(60) \times (0.56) + (100) \times (0.40) - (60) \times (0.94) - (20) \times (0.74)}{80} = 0.03$$

The equilibrium stages are stepped off between the operating lines and the equilibrium curve as indicated in the diagram. A total of 11 and a fraction equilibrium stages are required with feeds optimally introduced at the fifth and eighth stages from the top. These stages do not include the total condenser, which is not an equilibrium stage. If the side stream is withdrawn at the third stage from the top, it would have a composition of 72% mole acetone. The actual product rates would have to be adjusted to satisfy this composition:

$$F1X_{F1} + F2X_{F2} = DX_D + SX_S + (F1 + F2 - D - S)X_B$$

$$(60) \times (0.56) + (100) \times (0.40) = 0.94D + (20) \times (0.72) + (60 + 100 - D - 20) \times (0.03)$$

$$D = 60.44 \text{ kmol/h}$$

$$B = 60 + 100 - 20 - 60.44 = 79.56 \text{ kmol/h}$$

The stepped stages may be altered slightly if the operating lines are redrawn based on a side draw composition of 72% instead of the original 74% mole acetone. Alternatively, the reflux ratio may be adjusted such that the liquid draw from the third stage has a composition of 74% acetone. The steps necessary for this approach are as follows:

1. For the given values of $X_D = 0.94$ and $X_S = 0.74$, draw the top operating line with a slope that results in three stages between X_D and X_S.
2. Calculate the distillate and bottoms rates consistent with the given feed rates, side draw rate and the product compositions $X_D = 0.94$, $X_S = 0.74$, and $X_B = 0.03$.
3. Calculate the slopes of the other operating lines as described in the base case solution.
4. Step off the stages to determine the required number of trays and the optimum feed trays below the side draw.

6.3.1 ANALYTICAL METHOD

Example 6.7 demonstrates the mathematical procedure for solving binary distillation columns as described in Section 5.2.1.

EXAMPLE 6.7: SEPARATION OF BENZENE FROM TOLUENE USING THE ANALYTICAL METHOD

A column with partial condenser and reboiler is to be used for the separation of benzene (1) from toluene (2), giving a distillate with 0.95 mole fraction benzene and a bottoms product with 0.10 mole fraction benzene. The column will operate at 105 kPa pressure and a reflux ratio of 4. The feed, at 55°C and a flow rate of 100 kmol/h containing 45 mol% benzene and 55 mol% toluene, enters the column at the fifth theoretical stage from the top. The estimated average relative volatility (benzene relative to toluene) is assumed constant, and estimated at 2.41. Based on the column conditions and thermodynamic properties, the predicted q-value is 1.2. It is required to determine the number of theoretical stages below the feed to complete the separation.

SOLUTION

Given data:

Feed flow rate and composition: $F = 100$ kmol/h, $X_{F1} = 0.45$, $X_{F2} = 0.55$

Feed stage: $f = 5$

Reflux ratio: $R = 4$

Products specified benzene composition:

$Y_1 = 0.45$, $X_N = 0.10$ (The component subscripts are dropped)

Relative volatility: $\alpha = K_1/K_2 = 2.41$

By overall benzene material balance, $(100) \times (0.45) = 0.95D + 0.10(100 - D)$

Distillate rate, $D = 41.2$ kmol/h; Bottoms rate, $B = 58.8$ kmol/h

$$L_r = RD = 4 \times 41.2 = 164.8 \text{ kmol/h}$$

$$V_r = D + L_r = 41.2 + 164.8 = 206.0 \text{ kmol/h}$$

From Equation 5.12 with $j = 1$,

$$X_1 = Y_1/[\alpha - Y_1(\alpha - 1)] = Y_1/(2.41 - 1.41Y_1) = 0.8874$$

From Equation 5.13 with $j = 1$,

$$Y_2 = (L_r/V_r)X_1 + (D/V_r)Y_1 = (164.8/206.0)X_1 + (41.2/206.0)Y_1 = 0.8999$$

With $j = 2$, $X_2 = 0.7886$, $Y_3 = 0.8209$
With $j = 3$, $X_3 = 0.6554$, $Y_4 = 0.7143$
With $j = 4$, $X_4 = 0.5092$, $Y_5 = 0.3810$
At the feed stage, from Equation 5.16,

$$L_s = L_r + qF = 164.8 + 1.194 \times 100 = 284.2 \text{ kmol/h}$$

From Equation 5.17,

$$V_s = V_r + (q - 1)F = 206.0 + (1.2 - 1) \times (100) = 226.0 \text{ kmol/h}$$

From Equation 5.15,

$$100 \times 0.45 + 164.8 \times 0.5092 - 284.2 \times 0.3810 = 206.0 \times 0.5974 - 226.0Y_6$$

$$Y_6 = 0.4532$$

From Equation 5.12, with $j = 6$,

$$X_6 = Y_6/(2.41 - 1.41Y_6) = 0.2559$$

From Equation 5.14 with $j = 7$,

$$Y_7 = (L_s/V_s)X_6 - (B/V_s) = (284.2/226.0)(0.2559) - (58.8/226.0)(0.10) = 0.2958$$

From Equation 5.12 with $j = 7$,

$$X_7 = Y_7/(2.41 - 1.41Y_7) = 0.2958/(2.41 - 1.41 \times 0.2958) = 0.1484$$

With $j = 8$, $Y_8 = 0.1606$, $X_8 = 0.0735$.

The required number of theoretical stages below the feed (including the partial reboiler) is between 2 and 3. The calculations may be repeated with 8 total theoretical stages while adjusting the reflux to meet $X_N = 0.10$.

It should be noted that the calculated number of trays are theoretical stages that must be converted to actual trays. Once the column is constructed with the determined number of actual trays, feed and side draw locations, the operation of the column will be controlled to satisfy the composition specifications by adjusting such operating variables as product rates, reflux ratio, and condenser and reboiler duties.

NOMENCLATURE

B   Bottoms stream designation or molar flow rate
D   Distillate or overhead stream designation or molar flow rate
F   Feed stream designation or molar flow rate
H   Vapor molar enthalpy
h   Liquid molar enthalpy
L   Liquid molar flow rate in the column
R   Reflux ratio
V   Vapor molar flow rate in the column
X   Component mole fraction in the liquid
Y   Component mole fraction in the vapor

SUBSCRIPTS

B	Bottoms designation
D	Distillate designation
F	Feed designation
r	Rectifying section designation
s	Stripping section designation

PROBLEMS

6.1.   The K-values and enthalpies of components 1 and 2 in a binary mixture at 1 atm are given by the following equations:

$$\ln K_1 = 11.9 - 4000/T \text{ (K)}$$

$$\ln K_2 = 12.1 - 4333/T \text{ (K)}$$

Liquid enthalpies:

$$h_1 \text{ (kJ/kmol)} = 10300 + 188T \text{ (°C)}$$

$$h_2 \text{ (kJ/kmol)} = 11516 + 217T \text{ (°C)}$$

Vapor enthalpies:

$$H_1 \text{ (kJ/kmol)} = 44136 + 133.6T \text{ (°C)}$$

$$H_2 \text{ (kJ/kmol)} = 51690 + 167.0T \text{ (°C)}$$

Construct the $Y–X$ and $H–X$ diagrams for this binary. What assumptions are necessary for generating the enthalpy diagram from the given data alone?

6.2.   A binary stream at the rate of 1000 kmol/h containing 35% mole of component 1 (the lighter component) is to be separated in a distillation column to produce 95% component 1 in the distillate and 90% component 2 in the bottoms. The column will have a partial reboiler and a partial condenser, and will operate at 1 atm. It is proposed to utilize an existing hot process stream in the plant as a heat source for the reboiler, which limits the reboiler duty to 58×10^6 kJ/h. Using either the $Y–X$ diagram or the $H–X$ diagram, determine the number of theoretical trays required, the optimum feed location, the product rates, and the condenser duty. Assume feed thermal conditions that result in a saturated liquid at the feed tray pressure. Use thermodynamic data from Problem 6.1.

6.3.   The distillation column described in Problem 6.2 is to be designed with the provision that the hot process stream will supply only 46.4×10^6 kJ/h to the reboiler. It is proposed to supplement the column duty requirements by utilizing another process stream that can supply 23.2×10^6 kJ/h. However, this stream is at 95°C, and a temperature approach of 14°C is required between the process stream and the column temperature where the heat would be supplied. Determine the optimum locations of the heater and the feed and the total number of equilibrium stages. Calculate the condenser duty.

6.4. Two streams, each one being a binary mixture of components 1 and 2, are to be separated in a distillation column to produce a 95% mole component 1 distillate and a 5% mole component 1 bottoms. The flow rates and compositions of the feed streams are as follows:

	Mole Fraction	
Component	Stream 1	Stream 2
1.	0.40	0.70
2.	0.60	0.30
Flow rate (kmol/h)	600.0	400.0

The column will operate at 1 atm with a total condenser at the bubble point and a reflux ratio of 3.0. The thermodynamic data in Problem 6.1 may be used for this mixture. The feeds are assumed to be saturated liquids at feed tray conditions.

Two alternative schemes are proposed: (1) Mix the feed streams and send the mixture to a single point in the column; and (2) Feed each stream separately to different trays in the column. For each option calculate the required number of trays, the optimum feed locations, and the condenser and reboiler duties. Use either the Y–X or H–X diagram methods.

6.5. A distillation column was designed to separate a 1000 kmol/h binary mixture of 50% mole component 1 to produce a distillate of 95% mole component 1 and to recover in this product 90% of component 1 in the feed. The column was constructed to handle a liquid traffic of 2500 kmol/h. Due to upstream process changes, the column feed composition dropped to 40% mole component 1 at the same total rate of 1000 kmol/h. While maintaining the required 95% distillate composition and operating at optimum performance, what recovery of component 1 is achievable, and can the column handle the required liquid traffic? The column operates with a partial condenser at 100 kPa. The feed is a saturated liquid at feed tray conditions. Thermodynamic data given in Problem 6.1 may be used in this problem.

6.6. A column with eight equilibrium trays and a partial condenser and reboiler is used to separate a 60–40 mole percent mixture to produce a distillate containing 95% mole component 1. The feed may be introduced in either the fourth or seventh stage from the top. Assuming the feed location is optimum in both cases, calculate component 1 recovery in the distillate in each case. The feed enters the column as saturated liquid. Use thermodynamic data from Problem 6.1.

6.7. A distillation column is designed to separate a 70–30 mole percent mixture to produce a distillate product with 97.5% mole component 1 and a bottoms product with 20% mole component 1. The column has a partial condenser and reboiler and operates at 1 atm pressure. Determine the required number of theoretical trays and the optimum feed location at a reflux ratio of 1.5. Calculate the recovery of component 1 in the distillate. Calculate the recovery at the same distillate purity and reflux ratio if the feed is introduced three trays below the optimum feed tray. Use thermodynamic data from Problem 6.1.

6.8. A 100 kmol/h stream containing 60 mol% benzene (1) and 40 mol% toluene (2) is to be separated in a distillation column to produce a distillate containing 95 mol% benzene. An existing column has the equivalent of seven equilibrium stages plus a partial reboiler and a total condenser. Thus, stage 1 is a total condenser and stage 9 a partial reboiler. The feed, a saturated liquid, is introduced in stage 4. The column operates at a constant pressure of 110 kPa.

Use analytical calculations (the binary distillation algorithm) in what follows. Assume the relative volatility of benzene to toluene is constant at $\alpha_{12} = 2.5$.

a. Determine the reflux rate required to achieve a benzene recovery of 90% in the distillate.

b. For the same benzene purity in the distillate, what is the highest possible benzene recovery if the column can handle a maximum internal liquid rate of 200 kmol/h?

6.9. A stream containing 40 mole% carbon disulfide (1) and 60 mole% carbon tetrachloride (2) is fed as saturated liquid to a distillation column with a partial condenser. The distillate is 95 mole% carbon disulfide (CS_2), and 92.4% of the CS_2 in the feed is recovered in the distillate. Determine the following:

a. The distillate and bottoms rates and compositions if the feed rate is 100 moles/h.

b. The required number of trays and the optimum feed location, graphically using the X–Y diagram, if the reflux ratio (L/D) is 4 and the Murphree vapor efficiency is 80%.

The following are vapor–liquid equilibrium data at column pressure for the carbon disulfide—carbon tetrachloride binary:

X(CS)	0.05	0.10	0.20	0.30	0.40	0.50	0.60	0.70	0.80	0.90	0.95
Y(CS)	0.135	0.245	0.42	0.545	0.64	0.725	0.79	0.85	0.905	0.955	0.975

6.10. A stream of 60 kmol/h component 1 and 40 kmol/h component 2 is sent to a distillation column with a partial condenser and a partial reboiler. The feed thermal conditions are such that it is a saturated vapor at the feed tray conditions. The distillate rate is 40 kmol/h with a mole fraction of component 1 equal to 0.95. The reflux ratio is 3.0. Determine the required number of equilibrium stages, the optimum feed location, and the bottoms composition. The following equilibrium compositions are given for this binary at the column pressure:

X_1	0.05	0.10	0.20	0.30	0.40	0.50	0.60	0.70	0.80	0.90	0.95
Y_1	0.135	0.245	0.42	0.545	0.64	0.725	0.79	0.85	0.905	0.955	0.975

6.11. Solve Problem 6.8 graphically or analytically with the column having two feed streams: Stream 1, 80 kmol/h, 50% mole benzene, and 50% mole toluene, saturated liquid, sent to stage 3; Stream 2, 20 kmol/h, 30% mole benzene, and 70% mole toluene, saturated vapor, sent to stage 5. Use any other necessary data as given in Problem 6.8, and find the product compositions at a reflux ratio of 2 and a distillate rate of 50 kmol/h.

REFERENCES

Johnson, J. E. and D. J. Morgan, *Chemical Engineering*, July 8, p. 72, 1985.
Johnson, J. E. and D. J. Morgan, *Encyclopedia of Chemical Process Design*, New York, Marcel Dekker, 1986.

7 Multi-Component Separation
Conventional Distillation

The principles of multistage separation were applied in Chapters 5 and 6 mainly to binary mixtures. The objective was to formulate the theories and methods of solution and performance prediction for an idealized system. This model was used to study the effect of different parameters on the outcome of a separation process and also to quantitatively solve certain binary problems.

Most practical separation problems, however, involve multi-component mixtures, simply because mixtures occurring in nature are rarely binary. Certain industrial streams, the products of various processing steps, could include primarily two components to be further separated in downstream unit operations. Even such streams are in fact multi-component mixtures as a result of impurities carried along in the stream.

Calculating or predicting the performance of multi-component, multistage columns is carried out rigorously by setting up the equations that relate the various interacting variables and then solving the equations. Discussion of these methods in detail is deferred to Chapter 13. Although the mathematical models are accurate and rigorous, poring over a multitude of equations hardly helps the engineer determine which variables could affect which performance specifications and to what extent.

This chapter, as well as a few more that follow, is primarily concerned with a qualitative understanding of the factors that influence the separation process. The principles and ideas introduced in Chapters 2 through 6 and applied to idealized and binary systems are generalized here to multi-component separation.

The discussion in this chapter centers around multi-component distillation with conventional columns, that is, columns with a single feed and overhead and bottoms products. In general, such columns have a condenser and a reboiler, although the special cases of rectifiers and reboiled strippers are also considered.

7.1 CHARACTERISTICS OF MULTI-COMPONENT SEPARATION

A multi-component mixture is separated in a fractionation column into an overhead or distillate product that is enriched in the lighter components and a bottoms product that is enriched in the heavier components. The product compositions depend on the extent of fractionation (or separation) taking place inside the column and on the product rates.

Consider a feed mixture of components 1 through C with mole fractions Z_1 through Z_C. It is convenient to list the components in the feed in the order of decreasing volatility, such that component 1 has a lower boiling point than component 2, which has a lower boiling point than component 3, and so on. If the column has a sufficient number of stages and enough reflux to result in perfect fractionation, the product compositions are determined entirely by the product rates. If, for instance, the overhead rate is less than or equal to FZ_1 (where F is the molar feed rate), the overhead composition would be pure component 1. As the overhead rate is increased, a certain amount of component 2 starts flowing with the overhead. Since perfect fractionation is assumed, the overhead composition is determined by material balance alone. If the overhead (or distillate) rate is D, the following relationships hold for values of D between FZ_1 and $F(Z_1 + Z_2)$:

$$FZ_1 = DY_1$$

$$Y_2 = 1 - Y_1$$

where Y_1 and Y_2 are mole fractions of components 1 and 2 in the overhead. Similar relationships can be derived for the overhead composition as its rate is increased beyond $F(Z_1 + Z_2)$. Thus, in this model, the point in the component list where separation takes place shifts as the overhead rate changes.

With a finite number of stages, where fractionation is not perfect, the separation point also shifts with the product rate. The above relationships, however, are no longer accurate. Nevertheless, for a given overhead or bottoms flow rate, there generally exists a pair of components i and $i + 1$ with different volatilities, between which the separation takes place. Since component i is more volatile than component $i + 1$, most of i in the feed goes with the overhead, while most of $i + 1$ goes with the bottoms. It is also clear that the larger part of each of the components that are lighter than i flows with the overhead, while those heavier than $i + 1$ end up mostly in the bottoms. Component i is known as the *light key* component. It is the heaviest component that goes mostly in the overhead. Component $i + 1$ is the *heavy key* and is the lightest component that goes mostly in the bottoms.

It may be desired to separate components i and j that are not necessarily adjacent to each other on the volatility list. Between them there may be irrelevant components with intermediate volatilities. In this case, if i is lighter than j, then i is the light key and j is the heavy key. The components in between are *distributed components*; they may flow either mostly with the overhead or with the bottoms.

7.2 FACTORS AFFECTING SEPARATION

The separation into lighter and heavier products in a multistage column is accomplished as a result of a composition gradient that develops along the column. The upper stages have higher concentrations of lighter components and the lower stages have higher concentrations of heavier components. The composition gradient is maintained by the counterflow of liquid and vapor streams with different compositions passing each other and interacting on each stage. In conventional distillation columns, the liquid and vapor streams within the column are generated internally by condensing vapor in the condenser and reboiling liquid in the reboiler. In absorbers and strippers, the column liquid and vapor traffic is maintained by feeding external liquid and vapor streams at the top and bottom of the column, respectively.

As the composition changes from stage to stage in the column, so does the temperature. If the feed to a distillation column is a narrow-boiling mixture, the compositional gradient along the column is linked mostly to the temperature gradient. The predominant factor influencing the temperature and composition profile in this type of column is the temperature dependence of the vapor–liquid equilibrium coefficients (K-values). The contribution of the heat of vaporization/condensation to the temperature profile is secondary. In fact, in the idealized treatment of binary systems in previous chapters, the latent heat of vaporization was balanced out by the latent heat of condensation (Section 5.1). A simple, two-stage, binary model was also described which illustrates the phenomenon of a process that is entirely of a distillation nature (Section 3.2).

In columns or column sections where the liquid composition is considerably different from the vapor composition, the composition change in each phase from one stage to the next is not only the result of temperature variation. In this type of process the temperature variation along the column is influenced less by phase equilibrium relations than by latent heat or enthalpy balances. Such situations occur to some extent in wide-boiling distillation columns although they are more prevalent in absorbers and strippers. In these columns, no temperature gradient is necessary to generate the compositional gradient along the column. The composition gradient is brought about by streams with distinct compositions entering the column at different locations.

The difference between the thermal behavior in predominantly distillative and predominantly absorptive processes is a consequence of the different types of mass transfer in each case. In the former, mass is transferred from the liquid phase to the vapor phase and vice versa at approximately the same molar rate. The net material transfer between the phases is therefore small and the ratio of liquid flow to vapor flow (L/V) in a column section is nearly constant. In absorption or stripping columns, there is a net mass transfer in one direction and the L/V ratio is not constant.

In conclusion, the temperature profile in conventional distillation columns is the result of both phase equilibrium relations and enthalpy balances. In narrow-boiling mixtures, the phase equilibrium effect is generally more pronounced, while in wide-boiling mixtures, the enthalpy balances are more significant. The importance of the distinction between the two effects is twofold. First, different mathematical solution algorithms are better suited for each situation, as will be discussed in Chapter 13. Second, the understanding and prediction of column performance is enhanced when the two effects are recognized. Examples 7.1 and 7.2 illustrate the two cases.

EXAMPLE 7.1: SEPARATION OF A NARROW-BOILING MIXTURE

A mixture of methane, ethane, propane, and butane is to be split between ethane and propane using an 11 theoretical stage, conventional column operating at 2600 kPa. The feed, with the following composition, enters the column on the fourth stage from the top:

Component	Methane	Ethane	Propane	Butane
mol/h	120	460	27	5

TABLE 7.1

Results of Example 7.1

Stage	Temp (°C)	Tray Liq (kmol/h)	Tray Vap (kmol/h)	Methane Y	K	L/V
COND	−8	580	580	0.207	3.65	
2	−3	588	1160	0.132	3.70	0.50
3	−2	577	1168	0.121	3.74	0.50
4[a]	0	1272	1153	0.118	3.81	0.50
5	7	1259	1240	0.318E−1	3.91	1.02
6	15	1189	1227	0.836E−2	4.11	1.02
7	26	1120	1157	0.209E−2	4.38	1.03
8	39	1092	1088	0.492E−3	4.59	1.03
9	52	1097	1060	0.110E−3	4.66	1.03
10	62	1103	1065	0.244E−4	4.67	1.03
REB	71	32	1071	0.535E−5	4.68	1.03

[a]   Feed tray.

The column is solved by computer simulation at an overhead rate of 580 kmol/h and a reflux ratio of 1.0. The results are given in Table 7.1.

The vapor and liquid rates on each tray are the net rates leaving that tray. In particular, the vapor rate from the condenser (or stage number 1) is the distillate product rate, and the liquid rate from the reboiler (or the bottom stage) is the bottoms product rate. The L/V ratio is the ratio of liquid to vapor rates passing each other between trays; that is, L_j/V_{j+1}. The methane mole fraction in the vapor and its K-value are given on each tray.

Typical characteristics of a predominantly distillative process are evident in this example. The temperature rises steadily from stage to stage between the condenser and the reboiler. The K-values increase with temperature and the methane mole fraction decreases. The L/V ratio is very nearly constant in each section of the column.

EXAMPLE 7.2: STRIPPING OUT LIGHT COMPONENTS IN A DEMETHANIZER

The feed stream is a wide-boiling mixture with the following composition:

Component	kmol/h
Carbon dioxide	0.2
Nitrogen	0.5
Methane	21.7
Ethane	18.4
Propane	26.7
i-Butane	4.2
n-Butane	15.0
i-Pentane	3.3
n-Pentane	5.0
Hexane	2.5
Heptane	2.5

TABLE 7.2
Results of Example 7.2

Stage	Temp (°C)	Tray liq (kmol/h)	Tray vap (kmol/h)	Nitrogen Y	K	L/V
1[a]	−65	103	20.0	0.250E−1	26.2	5.15
2	−65	103	22.6	0.433E−2	26.1	4.54
3	−64	103	23.1	0.739E−3	26.1	4.46
4	−64	103	23.2	0.126E−3	26.1	4.45
5	−64	103	23.2	0.214E−4	26.1	4.45
6	−64	103	23.2	0.365E−5	26.1	4.45
7	−63	103	23.2	0.622E−6	26.3	4.45
8	−57	104	23.1	0.105E−6	26.9	4.46
9	−31	111	24.1	0.167E−7	27.9	4.31
REB	20	80	31.1	0.195E−8	26.5	3.57

[a] Feed tray.

The feed stream is a saturated liquid at the column pressure of 1750 kPa. The purpose of the column is to remove in the stripping section most of the methane and lighter components from the feed. The bottoms is the purified product. There is no need to purify the overhead from heavier components. Accordingly, no rectifying section is required and the feed is introduced at the top tray. No condenser is required, as the liquid feed serves as reflux.

This type of column is a special case of the conventional column and is described as a reboiled stripper. Because of the wide-boiling feed and the fact that the external feed is used as reflux, this column has mixed characteristics of distillation and absorption/stripping. The solution results are summarized in Table 7.2. In contrast to the distillation column of Example 7.1, the temperature profile in this column is fairly constant over a considerable number of trays. The K-values of nitrogen on these trays are also constant, although the nitrogen mole fraction changes noticeably from tray to tray. The L/V ratio goes through significant changes at both ends of the column.

7.3 SPECIFYING COLUMN PERFORMANCE

It has been established on previous occasions (Sections 3.2.2 and 5.2.1) that a conventional two-product distillation column with a fixed feed and configuration and operating at a given pressure has two degrees of freedom. The configuration is defined by the number of stages, type of condenser and reboiler, and the feed location. The feed is of fixed flow rate, composition, and thermal conditions.

The description rule introduced in Section 5.2.1 may be used to check the degrees of freedom or the number of variables which must be set to define the column operation. For an existing column, the variables that can be set by construction determine the column configuration. The variables that can be controlled by external means include those variables that define the feed, the column pressure, and the condenser and reboiler duties. With a fixed feed and a pressure usually controlled to maintain an acceptable column temperature level, two variables are left for the purpose of controlling the separation—the condenser and reboiler duties. The product rates are controlled to maintain liquid levels in the condenser and reboiler and as such are not independent variables. The condenser and reboiler duties are considered as independent variables: they may either be fixed at certain values or relaxed to allow two other variables to determine the column performance.

The column may be operated to meet various performance specifications within certain ranges. The variables that can be specified in multi-component separation include all the component compositions, rates, or recoveries in the two products as well as the product rates, properties, and temperatures, the reflux and boilup ratios, the condenser and reboiler duties, and the tray temperatures and liquid and vapor rates.

The two variables selected to define the column performance become the independent variables, and the others are calculated to satisfy the mass balances, energy balances, and equilibrium relations. Many of the parameters are interdependent to varying degrees, but the two selected as the independent variables must, at least, be independent of each other over certain ranges. In fact, the higher the degree of independence between the two specified variables, the wider the feasible ranges over which the column can operate and the easier for the specified values of the variables to be met.

The extent of interdependence among the different variables may be investigated by observing the surface that represents each variable as a function of the two "most independent" variables, namely, the reflux ratio and one of the product rates. Stated mathematically, any variable p_k may be expressed as

$$p_k = f(R, B)$$

where R is the reflux ratio and B is the bottoms rate. The dependence of p_k on either of the independent variables is measured by its partial derivatives with respect to R or B. The interdependence of any two parameters p_k and p_l on each other is similarly measured by the relative values of their partial derivatives with respect to the reflux ratio and bottoms rate.

It is impossible to express these derivatives in analytical form since the column equations can only be solved numerically. The interdependence of the various parameters is therefore investigated on the basis of numerical solutions for different column situations.

7.3.1 Variation in Dependent Variables with Reflux Ratio and Product Rate

The variables most often used for evaluating column performance include component fractions and recoveries in each product, the temperature of each product, and the condenser and reboiler duties. It is convenient to scale the variables in such a manner so as to provide a uniform basis for comparing their variation. The reflux is expressed as a ratio and is therefore dimensionless and could vary from zero to infinity. The product rates are expressed as fractions of the total feed, and each component recovery in each product is expressed as a fraction of the same component in the feed. These quantities could thus vary from 0 to 1. The component concentrations in the products are expressed as mole fractions and could also vary from 0 to 1. The overhead and bottoms temperatures are designated as T_D and T_B, respectively. The condenser and reboiler duties, Q_C and Q_R, are expressed on a per-mole-feed basis. Example 7.3 serves to illustrate the interdependence of the variables in a conventional column.

EXAMPLE 7.3: COLUMN VARIABLE INTERDEPENDENCE

A feed stream composition is defined as follows:

Component	kmol/h
Methane	2
Ethane	4
Propane	17
i-Butane	9
n-Butane	26
i-Pentane	12
n-Pentane	19
n-Hexane	11
Total	100

The thermal conditions of the feed are such that it is 50% vapor at the feed tray pressure. The column has 18 theoretical stages, including a partial condenser and a reboiler. The feed is introduced at the eighth stage from the top. The column pressure is fixed at 825 kPa.

Initially, the overhead product rate is fixed, and the variation in each column variable is investigated as a function of the reflux ratio.

The flow rate of the overhead or bottoms product determines roughly which components go mostly in the overhead and which ones in the bottoms. This also defines the key components where the separation takes place. In this example, an overhead rate of 50 kmol/h would include most of the methane, ethane, propane, isobutane, and n-butane. The bottoms product would include most of the hexane, n-pentane, and isopentane. The n-butane is, therefore, considered the light key component and the isopentane, the heavy key component. The product compositions at a reflux ratio of 1.0 are given in Table 7.3.

With the product rates fixed, the sharpness of separation is determined by the reflux ratio. (The column pressure and number of stages are fixed.) The reflux ratio mostly affects the product compositions of components with boiling points in the vicinity of the key components. Components that are either much lighter or much heavier than the key components are not significantly affected by variations in the reflux ratio. This may be seen by comparing the behavior of the distillate compositions $Y(C_3)$ and $Y(NC_4)$ at a distillate rate of 50 kmol/h (Figure 7.1).

As the distillate rate is changed, the split point shifts, and this may result in a change in the identity of the key components. In this example, at a distillate rate of 75 kmol/h, most of the isopentane and lighter components go in the overhead, while most of the n-pentane and heavier go in the bottoms (Table 7.3). Hence,

TABLE 7.3
Results of Example 7.3

Reflux Ratio = 1.0

		50		75	
		Distillate	Bottoms	Distillate	Bottoms
Component	Feed (kmol/h)	(kmol/h)	(kmol/h)	(kmol/h)	(kmol/h)
Methane	2.00	2.00	0.00	2.00	0.00
Ethane	4.00	4.00	0.00	4.00	0.00
Propane	17.00	16.99	0.01	17.00	0.00
i-Butane	9.00	7.65	1.35	9.00	0.00
n-Butane	26.00	17.12	8.88	25.98	0.02
i-Pentane	12.00	1.42	10.58	8.77	3.23
n-Pentane	19.00	0.82	18.18	8.22	10.78
n-Hexane	11.00	0.00	11.00	0.03	10.97
Totals	100.00	50.00	50.00	75.00	25.00

Distillate Rate (kmol/h) appears above the 50 and 75 column groups.

isopentane is the light key component, and n-pentane is the heavy key component. At this distillate rate, since the NC$_4$ boiling point is not close to the key components, its composition in the distillate is not sensitive to the reflux ratio. This is
not the case at a distillate rate of 50 kmol/h, where the NC$_4$ composition is quite
sensitive to the reflux ratio (Figure 7.1).

Figure 7.1 shows other variables plotted versus reflux ratio at a constant distillate
rate. A plot of B_r, the bottoms rate expressed as a fraction of the feed, is, of course,
constant at 0.5 when the overhead rate is 50 kmol/h. The condenser and reboiler

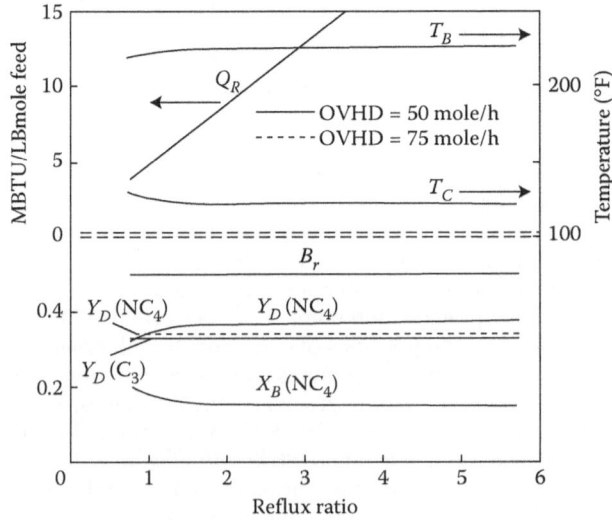

FIGURE 7.1   Dependence of column variables on the reflux ratio.

duties, Q_C and Q_R, vary considerably with the reflux ratio as may be expected due to the variation in condensation and boilup rates as the reflux ratio varies. Since the product compositions do not vary significantly at a constant product rate, the product temperatures also do not vary much with reflux ratio. Fine-tuning of the compositions of key components and adjacent components by adjusting the reflux ratio does not affect the product temperatures to a great extent.

If the number of stages and reflux ratio are held constant, the fractionation attainable between the products is fixed. If the overhead rate is varied, the components where the separation takes place change. In general, most variables (except the condenser and reboiler duties) are more dependent on the product rates than on the reflux ratio. Figure 7.2 presents plots of a number of variables as a function of the overhead rate at a fixed reflux ratio of 1.0.

As the overhead rate is increased, the concentration of light components in the overhead goes down (Y_D for C_1 in Figure 7.2). For heavy components, the concentration goes up (not shown). Intermediate components go through a maximum, which determines the overhead rate at which the column should be operated to maximize the concentration of any of these components. For example, if the objective is to maximize the concentration of NC_4 in the overhead, the overhead rate should be around 65 kmol/h.

It may be desired to meet a certain component concentration, say Y_D (NC_4) = 0.38, which constitutes one performance specification. To meet this concentration, either the reflux ratio or the overhead rate (or both) would have to be varied. Figure 7.2 indicates that, if the reflux ratio is specified at 1.0, the NC_4 specification may be met by varying the overhead rate. It is also observed that two solutions are possible—one at an overhead rate of 58 kmol/h and another at an overhead rate of 71 kmol/h.

The variation in some of the other parameters with overhead rate at a constant reflux ratio is shown in Figure 7.2. The significance of these relationships is in determining the feasibility of specifying the column performance in terms of a given pair of variables. This question is explored further in the following section.

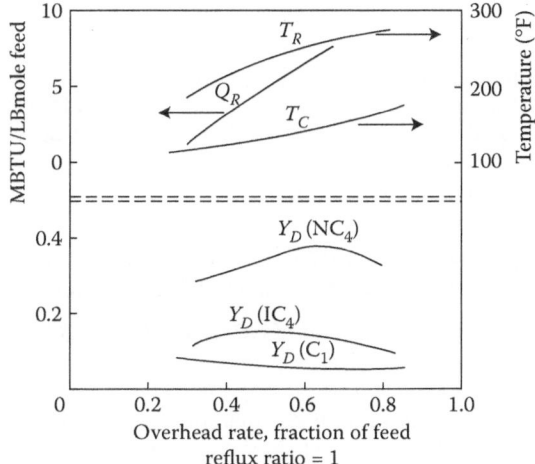

FIGURE 7.2 Dependence of column variables on the product rate. Reflux ratio = 1.0.

7.3.2 Parameter Feasible Ranges

Studying the interdependence between column variables is important when attempting to select parameter pairs for specifying the performance of a conventional column. The reflux ratio and product rate are generally considered the most independent pair. The other extreme may be illustrated by a situation where both product rates are specified. These are two-column variables that are totally dependent since specifying one of them fixes the other by simple material balance. Once one product rate is known, no additional information about the column performance is acquired by providing the other product rate. This fact is indicated by the horizontal line for the bottoms rate shown in Figure 7.1.

For any two variables, it is possible to construct a diagram that represents the feasible area on a plane defined by the two variables as coordinates. If the two product rates are the coordinates, the area is reduced to a straight line, that is, the feasible area is zero. The feasible area varies for different variable pairs: the larger the area, the more independent the variables. The process discussed in Example 7.3 is examined in Examples 7.4 through 7.8, each with a different pair of variables selected as the independent variables.

EXAMPLE 7.4: INDEPENDENT VARIABLES—REFLUX RATIO AND OVERHEAD RATE

At a given product rate, the reflux ratio may be varied quite independently by changing the vaporization rate in the reboiler and the condensation rate in the condenser, that is, by changing the condenser and reboiler duties. This represents internal circulation within the column and is thus independent of the product rate. Nevertheless, the two quantities are not entirely independent; if one variable is fixed, the other may vary only within a certain feasible range. Outside of this range, the vapor or liquid phase may dry up on certain column trays, or one of the products may vanish.

Figure 7.3 shows the feasible region when the independent variables are the distillate rate and the reflux ratio. The region is fairly large, underlining the high degree of independence between the two variables. The feasible range depends in part on the feed thermal conditions. At low distillate rates, a minimum boilup rate (and, hence, reflux ratio) is required to keep the vapor from drying up in the stripping section. At intermediate distillate rates, both the vapor in the stripping section and the liquid in the rectifying section start drying up below a certain reflux ratio. At higher distillate rates, the liquid in the rectifying section starts drying up below a certain reflux ratio.

EXAMPLE 7.5: INDEPENDENT VARIABLES—NC_4 COMPOSITION IN THE OVERHEAD AND REFLUX RATIO

Referring to Table 7.3, the composition of NC_4 in the feed is 26%. It may be desirable either to maximize or minimize its concentration in the distillate. For a given reflux ratio, this may be achieved by varying the distillate rate. The mole fraction of NC_4 in the distillate will always be less than the fraction of NC_4 in the portion of the feed that includes NC_4 and the lighter components, or

$$Y_D(NC_4) < 26/(2 + 4 + 17 + 9 + 26) < 0.448$$

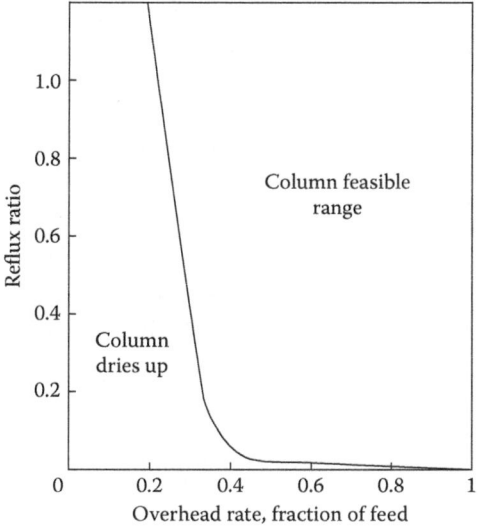

FIGURE 7.3 Reflux ratio—product rate feasible region.

The range between the minimum and maximum values of Y_D of NC_4 increases with increasing reflux ratio as each approaches an asymptote at total reflux. At lower reflux ratios, the degree of fractionation is small, and thus the range between the minimum and maximum feasible values of Y_D of NC_4 is narrow. The column feasible region is plotted on a Y_D versus reflux ratio diagram in Figure 7.4. Within this region the two variables are independent of each other.

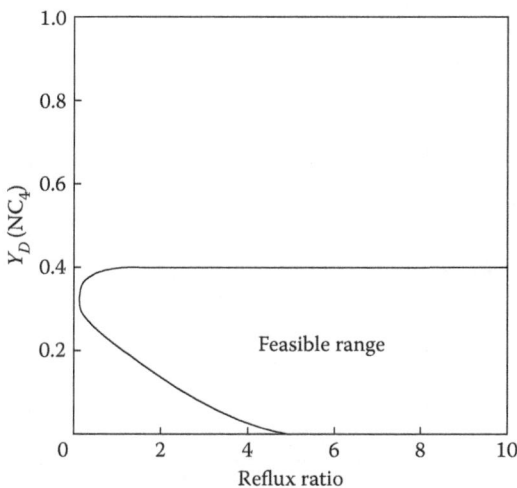

FIGURE 7.4 Reflux ratio—Y_D (NC_4) feasible region.

EXAMPLE 7.6: INDEPENDENT VARIABLES—NC$_4$ COMPOSITION IN THE OVERHEAD AND OVERHEAD RATE

Fixing the product rates generally determines the approximate range of compositions of the products. Thus, for a given distillate rate, the composition of NC$_4$, for instance, can vary within a relatively small range. Conversely, if Y_D of NC$_4$ is specified, the overhead rate that can satisfy this specification may vary within a limited range. Therefore, the two variables, distillate rate and Y_D of NC$_4$, are independent variables within a limited feasible region as shown in Figure 7.5.

EXAMPLE 7.7: INDEPENDENT VARIABLES—NC$_4$ COMPOSITION IN THE DISTILLATE AND IC$_5$ COMPOSITION IN THE BOTTOMS

The interdependence of these two variables is best evaluated in terms of their dependence on the distillate rate and reflux ratio. Since the two components have boiling points that are not very different and both are of intermediate volatilities compared to the feed mixture, the two components are expected to be distributed between the distillate and bottoms over a considerable range of product rates. Within this range, both Y_D of NC$_4$ and X_B of IC$_5$ may be varied by adjusting both the distillate rate and the reflux ratio. The feasible region within which Y_D of NC$_4$ and X_B of IC$_5$ can vary independently is shown in Figure 7.6.

EXAMPLE 7.8: INDEPENDENT VARIABLES—C$_1$ COMPOSITION IN THE DISTILLATE AND NC$_6$ COMPOSITION IN THE BOTTOMS

This is an extreme example where the designated compositions are independent only within a very narrow region. Most of the methane goes in the distillate and most of the hexane goes in the bottoms over the greater part of the range of distillate rates. The compositions of methane in the distillate and hexane in the bottoms are primarily the functions of the distillate rate. The reflux ratio has a minor effect and only within certain distillate rate ranges. Since both Y_D of C$_1$ and X_B of NC$_6$ are primarily dependent on one variable, they are mutually independent within a small feasible range, as shown in Figure 7.7.

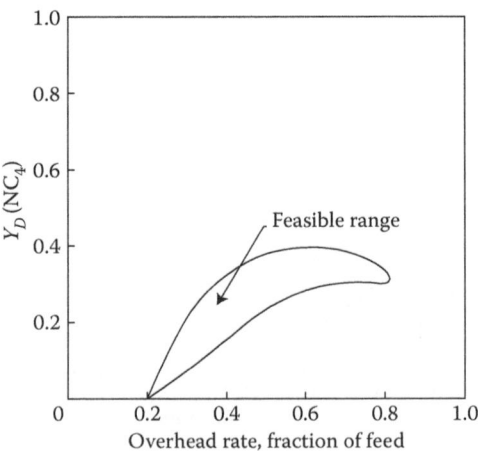

FIGURE 7.5   Product rate—Y_D (NC$_4$) feasible region.

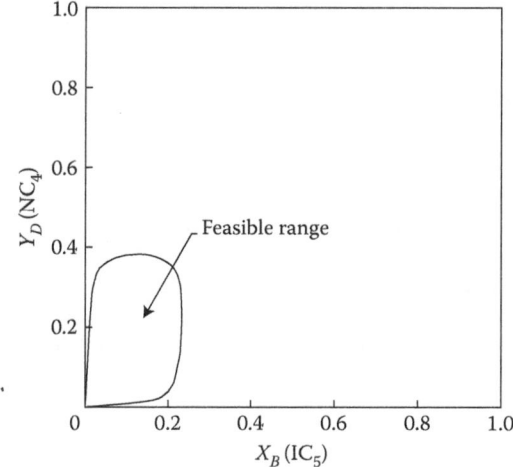

FIGURE 7.6 Y_D (NC$_4$)—X_B (IC$_5$) feasible region.

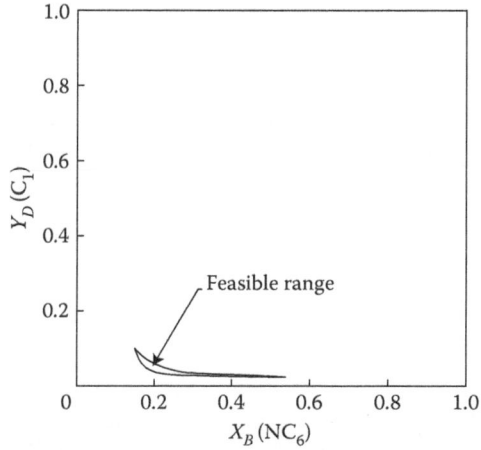

FIGURE 7.7 Y_D (C$_1$)—X_B (NC$_6$) feasible region.

7.3.2.1 Product Temperature as the Independent Variable

Since a column product is at its dew point if it is vapor or at its bubble point if it is liquid, it follows that the product temperature is directly related to the product composition. The product composition was shown to be in general mostly dependent on the flow rate, and the composition fine tuning to be dependent on the reflux ratio. Hence, also the product temperature is, for the most part, a function of its flow rate and, to a lesser degree, of the reflux ratio. Limited independence is therefore expected between a product temperature and its flow rate, while considerable independence is expected between a product temperature and the reflux ratio. Examples 7.9 through 7.11, based on the problem defined in Example 7.3, are presented to illustrate the relationship between the product temperature and the column performance variables.

EXAMPLE 7.9: INDEPENDENT VARIABLES—OVERHEAD TEMPERATURE AND REFLUX RATIO

As the overhead (distillate) product rate and reflux ratio are independent of each other over a wide region (Example 7.4), so are the overhead temperature and reflux ratio. At a given reflux ratio, the overhead temperature may be raised or lowered by increasing or decreasing the overhead rate. The practical limits are when the bottoms, overhead, or parts of the column begin drying up (vapor or liquid). The range between minimum and maximum possible temperatures is wider at higher reflux ratios. The feasible region where the overhead temperature and reflux ratio are mutually independent is shown in Figure 7.8.

EXAMPLE 7.10: INDEPENDENT VARIABLES—OVERHEAD TEMPERATURE AND OVERHEAD RATE

Once the overhead rate is specified, the overhead temperature may be controlled to a certain degree by varying the reflux ratio. The reflux ratio determines the fractionation and, hence, the product compositions and temperatures. However, the composition changes induced by reflux ratio variation are of a "fine-tuning" nature and do not greatly influence the temperature. Figure 7.9 shows the feasible region within which the overhead rate and temperature may be varied independently.

EXAMPLE 7.11: INDEPENDENT VARIABLES—OVERHEAD TEMPERATURE AND BOTTOMS TEMPERATURE

The overhead and bottoms temperatures are determined by the product compositions, which in turn are controlled by the product rate and reflux ratio. A

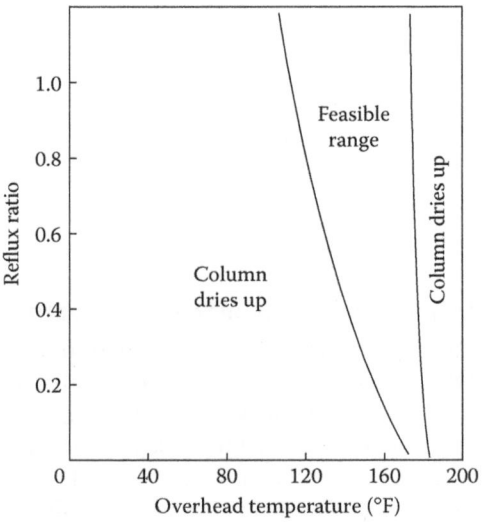

FIGURE 7.8   Reflux ratio–overhead temperature feasible region.

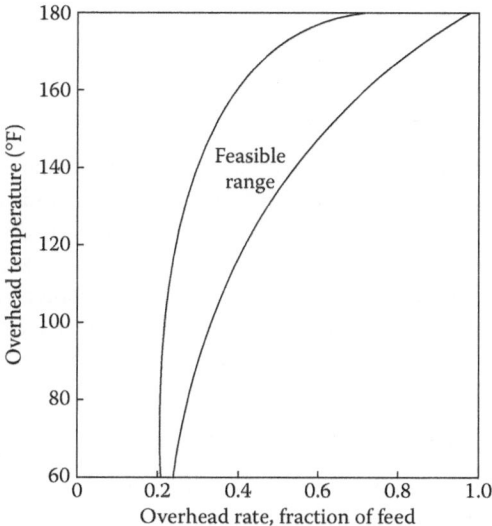

FIGURE 7.9   Overhead rate–overhead temperature feasible region.

wide feasible region is expected (Figure 7.10) in which the overhead and bottoms temperatures may vary independently. The lines of constant overhead rates and constant reflux ratios may be used to determine the values of these variables corresponding to a given overhead and bottoms temperature pair. As the reflux ratio is increased at a constant product rate, the difference between the overhead and bottoms temperatures increases as a result of better fractionation. As the overhead

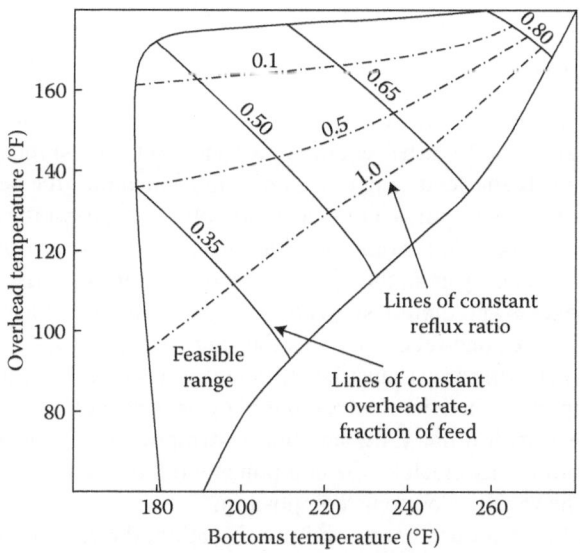

FIGURE 7.10   Overhead temperature–bottoms temperature feasible region.

rate is increased at a constant reflux ratio, the temperatures of both products go up since the concentration of heavier components in both products goes up.

7.4 NUMBER OF TRAYS AND FEED LOCATION

The number of trays in each column section (rectifying or stripping) is determined by the required fractionation in that section at a given L/V ratio. This ratio, which is directly related to the reflux ratio, is usually limited by practical considerations such as economically acceptable reboiler and condenser duties. Also, the reflux ratio is limited by tray hydraulics considerations, discussed in Chapters 14 and 15. Column flooding can occur if the vapor or liquid velocities become excessive. The total column trays and feed location are set once the number of trays in each section is known.

7.4.1 MINIMUM REFLUX AND MINIMUM TRAYS

For a given L/V ratio in a column section, better fractionation is generally achieved as the number of trays in that section is increased. Above a certain number of trays the composition remains unchanged from tray to tray at certain points in the column. These "pinch points" may be observed on a binary $Y–X$ diagram if the operating line touches the equilibrium curve. The same situation can occur in a multi-component system. At a given reflux ratio, there exists a certain number of trays above which the separation cannot be improved. Therefore, a minimum reflux ratio is required to meet a specified separation. If the column is operated at that minimum, an infinite number of trays would be required to achieve that separation. Also, by analogy to binary distillation, a minimum number of trays is required for a given separation. At the minimum trays, total reflux is necessary in order to meet the specified separation.

7.4.2 FEED LOCATION

For a given total number of trays, the optimum feed location is that which minimizes the reflux ratio (and therefore the reboiler and condenser duties) required to deliver a given separation. If the feed is not at its optimum location, the section with the reduced number of trays requires a higher reflux ratio to perform the same fractionation. The higher reflux ratio becomes too high for the other, over-trayed section.

In a multi-component system, the optimum feed location depends on the light and heavy key components and their desired concentrations in the products. A feed location that is optimum for one set of specifications may be a poor selection for another. The number of rectifying trays must be sufficient to remove from the overhead as much of the components heavier than the light key as is needed to meet the required overhead composition. Similarly, the number of stripping trays must be sufficient to strip from the bottoms as much of the components lighter than the heavy key as is needed to meet the required bottoms composition.

When the feed is introduced at its optimum location, the rectifying and stripping sections are balanced with regard to the amount of fractionation taking place in each.

Uniform fractionation is characterized by even variation from tray to tray of the separation parameter, defined as the logarithm of the ratio of the light to heavy key concentrations in the liquid on a given tray:

$$S = \ln(X_{lk}/X_{hk})$$

The selection of the feed tray and its effect on the column performance is examined in Examples 7.12 and 7.13.

EXAMPLE 7.12: DEBUTANIZER COLUMN

The mixture of Example 7.3 is to be separated between n-butane and isopentane in an 18 theoretical stage column. The separation specifications are given:

Mole fraction of NC_4 (light key) in the bottoms: 0.04
Mole fraction of IC_5 (heavy key) in the distillate: 0.018

Assuming these specifications require approximately equivalent degrees of stripping and rectification, suggests that the feed should be placed around the middle of the column. If the feed is placed on the tenth tray from the top, the required reflux ratio calculated by simulation is 1.5.

If the feed is placed on the fourth tray from the top, there would be only four rectifying trays and, therefore, a higher reflux ratio, calculated at 2.0, would be required to obtain the same rectification. The stripping section becomes over-trayed, with several trays showing little change in the separation parameter. If the feed tray is too low, the rectifying section becomes over-trayed. Figure 7.11 is a schematic plot (not to scale) of the separation parameters for the different feed tray locations.

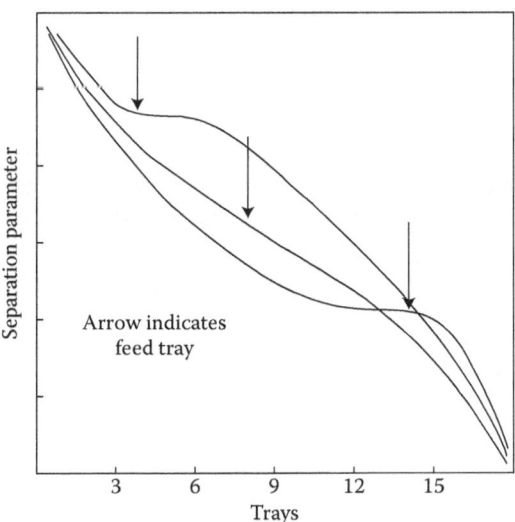

FIGURE 7.11   Schematic plots of the separation parameter.

EXAMPLE 7.13: DEPROPANIZER COLUMN

The same mixture as above is to be separated into propane and lighter components in the distillate and isobutane and heavier components in the bottoms, using
an 18-stage column. The main purpose of the column is to strip as much propane
and lighter components as possible from the bottoms, while tolerating some loss
of isobutane and heavier components in the distillate. Hence, a large number
of stripping trays is required, and a small number of rectifying trays should be
adequate. The feed should therefore be placed on a tray in the upper part of the
column, such as tray 5 or 6.

7.4.3 EFFECT OF FEED THERMAL CONDITIONS

The feed thermal conditions affect the L/V ratio in both column sections. For a given
reflux ratio, the higher the feed temperature, the lower the L/V ratio. Since better
rectification is achieved at higher L/V ratios and better stripping is achieved at lower
L/V ratios, it follows that a higher feed temperature requires more rectifying trays
and fewer stripping trays to make the same separation. Thus, a hotter feed should
be fed at a lower tray and a colder feed should be fed at a higher tray in the column.

7.4.4 RECTIFIERS AND REBOILED STRIPPERS

A rectifier is a column with no stripping section, that is, the feed is sent to the reboiler
or bottom tray. The feed could either be sent directly to the reboiler or, if sufficiently
preheated, it could be sent to the bottom tray and no reboiler would be necessary.
The purpose of such a column is to produce a distillate product that is enriched in
lighter components from which the heavier components are refluxed in a condenser
back to the column. Since no stripping section is used, a certain loss of light components in the bottoms should be expected.

A reboiled stripper is the opposite of a rectifier in that it has no rectifying section. The feed is sent either directly to the condenser or, if sufficiently cold, to the
top tray, requiring no condenser. The liquid portion of the feed acts as reflux. The
main purpose of this type of column is to produce a bottoms product that is enriched
in heavier components and stripped of the lighter ones. With no rectifying trays, a
reboiled stripper could result in some heavies being lost in the overhead.

Note that in reboiled strippers without a condenser or in rectifiers without a
reboiler, there is only one degree of freedom or one independent variable. In a rectifier, for instance, varying the condenser duty directly affects both the reflux ratio and
the overhead rate. These two variables can no longer be varied independently and,
therefore, only one performance specification may be made.

NOMENCLATURE

B Bottoms stream designation or molar flow rate
D   Distillate or overhead stream designation or molar flow rate
F Feed stream designation or molar flow rate
K Vapor–liquid equilibrium distribution coefficient

L Tray liquid molar flow rate
p_k Column variable
Q Heat duty per unit time or per mole of feed
R Reflux ratio
S Separation parameter
T Temperature
V Tray vapor molar flow rate
X Component mole fraction in the liquid
Y Component mole fraction in the vapor
Z Component mole fraction in the feed

SUBSCRIPTS

B Bottoms designation
C Condenser designation
D Distillate designation
hk Heavy key designation
i Component designation
j Tray designation
lk Light key designation
R Reboiler designation

PROBLEMS

7.1. How many variables must be specified in order to define the performance of an existing single-feed, two-product column with a partial condenser (vapor distillate only), a reboiler, and a fixed pressure profile? The feed rate, composition, and thermal conditions are also fixed. How would you conceptually control the column operation?

7.2. An existing column with a partial condenser and a reboiler is used to separate a given mixture of methane, ethane, propane, and butane into a distillate product containing most of the methane, and a bottoms product containing the remaining components. The column pressure profile is fixed at given values. It is required to calculate the condenser and reboiler duties. Which of the following specifications more reliably determine the required duties?

   a. Distillate rate and reflux ratio.
   b. Concentration of propane in the bottoms and methane in the distillate.
   c. Bottoms rate and butane concentration in the bottoms.
   d. Methane concentration in the bottoms and ethane concentration in the distillate.
   e. Ethane recovery in the bottoms and reflux ratio.

7.3. A stream containing components A, B, and C is separated in a distillation column into distillate and bottoms products with the indicated component recoveries. (The components are listed in order of decreasing volatility).

	Distillate Recoveries (%)	Bottoms Recoveries (%)
A	97	3
B	55	45
C	2	98

The bottoms rate will be set at a fixed rate. The reflux rate will be manipulated to control the recovery of either component B or C in the distillate. Which of these recoveries would you recommend as a controller input?

7.4. How many variables must be specified in order to define the performance of an existing single-feed, two-product column with a total condenser and a reboiler, and a fixed pressure profile? The feed rate, composition, and thermal conditions are fixed. How would you conceptually control the column operation?

7.5. A portion of a constant composition stream at a given temperature and pressure is used as the only feed to an existing distillation column. The column has a partial condenser and reboiler, and its only products are the bottoms and vapor distillate. What assumptions must be made to determine the degrees of freedom? Describe a conceptual control scheme to control the column pressure, product purities, and reflux rate.

7.6. A reboiled stripper column is to be designed to separate a feed stream to produce an overhead and a bottoms product. The feed is of fixed rate, composition, and thermal conditions and will be sent to the top of the column. The column will have a reboiler, but no condenser. Determine the number of degrees of freedom.

The column will be used for removing methane and nitrogen from a mixture of nitrogen and hydrocarbons ranging from methane to hexane. What specifications could be used to define the column performance once it is built?

7.7. In a vapor recompression process, heat for the reboiler of a distillation column is supplied by the vapor distillate. The temperature of the vapor distillate is raised by compressing it isentropically, and the heated vapor is used as the heat source for the reboiler. If this is the only heat source for the column, and the column feed is of fixed flow rate, composition, and thermal conditions, determine the number of degrees of freedom for an existing column. What control loops can be used to control the column performance?

7.8. An existing 18-stage column is equipped with feed tray locations at stages 4, 7, 10, 13, counted from the top. For the feed stream given in Example 7.2, suggest recommended feed locations for the following separation objectives: (1) nitrogen/methane, (2) ethane/propane, (3) n-butane/i-pentane, and (4) n-pentane/hexane.

8 Absorption and Stripping

The distinction was made in Chapters 3 and 7 between multistage separation processes where the composition gradient from stage to stage is primarily caused by temperature gradients, as in distillation, and those where the composition gradient is the result of external feeds having compositions and average boiling points that are considerably different from the main feed to the column. Such is the case in absorption and stripping. In many situations both effects come into play.

This chapter is concerned with the performance of simple absorbers and strippers, defined as countercurrent vapor–liquid multistage separation columns, with a liquid feed at the top stage and a vapor feed at the bottom stage. A vapor product comes off the top stage and a liquid product off the bottom stage. No reboilers or condensers are used and no heat is assumed to cross the column walls (although, for the purpose of studying the thermal effects of absorption and stripping, and for their potential benefit, heat sources or sinks may be included).

In absorption, net mass is transferred from the vapor to the liquid, resulting in the recovery of certain components from the vapor feed by the liquid. In stripping, net mass is transferred from the liquid to the vapor, resulting in the recovery of certain components from the liquid feed by the vapor. The end effect is that the main feed (vapor in absorbers, liquid in strippers) is separated, part of it going with the vapor product and the other part with the liquid product. As in distillation, the heaviest component going mostly with the vapor is the light key, and the lightest component going mostly with the liquid is the heavy key. In absorbers or strippers, however, it is mainly the heavy key or the light key, respectively, that is of interest. Thus, in these columns, reference is usually made to one key component only.

The factors that affect an absorption column or a stripping column performance include the feed compositions and thermal conditions, the ratio of liquid to vapor feed flow rates, the column temperature and pressure, and the number of stages.

8.1 THERMAL EFFECTS

The basics of absorption and stripping were discussed in Section 3.3, in terms of ternary single- or two-stage systems. One characteristic feature of these processes is the large difference between the average boiling points of the liquid and vapor in the column. A key component—or group of components that include the key component—having an intermediate boiling point is transferred either from the vapor to the liquid (absorption) or vice versa (stripping). On each stage the intermediate components distribute themselves between the vapor and liquid until their fugacities in both phases are equal or until $Y_i = K_i X_i$, where K_i is the vapor–liquid equilibrium distribution coefficient and Y_i and X_i are the mole fractions in the vapor and liquid of component i. In ideal solutions, this relationship simplifies to Raoult's law

(Chapter 1), $PY_i = p_i^0 X_i$, where P is the stage pressure and p_i^0 is the vapor pressure of component i at the stage temperature.

Mass transfer takes place in order to achieve phase equilibrium, and this mass transfer primarily involves the intermediate components. Because of the large difference between the boiling points of the main components in the liquid and vapor, these components experience little mass transfer between the phases. Referring to Raoult's law, the vapor pressure of the light components is too high to allow them to move in any appreciable amounts to the liquid. (If p_i^0 is large, X_i must be small to keep their product equal to the partial pressure of i, PY_i). Conversely, the vapor pressure of the heavy components is too low to allow it to be transferred significantly to the vapor. Hence, in absorption or stripping, net mass is transferred from one phase to the other. This is in contrast to distillation, where mass transfer between the phases takes place in both directions.

In absorption, the transfer of molecules from the vapor to the liquid is a condensation process that is accompanied by the release of an amount of heat equivalent to the latent heat of condensation of the components being absorbed. If the process is adiabatic, where no heat crosses the system boundaries, the heat released by absorption is converted to sensible heat, resulting in a temperature rise. This thermal effect is reversed in stripping since the stripped components are transferred from the liquid state to the vapor state. The latent heat of vaporization is responsible for a temperature drop in adiabatic stripping processes.

The temperature variation in an absorption or stripping column is a side effect and is not a required contributor to separation, as in distillation. In fact, in situations where the temperature variation is excessive, the column may have to be cooled or heated in order to counter the absorption/stripping thermal effects. The column may be operated isothermally, with all stages maintained at the same temperature, by applying the right amounts of heater/cooler duties.

The thermal effects of absorption and stripping are demonstrated in the following example using a single equilibrium stage, that is, a flash operation with two feeds.

EXAMPLE 8.1: SINGLE-STAGE ABSORPTION AND STRIPPING OF PROPANE

Propane (the key component) is recovered in one case from a nitrogen–propane gas mixture by absorption with a liquid consisting mostly of decane. In another case propane is recovered from a decane–propane solution by stripping it with a gas that is mostly nitrogen. All the feed streams are maintained at a temperature of 24°C and a pressure of 1725 kPa. The absorption or stripping takes place in an insulated vessel maintained at 1725 kPa, where the vapor and liquid feeds mix, reach phase equilibrium, then separate into vapor and liquid products. The feed and product compositions and thermal conditions are given in Table 8.1.

The process involves no pressure change; hence, any temperature change is solely the result of mixing and separation of liquid and vapor. Table 8.1A shows the liquid product to be richer in propane than the liquid feed. Propane absorbed from the gas was transferred from the vapor phase to the liquid phase. A small amount of nitrogen was also absorbed, and a small amount of decane was vaporized in the process. The vaporized decane is negligible compared to the propane that was condensed. Overall, the liquid product flow rate is larger than the liquid feed rate. Net condensation took place, which resulted in a temperature rise of about 4°C.

TABLE 8.1
Results of Example 8.1

Component	Gas Feed (kmol/h)	Liquid Feed (kmol/h)	Gas Product (kmol/h)	Liquid Product (kmol/h)
A. Adiabatic Absorption				
N_2	100.0	0.0	97.6	2.4
C3	14.9	0.1	5.3	9.7
NC10	0.0	100.0	0.0	100.0
Totals	114.9	100.1	102.9	112.1
Temperature (°C)	24.0	24.0	28.0	28.0
Pressure (kPa)	1725	1725	1725	1725
B. Adiabatic Stripping				
N_2	100.0	0.0	97.6	2.4
C3	0.1	14.9	4.8	10.2
NC10	0.0	100.0	0.0	100.0
Totals	100.1	114.9	102.4	112.6
Temperature (°C)	24.0	24.0	21.6	21.6
Pressure (kPa)	1725	1725	1725	1725
C. Isothermal Absorption				
N_2	100.0	0.0	97.6	2.4
C3	14.9	0.1	5.0	10.0
NC10	0.0	100.0	0.0	100.0
Totals	114.9	100.1	102.6	112.4
Temperature (°C)	24.0	24.0	24.0	24.0
Pressure (kPa)	1725	1725	1725	1725
Duty = −67,000 kJ/h				
D. Isothermal Stripping				
N_2	100.0	0.0	97.6	2.4
C3	0.1	14.9	5.0	10.0
NC10	0.0	100.0	0.0	100.0
Totals	100.1	114.9	102.6	112.4
Temperature (°C)	24.0	24.0	24.0	24.0
Pressure (kPa)	1725	1725	1725	1725
Duty = 35,000 kJ/h				

Table 8.1B summarizes the results of the stripping process. A certain amount of propane was transferred from the liquid to the vapor and a smaller amount of nitrogen went in the opposite direction. There was net vaporization since the vapor product rate is larger than the vapor feed rate. A temperature drop of about 2°C resulted.

The processes are carried out again under isothermal conditions. A heat source or sink is included to maintain the vessel temperature at 24°C. Tables 8.1C and 8.1D indicate that better absorption and stripping are achieved isothermally. The cost, however, is a cooling duty of 67,000 kJ/h for absorption and a heating duty of 35,000 kJ/h for stripping.

8.2   LIQUID-TO-VAPOR RATIOS

In distillation columns, a higher reflux ratio means higher liquid-to-vapor traffic, or L/V ratio, in the rectifying section and a lower L/V ratio (higher V/L ratio) in the stripping section. For a given number of stages, the separation is enhanced by raising the reflux ratio, with maximum separation obtained at total reflux. In absorbers the separation of the intermediate components from the gas feed is also improved with higher L/V ratios. Similarly, in strippers the separation of intermediate components from the liquid is improved with higher V/L ratios. Normally in absorbers the gas feed is at a constant rate, and the L/V ratio is varied by varying the liquid feed or solvent rate. In strippers the liquid feed is usually fixed, and the V/L ratio is varied by varying the stripping gas rate.

The limiting condition of total reflux in distillation is approached when most of the overhead is refluxed back to the top of the column and most of the bottoms is reboiled and sent back to the bottom of the column. In absorbers and strippers, the concept of total reflux does not apply since the liquid reflux (solvent) or stripping gas are external feeds. The rates of these feeds may be increased indefinitely, whereupon the product compositions reach asymptotic values. The required feed rate for a given number of stages is determined to satisfy certain product composition specifications. Obviously, the column size sets practical limitations on the allowable flow rate ranges. Moreover, as the liquid solvent or stripping gas rates are increased, larger quantities of the gas or liquid feeds get carried along with the intermediate components to be recovered. This translates into poor separation with the existing number of stages, and a larger number may be required.

Since absorption and stripping processes involve mostly one way mass transfer from one phase to the other, the L/V ratio in absorbers and strippers varies considerably from stage to stage. In absorbers the L/V ratio usually varies most appreciably in the bottom stages where most of the absorption takes place. Any thermal effects causing a temperature rise are also most noticeable in these stages. In the upper stages the final purification takes place, which involves somewhat less mass transfer and temperature change. In strippers most of the L/V and temperature variations occur in the upper stages where most of the stripping takes place, while the final purification occurs in the bottom stages.

Example 8.2 examines the variation in the L/V ratio, temperature, and key component composition in an absorber column and the effect of the relative feed rates on the performance of the column.

EXAMPLE 8.2: ABSORPTION OF PROPANE BY NC$_{12}$

Dodecane (NC$_{12}$) is the solvent or lean oil used to recover propane and heavier components from a rich gas mixture by absorption. The initial conditions of both feed streams are fixed at 38°C and 2760 kPa. Under these conditions the lean oil is a subcooled liquid, and the rich gas is a superheated vapor. The column has ten theoretical stages and is maintained at 2760 kPa.

The lean oil enters the column at the top, and the rich gas at the bottom. The product streams are the rich oil leaving the bottom of the column and the lean gas leaving the top. The flow rates, compositions, and thermal conditions of all four streams are given in Table 8.2. Most of the nitrogen, methane, and ethane, which are the components lighter than the key component, leave the column with

TABLE 8.2
Stream Data for Example 8.2

Component	Rich Gas (kmol/h)	Lean Oil (kmol/h)	Lean Gas (kmol/h)	Rich Oil (kmol/h)
N_2	20.0	0.0	19.4	0.6
C1	800.0	0.0	722.8	77.2
C2	80.0	0.0	48.5	31.5
C3	70.0	0.0	3.3	66.7
NC4	20.0	0.0	0.0	20.0
NC5	10.0	0.0	0.0	10.0
NC12	0.0	500.0	0.1	499.9
Totals	1000.0	500.0	794.1	705.9
Temperature (°C)	38.0	38.0	41.2	47.7
Pressure (kPa)	2760	2760	2760	2760

the vapor product. Most of the propane (the key component) and heavier components leave with the liquid product. The separation of the gas feed thus takes place between the ethane and propane. By far, most of the mass transfer is from the vapor to the liquid, as only 0.1 mole of lean oil (dodecane) is vaporized and becomes part of the vapor product.

Table 8.3 shows the column profiles of temperature, L/V ratio, and the K-value and vapor concentration of propane. The profiles are also plotted in Figure 8.1. The temperature rise due to absorption is highest at the bottom stage where the liquid leaves at about 10°C above the inlet temperature of both feeds. The column temperature rise tapers off at the upper stages where the absorption rates are lower. The L/V ratio goes up progressively from the top of the column down as more material is transferred from the vapor to the liquid.

Going up the column, the vapor is at equilibrium with leaner oil (dodecane with lower concentrations of propane and other dissolved gases). Hence, the propane concentration in the vapor is lower in the upper column stages. The K-values of propane are higher in the lower column stages because the temperature is

TABLE 8.3
Column Profiles for Example 8.2

Stage	Temperature (°C)	L/V	$Y(C3)$	$K(C3)$
1 (top)	41.2		0.0042	0.574
2	41.9	0.671	0.0086	0.584
3	42.5	0.675	0.0137	0.592
4	43.1	0.678	0.0193	0.599
5	43.8	0.680	0.0255	0.607
6	44.4	0.682	0.0323	0.615
7	45.2	0.685	0.0396	0.624
8	46.2	0.687	0.0471	0.636
9	47.3	0.690	0.0547	0.649
10 (bottom)	47.7	0.694	0.0620	0.656

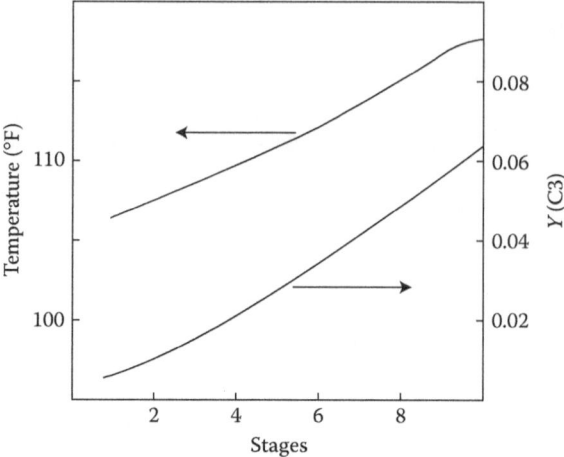

FIGURE 8.1   Absorber column profiles.

higher. If external cooling were used to keep the temperature from rising and thereby keep the *K*-values down, better absorption would be expected.

To investigate the effect of the *L/V* ratio on the separation, the lean oil rate is varied from 25 to 5000 kmol/h at a fixed rich gas feed rate of 1000 kmol/h. Table 8.4 lists the recoveries in the rich oil and concentrations in the lean gas of several

TABLE 8.4

Effect of Lean Oil Rate on Column Performance (Example 8.2)

| Lean Oil Rate (kmol/h) | Percent Recovery in Rich Oil | | | | |
	C1	C2	C3	NC4	NC5
25	0.73	3.16	9.19	26.50	69.50
100	2.24	9.19	25.11	63.60	99.40
300	5.99	24.20	65.76	99.90	100.00
600	11.47	47.37	98.88	100.00	100.00
750	14.19	59.40	99.86	100.00	100.00
1000	18.71	78.14	99.99	100.00	100.00
5000	94.83	100.00	100.00	100.00	100.00
	Mole Fraction in Overhead Vapor				
25	0.816	0.080	0.065	0.015	0.003
100	0.837	0.078	0.056	0.008	0.0
300	0.879	0.071	0.028	0.0	0.0
600	0.919	0.055	0.001	0.0	0.0
750	0.930	0.044	0.0	0.0	0.0
1000	0.947	0.025	0.0	0.0	0.0
5000	0.738	0.0	0.0	0.0	0.0

Feed gas rate = 1000 kmol/h.

components. The recovery is the percentage of a component in the rich gas feed that is recovered in the rich oil. As implied by the definition of key components in absorbers, the key component is that which has the lowest recovery above 50%. At low lean oil rates, only the heavier components' recoveries are over 50%. As the lean oil rate is increased, the recovery of all components goes up and the key component changes. Table 8.4 indicates that, at a lean oil rate of 25 kmol/h, the key component is pentane. At 100 kmol/h, the key component is butane. At 300 and 600 kmol/h, it is propane, and so on. The required lean oil rate is thus determined in part by where the separation of the rich gas should take place. A closer look at the effect of the liquid solvent feed rate (or stripping gas rate) on the column performance is deferred to Section 8.4.

8.3   NUMBER OF STAGES

The component separation in absorbers or strippers depends both on the number of stages in the column and on the ratio of liquid-to-vapor feed rates. This ratio is bracketed on the basis of key component selection (Section 8.2). Most of the mass transfer of different components from one phase to the other usually takes place at different stages in the column. The transfer of the key component generally takes place over a larger number of stages than the other components. This implies that the number of stages affects mostly the key component recovery.

Consider, for instance, an absorber with a small number of stages operating with the feeds at a fixed ratio and resulting in certain component recoveries. If more stages are added to the column, the recovery of most components will go up. Beyond a certain number of stages, the heavier components are recovered in their entirety, while the liquid becomes saturated with the lightest components. As the number of stages is further increased, the recovery of the lightest components is not affected, although more of the heavier components (still lighter than the key component) are recovered. If more stages are added, these components will, in turn, reach saturation, at which point additional stages do not affect their recovery. Depending on the ratio of feeds and on the vapor–liquid equilibrium data, further increase in the number of stages will either saturate the liquid with the key component or will result in 100% recovery of this component. In either case, additional stages are not warranted.

With a fixed number of stages, the liquid-to-vapor feed ratio is adjusted for fine tuning of one of the component's (usually the key) recovery or concentration in the bottoms or overhead. Note that any of these variables determines the rest since they are all related by the column equations.

EXAMPLE 8.3: EFFECT OF THE NUMBER OF STAGES ON THE ABSORPTION OF LIGHT HYDROCARBONS

The process described in Example 8.2 is used to study the effect of the number of stages on the absorption characteristics. The rich gas feed rate is fixed at 1000 kmol/h, and the lean oil rate (NC_{12}) is selected so that the key component is propane. Table 8.4 reveals that this condition can be met with lean oil rates between 300 and 600 kmol/h.

Tables 8.5 and 8.6 represent the recoveries of ethane, propane, and n-butane for different numbers of stages. Propane is the key component, ethane is the next

TABLE 8.5
Effect of Number of Stages on Absorber Performance

Number	Recovery in Bottoms, Percent of Feed		
of Stages	Ethane	Propane	n-Butane
2	23.58	54.39	85.64
5	23.98	63.01	97.88
10	23.95	64.74	99.89
20	23.95	64.94	100.00
30	23.95	64.94	100.00
40	23.95	64.94	100.00

Lean oil rate = 300 kmol/h (Example 8.3).

TABLE 8.6
Effect of Number of Stages on Absorber Performance

Number	Recovery in Bottoms, Percent of Feed		
of Stages	Ethane	Propane	n-Butane
2	42.94	78.21	95.69
5	47.07	93.68	99.93
10	47.29	98.88	100.00
20	47.29	99.96	100.00
30	47.29	100.00	100.00
40	47.29	100.00	100.00

Lean oil rate = 600 kmol/h (Example 8.3).

lightest (light key), and n-butane is the next heaviest component. Table 8.5 gives the results for a lean oil rate of 300 kmol/h, and Table 8.6 for 600 kmol/h. The recoveries are plotted in Figures 8.2 and 8.3.

At each lean oil rate, the recoveries reach a constant value after a certain number of stages. At a lean oil rate of 300 kmol/h, propane reaches a constant recovery of 64.94% after 20 stages and, at a lean oil rate of 600 kmol/h, it reaches 100% recovery with 30 stages. The ethane recovery reaches a constant value after five stages at a lean oil rate of 300 kmol/h and after eight stages at a lean oil rate of 600 kmol/h. N-butane is totally absorbed after 20 stages at a lean oil rate of 300 kmol/h and after 10 stages at a lean oil rate of 600 kmol/h.

8.4   PERFORMANCE SPECIFICATIONS

An absorber or stripper is designed to meet certain process requirements, such as a specified recovery of desirable components or a specified maximum concentration of an impurity in a given product. In the design phase, the engineer has at his or her disposal a number of parameters that may be manipulated to meet the process

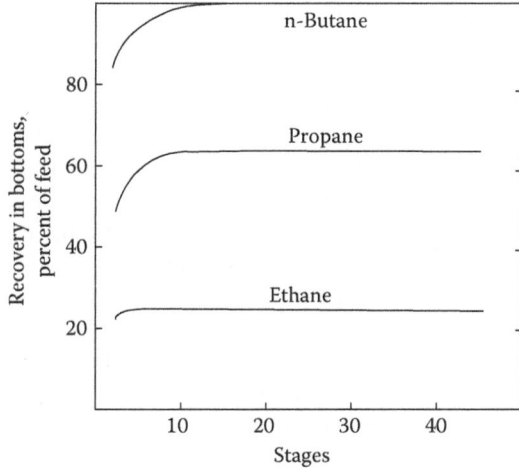

FIGURE 8.2 Effect of number of stages on absorber performance. Lean oil rate = 300 mol/h (Example 8.3).

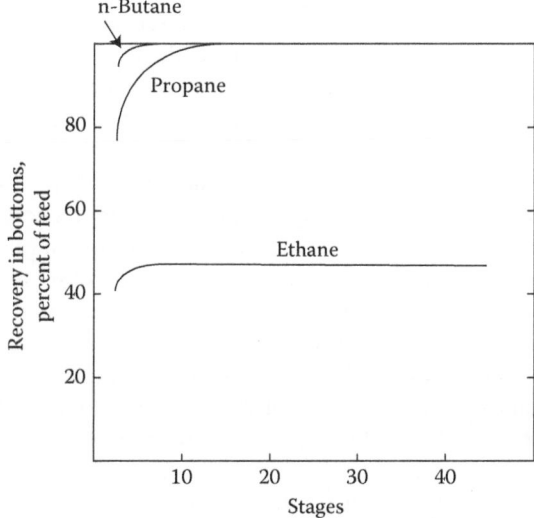

FIGURE 8.3 Effect of number of stages on absorber performance. Lean oil rate = 600 mol/h (Example 8.3).

specifications. The engineer can decide on such parameters as column size, number of stages, feed conditions, operating pressure, whether or not heaters or coolers will be used, and so on. Once a column is built, the ability to control its performance is greatly restricted. In fact, the degrees of freedom are reduced to zero for an absorber or stripper that has a fixed number of stages, is operated adiabatically at a fixed pressure, and whose feeds are of fixed flow rates, compositions, and thermal conditions

(Section 3.3.3). The product rates and compositions in such a column are determined by the energy and material balances and the vapor–liquid equilibrium relations. Under these circumstances the operator has no control over the column performance.

The number of independent variables required to define the operation of an absorber or stripper may also be determined by applying the description rule, stated in Section 5.2.1. The number of trays or the column height is set by construction and may, in the design phase, be used as design variables. Since, by definition, the feeds are introduced at the top and bottom of the column, the feed locations are not variable. The feed compositions and thermal conditions are set outside the column region and are therefore beyond the operator's control. The operator can, however, control the valves on the two feeds and the two products. One of these four valves, usually the bottoms product valve, cannot be controlled independently since it must be set at steady state such as to maintain the required liquid level in the bottom of the column. The overhead valve is usually used to control the column pressure. The two feed valves may be controlled independently; one controls the main process stream rate and the other controls the solvent or stripping gas flow rate.

The parameter used as the control variable to achieve performance specifications is the absorbent rate in absorbers or the stripping gas rate in strippers. With one feed rate allowed to vary, this type of column has one degree of freedom; hence, one specification would be required to define its operation.

If the variable absorbent or stripping gas feed rate in an absorber or stripper is considered as the independent variable, the dependent variables would include product rates and compositions, component recoveries, product temperatures, and so on. When one of these variables is specified, the absorbent or stripping gas stream rate must be calculated to satisfy the specification. The ability of the stream rate to satisfy a given variable specification depends on the relationship between that variable and the stream rate. Each variable may be specified within a certain feasible range, determined by the dependence of the parameter on the stream rate.

As discussed in Section 8.2, the identity of the key component in an absorber- or stripper-type separation depends on the ratio of the liquid-to-gas feed rates. Within any given range of liquid-to-gas feed ratios, the component whose concentration in the product stream is most strongly dependent on the feed rates ratio is the key component. The other component concentrations are less dependent on this ratio, and the farther their boiling points are from the key component boiling point, the less the dependence. Example 8.4 illustrates the different ways of specifying an absorber and also considers the feasible ranges of various performance specifications.

EXAMPLE 8.4: ABSORBER PERFORMANCE
SPECIFICATIONS AND FEASIBLE RANGES

This example is an expansion on Example 8.2, the dodecane absorber. The recovery in the rich oil and the mole fractions in the overhead in Table 8.4 are plotted versus the lean oil rate for a number of components in Figure 8.4.

If one variable is specified, the lean oil rate and the column performance are determined. For instance, if the mole fraction of methane in the overhead is specified at 0.9, the lean oil rate required to meet this specification is 440 kmol/h.

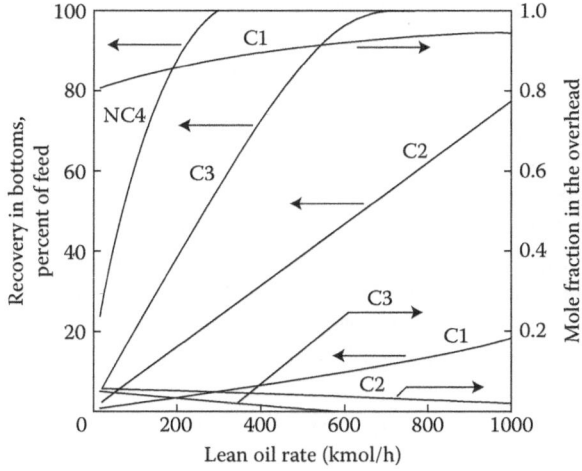

FIGURE 8.4 Effect of lean oil rate on column performance. Feed gas rate = 1000 mol/h (Example 8.4).

At this lean oil rate, the mole fraction of ethane in the overhead is 0.06, the recovery of methane in the liquid is 8%, that of ethane 35%, propane 84%, n-butane 100%, and so on (Figure 8.4).

Each variable may be specified within a certain feasible range. The methane mole fraction in the overhead may be specified at any value between 0.80 and 0.96. Outside this range, the specification cannot be met regardless of the lean oil rate. Mathematically, no solution to the problem exists.

A glance at Figure 8.4 shows that the feasible ranges for some variables are larger than other variables. For instance, the recovery of propane in the bottoms may be specified at practically any point within the entire range from 0 to 100%. It is essential, however, for defining the column performance, not only for the specification to be feasible, but for a unique solution to exist. If, for instance, the propane recovery in the bottoms is specified at 100%, no unique solution exists since any lean oil rate over about 800 mol/h satisfies the specification.

Within different lean oil rate regions, different specifications provide "more unique" definitions of the column performance. For instance, at a lean oil rate of around 700 kmol/h, the key component is ethane. Therefore in this region the column performance is better defined by specifying the ethane recovery or concentration than, say, the methane concentration in the overhead. This is because at about this lean oil rate, the dependence of the ethane recovery or concentration on the lean oil rate is stronger than that of the methane concentration in the overhead. At lower lean oil rates, propane becomes the key component; hence, its concentration or recovery would be a better variable for defining the column performance.

8.5 GRAPHICAL REPRESENTATION

If certain assumptions are made, the absorption or stripping process may be described graphically in a manner similar to the McCabe–Thiele representation of binary distillation. As in binary distillation, it is emphasized here that the graphical representation is not a substitute for a rigorous solution but rather a tool that can be used in conjunction with it to obtain a visual appreciation of the effect of the various parameters on the performance of the column.

The graphical method assumes an idealized absorption/stripping model that is defined in terms of essentially three components or groups of components: a liquid, a gas, and a distributed component or solute. The liquid is assumed not to vaporize (i.e., it has a very low K-value), and the gas is assumed not to dissolve in the liquid (i.e., it has a very high K-value). The distributed component is the key component to be absorbed by the liquid or stripped by the gas. It distributes itself between the two phases to satisfy vapor–liquid equilibrium criteria (Chapter 1).

Another simplification in the idealized model is the exclusion of energy balances and the heat of absorption or stripping. The tray temperatures must be assumed or determined independently. An isothermal column may be assumed with the implication that heat sources or sinks are available for maintaining a constant temperature.

Since the liquid and gas components do not change phases, their flow rates remain constant in the column. It follows that, if the distributed component compositions are expressed as mole ratios to the liquid or gas instead of mole fractions, the column operating line becomes a straight line.

The operating line equation is derived from a material balance on the distributed component around trays 1 through j, counting from the top:

$$L'X_0' + V'Y_{j+1}' = L'X_j' + V'Y_1'$$

or

$$Y_{j+1}' = (L'/V')(X_j' - X_0') + Y_1'$$

where L' and V' are liquid and vapor molar flow rates on a solute-free basis. Since the flow rates are constant, this equation represents a straight operating line. The mole ratios of the distributed component to the liquid or gas, X' or Y', are related to the mole fractions as follows:

$$X' = \frac{X}{1 - X}$$

$$Y' = \frac{Y}{1 - Y}$$

The component subscript is dropped, as reference is made exclusively to the distributed component. The subscripts in the operating line equation refer to the tray number, with X_0' referring to the external liquid feed composition and Y_1' to the overhead vapor composition.

Since the operating line equation must be plotted on an $Y'-X'$ diagram to produce a straight line, the equilibrium data must be converted to Y' versus X' points to generate the equilibrium curve on the same diagram. The operating line relates compositions (expressed as mole ratios) of vapor and liquid streams counter-flowing between stages j and $j + 1$. In absorbers, Y'_{j+1} is greater than Y'_j and, therefore, the operating line is above the equilibrium curve. The opposite is true for strippers.

Example 8.5 illustrates the graphical solution of an absorber problem.

EXAMPLE 8.5: ABSORPTION OF SO$_2$ FROM NITROGEN BY WATER

A countercurrent absorber is used to lower the sulfur dioxide contained in a nitrogen–SO$_2$ gas mixture to acceptable levels by contact with water as the absorbent. The inlet gas is 3% mole SO$_2$ and 97% mole nitrogen. The concentration of SO$_2$ in the treated gas must be brought down to 1% mole. The column is maintained at 30°C and 200 kPa.

It is required to calculate the water-to-gas ratio that would achieve the desired separation at different numbers of stages.

SOLUTION

As in Example 6.5 (the demethanizer column), this example is intended to provide graphical visualization based on a rigorous computer column simulation. The equilibrium curve shown in Figure 8.5 may be constructed based on equilibrium vapor and liquid compositions from the simulation, converted from mole fractions to mole ratios.

The lower end of the operating line corresponds to the conditions at the top of the column where the inlet liquid is pure water (SO$_2$ mole ratio $X'_0 = 0$), and the overhead gas has an SO$_2$ mole ratio $Y'_1 = 0.01$. (At these low concentrations the mole fractions and mole ratios are almost identical). The top of the operating line corresponds to the column bottom, where the concentration of the inlet gas is given as $Y'_{N+1} = 0.03$. Thus, the intersection of the operating line with a horizontal line drawn through $Y'_{N+1} = 0.03$ marks the upper end of the operating line. Its slope, based on SO$_2$ material balance, is equal to L'/V', the liquid-to-gas ratio. The X' coordinate of the top of the operating line, X'_N, corresponds to the SO$_2$ concentration in the water leaving the column. The number of stages required at a given liquid-to-gas ratio is determined by stepping off equilibrium stages between the operating line and the equilibrium curve (Figure 8.5).

The operating line shown in Figure 8.5 is constructed such as to meet the absorption specifications with three equilibrium stages. This is accomplished by trial and error, with the lower end of the operating line fixed at $X'_0 = 0$ and $Y'_1 = 0.01$, and its upper end moved along the $Y'_4 = 0.03$ line to a point where the number of stepped stages is three. The liquid to vapor ratio is calculated from the slope of the operating line:

$$\frac{L'}{V'} = \frac{Y'_4 - Y'_1}{X'_3 - X'_0} = \frac{0.03 - 0.01}{0.00156 - 0} = 12.82$$

The stage compositions are read from the diagram. The procedure may be repeated for different numbers of stages to determine the relationship between the number of stages and the required liquid-to-gas ratio.

It can be seen from the diagram that the higher the liquid-to-gas ratio, the fewer the number of stages required for the separation. The concentration of SO$_2$

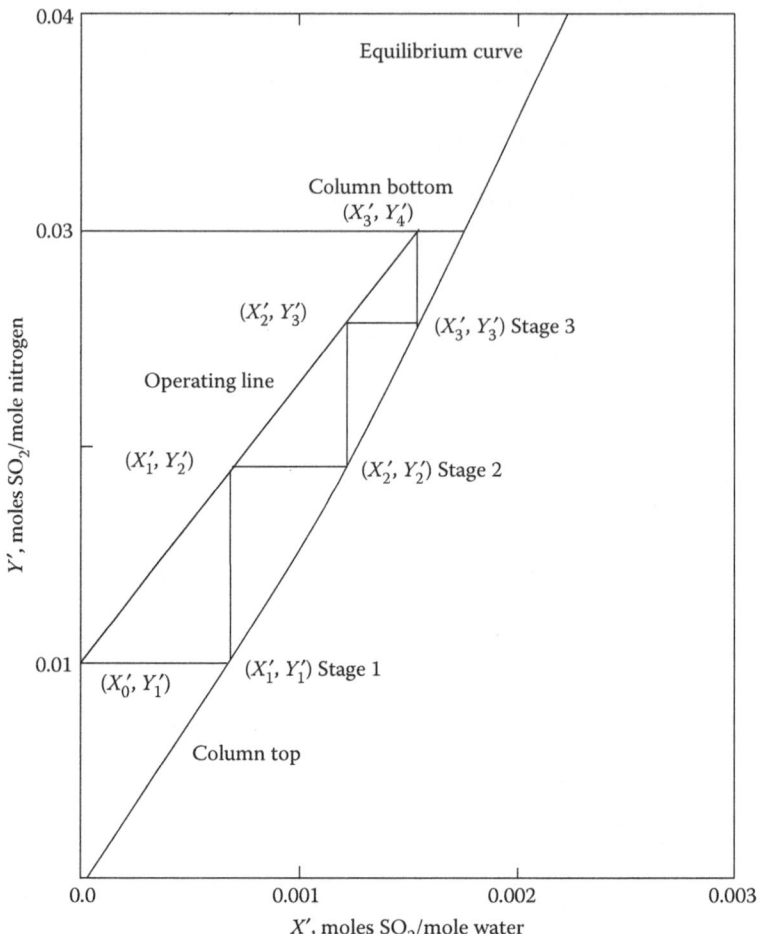

FIGURE 8.5   Absorption of SO_2 from nitrogen by water.

in the water effluent also drops as the water-to-gas ratio increases. As the L'/V' ratio is lowered, the operating line moves closer to the equilibrium curve, and the number of stages increases. If the upper end of the operating line is moved along the horizontal line through Y'_{N+1} toward the equilibrium curve, a point is reached where the operating line touches the equilibrium curve. At these conditions, the L'/V' ratio is at its minimum and the number of stages becomes infinite. If L'/V' is lowered below the minimum ratio, the operating line would cross the equilibrium curve, and the SO_2 would start moving from the liquid to the gas, as in a stripping process.

As in distillation, the column design should avoid pinch conditions; that is, the column should be designed to handle adequate water flow rates so that the operating line would not get too close to the equilibrium curve. The decision on the number of stages versus the liquid-to-gas ratio is an economical one, where the capital investment in column construction is weighed against operating (as well as capital) costs involved in pumping the required amounts of water, and so on.

8.6 ANALYTICAL SOLUTION

As discussed in Chapters 5 and 6, mathematical solution of the column, while still based on simplifications as in the graphical method, can supplement that approach by allowing fast and easy evaluation of the effect of the various column process parameters.

The mathematical method can be applied to absorbers and strippers if equations are available for the operating line and the equilibrium curve. The calculations alternate between the equilibrium equation and the operating line equation, covering the column from one end to the other. If the number of stages is given, a trial and error procedure may be required, as is the case in the graphical method. Better accuracy can be expected in the mathematical method than the graphical, where the accuracy may be limited by the graphics.

Example 8.6 presents one possible application of the mathematical method.

EXAMPLE 8.6: ABSORPTION OF ETHYL ALCOHOL FROM CO_2 USING WATER AS THE ABSORBENT

An absorber column is to be designed to lower the concentration of ethyl alcohol in a CO_2 stream using water as the absorbent. The ethyl alcohol is assumed to be the only component transferring between the vapor and liquid phases, so that the flow rates of ethyl alcohol and water are constant on a solute-free basis. The column pressure and temperature are held constant at 1 kPa and 30°C, at which conditions the K-value of ethyl alcohol is 0.60. The inlet CO_2-ethyl alcohol flow rate is 200 kmol/h at 2 mol% ethyl alcohol. The inlet absorbent rate is 160 kmol/h pure water. Find the smallest number of absorber equilibrium stages that will bring the ethyl alcohol concentration in the effluent gas stream down to 0.1 mol%.

SOLUTION

The inlet water rate and alcohol concentration are $L_0 = L_0' = 160$ kmol/h, $X_0 = X_0' = 0$, where (') indicates solute-free basis and mole ratio.

The inlet vapor stream enters the column at the bottom stage, N, at flow rate $V_{N+1} = 200$ kmol/h. The ethyl alcohol flow rate at the inlet is $0.02 \times 200 = 4$ kmol/h. The solute-free vapor flow rate is $V' = 200 - 4 = 196$ kmol/h, and the mole ratio $Y_{N+1}' = 4/196 = 0.020408$.

The vapor leaves the column at the same solute-free rate of 196 kmol/h. The alcohol concentration in the vapor leaving the column is $Y_1 = 0.001 = $ EtOH/(EtOH + 196), so that EtOH = 0.001(EtOH + 196); EtOH(1 − 0.001) = 0.001 × 196; EtOH = 196 × 0.001/0.999 = 0.1962; $Y_1' = 0.1962/196 = 0.001001$.

The solute-free liquid leaving the column is $L' = 160$ kmol/h. By material balance, the ethyl alcohol leaving the column with the water is $4 - 0.1962 = 3.8038$ kmol/h, at a mole ratio $X_N' = 3.8038/160 = 0.02377$.

In summary, $L' = 160$ kmol/h, $X_0' = 0$, $X_N' = 0.02377$

$$V' = 196 \text{ kmol/h}, \quad Y_1' = 0.001001, \quad Y_{N+1}' = 0.020408$$

The equilibrium equation, $K = Y/X = [Y'/(1 + Y')]/[X'/(1 + X')] = 0.60$. Rearranging, for any stage j,

$$X_j' = Y_j'/(0.60 - 0.40Y_j')$$

The operating line equation, by ethyl alcohol material balance applied to a column section starting at the inlet to the top stage (0) and ending at stage j is expressed as

$$Y'_{j+1} = Y'_1 + (L'/V')(X'_j - X'_0) = Y'_1 + (160/196)(X'_j - 0) = 0.001001 + 0.8163265\ X'_j$$

The results are tabulated below:

J	Operating Line, Y'_j	Equilibrium, X'_j
1.	0.001001 (given)	0.001669
2.	0.002363	0.003945
3.	0.0042214	0.0070555
4.	0.0067606	0.0113187
5.	0.010241	0.0171857
6.	0.015030	0.0253004
7.	0.021654	0.0366186

The computations indicate that based on the given specifications, a vapor stream with an ethanol mole fraction of $0.021654/1.021654 = 0.0212$ would require an absorber with six equilibrium stages to bring down the ethanol concentration in the effluent gas to 0.10 mol%.

NOMENCLATURE

K Vapor–liquid equilibrium distribution coefficient
L Liquid molar flow rate
L' Liquid molar flow rate on a solute-free basis
P Stage or column pressure
p^0 Component vapor pressure
V Vapor molar flow rate
V' Vapor molar flow rate on a solute-free basis
X Component mole fraction in the liquid
X' Mole ratio of a component in the liquid to the solute-free liquid
Y Component mole fraction in the vapor
Y' Mole ratio of a component in the gas to the solute-free gas

SUBSCRIPTS

i Component designation
j Tray designation

PROBLEMS

8.1. How many variables must be specified in order to define the performance of an existing absorber with a liquid feed at the top and a vapor feed at the bottom? The column pressure and the flow rates, compositions, and thermal conditions of both feed streams are fixed.

8.2. A column is designed to remove lighter components from a liquid feed stream by stripping with steam. The feed stream is sent to the top tray, and is at constant rate, composition, and pressure, but its temperature may be controlled. The stripping steam, going to the bottom tray, is at constant temperature and pressure, but its rate may be controlled. How many degrees of freedom exist during the design phase? If the separation is a function of the ratio of overhead rate to bottoms rate, the column bottom temperature, and the column pressure, what control loops would you use to control the separation?

8.3. An absorber column is used to remove most of the propane and heavier components from a gas stream containing methane, ethane, propane, butane, and pentane by treating with lean oil. The gas rate is constant and the lean oil rate is to be calculated to satisfy a performance specification. Which of the following specifications better defines the lean oil rate?
    a. The butane recovery in the bottoms.
    b. The propane concentration in the overhead.
    c. The overhead rate.
    d. The bottoms rate.
    e. The column temperature.
    f. The column pressure.

8.4. The acetone concentration in an effluent air stream is to be reduced from 1% mole to 0.05% mole by scrubbing with water in a countercurrent absorber. Assuming Henry's law holds at these concentrations, and that it may be expressed as $Y = 2.45X$, where X and Y are equilibrium acetone mole fractions in the liquid and gas, calculate the minimum water rate required for an air–acetone mixture rate of 100 kmol/h. How many equilibrium stages would be required if the water rate is twice the minimum? What is the concentration of acetone in the water leaving the absorber? Assume water vapor in the gas and air dissolved in the water are negligible. Solve this problem graphically.

8.5. A countercurrent stripper column with three equilibrium stages is to be used for removing 90% of a component dissolved in a liquid stream by stripping with an inert gas. The liquid feed rate is 100 kmol/h, and the mole fraction of the dissolved component is 0.01. This component is also present in the stripping gas at a mole fraction of 0.0001. Calculate the required stripping gas rate. Assume Henry's law, given in the form $Y = 2X$, holds for the dissolved component. The stripping gas may be assumed immiscible in the liquid and the liquid essentially nonvolatile. What is the composition of the stripping gas as it leaves the column?

8.6. Effluent air containing 0.1 mole% H_2S is to be scrubbed with a caustic solution in a counterflow absorber to lower the H_2S concentration in the air to 0.01 mole%. The equilibrium concentrations of H_2S in the air and the liquid follow the relationship $Y = 3X$, where X and Y are equilibrium mole fractions of H_2S in the liquid and air, respectively. It may be assumed that water vapor in the air and air dissolved in the liquid are negligible. Determine the required number of equilibrium stages if the liquid rate is three times the minimum.

8.7. A countercurrent absorber is used to lower the SO_2 content in air by scrubbing with water. It may be assumed that the amounts of water vaporized and air dissolved in the water are negligible. The mole ratios of SO_2 in the inlet streams and the specified values in the outlet streams are as shown below. Calculate the flow rate of water per 100 kmol/h air.

	Inlet Streams		Outlet Streams	
	Water	Air	Water	Air
SO_2	$X' = 0.005$	$Y' = 0.040$	$X' = 0.020$	$Y' = 0.015$

If the equilibrium SO_2 distribution between water and air is given as $Y' = 1.6X'$, calculate the minimum water flow rate that would meet the SO_2 specification in the outlet air.

8.8. A countercurrent absorber is used to lower the sulfur dioxide content in a nitrogen–SO_2 gas stream to acceptable levels by contact with water as the absorbent. The inlet gas is 2.4 mol% SO_2, and in the treated gas it must be brought down to 1 mole%. The inlet gas rate is 50 kmol/h on a SO_2-free basis, and the inlet water rate is 500 kmol/h. The K-value of SO_2 is $K = Y/X = 15.8$. The water and nitrogen are assumed to undergo no phase transfer, and their flow rates may be assumed constant. Derive the operating line equation based on material balance around an envelope that includes the column top. Determine the required number of equilibrium stages using the 'simulated' graphical method, by alternating the computations between the operating line and the equilibrium curve.

8.9. Solve Problem 8.4 numerically.

9 Complex Distillation and Multiple Column Processes

Complex distillation may be defined as a multistage vapor–liquid separation process that includes one or more of the following features: multiple feeds, side draws, pumparounds, and side heaters or coolers.

Absorbers and strippers are examples of multiple feed columns. In general, in multiple feed columns, the feeds are of different compositions and/or thermal conditions and are fed to different column trays. In certain processes, multiple products are taken from the column, each from a different tray, each with a distinct composition corresponding to the draw tray fluid composition.

Side coolers or heaters are used for redistributing the heat load along the column in order to control liquid and vapor traffic or L/V ratios in the various sections of the column. Pumparounds may be used as the means of transferring heat between the side heaters or coolers and the column.

Multistage complex columns often perform the functions of a number of columns combined in one. In analyzing the performance of complex columns, it is helpful to consider the column as made up of column sections, single equilibrium stages, and stream splitters.

A column section is a vertical vapor–liquid countercurrent adiabatic column section with no heaters or coolers, and whose only feeds and products are a liquid feed and a vapor product at the top, and a vapor feed and a liquid product at the bottom. The feeds to the column section are considered fixed or determined by their points of origin. Assuming the number of stages and the pressure in the column section are also fixed, this unit has no variables that can be controlled independently and therefore has zero degrees of freedom.

Another module in the column makeup is the equilibrium stage. In general, this unit can receive multiple feed streams and can have heat transferred in or out. Its products can be liquid or vapor, or liquid and vapor at equilibrium. At a fixed pressure and with feed streams determined by their points of origin, the heat duty is the only variable that can be controlled independently in this unit, giving it one degree of freedom.

A stream splitter is also a unit that could be a column building block. It is a module that represents the splitting of a stream into two products having the same composition and thermal conditions as the inlet stream. The split ratio can be controlled independently, so this unit has one degree of freedom.

As a simple example, an absorber is itself one column section with zero degrees of freedom if the feeds and the column pressure, as well as the number of stages, are fixed, and if it has no heaters or coolers. As another example, a conventional distillation column may be considered as made up of three equilibrium stages and two column sections. One equilibrium stage with no heat transfer represents the feed tray; it has no degrees of freedom. Two column sections represent the rectifying and stripping sections, with no degrees of freedom. A partial condenser and a reboiler each is an equilibrium stage with heat transfer, allowing two degrees of freedom. For a partial condenser with vapor and liquid products, and for a total condenser, a splitter is used to divide the condensed liquid into a liquid distillate and a reflux stream. This unit has one degree of freedom, giving the column a total of three degrees of freedom. Degrees of freedom of complex columns may be analyzed in a similar manner.

The separation brought about in a given column section depends on the number of stages and the L/V ratio in that section. The L/V ratio is a function of the feeds and product rates, heater and cooler duties, and pumparounds associated with the section. The fractionation may be of a distillative or an absorptive or stripping nature (Chapters 7 and 8), although one effect or the other may be predominant in different situations. The various features of complex distillation are described in this chapter.

The separations that cannot be achieved in a single column are carried out in multiple columns, frequently interconnected in various complex configurations aimed at optimizing the process while meeting the separation specifications. Multiple column processes are also discussed in this chapter.

9.1 MULTIPLE FEEDS

When more than one stream is fed to a column, it is possible to distinguish between two types of feed: one as a stream to be separated and the other as a stream intended to bring about or enhance the separation of another stream. The gas feed to an absorber and the liquid feed to a stripper are of the former type, while the absorbent and the stripping gas are of the latter. Multiple feed columns may be grouped in the following categories: columns with no condenser or reboiler, columns with a reboiler only, columns with a condenser only, and columns with both a condenser and a reboiler. The first category includes simple absorbers and strippers, covered in Chapter 8. The other categories are discussed below. Different feed types are considered and the factors that determine the optimum feed tray are examined.

9.1.1 COLUMNS WITH A REBOILER AND NO CONDENSER

Since these columns have no condenser, one of the feeds provides external reflux and must be fed at the top of the column. The external reflux is either a high-boiling liquid, such as an oil absorbent, or a stream that is sufficiently cold to maintain a liquid reflux down the column. Obviously, no rectification takes place on the top feed and a certain fraction of its components that are volatile enough at the top tray temperature will exit the column with the overhead. The trays above and below the main feed make up the rectifying and stripping sections for that stream. The rectifying section uses a mass separating agent (the top feed or external reflux) to remove the heavier components from the overhead, while the stripping section uses an energy separating agent (the reboiler) to remove the lighter components from the bottoms. The top feed rate mainly controls the recovery of the heavy key component in the bottoms and the purity of the light key component in the overhead, while the reboiler duty mainly controls the recovery of the light key in the overhead and the purity of the heavy key in the bottoms.

Two-feed columns with a reboiler usually have one or two degrees of freedom, depending on whether only the reboiler duty or both the reboiler duty and one of the feed rates are variable. Other column parameters—namely, the number of trays, the feed tray locations, the feed compositions and thermal conditions, and the column pressure—are assumed fixed in an existing column. If one or two of the variable parameters are not fixed, the column operation must be defined by setting one or two performance specifications, such as product rates and compositions. The number of variables required to define the operation of this type of column may be verified by applying the description rule (Section 5.2.1). For an existing column with fixed-feed compositions and thermal conditions, the variables that can be controlled independently by external means include the two feed rates, one product rate, and the reboiler duty. One feed rate is usually determined based on design throughput, and the overhead rate is used to control the column pressure, leaving potentially two variables to control the separation.

Depending on the difference between the average boiling points of the feeds, the column, and mainly its rectifying section, may have either stronger distillative or absorption characteristics. Examples 9.1 and 9.2 apply to these two cases.

EXAMPLE 9.1: DEMETHANIZER COLUMN

The separation of light hydrocarbon gases into methane and lighter components as one product and ethane and heavier components as another is typically carried out in expander plants. In a simplified version of this process, the high-pressure feed gas is cooled, then flashed and separated in a high-pressure separator into vapor and liquid streams. The vapor is expanded in a turbo-expander, a process that drops its temperature and partially liquefies it. The liquid from the flash drum is throttled through a valve down to about the same pressure as the expander discharge. The adiabatic pressure drop across the valve lowers the stream temperature, although this temperature drop is smaller than that across the expander. The expander discharge is thus at a lower temperature than the valve effluent.

The liquid from the expander is fed to the top of the demethanizer column as external reflux, and the valve outlet stream is fed to an intermediate tray. In this example, the column has 12 theoretical trays and a reboiler (a total of 13 theoretical stages). As shown in Figure 9.1, one stream goes to tray 1 and the other to tray 5, counting trays from the top of the column. Streams 3 and 4 are the overhead and bottoms products, respectively. The column and the reboiler operate at a constant pressure of 2900 kPa. The compositions, flow rates, and thermal conditions of the streams are given in Table 9.1. The process requirement is that the bottoms product should have a maximum of 0.001 mole fraction methane.

Since this column has a reboiler but no condenser, and if the feeds are held at constant rates, one degree of freedom remains if the reboiler duty is variable. One specification is therefore required to define the column performance: the methane

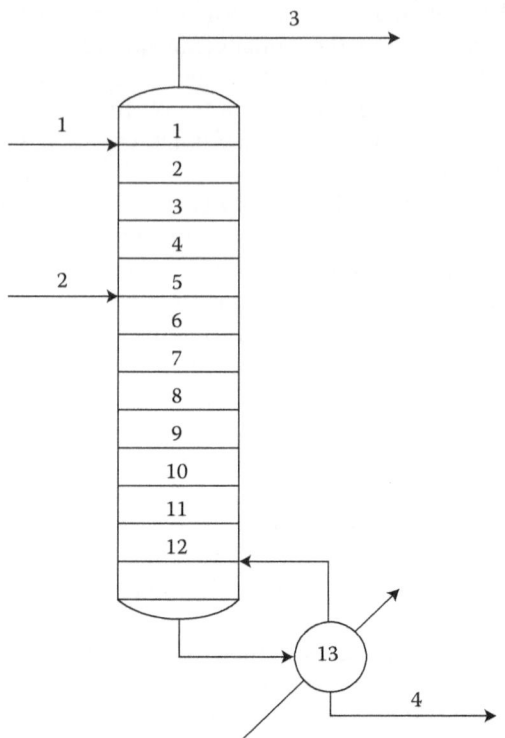

FIGURE 9.1 Demethanizer column (Example 9.1).

TABLE 9.1
Stream Data for Example 9.1

| | Mole Fraction | | | |
Component	Stream 1 Upper Feed	Stream 2 Lower Feed	Stream 3 Overhead	Stream 4 Bottoms
N_2	0.002	0.004	0.0049	0.0000
CO_2	0.050	0.030	0.0376	0.0285
Methane	0.580	0.640	0.8708	0.0010
Ethane	0.250	0.170	0.0824	0.4630
Propane	0.080	0.085	0.0039	0.2865
i-Butane	0.020	0.015	0.0003	0.0568
n-Butane	0.010	0.020	0.0001	0.0617
i-Pentane	0.005	0.010	0.0000	0.0309
n-Pentane	0.002	0.010	0.0000	0.0283
Heptane	0.001	0.016	0.0000	0.0433
Flow rate (kmol/h)	1000	3000	2870	1130
Temperature (°C)	−73	−46	−63	41
Pressure (kPa)	2900	2900	2900	2900

concentration in the bottoms. The reboiler duty required to meet this specification is to be determined.

Design Considerations

Since no condenser is used, the colder feed, serving as external reflux, must be fed to the top tray. With no rectifying trays available above this feed, some of the heavy key component in this feed is "lost" in the overhead. In this example ethane is the heavy key (methane is the light key). Table 9.1 shows that, of the 250 kmol ethane in the upper feed, over 230 mol go overhead.

The distribution of components in the main feed (stream 2) between the overhead and bottoms is consistent with the desired heavy key/light key separation. Most of the ethane in this stream goes with the bottoms and most of the methane with the overhead. Overall, this column may be considered not highly efficient from the standpoint of recovering ethane in the bottoms product but is very efficient in purifying it from the methane (hence its designation as a demethanizer).

In selecting the feed tray for the main feed, it is desirable to maximize the number of stripping trays in order to minimize the reboiler duty required to achieve the specified methane stripping. However, the feed tray should not be too high. If it is, an excessive amount of ethane is lost in the overhead because of an insufficient number of rectifying trays. Table 9.2 shows reboiler duty requirements and ethane recoveries in the bottoms for different feed tray locations. The methane mole fraction in the bottoms is fixed in all cases at 0.001. The results indicate that the optimum feed tray is the third tray from the top.

Column Characteristics

Even though the column has no condenser and the reflux is from an external feed, the column, including its rectifying section, behaves more like a distillation column than an absorber. The reason is that the two feeds—the external reflux and the main feed—have similar compositions and average boiling points. Also, the vapor and liquid compositions on each tray are not much different from each other so that mass transfer takes place both ways between the phases. As a result, the L/V ratio variation from tray to tray is fairly small in each column section, and the temperature rises steadily from tray to tray going down the column. The column profiles are shown in Table 9.3.

TABLE 9.2
Effect of Feed Tray Location on Reboiler Duty and Ethane Recovery (Example 9.1)

Feed Tray for Stream 2	Reboiler Duty (kJ/h)	Ethane Recovery in Bottoms (kmol/h)
1	15.6E6	513
2	14.3E6	528
3	14.2E6	534
4	14.3E6	533
5	14.6E6	523
7	16.2E6	453

TABLE 9.3
Column Profiles for Example 9.1

Tray	Temperature (°C)	Liquid Flow (kmol/h)	Vapor Flow (kmol/h)	L/V
1[a]	−63	850	2870	
2	−56	821	2720	0.313
3	−53	791	2691	0.305
4	−51	698	2660	0.297
5[a]	−41	1844	686	1.017
6	−28	1932	714	2.584
7	−14	2076	801	2.411
8	−2	2233	946	2.195
9	7	2357	1102	2.026
10	13	2442	1227	1.921
11	18	2484	1312	1.862
12	24	2433	1354	1.835
13	41	1130	1303	1.868

[a] Feed trays.

EXAMPLE 9.2: REBOILED ABSORBER

The purpose of this column is to recover propane from a light gas mixture by absorption with a lean oil stream. In addition, the stream containing the recovered propane (the rich oil stream) is to be stripped of some of the lighter gases using a reboiler. The column has thus two functions and its full designation would be a reboiled stripper absorber. A condenser could be used instead of an external reflux, making it a distillation column. The benefit gained by using an external absorbent is that the absorber operates at a moderate temperature, whereas a condenser would have to be held at a considerably lower temperature and/or higher pressure in order to condense the propane. The drawback of using an absorber is the additional downstream processing required to separate the propane from the oil.

The column has two feeds, as in Example 9.1, but in this application the feeds are of very different average boiling points, and they enter the column at about the same temperature. The oil feed has the higher boiling point and is fed at the top of the column as external reflux, while the gas mixture is fed at an intermediate tray. The top section of the column between the two feeds is the absorption section where most of the propane recovery takes place. The lower section, between the gas feed and the reboiler, is the reboiled stripping section where ethane and lighter components are stripped off as in conventional columns.

A schematic of the column is shown in Figure 9.2 and the stream data are given in Table 9.4. The column has nine theoretical trays and a reboiler and operates at 1035 kPa. The rich gas is fed on tray 7, allowing for seven absorption trays and leaving trays 8 and 9 and the reboiler for stripping. The oil feed is a relatively high-boiling stream, and by far most of it flows down the column as reflux with very little being lost in the overhead. The main feed is separated between ethane and propane, the light and heavy key components.

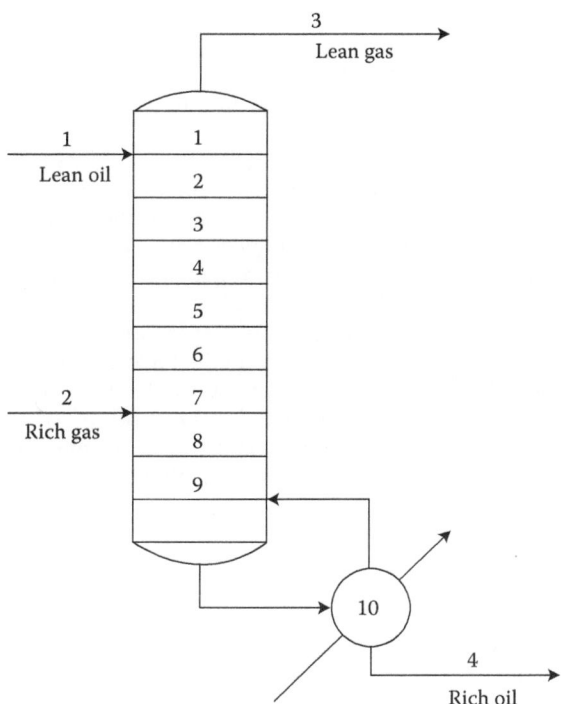

FIGURE 9.2 Reboiled absorber (Example 9.2).

TABLE 9.4
Stream Data for Example 9.2

	Mole Percent			
Component	Stream 1 Lean Oil	Stream 2 Rich Gas	Stream 3 Lean Gas	Stream 4 Rich Oil
N_2	0	2.50	6.67	0
CO_2	0	10.00	26.04	0.15
Methane	0	20.00	53.26	0.03
Ethane	0	2.50	5.35	0.30
Propane	0	65.00	8.68	37.23
Dodecane	100.00	0	0.004	62.29
Flow rate (kmol/h)	1033	1000	374	1659
Temperature (°C)	24	24	31	84
Pressure (psia)	1035	1035	1035	1035

The column profiles are given in Table 9.5, showing the temperature, liquid and vapor flow, L/V ratio, and ethane and propane mole fractions in the liquid on each tray. The profiles indicate that the upper six trays act as typical absorption trays with a rise in temperature, peaking on tray 6, due to the heat of absorption and also due to hot vapors from the reboiler. The L/V ratio varies steadily from tray to tray in the absorption section, underscoring the one-way mass transfer of propane from the vapor to the liquid. In the stripping section, the temperature changes quite steeply, while L/V starts leveling off at the bottom trays. This is because the stripping is accomplished with a reboiler rather than an external stripping medium. The column may thus be viewed as a hybrid of an absorber column and the stripping section of a distillation column. The ethane and propane concentration profiles in Table 9.5 demonstrate the absorption of propane in the upper column section and the stripping of ethane in the lower section.

Assuming a fixed-configuration, fixed-pressure column, there could still be two independent operating variables: the reboiler duty and the oil feed rate. These two parameters may be varied independently to meet the specified propane recovery and ethane rejection in the bottoms. Although all the parameters are

---

TABLE 9.5
Column Profiles for Example 9.2

Tray	Temperature, (°C)	Liquid Flow (kmol/h)	Vapor Flow (kmol/h)	L/V
1[a]	31	1234	374	
2	38	1401	576	2.14
3	46	1572	742	1.89
4	50	1693	914	1.72
5	52	1758	1034	1.64
6	53	1793	1100	1.60
7[a]	51	1856	1134	1.58
8	56	1985	197	9.42
9	60	2074	327	6.07
10	84	1659	415	5.00

	Mole Fraction in Liquid	
Tray	Ethane	Propane
1[a]	0.0156	0.0838
2	0.0178	0.1937
3	0.0144	0.2910
4	0.0106	0.3505
5	0.0081	0.3801
6	0.0068	0.3946
7[a]	0.0064	0.4152
8	0.0074	0.4578
9	0.0061	0.4890
10	0.0030	0.3723

[a] Feed trays.

TABLE 9.6
Effect of Feed Tray Location (Example 9.2)

Feed Tray	Reboiler Duty (kJ/h)	Lean Oil Rate (kmol/h)
5	19.5E6	1071
6	17.9E6	1040
7	17.2E6	1033
8	17.5E6	1058
9	19.3E6	1143

Specifications: Propane recovery in bottoms: 95%; ethane mole fraction in bottoms: 0.003.

interdependent to varying degrees, it is logical to visualize the lean oil rate as primarily affecting the propane recovery, and the reboiler duty mainly determining the ethane rejection. The process requirements for this example are to recover in the bottoms 95% of the propane in the feed and to bring the ethane mole fraction in the bottoms down to 0.003. To meet these specifications, a reboiler duty of 17.2E6 kJ/h and a lean oil rate of 1033 kmol/h are required.

FEED TRAY

The reboiler duty and oil feed required to meet the process specifications correspond to a fixed gas feed location on tray 7. If the feed trays were to be altered while meeting the same column performance specifications, the reboiler duty and oil rate requirements would, in general, take new values. A higher feed tray implies fewer absorption trays and, therefore, a higher oil rate would be required to meet the recovery specification. A higher feed tray also means more stripping trays are available. Two conflicting factors could affect the reboiler duty: (1) A lower duty is needed since more stripping trays are available, or (2) a higher duty is needed since the reboiler must handle a higher liquid flow caused by the higher oil rate. The opposite is true if the feed tray is lowered. If the second factor is controlling, an optimum feed location would exist that minimizes both the oil rate and reboiler duty. If the first factor is controlling, that is, a higher feed tray requires a smaller reboiler duty, then the feed location may be optimized to minimize either the oil rate or the reboiler duty or some function (such as cost) associated with both variables.

In this example, both the reboiler duty and oil rate are minimized when the feed is on the seventh tray. The reboiler duty and oil rate requirements are summarized for several feed trays in Table 9.6.

9.1.2 COLUMNS WITH A CONDENSER AND NO REBOILER

With no reboiler, an external feed, at least partially vapor, is necessary at the bottom tray to act as an external stripping medium. This feed could be the only one, in which case the column may be designated as a rectifier (Section 7.4). Such a column would perform the function of separating the feed into an overhead and a bottoms product, where the heavies are removed from the overhead product to a specified level. No particular concern is placed on the amount of light components lost in the bottoms.

In this section, consideration is given to columns where the bottom feed serves primarily as a stripping medium and where the main feed to be separated is placed on an intermediate tray. Although the stripping stream must be at least partially vapor, the main feed could be at any thermal condition. The stripping stream should be sufficiently hot or of low enough boiling point to maintain vapor traffic throughout the column. Steam is commonly used as the stripping medium because of its relatively low cost and ease of its downstream separation from the stripped gases by condensation and water decantation.

In some situations, external stripping is favored over reboiled stripping, just as in certain situations external reflux is favored over a condenser. Stripping with an external stream can take place at lower temperatures than would exist in a reboiler. Reboiled stripping requires high temperatures in order to drive up the K-values, to cause lighter components to cross from the liquid to the vapor. In external gas stripping, the K-values and temperature need not be raised since the transfer of light components from the liquid to the vapor takes place as a result of lowering their partial pressure or vapor mole fraction by introducing a lighter component (the stripping medium). As the lighter components' vapor mole fraction is lowered, their liquid mole fraction must also come down to maintain the vapor–liquid equilibrium constant. The result is that some of the light and intermediate components transfer from the liquid to the vapor.

Examples of steam stripping include stripping of light components from crude oil or regenerating lean oil from rich oil. Using a reboiler in such applications could require unacceptably high temperatures. Example 9.3 compares two columns that perform the same function, one using a reboiler and the other using steam stripping.

Columns with a condenser and an external stripping stream include the usual rectifying section between the condenser and the main feed as well as the stripping section between the two feeds. A fixed-configuration, fixed-pressure column of this kind has two independent operating parameters or two degrees of freedom: the condenser duty and the stripping stream flow rate. If the two parameters are variable, two process specifications would be required to define the column performance, such as the recovery of one component and the rejection of another in one of the products.

EXAMPLE 9.3: RICH OIL STRIPPER

An oil stream contains heavy oil components decane (n-$C_{10}H_{22}$), undecane (n-$C_{11}H_{24}$), dodecane (n-$C_{12}H_{26}$), and tridecane (n-$C_{13}H_{28}$), plus absorbed components pentane and hexane. The stream is to be separated to recover the pentane and hexane and also to regenerate the lean oil. A 10-stage column operating at 170 kPa is to be used for this purpose. The column is equipped with a partial condenser (first stage) and may either be stripped with steam entering the tenth stage or with a reboiler as the tenth stage. The rich oil enters the column on the fifth stage from the top as a slightly subcooled liquid at 120°C and 170 kPa. Superheated steam at 200°C and 170 kPa is used for stripping. The schematic of the columns is shown in Figure 9.3, and the stream data are given in Table 9.7.

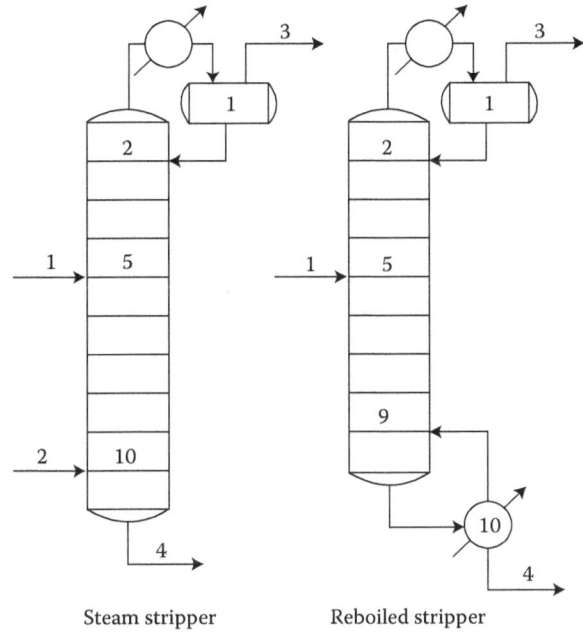

FIGURE 9.3   Steam stripper and reboiled stripper (Example 9.3).

TABLE 9.7
Stream Data for Example 9.3

| | | | Stream 3 Overhead Vapor | | Stream 4 Stripped Oil | |
	Stream 1 Rich Oil	Stream 2 Steam	Steam	Reboiled	Steam	Reboiled
			kmol/h			
H₂O	0.00	330	3.88	0.00	1.24	0.00
Pentane	13.20		13.20	13.20	0.00	0.00
Hexane	8.58		8.58	8.58	0.00	0.00
Decane	8.25		0.01	0.20	8.24	8.05
Undecane	34.00		0.00	0.02	34.00	33.98
Dodecane	33.00		0.00	0.00	33.00	33.00
Tridecane	2.97		0.00	0.00	2.97	2.97
Totals	100.00	330	25.67	22.00	79.45	78.00
Temperature (°C)	120	200	66	85	122	224
Pressure (kPa)	170	170	170	170	170	170

The process specifications require a minimum recovery of 99% for both the stripped hydrocarbons and the regenerated oil. At a reflux rate of 50 kmol/h and a bottoms rate of 78 kmol/h, both columns surpass the two specifications. The column profiles, however, are different in the two cases, as shown in Table 9.8. Much of the stripping steam condenses in the column and is withdrawn as decanted water. The decanted water is not shown in Table 9.8, and this accounts for the apparent material imbalance on some of the trays. A significant difference between the two columns is the temperature profile, which is substantially lower in the steam-stripped column. Also, the temperature range in the reboiled column is wider than that of the stripped column, which is consistent with the distillative nature of the former and the external stripping characteristics of the latter. The steam-stripping column requires a much lower L/V ratio (higher V/L ratio) than the reboiled column because the vapor flow in the steam column includes the stripping steam. This is a factor that must be taken into account in selecting the column type for the process. The vapor flow affects column hydraulics, such as flooding conditions and pressure drop, which are discussed in Chapters 14 and 15.

TABLE 9.8
Column Profiles for Example 9.3

Tray	Temperature (°C)	Liquid Flow (kmol/h)	Vapor Flow (kmol/h)	L/V
		Steam Stripping		
1	66	50.0	25.7	
2	79	16.6	400.6	0.125
3	94	14.4	367.2	0.045
4	105	13.2	365.0	0.039
5[a]	110	95.4	363.8	0.036
6	112	95.1	345.9	0.276
7	115	96.4	345.7	0.275
8	118	97.5	347.0	0.278
9	119	98.1	348.1	0.280
10[a]	122	79.4	348.6	0.281
		Reboiled Stripping		
1	85	50.0	22.0	
2	149	44.3	72.0	0.694
3	182	49.8	66.3	0.668
4	192	50.1	71.8	0.694
5[a]	198	202.2	72.1	0.694
6	213	228.0	124.2	1.628
7	217	232.3	150.0	1.520
8	218	232.4	154.3	1.506
9	221	231.6	154.4	1.505
10	224	78.0	153.6	1.508

[a] Feed trays.

9.1.3 COLUMNS WITH A CONDENSER AND A REBOILER

In many applications, different process streams that contain essentially the same components must undergo similar separation specifications. The process streams may have different flow rates, compositions, and/or thermal conditions. Each stream could be separated in a different column or the different streams could be first mixed together and then separated into the desired products in one column. Alternatively, the streams could be fed directly to the multistage column without prior mixing. Since the optimum feed tray depends on the feed's composition and thermal conditions, each stream is, in general, fed to a different tray.

Excluding cases where some of the feed streams serve as absorption or stripping media, a multi-feed column requires a condenser and a reboiler to generate internal reflux and boilup for fractionation, just as in single-feed conventional columns. For a fixed total number of trays, the feed locations should be optimized for a specified separation to minimize the reflux and boilup ratios or the condenser and reboiler duties. The column section below the lowest feed is the stripping section, and that above the highest feed is the rectifying section. In the column sections between the feeds, the L/V ratio, and hence the fractionation, is under the combined effect of the reflux and boilup and the external feeds. The feeds should thus be located to take full advantage of their different compositions and thermal conditions to enhance the separation, while at the same time sufficient stripping trays and rectifying trays should be allowed. All these factors are interrelated and highly complex and must be incorporated in the overall column solution. The toluene–xylene column described below provides a qualitative insight into the performance of a multi-feed column.

EXAMPLE 9.4: TOLUENE–XYLENE COLUMN

Two mixtures of toluene and xylene are separated into toluene and xylene using one column consisting of 36 theoretical stages, including a partial condenser and a reboiler and operating at a pressure of 140 kPa. The two feed streams are at the same temperature and pressure but are of different flow rates, compositions, and phase (Table 9.9). The process requirements are a toluene recovery of 99.9% in the overhead and a xylene recovery of 96.2% in the bottoms. Recoveries are based on total feed.

TABLE 9.9
Stream Data for Example 9.4

Component	Feed 1 Vapor	Feed 2 Liquid	Overhead	Bottoms
	kmol/h			
Toluene	640.00	80.00	719.27	0.73
Xylene	160.00	120.00	10.53	269.47
Totals	800.00	200.00	729.80	270.20
Temperature (°C)	135	135	122	151
Pressure (kPa)	140	140	140	140

Since the system is binary, the two recovery specifications determine the total product rates by overall material balance. With the total number of stages fixed, the reflux ratio required to achieve the desired separation depends on the feed locations. It is desirable to operate the column at the lowest possible reflux ratio in order to minimize the condenser and reboiler duties.

The optimum feed location for a single-feed binary stream may be estimated graphically using a Y–X diagram (Chapters 5 and 6). The complexity of graphical estimation increases steeply with the number of feeds and is infeasible for multi-component mixtures. Taken separately, the optimum tray for feed 1 would, on the one hand, be higher than that of feed 2 because feed 1 is richer in the lighter component. On the other hand, feed 1 optimum tray would be lower than feed 2 optimum tray because feed 1 is a vapor and feed 2 is a liquid. Generalizing the graphical binary concepts to multi-component systems, streams are fed at upper trays if they are lighter in composition and/or if they have a higher liquid fraction or are subcooled to a lower temperature, and vice versa. In this example, if the streams are fed separately, the phase condition appears to be a more predominant factor than the composition, and feed 1 optimum tray is lower than feed 2 optimum tray. Table 9.10 shows the variation in the reboiler duty with the feed tray for each of streams 1 and 2 when fed separately to the column. The optimum for feed 1 is tray 18 and that for feed 2 is tray 11.

When both streams are fed to the same column, the above considerations still apply, but their relative importance may vary. If one of the streams is predicted to have a higher optimum feed tray than the other on account of both its composition and thermal conditions, then the optimum feed trays in the two-feed column would correspond to the same relative locations as with the streams fed separately. However, if the two factors are contradictory, as in this example, the optimum feed locations in the two-feed column may or may not correspond to the relative single-feed optimum locations. In the two-feed toluene–xylene column, the reboiler duty is minimized when feed 1 is higher than feed 2, which is contrary to the single-feed optimum feed trays. With the two-feed column, the

---

TABLE 9.10
Reboiler Duty for a Single-Feed
Column (Example 9.4)

	kJ/h	
Feed Tray	**Feed 1**	**Feed 2**
8		9.00E6
10	28.6E6	8.80E6
11	25.4E6	8.71E6
13	23.8E6	8.80E6
14	22.9E6	
15		9.15E6
16	22.3E6	
18	22.1E6	
19	22.2E6	
20	22.3E6	

TABLE 9.11
Reboiler Duty for a Multiple-Feed Column
(Example 9.4)

Feed 1 Tray	Feed 2 Tray	Reboiler Duty (kJ/h)
10	20	31.75E6
11	22	29.59E6
15	22	28.42E6
16	22	28.56E6
16	24	28.60E6

compositional effect on the feed location outweighs the phase effect. Table 9.11 lists reboiler duties for several combinations of feed trays. The optimum corresponds to feed 1 on tray 15 and feed 2 on tray 22.

9.2 MULTIPLE PRODUCTS

Column products other than the distillate and bottoms may be taken out as liquid or vapor side draws from any tray. Multiple product columns are used in certain applications as a one-step process for separating multi-component streams into a number of products with different compositions. A multi-product column could, therefore, perform the equivalent task of several columns. In many situations the separation realized in the multi-product column is only a crude separation to be followed by further downstream processing.

The compositions of the side products depend on their location in the column, in accordance with the composition profile that exists along the column. A small liquid product taken from a column tray has the same composition as the undisturbed liquid on that tray. As the product rate varies, the column L/V ratio and other column parameters also vary. New steady-state conditions are re-established with a new column composition profile. The side product composition is therefore also dependent on its flow rate.

9.2.1 COLUMN SECTIONS

In a multi-component column, the concentration of each component peaks at a certain tray, as shown schematically in Figure 9.4. The lighter components reach their peak at the upper trays, and the heavier components at the lower trays. Side products may be drawn at the concentration peak trays or at other trays to obtain products with different compositions. The set of trays between two adjacent products constitutes a column section, and the fractionation between the two products depends on the number of trays and the L/V ratio in that section. The L/V ratio is a function of the column reflux ratio and the side product flow rates. The separation of components between two products defining a column section identifies a light key and a heavy key component for that section.

Multi-product columns consist of the basic modules: column sections, equilibrium stages, and stream splitters. Conceptually, each side product is created by a

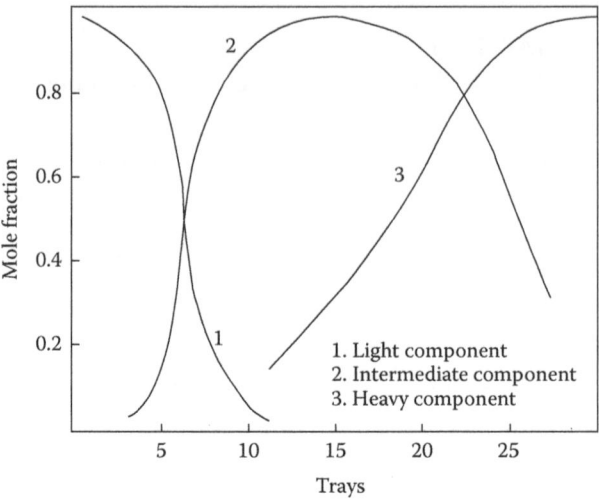

FIGURE 9.4 Composition profiles in a multi-component column.

splitter module between two sections; and since a splitter has one degree of freedom, each side product adds one degree of freedom to the column. While the separation pattern, or the key components in each section, is determined by the product rates and relative locations of the feeds and products, the separation sharpness is determined by the L/V or reflux ratio.

9.2.2 DEGREES OF FREEDOM

Each side product provides one additional independent column variable. To define the column performance, the flow rate of each side product must be known. Alternatively, a side product flow rate may be allowed to vary in order to meet a performance specification such as the concentration of a component in that product. The side product flow rate becomes a dependent variable which must be calculated to satisfy the performance specification. It has been established in Chapter 7 that a fixed-feed, fixed-configuration, fixed-pressure column with a partial condenser (having only a vapor distillate) and a reboiler has two degrees of freedom. Two variables, such as the condenser and reboiler duties, may be varied independently. Each side product adds to the column one degree of freedom. Hence, a column as defined above with S side products has $2 + S$ degrees of freedom. The duties and side product flow rates can each be varied independently, allowing $2 + S$ performance specifications. This conclusion can be reached by applying the description rule since each additional product rate can be controlled independently by external means.

9.2.2.1 Modular Representation

The number of degrees of freedom can also be determined by representing the column using the basic units: column section, equilibrium stage, and stream splitter.

These units are assumed to be at fixed pressure, determined independently based on material flow and equipment considerations. The feed rates, composition, and thermal conditions are either fixed or determined by upstream conditions, and therefore do not contribute to the degrees of freedom.

The column section is a vertical multistage vapor–liquid countercurrent adiabatic column, with a liquid feed and a vapor product at the top, and a vapor feed and a liquid product at the bottom. At a constant pressure this unit has no variable that can be controlled independently, and therefore has zero degrees of freedom.

The equilibrium stage can receive multiple feed streams and can have a heat duty (transferred in or out). Its products can be vapor or liquid, or vapor and liquid at equilibrium. The heat duty can be controlled independently; therefore at constant pressure this unit has one degree of freedom.

The stream splitter splits one feed stream into two products having the same composition and thermal conditions as the feed. The split ratio may be controlled independently, giving this unit one degree of freedom.

9.2.2.2   General Column Performance Considerations

Once a column has been designed and built, the total number of trays and the feed and product locations are fixed. The column pressure may be varied within certain limits but is usually dictated by other process considerations, such as condenser cooling medium temperature and reboiler temperature, and so on. Assuming a fixed column pressure, the operating variables that are available for meeting product specifications, such as purities or recoveries, are the reflux ratio and the product rates. Higher reflux ratios generally improve fractionation throughout the column. If the fractionation between two products appears adequate (i.e., sufficient number of trays and appropriate L/V ratio in the column section between the two products), but the products are not split at the desired point, then the product rates must be varied to meet their composition specifications.

EXAMPLE 9.5: DEGREES OF FREEDOM ANALYSIS USING COLUMN MODULES REPRESENTATION

Desalted hot crude oil feed is sent to a preflash column where it is separated into gas, gasoline, naphtha, and topped crude as shown in Figure 9.5. Using column modules representation, determine the number of degrees of freedom for this operation. What specifications may be used to define the column performance? The crude feed is of fixed composition and thermal conditions. The stripping steam thermal conditions are fixed, but its flow rate may be varied. The column pressure is independently fixed.

Solution

Figure 9.5 shows the distillation column and its modular representation. The feeds to the modules are determined by upstream modules or units and are therefore considered fixed. External feeds may or may not be variable. In this example the stripping steam rate is variable, adding one degree of freedom to the process.

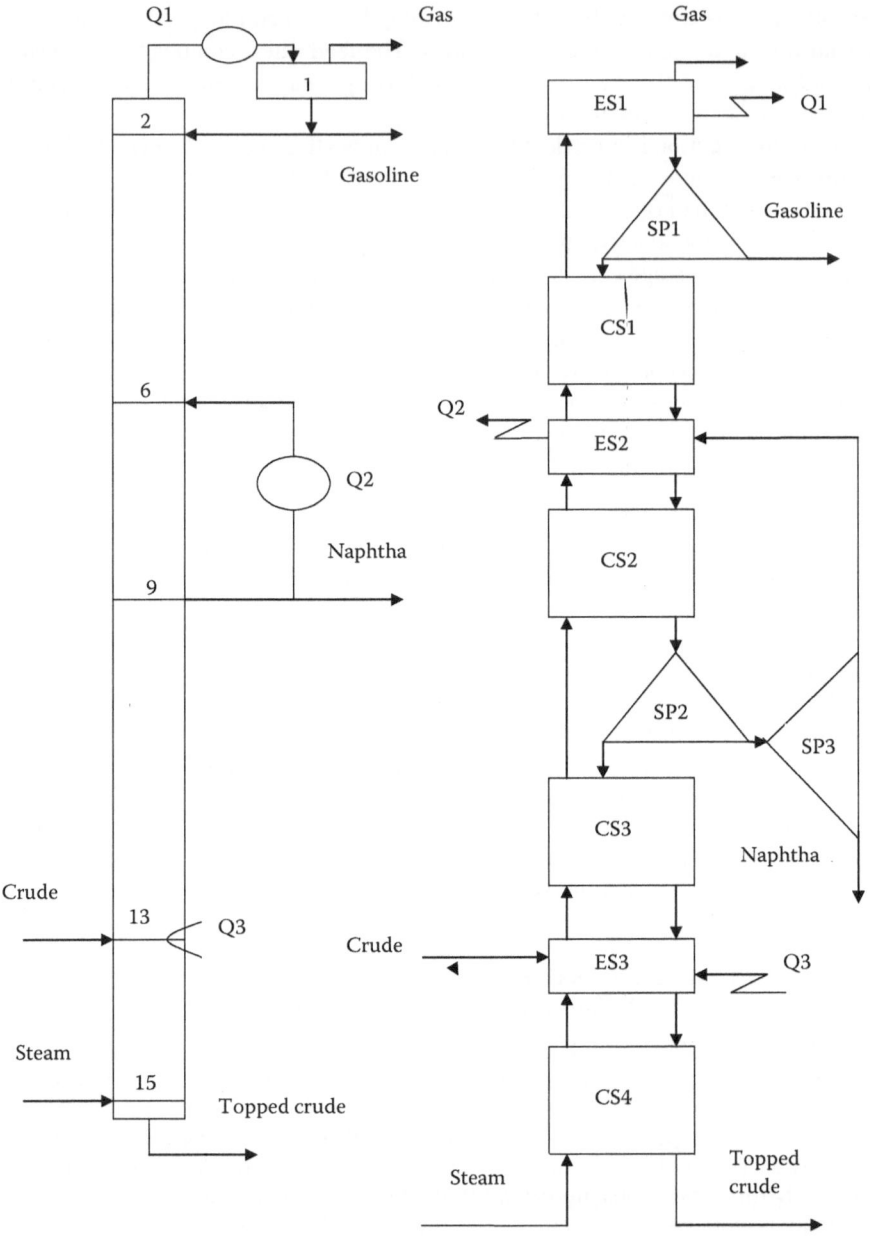

FIGURE 9.5 Column modules representation (Example 9.5).

Equilibrium stages with heat transfer and a fixed pressure provide one degree of freedom each. The feed heater is modeled as a heater on the feed tray. Splitters have one degree of freedom, the split ratio. Column sections have zero degrees of freedom.

Column Part	Module	Degrees of Freedom
Condenser	ES1, equilibrium stage	1, duty Q1
Reflux splitter	SP1, splitter	1, split ratio
Trays 2–5	CS1, column section	0
Pumparound return tray 6	ES2, equilibrium stage	1, duty Q2
Trays 7–9	CS2, column section	0
Liquid draw splitter	SP2, splitter	1, split ratio
Pumparound splitter	SP3, splitter	1, split ratio
Trays 10–12	CS3, column section	0
Feed tray 13	ES3, equilibrium stage	1, duty Q3
Trays 14–15	CS4, column section	0
Variable rate stripping steam		1, steam rate

Degrees of Freedom   7

Each variable corresponding to a degree of freedom may be coupled with a performance specification. Following is a possible set of specifications:

Variable	Specification
Q1	Condenser temperature
SP1 split ratio	Gasoline product rate
Q2	Tray 5 liquid rate
SP2 split ratio	Pumparound rate
SP3 split ratio	Naphtha product rate
Q3	Tray 12 liquid rate
Stripping steam rate	Bottoms temperature

9.2.3 PARTIAL AND TOTAL CONDENSERS

The condenser is the stage where overhead vapors are condensed and liquid is returned to the top of the column as reflux. The condenser is partial if only part of the vapor is condensed and refluxed and the remainder leaves the condenser as vapor distillate. This type of condenser adds one equilibrium stage to the column trays because it holds a vapor phase and a liquid phase at equilibrium with each other. A total condenser is one where the entire overhead vapor is condensed (cooled to the bubble point temperature or subcooled to a lower temperature), part of the condensate is returned as reflux, and the remaining part is taken as liquid distillate. This type of condenser does not count as an equilibrium stage because no vapor–liquid separation takes place in it. The liquid distillate composition is identical to the composition of the vapor leaving the column top tray.

A third type of condenser is often used, one which is a partial condenser but where only part of the condensate is refluxed and the remaining part is drawn as liquid distillate along with the vapor distillate. This type of condenser constitutes an equilibrium stage. The two overhead products have distinct compositions, being the equilibrium vapor and liquid phases. A column with vapor and liquid overhead products may be regarded as a special case of a multi-product column since the liquid overhead is essentially a side product from the condenser, or top stage.

The partial condenser with a vapor product and all the liquid being refluxed is represented by an equilibrium stage. It adds to the column one degree of freedom: the condenser heat rejection rate.

The total condenser consists of a stage with heat rejection, and a splitter that splits the condensed vapor into a reflux and a liquid product. The stage heat rejection rate is one degree of freedom, and the splitter split ratio is another. Hence, a total condenser adds two degrees of freedom to the column.

The partial condenser with vapor and liquid products is represented by an equilibrium stage and a splitter, each of these units having one degree of freedom: The heat rejected at the equilibrium stage and the splitter split ratio. This type condenser also adds two degrees of freedom to the column.

9.2.3.1   Performance of Multi-Product Columns

The factors influencing the performance of multi-product columns are analyzed in Examples 9.6 and 9.7.

EXAMPLE 9.6: SEPARATION OF LIGHT HYDROCARBONS

A stream containing ethane, propane, butane, and pentane is to be separated into four products, each containing mainly one of the four components. A single column with 30 theoretical stages, including a partial condenser and a reboiler, is to be used for this purpose. The column pressure is set at 200 psia. The stream compositions and thermal conditions are given in Table 9.12.

TABLE 9.12
Stream Data for Example 9.6

Component	Feed	Distillate	Side Draw 1	Side Draw 2	Bottoms
			kmol/h		
		Reflux Ratio = 8			
C2	12.00	8.75	2.34	0.91	0.00
C3	48.00	3.25	35.60	8.82	0.33
NC4	25.00	0.00	10.05	9.52	5.43
NC5	15.00	0.00	0.01	5.75	9.24
Totals	100.00	12.00	48.00	25.00	15.00
Temperature (°C)	65.6	4.4	42.8	68.3	117.9
P (kPa)	1380	1380	1380	1380	1380
		Reflux Ratio = 16			
C2	12.00	9.82	1.66	0.53	0.00
C3	48.00	2.18	39.77	6.01	0.02
NC4	25.00	0.00	6.57	16.93	1.50
NC5	15.00	0.00	0.00	1.53	13.48
Totals	100.00	12.00	48.00	25.00	15.00
Temperature (°C)	65.6	−2.3	41.4	73.9	134.9
P (kPa)	1380	1380	1380	1380	1380

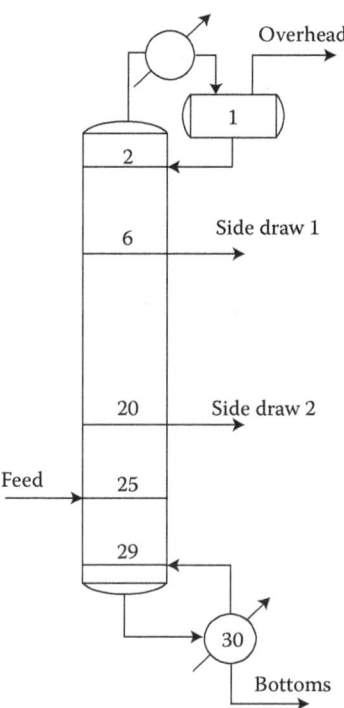

FIGURE 9.6   Multi-product column (Example 9.6).

The relative locations of the products along the column may be estimated on the basis of total reflux shortcut calculations, discussed in Chapter 12. The side draw trays are fixed in this example at the locations indicated in Figure 9.6. The feed location is fixed at tray 25 from the top. The effect of changing the feed location on the column performance is also considered.

COLUMN SPECIFICATIONS

The column has four products, two of which are side draws. Since it also has a partial condenser and a reboiler, it has a total of 2 + S, or 4, degrees of freedom and, therefore, four performance specifications are required to define its operation.

One possible set of specifications would include the flow rates of three of the products and the column reflux ratio. The way the components are split among the products, the identity of the key components in each section are determined by the product rates, and the locations of the feed and products. In this example the products are specified at flow rates that correspond to the component flow rates in the feed. Thus, the distillate is specified at 12 kmol/h, the upper side draw at 48 kmol/h, and the lower side draw at 25 kmol/h. By overall material balance, the resulting bottoms flow rate is 15 kmol/h.

The column is expected to recover most of the ethane in the distillate, the propane in the upper side draw, the butane in the lower side draw, and the pentane in the bottoms. Ethane is the light key in the upper section, propane is the heavy key in the upper section and the light key in the middle section, butane is the heavy key in the middle section and the light key in the lower section, and pentane is the heavy key in the lower section. The distinction between rectification and stripping is relative in a multi-product situation. In the upper section, for instance, ethane is stripped from the upper side draw and propane is separated by reflux from the distillate. Similarly, propane is stripped from the lower side draw, butane is separated from the upper side draw by liquid reflux, butane is stripped from the bottoms, and pentane is separated from the lower side draw by reflux.

The fourth column specification is the reflux ratio, which controls the quality of separation between the products. The reflux ratio sets the L/V ratios in the column. These ratios tend to be fairly constant within each section, depending on the distillation characteristics of the mixture. They are different, though, in each section because the side draws alter the fluid flow in the column. If, for instance, a side draw is a liquid, the liquid flow on the trays below the draw tray is smaller than that on the trays above it. In the feed section, the L/V ratio is generally different above and below the feed tray, depending on the feed thermal conditions.

If all other parameters are held constant, a higher reflux ratio results in better fractionation throughout the column and better separation between the products. Table 9.12 summarizes the stream data at reflux ratios of 8 and 16.

Feed Location

If the feed was placed in the column top section, it would be separated in that section into ethane and the rest of the components, which would be separated into the other products in the lower sections. If the feed was placed in the middle section, it would be separated between propane and butane in that section. The upper and lower sections would then make the separation between ethane and propane, and butane and pentane, respectively. The feed could be placed in the lower section, where the separation would take place between butane and pentane. Further separation into the lighter products would take place in the upper sections.

If the feed is introduced in the upper section, at a given reflux ratio some of the heavier components escape in the distillate and upper side draw due to inadequate rectification. If, at the same reflux ratio, the feed is introduced in the middle section, an appreciable degree of rectification takes place between the feed and the upper side draw, resulting in better recoveries of the desired components in the lower products at the expense of slightly lower purity of the distillate. Putting the feed in the lower section enhances the purity of the bottoms although less stripping of the light components takes place, causing part of them to escape in the lower side draw. In general, for given product rates and reflux ratio, the products that are closer to the feed tend to have higher purities. The other factors that affect the column performance should be included in the search for an optimum feed tray. The product compositions at three different feed locations are summarized in Table 9.13.

Effect of Product Rates

With a fixed configuration and reflux ratio, each product composition may be controlled independently by varying its draw rate. The overhead contains primarily ethane, while heavier components are refluxed. Ethane is the lightest component; therefore, any impurities in this product are heavier components. Lowering the overhead rate should improve its purity by decreasing the concentration of heavier components. The obvious penalty is a lower ethane recovery in this product. Since the fractionation is fixed, the overhead rate must be determined to achieve some balance between purity and recovery. Table 9.14 indicates an ethane recovery in

TABLE 9.13
Effect of Feed Location (Example 9.6, Reflux Ratio = 16)

	kmol/h			
Component	Distillate	Side Draw 1	Side Draw 2	Bottoms
	Feed on Tray 4 (Upper Section)			
C2	10.62	1.37	0.00	0.00
C3	1.37	34.78	11.84	0.01
NC4	0.01	7.93	10.61	6.46
NC5	0.00	3.92	2.55	8.53
	Feed on Tray 10 (Middle Section)			
C2	10.13	1.87	0.00	0.00
C3	1.87	44.74	1.38	0.00
NC4	0.00	1.36	19.66	3.98
NC5	0.00	0.03	3.95	11.02
	Feed on Tray 25 (Lower Section)			
C2	9.82	1.66	0.53	0.00
C3	2.18	39.77	6.01	0.02
NC4	0.00	6.57	16.93	1.50
NC5	0.00	0.00	1.53	13.48

TABLE 9.14
Effect of Product Rates (Example 9.6, Reflux Ratio = 16)

		Mole Fraction			
	kmol/h	C2	C3	NC4	NC5
Overhead	12	0.818	0.182	0.000	0.000
Side draw 1	48	0.035	0.828	0.137	0.000
Side draw 2	25	0.021	0.241	0.676	0.062
Bottoms	15	0.000	0.002	0.101	0.897
Overhead	10	0.912	0.088	0.000	0.000
Side draw 1	48	0.046	0.824	0.130	0.000
Side draw 2	25	0.027	0.300	0.610	0.063
Bottoms	17	0.000	0.005	0.206	0.789
Overhead	13	0.773	0.227	0.000	0.000
Side draw 1	47	0.031	0.840	0.129	0.000
Side draw 2	25	0.019	0.222	0.704	0.055
Bottoms	15	0.000	0.001	0.090	0.909
Overhead	12	0.819	0.181	0.000	0.000
Side draw 1	45	0.035	0.868	0.097	0.000
Side draw 2	28	0.021	0.241	0.682	0.056
Bottoms	15	0.000	0.002	0.102	0.896

the overhead of $12 \times 0.818 = 9.82$ mol/h at a purity of 0.818 mole fraction or a recovery of $10 \times 0.912 = 9.12$ mol/h at a purity of 0.912 mole fraction.

An intermediate product such as the upper side draw, which is mostly propane, may contain impurities from components both lighter and heavier than the main component. The upper side draw is at the same time the bottom product of the top column section and the top product of the second section. The fractionation in the upper section determines to what extent ethane is stripped off from the propane product, and the fractionation in the second section determines to what extent butane is removed from the propane product. Again, in this situation since the number of stages in each section and the reflux ratio are all fixed, the fractionation is fixed. The propane recovery and purity depend mostly on its flow rate and on the flow rates of the adjacent products above and below it. If the propane product contains too much ethane, its flow rate should be cut back and the overhead rate increased. If the propane product contains too much butane, its flow rate should be cut back and the lower side draw rate increased. Table 9.14 summarizes the purities of components in different products at different flow rates. The recoveries can also be calculated from Table 9.14. The dependence of the other products' compositions on their rates may be analyzed in a similar manner.

EXAMPLE 9.7: AMMONIA–ACETONE–WATER COLUMN

A distillation column operating at 500 kPa is used to recover pure ammonia in the distillate from a water–acetone–ammonia solution. The column has 18 theoretical trays, a total condenser, and a reboiler; the feed is introduced on the tenth tray from the top. In the same column, crude separation is to be made between the acetone and the water by taking a liquid side stream. The feed, which is superheated vapor at 200°C and 500 kPa before entering the column, has the following component rates:

Component	Feed Rate (kmol/h)
Ammonia	4000
Acetone	200
Water	500
Total	4700

It is required to determine the best draw tray for the side stream and the effect of the draw rate on the separation.

The column is solved by computer simulation with vapor–liquid equilibrium calculations based on the van Laar liquid activity coefficient equation (Chapter 1). Initially the column is solved without a side draw to determine the composition profiles in the column. The column is solved at a reflux ratio of 1 and a bottoms rate of 800 kmol/h. The composition profiles with no side draw are shown in Figure 9.7. The trays are numbered from the top down, with the condenser as number one. A total of 20 trays are shown, which include the condenser and the reboiler.

The results indicate that the column is overtrayed, especially in the rectifying section, since almost pure ammonia is obtained only two or three trays above the feed. Although the bottoms rate of 800 kmol/h ensures a high-purity ammonia in

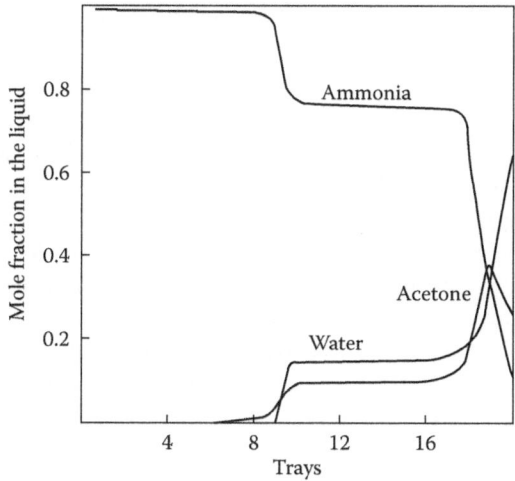

FIGURE 9.7   Composition profiles in the ammonia–acetone–water column (Example 9.7).

the overhead, it does that at the expense of ammonia recovery since much of it is forced down the column and leaves with the bottoms.

The acetone concentration peaks at tray 19, or one stage above the reboiler. This would be a logical place to put the side draw. As the side draw rate is increased, its acetone concentration goes down since more and more water gets drawn with the acetone. The variation in the acetone concentration with the side draw rate is shown in Figure 9.8.

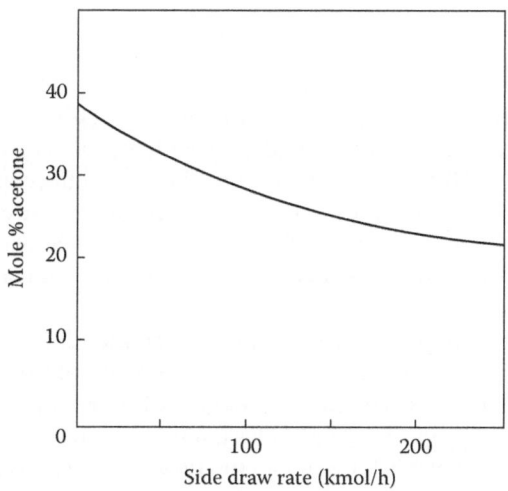

FIGURE 9.8   Effect of side draw rate on acetone concentration (Example 9.7).

9.3   SIDE HEATERS/COOLERS AND PUMPAROUNDS

Distillation columns and other multistage columns generally require heat addition and/or removal, which normally take place at the reboiler and/or condenser.
For a variety of reasons, it may be desirable in certain types of columns to add or
remove heat at intermediate trays aside from the condenser and reboiler. Examples
of such columns include demethanizers that utilize a side reboiler alongside the bottom reboiler, multi-product columns where intermediate condensers are associated
with some of the side products, and absorber intercoolers used for partial removal of
the heat of absorption. The method most commonly employed for exchanging heat
between a column tray and a heat source or sink is the pumparound, where a fluid is
drawn from the tray, sent to a heat exchanger, then pumped back to the column. The
following sections pertain to the various applications of side heaters and coolers and
the different types of pumparounds.

9.3.1   APPLICATIONS

The purposes of side heaters and coolers may be categorized as follows: utilization
of heat sources or sinks at different temperature levels, removal of heat of absorption, and control of vapor and liquid flows in the column.

9.3.1.1   Temperature Levels

Whenever feasible, processes are designed for maximum energy integration aimed at
recovering heat from process streams to the extent that is technically and economically advantageous. For instance, a hot process stream may be a suitable heat source
for a column reboiler, while a relatively cold stream may be used as the heat sink for
the condenser. The net outcome is that the hot and cold streams are brought closer to
their respective target temperatures, thereby resulting in utility savings in both the
column and process streams. It happens quite frequently, however, that the temperature levels of the process streams may not be right for heat exchange at the reboiler
or condenser temperatures. For instance, consider the heat removal in a column to
be done mostly at the condenser, using cooling water at 30°C. It may be desired to
supplement the heat removal in the condenser with a process stream that is to be
heated, say from 50 to 60°C. This stream cannot be used at the condenser if the condenser temperature must be maintained at 38°C. The process stream could be used
for heat removal at a lower tray where the temperature is high enough to allow heat
exchange with the 50°C process stream. A side cooler or condenser could be used
for this purpose.

The use of process streams to supplement the reboiler duty using side reboilers
is similar in principle. In demethanizers, for example, one process stream may
be used as a heat source for the bottom reboiler, while another stream, at a lower
temperature, may be used as a side reboiler a few trays above the bottom reboiler,
where the temperature level is low enough to allow heat exchange with the process
stream.

The implications of side heaters or coolers on the column performance are discussed later in this section.

9.3.1.2 Heat of Absorption

The effect of heat of absorption on the performance of absorbers was discussed in Chapter 8. Under certain circumstances, it may be necessary for improving the absorber efficiency to counteract the effect of the heat of absorption with heat removal using side coolers on certain trays along the column.

9.3.1.3 Column Vapor and Liquid Flows

Side coolers or condensers are commonly used in multi-product columns to prevent excessive variations in the vapor and liquid flow profiles along the column. The multi-product column described in Section 9.2 had, for simplicity, only one condenser and a reboiler. If the feed in this column is introduced at its lower part, the vapor and liquid rates would continuously increase above the feed to the top of the column. A similar situation would exist in crude columns if no side coolers were used.

The variations in the column flow profiles are caused by the liquid side draws. The liquid flow at the top of the column is the reflux rate. Each liquid draw removes a portion of the column liquid flow. With a reduced liquid flow, the vapor flow also decreases since the L/V ratio remains unchanged, assuming other factors are constant. The column flow profiles can be manipulated by redistributing the condenser duty among side coolers along the column.

If a condenser or side cooler is placed on a draw tray or right below it, a portion of the rising vapor condenses. This causes the liquid flow, and consequently the vapor flow, below the side draw to rise, which tends to equalize the profiles above and below the draw tray. The side cooler duty may be adjusted to obtain the desired profiles. In this manner, maintaining the desired liquid and vapor flows in the lower sections of the column does not require unnecessarily high flows in the upper sections.

Each side cooler or heater placed on the column adds one degree of freedom to the column. By varying the side cooler or heater duties, it is possible to meet certain performance specifications such as liquid or vapor profiles or side product purities. The side condensers can enhance the side product purities because better fractionation may be achieved by manipulating the L/V ratio below the side draw.

Example 9.8 examines the effect of side coolers on the column performance.

EXAMPLE 9.8: COLUMN SIDE COOLERS

The liquid and vapor profiles in the multi-product column of Example 9.6 are shown in Table 9.15 for the case of a single condenser at the top stage. The feed is on theoretical tray 25 counting from the condenser, and the reflux ratio is 16. As Table 9.15 indicates, the liquid and vapor profiles increase from the feed tray to the condenser with sizable changes at each draw tray.

Keeping the reflux ratio at 16, additional condensers are placed as heat sinks on trays 7 and 21, just below the draw trays, to maintain fairly uniform liquid and vapor profiles along the column. The new profiles are given in Table 9.16.

The extra side coolers result in better fractionation for the column products even though the reflux ratio is kept the same. The new product compositions are

TABLE 9.15

Column Profiles: No Side Coolers (Example 9.8, Reflux Ratio = 16)

Tray	Temperature (°C)	Liquid Flow (kmol/h)	Vapor Flow (kmol/h)	Product (kmol/h)	Duty (kJ/h)
1	−2	192	12	12	−2.53E6
2	14	185	204		
3	27	187	197		
4	33	188	199		
5	37	186	200		
6	42	130	198	48	
7	48	123	190		
8	56	118	183		
9	62	116	178		
10	66	115	176		
11	69	115	175		
12	71	115	175		
13	71	115	175		
14	72	115	175		
15	72	115	175		
16	72	115	175		
17	72	114	175		
18	73	114	174		
19	73	113	174		
20	74	85	173	25	
21	76	83	170		
22	79	80	168		
23	82	77	165		
24	84	76	162		
25 Feed	96	120	102		
26	109	122	105		
27	119	125	107		
28	126	127	110		
29	131	129	112		
30	135	15	114	15	1.94E6

shown in Table 9.17 and should be compared with the compositions in the upper part of Table 9.14. With the side coolers, a lower reflux ratio is required to obtain a separation equivalent to the case with a single condenser. Table 9.17 also summarizes the product compositions at a reflux ratio of 10 with side coolers. The compositions at this reflux ratio are comparable to those at the higher reflux ratio of 16 (Table 9.14). The result of adding side coolers is that the condenser duty is redistributed among the top and side condensers. The variations in the column profiles are diminished, and the actual vapor and liquid flows are reduced in the top column section since the reflux ratio is lower. It is highly preferable to design a column with a mostly uniform diameter throughout than to have to design the upper sections with larger diameters to handle the higher fluid flows.

TABLE 9.16
Column Profiles with Side Coolers (Example 9.8, Reflux Ratio = 16)

Tray	Temperature (°C)	Liquid Flow (kmol/h)	Vapor Flow (kmol/h)	Product (kmol/h)	Duty (kJ/h)
1	−3	192	12	12	−2.53E6
2	14	185	204		
3	26	187	197		
4	33	188	199		
5	37	186	200		
6	41	132	198	48	
7	46	169	192		−0.66E6
8	54	161	229		
9	62	158	221		
10	67	156	218		
11	71	156	216		
12	73	156	216		
13	74	157	216		
14	75	157	217		
15	76	157	217		
16	76	157	217		
17	76	157	217		
18	76	156	217		
19	77	155	216		
20	77	128	215	25	
21	79	165	213		−0.66E6
22	83	165	250		
23	87	155	245		
24	91	150	240		
25 Feed	103	195	177		
26	116	200	180		
27	124	203	184		
28	130	206	188		
29	134	209	191		
30	137	15	194	15	3.26E6

9.3.2 PUMPAROUNDS

The addition or removal of heat in side heaters or coolers is commonly carried out by drawing a fluid from a column tray, pumping it through a heat exchanger, then returning it to the column. Different pumparound configurations may be envisioned for circulating the liquid or the vapor. In general, it is the liquid that is pumped for exchanging heat, either by subcooling it for heat removal or by partially vaporizing it for heat addition. The pumparound action itself, even with no heat exchange, can affect the column performance. Different types of liquid pumparounds with no associated heat exchange are considered next.

TABLE 9.17
Product Compositions with Side Coolers (Example 9.8)

	kmol/h	kJ/h	Mole Fraction			
			C2	C3	NC4	NC5
		Reflux Ratio = 16				
Overhead	12		0.823	0.177	0.000	0.000
Condenser		−2.53E6				
Side draw 1	48		0.035	0.852	0.113	0.000
Side cooler 1		−0.66E6				
Side draw 2	25		0.017	0.199	0.741	0.043
Side cooler 2		−0.66E6				
Bottoms	15		0.000	0.001	0.070	0.929
Reboiler		3.26E6				
		Reflux Ratio = 10				
Overhead	12		0.770	0.230	0.000	0.000
Condenser		−1.61E6				
Side draw 1	48		0.045	0.804	0.151	0.000
Side cooler 1		−0.66E6				
Side draw 2	25		0.024	0.266	0.658	0.052
Side cooler 2		−0.66E6				
Bottoms	15		0.000	0.001	0.086	0.913
Reboiler		2.31E6				

Either the entire amount of the liquid on a tray or part of it may be taken out and returned to the tray directly below it. The combined liquid flow (the pumparound and the liquid flowing directly to the tray below) would be the same as the liquid flow down to the tray below had there been no pumparound. From the standpoint of equilibrium stages, the column would perform exactly as though the pumparound did not exist although the tray efficiency may be affected due to the different flow pattern. If the liquid is returned several trays below the draw tray, the trays between the draw tray and the return tray are bypassed by the pumparound liquid. The amount of fractionation in that column section is therefore reduced. This pumparound thus tends to lower the overall number of effective trays in the column.

In another type of pumparound, the liquid drawn from one tray is returned a number of trays above the draw tray. This process also tends to lessen the effective number of trays and fractionation in that column section because of back-mixing higher-boiling liquid from the lower tray with lower-boiling liquid in the upper tray.

The above indicates that, while heat removal in the upper sections of multi-product columns improves fractionation, the means for heat removal could undermine it. The net outcome depends on the particular situation, but, in general, pumparounds should be designed to minimize any negative effect they might have on fractionation. For instance, the number of trays between the draw tray and return tray should be kept at a minimum. If a downward pumparound is used and no trays are bypassed between the draw tray and the return tray, no loss in fractionation is incurred.

The amount of heat exchanged in pumparounds depends on the pumparound rate and its temperature change across the heat exchanger. In downward pumparounds, the pumparound rate is limited by the liquid flow in the column. Since the temperature change is also limited by the temperature level of the available cooling or heating medium, the rate of heat exchange in downward pumparounds is subject to practical limitations that could be restrictive. In upward pumparounds, the pumparound rate could, in principle, be increased indefinitely by increasing liquid circulation in the column between the draw tray and the return tray. Of course, this also has its practical limits imposed by column hydraulics considerations, discussed in Chapters 14 and 15. In general, higher heat exchange rates are achievable with upward pumparounds than with downward pumparounds.

9.4   MULTIPLE COLUMN PROCESSES

While multi-product columns could be used to separate mixtures into products with distinct compositions, sharper separations usually require processing with multiple columns to achieve the desired splits and purities. A possible scheme would be to break down the separation process into a forward flowing series of single-feed, two-product distillation columns, each of which makes the separation between two key components. In a more complex multi-column approach, the process may include recycle streams, external separation-enhancing streams, stream splits, recombination of streams, energy integration, and so on, all aimed at achieving an economic optimum.

It is evident that a host of alternative arrangements of columns could be used for making the same set of separation specifications, with the number of arrangements escalating steeply as the number of components to be separated increases. Finding the optimum thus becomes a truly challenging task. Although computer simulation is the most efficient method for final evaluation and comparison of different plausible options, narrowing down the field on the basis of some systematic, logical approach is highly desirable for saving both computer time and human hours. Unfortunately, no systematic method for evaluating complex, multi-column separation processes seems to exist, although workable heuristic rules are available for selecting the better sequences among the simple, forward-flowing processes.

For processes belonging to this category and involving only single-feed, two-product distillation columns, a single sequence exists for separating a binary:

$$AB \rightarrow A + B$$

That is, the mixture AB is fed to a distillation column; the lighter component, A, is recovered in the distillate; and the heavier component, B, is recovered in the bottoms. (Components are arranged in the order of decreasing volatility.) Two sequences are possible for separating a ternary:

$$ABC \rightarrow A + BC \rightarrow A + B + C$$
$$ABC \rightarrow AB + C \rightarrow A + B + C$$

Two columns are needed in either of these sequences: the first to recover one of the components in the ternary and the second to separate the remaining two. For a four-component mixture, five alternative sequences could be used, each requiring three distillation columns:

$$ABCD \rightarrow A + BCD \rightarrow A + B + CD \rightarrow A + B + C + D$$
$$ABCD \rightarrow A + BCD \rightarrow A + BC + D \rightarrow A + B + C + D$$
$$ABCD \rightarrow AB + CD \rightarrow A + B + C + D$$
$$ABCD \rightarrow ABC + D \rightarrow A + BC + D \rightarrow A + B + C + D$$
$$ABCD \rightarrow ABC + D \rightarrow AB + C + D \rightarrow A + B + C + D$$

Sequences could be listed in this manner for any number of components.

For these simple processes, certain rules can be used to select a number of preferred sequences from which an optimum could be chosen. The rules are not absolute and will generally generate several options. Many factors that influence the final decision are specific to the particular process and its economics, and so on, and are not considered in the selection rules. The final selection should be based on rigorous simulation models of the chosen sequences.

According to the rules proposed by Seader and Westerberg (1977), the components in the feed are arranged by decreasing volatility and two sets of parameters are checked: the relative volatilities of adjacent pairs and the molar percentages of the components in the feed. The relative volatility, α_{ij}, of component i with respect to component j is used as an indicator of how readily they could be separated and is defined as the ratio of their vapor–liquid equilibrium coefficients:

$$\alpha_{ij} = K_i/K_j$$

The separation sequence is determined according to the following rules:

1. Where the relative volatilities vary widely, splits are sequenced in the order of decreasing relative volatility.
2. Where the relative volatilities do not vary widely, the removal of components is sequenced in the order of their molar percentages in the feed, starting with highest, if these percentages vary widely.
3. If neither relative volatilities nor molar percentages vary widely, the components are separated in a direct sequence of columns; that is, the lightest component is separated as the distillate product of the first column, the bottoms is sent to the next column where the next heavier component is separated as distillate product, and so forth.

In Example 9.9, separation strategy is considered for producing pentanes from a naphtha stream. The above heuristic rules are taken into account as well as other factors that are specific to the problem.

EXAMPLE 9.9: SEPARATING PENTANES FROM NAPHTHA

A mixture of primarily C5 components is to be separated from a naphtha stream containing hydrocarbons ranging from C3 to C10. It is proposed to utilize two existing columns to carry out the separation, one having 18 theoretical stages and the other, five. It is required to investigate alternative configurations for the process.

The feed composition, along with approximate average relative volatilities for the adjacent components over the expected range of conditions, is given in Table 9.18. The required separation is not between individual components but rather between three groups of components: those lighter than the pentanes (C3 through NC4), the pentanes (IC5 and NC5), and those heavier than the pentanes (NC6 through NC10). Consequently, the splits that matter are the NC4/IC5 and the NC5/NC6 splits. Table 9.18 indicates that the relative volatility of the NC5/NC6 pair is somewhat higher than that of the NC4/IC5 pair. That is, the NC5/NC6 split is the easier one.

TABLE 9.18
Naphtha Feed and Component Relative Volatilities (Example 9.9)

	kmol/h	α_{ij}
C3	148.2	
		1.75
IC4	91.0	
		1.18
NC4	239.2	
		1.79
IC5	135.2	
		1.19
NC5	174.2	
		2.29
NC6	218.4	
		1.25
Benzene	65.0	
		1.12
Cyclohexane	101.4	
		1.45
NC7	288.6	
		1.21
Toluene	426.4	
		2.16
NC10	712.4	
Total	2600.0	

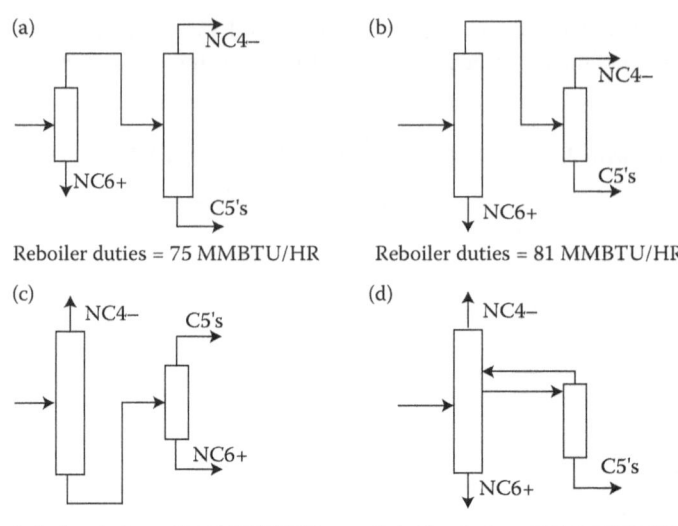

(a) NC4– NC6+ C5's
Reboiler duties = 75 MMBTU/HR

(b) NC4– C5's NC6+
Reboiler duties = 81 MMBTU/HR

(c) NC4– C5's NC6+
Reboiler duties = 84 MMBTU/HR

(d) NC4– NC6+ C5's
Reboiler duties = 77 MMBTU/HR

FIGURE 9.9 Alternative configurations for naphtha splitting (Example 9.9), (a) Reboiler duties = 75 MMBTU/HR, (b) Reboiler duties = 81 MMBTU/HR, (c) Reboiler duties = 84 MMBTU/HR, and (d) Reboiler duties = 77 MMBTU/HR.

Although the gap between the two key relative volatilities is not too wide, the first heuristic rule suggests that the first split should be between NC5 and NC6. Thus, the NC6 and heavier components would be removed in the bottoms of the first column. The distillate would then be separated in the second column into NC4 and lighter components in the distillate and the C5s in the bottoms (Figure 9.9a).

Another factor that should be considered in this application is the fact that the separation is to be carried out in two existing columns of different sizes. Since the easier separation requires fewer trays, the logical choice would be to use the smaller column for the initial split and the larger column for recovering the C5s. Figure 9.9 shows some of the possible configurations for the separation. Figure 9.9d shows a side stripper arrangement which differs from the rest in that it separates the pentanes as a side draw from the main column. This stream is then sent to the smaller column, which is used as a stripper to remove some of the lighter components in the overhead. The overhead stream is recycled back to the main column to recover some of the pentanes carried over.

Comparing the different options requires an economic analysis, although a good approach would be to compare the total reboiler duties required in each configuration to achieve similar separations and recoveries. The duty requirements for the different configurations are obtained from simulation and are shown in Figure 9.9. Based only on this criterion, the results appear to support the heuristic rules, which favor option A. This is followed closely by the recycle scheme, option D, which is not considered in the heuristic rules.

NOMENCLATURE

K_i Vapor–liquid equilibrium distribution coefficient
L Liquid molar flow rate in the column

S   Side product
V   Vapor molar flow rate in the column
α_{ij} Relative volatility of component *i* with respect to *j*

SUBSCRIPTS

i   Component designation
j   Component designation

PROBLEMS

9.1. An existing column has a reboiler, no condenser, and two feeds, one of which is at the top of the column. The top feed is of fixed temperature, pressure, and composition, but its rate is variable. The other feed is of fixed rate, composition, and pressure, and flows through a heat exchanger upstream of the column so that its temperature may be controlled before entering the column. How many independent variables are required to define this process? If the key components are ethane and propane, describe conceptual control loops to control the separation and performance of this process.

9.2. Two light hydrocarbon streams are to be separated into propane-rich and butane-rich products using one distillation column. Assuming the streams are of fixed rates, compositions, and thermal conditions, and that the column pressure is predetermined, how many degrees of freedom are available at the design phase if

 a. The streams are mixed and fed at the same tray.

 b. Each stream is fed separately to the column.

     Given the following feed stream definitions, what would be the recommended relative feed locations if they are fed separately?

	Feed 1	Feed 2
	kmol/h	
Ethane	15	25
Propane	65	82
Butane	37	77
Pentane	13	16
Temperature (°C)	10	24
Pressure (kPa)	680	680

9.3. A column has two feeds, a partial condenser, a partial reboiler, a vapor distillate, a bottoms product, and a side draw. The feeds are of fixed flow rates, compositions, and thermal conditions.

 a. How many degrees of freedom exist during the design phase?

 b. How many degrees of freedom exist once the column is built?

9.4. A distillation column is to be designed to separate a stream into two products, and to remove light impurities. A partial condenser is proposed, where the impurities would be removed in the vapor distillate, and the liquid distillate and bottoms would be the main column products.

Preliminary simulation calculations indicate that too much of the distillate product would be lost with the impurities, and the distillate product would contain too many impurities. How would you modify the column design to improve its performance?

9.5. A distillation column with a total condenser and a reboiler separates a stream into a distillate product, heavy residuals in the bottoms, and a main intermediate product as a side draw several trays above the reboiler. For an existing column, with a feed of fixed rate, composition, and thermal conditions, determine the number of variables required to define the column operation. Assume the column pressure is set, and no heat transfer occurs except in the condenser and reboiler. How would you control the separation in the column?

9.6. It is proposed to utilize a hot process stream as a heat source to supplement the reboiler duty of an existing distillation column. The heat from the process stream would be added to the column via a pumparound, a few trays above the reboiler. How many additional degrees of freedom would result from the proposed modification?

9.7. A column is designed for f feeds, a partial condenser with only a vapor product, a reboiler, a bottoms product, s side draws, and h pumparound heaters/coolers. Assuming the feeds are of fixed rates, compositions, and thermal conditions, determine the number of independent variables required to uniquely define the column during design, and its operation after construction.

9.8. A distillation column side draw constitutes the main product of the column, where lighter impurities are removed in the distillate and heavier impurities are removed in the bottoms. Because of anticipated variations in the feed composition the column is equipped with a side cooler just below, and a side heater just above the side draw to improve operating flexibility. What potential operating problems could the side cooler and heater help alleviate? What specifications would you use to define the column operation?

9.9. A stream containing components A, B, C, D, E, and F is to be separated into these six components using five two-product ordinary distillation columns. The composition of the feed and the relative volatilities of the components are listed below, with the components arranged by decreasing volatility. What separation sequence would you recommend for this process?

	Mole Percent in Feed	Relative Volatility
A	5	
		1.6
B	17	
		1.5
C	35	
		2.4
D	14	
		1.7
E	16	
		1.3
F	13	

9.10. How many operating degrees of freedom exist for the column configuration shown in Figure 9.9d? The main column has a reboiler and a partial condenser with only a vapor distillate. The sidestripper has a reboiler but no condenser. The external feed is of fixed rate, composition, and thermal conditions.

9.11. A stream containing benzene, toluene, and biphenyl is to be separated in a distillation column to produce purified benzene in the distillate. The separation will take place in an existing column with a total condenser, a partial reboiler, and several optional feed locations. The feed stream is of fixed flow rate, composition, and thermal conditions. The entire feed may be introduced at any one of the available feed trays, but may not be split and introduced at more than one feed tray. The condenser pressure is controlled by an inert gas flowing in and out of the reflux drum. Using column modules representation, determine the degrees of freedom for this operation, and recommend a set of specifications to define the column performance.

9.12. Two distillation columns are thermally coupled, with the side column receiving vapor and liquid from the main column as shown in the diagram. These are existing columns with fixed feed and draw locations. The side column provides initial fractionation of the feed, thereby allowing the main column to complete the separation with its existing trays. This column has a total condenser and a partial reboiler. The external feed is of fixed rate, composition and thermal conditions. The pressure in the columns is determined independently by the tray hydraulics. Determine the degrees of freedom of this system by representing it using basic modules, and suggest variables you would specify to define the column performance.

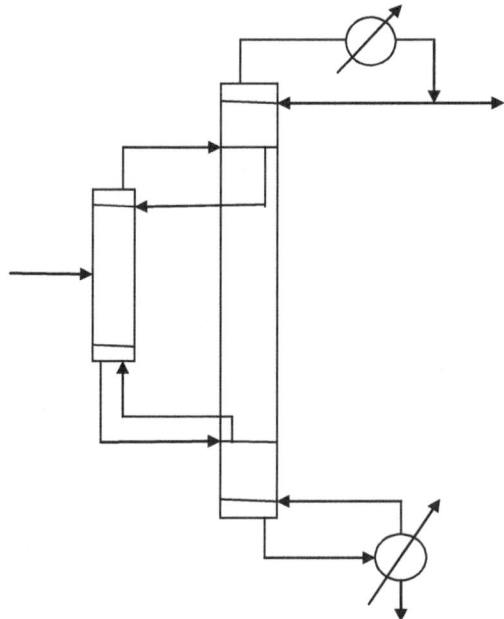

9.13. A distillation column with an attached side stripper is used to separate a feed stream F into products A, B, and C, as shown in the diagram. The main column has a total condenser and a partial reboiler, and the side stripper has a partial reboiler. The columns are existing units with fixed number of trays and feed and draw locations. The external feed, a product of an upstream unit, is of fixed flow rate, composition, and thermal conditions. The pressure in the columns is determined independently and is not available as a variable. Using basic modules representation, determine the degrees of freedom of this system. What variables would you specify to define the performance of these columns?

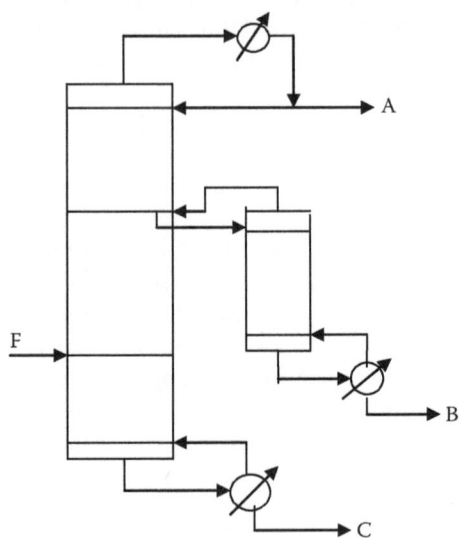

REFERENCE

Seader, J. D. and A. W. Westerberg, *AIChE J.*, 23, 951, 1977.

10 Special Distillation Processes

Mixtures exhibiting nonideal solution behavior present both challenges and opportunities in connection with separation processes. Azeotropes cannot be separated by ordinary distillation, yet the formation of azeotropes itself may be used as a means for carrying out certain separations. The formation of two liquid phases in a column may complicate the separation process; however, the coexistence of liquid phases with distinct compositions provides one more separation tool. Chemical reactions concurrent with distillation may be used either to enhance the separation or to perform both the reaction and the separation in one process.

This chapter covers several special distillation processes, including azeotropic and extractive distillation, three-phase distillation, and reactive distillation. As in other multistage separation processes, rigorous solution methods begin by setting up the material and energy balance equations, the phase equilibrium relations, and chemical reaction equations (if reactive processes are involved). Currently, the discussion centers around understanding these processes from the standpoint of performance analysis, assisted by graphical representation when applicable. Rigorous solution methods are discussed in Chapter 13.

10.1 AZEOTROPIC AND EXTRACTIVE DISTILLATION

The formation of azeotropes due to deviations from Raoult's law was discussed in Section 1.3. An azeotrope is a mixture that, at a given pressure (the azeotropic pressure), boils at a constant temperature (the azeotropic temperature) and has the same composition (the azeotropic composition) in the equilibrium vapor and liquid phases. Homogeneous azeotropes are those that form one liquid phase at equilibrium with the vapor; heterogeneous azeotropes are those that form two liquid phases at equilibrium with each other and the vapor.

Homogeneous azeotropes can be either minimum boiling, in which the boiling point of the azeotrope is lower than any of those constituents, or maximum boiling, where the boiling point of the azeotrope is higher than any of those constituents. Heterogeneous azeotropes are invariably minimum boiling. Azeotropes can be multi-component although those most commonly encountered in practice are either binary or ternary.

The absence of an interphase compositional differential makes the separation of azeotropes into their constituent components impossible by conventional vapor–liquid separation processes. The azeotropic pattern may be altered by adding an external component—an entrainer—that breaks the original azeotrope while forming new ones with the feed components. The process may be designed so that the new azeotropes may be separated from individual components or from other azeotropes by what is known as azeotropic distillation. Azeotropic distillation is not limited to the separation of azeotropes but is also used for separating close boilers that are difficult to be separated by conventional distillation.

Extractive distillation is another process used for separating azeotropes or close boilers. In this process, a solvent is added that alters the relative volatilities of the feed components through its preferential affinity for one or more of the components over the others.

In analyzing the complex processes of azeotropic and extractive distillation, it is convenient to represent the system as a ternary that includes the two key components to be separated and the entrainer or solvent. Real mixtures are, of course, normally multi-component, but focusing on the key components and the separating agent helps in obtaining a basic grasp of the problem. A qualitative understanding of the operation is important for preliminary evaluations aimed at selecting the entrainer or solvent, estimating the amount of entrainer or solvent, the approximate number of trays required, and the feed location, as well as considering any associated ancillary operations that may be essential for the process. With this background, the use of computer programs for accurate simulation of the process becomes highly effective.

The terms *entrainer* and *solvent* are commonly used interchangeably to refer to the separating agent used to enhance the separation of close boilers or azeotropes by azeotropic or extractive distillation. For consistency, the term *entrainer* will be used to designate the azeotropic distillation agent and *solvent*, the extractive distillation agent.

In azeotropic distillation, when two components are difficult to be separated due to the proximity of their boiling points or because they form an azeotrope, an entrainer is added that forms an azeotrope with at least one of the components to be separated. The entrainer must be selected so that the azeotropes that are formed have sufficiently spaced boiling points to facilitate their separation by distillation. Moreover, it must be verified that the azeotropes that are formed can themselves be separated by relatively straightforward and economical means. The number of suitable entrainers for a given situation is, therefore, quite limited, especially when other factors in the selection process are taken into account such as cost, stability, and safety.

The entrainer is usually introduced to the column with the feed or may be added in the column as reflux.

In extractive distillation the solvent attracts one of the components to be separated more strongly than the other on account of their different chemical structures. The solvent has a markedly higher boiling point than the components in the feed and tends to lower the volatility of one of the components preferentially. This component leaves the column at the bottoms with the solvent. Recovering the solvent from the bottoms is relatively easy because of its low volatility compared to the component to be separated.

Hydrogen bonding is one contributor to the preferential attraction between the solvent and certain components. Similarity of the chemical structure is another. For instance, the separation of benzene and cyclohexane, two close boilers that also form an azeotrope, can be achieved by extractive distillation using phenol as the solvent. Benzene is more strongly attracted to phenol because both contain an aromatic ring, whereas cyclohexane does not. Therefore, in this process, benzene is taken in the column bottoms with the phenol, and cyclohexane is recovered in the column overhead.

The choice of a selective solvent is easier the more the components to be separated differ in their chemical structure. It would be difficult or impossible, for instance, to find a selective solvent for the separation of stereoisomers. Nevertheless, the restrictions on extractive distillation solvents are less severe than those on azeotropic distillation entrainers, because the solvent recovery problem is virtually nonexistent due to the wide gap between the boiling points of the solvent and the components to be separated.

Due to its low volatility, the solvent in extractive distillation is introduced near the top of the column, several trays above the main feed tray. This is to ensure adequate concentration of the solvent on most trays, thereby maintaining the desired relative volatility throughout the column.

Various processes are used for separating components that are difficult or impossible to be separated by conventional distillation. Whether the difficulty of separation arises from the components' close boiling points or their tendency to form azeotropes, the separation processes must take into account the complex vapor–liquid equilibrium relationships of the system. The system to be considered involves both the components to be separated and the separating agent that, in one way or another, enhances the desired separation. The vapor–liquid equilibria of such mixtures is highly nonideal, and it is precisely this nonideality that is capitalized on to bring about the separation.

In view of the complexity of these processes, they are typically examined on a case-by-case basis with regard to selection of the separating agent, configuration of the column(s), process conditions, and so on. Nevertheless, certain general principles apply that allow categorizing the methods as described below. The processes are discussed generically, and actual applications are cited as examples for each process.

10.1.1 Separating Azeotropes with Pressure-Sensitive Composition

This is one case in which the separation of azeotropes may be carried out without the use of a separating agent. The method applies to minimum- or maximum-boiling homogeneous azeotropes that are characterized by an appreciable variation in the azeotropic composition with pressure. The effect of pressure on the azeotropic composition may be estimated for a binary, provided the activity coefficients can be adequately represented by one of the activity coefficient equations (Section 1.3.4).

The $Y–X$ and $T–X$ diagrams of a typical minimum-boiling, homogeneous, binary azeotrope are shown in Figure 10.1, where compositions are given as mole fractions of component A, the lighter (lower boiling) component to be separated from component B. The diagrams are given at two pressures, showing the directional effect of the pressure on the azeotropic composition.

The separation process employs two columns at two different pressures to take advantage of the pressure effect. The details of the process may vary from case to

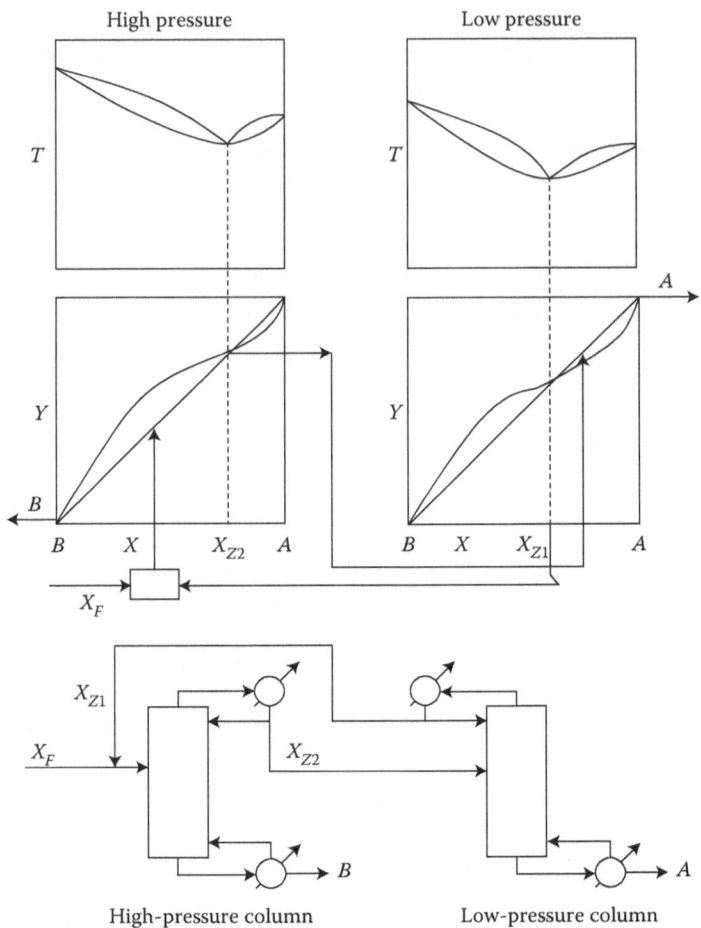

FIGURE 10.1 Separating azeotropes with pressure-sensitive composition.

case, depending in part on the feed composition relative to the azeotropic composition at the two column pressures. Assume the column pressures are such that the feed composition, X_F, is between the azeotropic compositions, X_{Z1} and X_{Z2}, at the lower and higher column pressures, respectively. Sending the feed to the lower pressure column would produce relatively pure component A in the bottoms and the azeotrope in the overhead. Sending the feed to the higher pressure column would produce relatively pure component B in the bottoms and the azeotrope in the overhead.

Figure 10.1 shows a schematic for one possible process. The feed is mixed with the overhead from the low-pressure column and then fed to the high-pressure column, where component B is separated in the bottoms and the higher pressure azeotrope is removed in the overhead. The overhead is sent to the lower pressure column, where the azeotropic composition is leaner in component A than the feed to this column. This column thus separates component A in the bottoms and the lower pressure azeotrope in the overhead. To complete the cycle, this overhead stream is mixed with fresh feed and sent to the high pressure column. The combined composition is leaner in component A than the fresh feed, thus ensuring that the combined feed is to the left of the higher pressure azeotropic composition.

The column sequence could be reversed and other configurations could be laid out for different feed compositions. For instance, if $X_F < X_{Z1}$, then the feed must go to the high-pressure column first so that the resulting azeotropic composition will be on the other side of the lower-pressure azeotropic composition. If $X_F > X_{Z2}$, the feed must first be sent to the lower-pressure column. The final selection of a particular configuration among possible alternatives is best made on the basis of computer simulation of the various options to determine the economic optimum.

The concept is the same for separating maximum-boiling, homogeneous, binary azeotropes by using two columns at different pressures. In this case, the pure components are recovered as overhead products since they are lighter than the azeotrope.

Examples of pressure-sensitive, homogeneous, binary azeotropes that are amenable to separation by processes as described above are given in Table 10.1, extracted from Othmer (1963).

TABLE 10.1
Homogeneous Azeotropes with Pressure-Sensitive Compositions

	P (kPa)	T (°C)	Composition (1st comp)
Minimum-Boiling Binaries			
n-Propyl alcohol/water	98.64	87	71.7 wt.%
	790.47	151	73.3 wt.%
Acetonitrile/water	20.00	34	72.5 mole%
	101.31	77	84.5 mole%
Ethyl alcohol/water	13.33	34.2	89.0 mole%
	193.28	95.3	99.6 mole%
Methyl alcohol/benzene	26.66	26	33.1 wt.%
	1466.30	149	63.0 wt.%
Methyl alcohol/MEK	13.33	18.4	52 wt.%
	271.93	92.3	85 wt.%
Acetone/water	342.58	95.8	78.0 mole%
	1375.66	155.8	96.5 mole%
Ethyl acetate/ethyl alcohol	3.33	−1.4	61.13 wt.%
	196.68	91.4	87.15 wt.%
Maximum-Boiling Binaries			
HCl/water	6.66	48.7	19.3 wt.%
	161.29	123	23.5 wt.%
HBr/water	13.33	74.12	47.03 wt.%
	159.96	137.34	49.80 wt.%
MEK/water	26.66	39.9	65.4 mole%
	101.31	73.3	72.2 mole%

10.1.1.1 Graphical Representation

Vapor–liquid equilibrium diagrams of homogeneous binary azeotropes may be viewed as consisting of two parts, each resembling a regular binary diagram. For instance, the Y–X diagrams in Figure 10.1 consist of a curve between the heavy component and the azeotrope and another one between the azeotrope and the light component. Each of these curves resembles a regular binary Y–X curve, although the second one lies below the diagonal. This is because the azeotrope, with its lower boiling point, is to the left of the higher boiling component A.

The significance of this analogy to regular binary systems is that the graphical techniques of binary distillation discussed in Chapters 5 and 6 may be applied to distillation columns used for separating a pure component from an azeotrope. The effect of such parameters as number of trays, reflux ratio, and feed location may be studied, at least qualitatively, using these graphical techniques for preliminary column design and for determining performance trends. Once a fundamental understanding of the process is achieved, a computer model can be set up to carry out the detailed rigorous calculations.

Equilibrium curve sections, each on the Y–X diagram to the left and right of the azeotrope, are used for a different column. Operating lines drawn on one section cannot be extended across the azeotropic point to the other section. The section of the equilibrium curve that lies below the diagonal may be redrawn above the diagonal if the azeotrope and the heavy pure component places are switched by expressing the compositions in terms of the heavier rather than the lighter component.

For each equilibrium curve section, a column may be designed with the required number of trays and reflux ratio to achieve the desired component purity at one end of the column and the azeotropic composition at the other.

EXAMPLE 10.1: ETHANOL/BENZENE SEPARATION

Ethanol and benzene form minimum-boiling azeotropes with different compositions at different temperatures and pressures. A 100 kmol/h stream at 60°C and 100 kPa, containing 35 mole% ethanol and 65 mole% benzene, is to be separated by capitalizing on the pressure sensitivity of the azeotropes. It is required to prepare a preliminary design for the process, showing the main units and flow sequence, and the various stream rates and compositions. Both the ethanol and the benzene should be separated to 99 mole% purity.

SOLUTION

The boiling points of ethanol and benzene at 100 kPa are 78.4°C and 80.1°C, respectively, and at 27 kPa they are 48.0°C (ethanol) and 42.3°C (benzene). The azeotropic temperature at 100 kPa is 68°C and the composition is 44.8 mole% ethanol. At 27 kPa the azeotropic temperature is 35°C and the composition is 36.0 mole% ethanol.

The feed stream has a lower concentration of ethanol than the azeotrope at 100 kPa. If the feed is sent first to a 100 kPa pressure column, the bottoms will be essentially pure benzene at ~80°C (corresponds to point B in Figure 10.1) and the distillate will be close to the azeotropic composition of 44.8 mole% ethanol. This stream is then fed to a 27 kPa pressure column. Its ethanol concentration is higher than the azeotropic composition at this pressure, 36.0 mole% ethanol. The bottoms of this column will be essentially pure ethanol at ~48°C (corresponds to point A in Figure 10.1) and the distillate will be close to the azeotropic composition of 36.0 mole% ethanol. This stream is recycled and combined with the main feed.

The following material balance calculations are carried out to determine the stream flow rates consistent with the required compositions. Designate the feed stream as F, the first column (high pressure) bottoms and distillate as B_1 and D_1, and those of the second column (low pressure) as B_2 and D_2, respectively. Stream D_2 is recycled and combined with F. The first column distillate composition will be taken at 44.4 mole% ethanol (slightly below the azeotrope) and that of the second column at 36.4 mole% ethanol (slightly above the azeotrope). The specifications are as follows:

	Feed F	D_1	B_1	D_2	B_2
1. Ethanol	$X_{F1} = 0.35$	$Y_{11} = 0.444$	$X_{11} = 0.01$	$Y_{21} = 0.364$	$X_{21} = 0.99$
2. Benzene	$X_{F2} = 0.65$	$Y_{12} = 0.556$	$X_{12} = 0.99$	$Y_{22} = 0.636$	$X_{22} = 0.01$

Ethanol material balance on column 1:

$$FX_{F1} + D_2Y_{21} = D_1Y_{11} + B_1X_{11}$$
$$100 \times 0.35 + 0.364D_2 = 0.444D_1 + 0.01B_1$$

Benzene material balance on column 1:

$$FX_{F2} + D_2Y_{22} = D_1Y_{12} + B_1X_{12}$$
$$100 \times 0.65 + 0.636D_2 = 0.556D_1 + 0.99B_1$$

Ethanol material balance on column 2:

$$D_1Y_{11} = D_2Y_{21} + B_2X_{21}$$
$$0.444D_1 = 0.364D_2 + 0.99B_2$$

Benzene material balance on column 2:

$$D_1Y_{12} = D_2Y_{22} + B_2X_{22}$$
$$0.556D_1 = 0.636D_2 + 0.01B_2$$

Solving these equations, $B_1 = 65.306$ kmol/h, $B_2 = 34.694$ kmol/h

$$D_1 = 271.45 \text{ kmol/h}, D_2 = 236.75 \text{ kmol/h}$$

It may be worthwhile to check the outcome if the column sequence is reversed. Sending the feed initially to the low-pressure column, the bottoms will be essentially pure benzene and the distillate would be close to the 36 mole% ethanol azeotrope. If this stream is sent to the high-pressure column, with its ethanol concentration lower than the high-pressure azeotrope (44.8 mole% ethanol), the bottoms would again be essentially pure benzene and the distillate would be the azeotrope. Therefore this setup would not produce the required separation.

10.1.2 SEPARATING HETEROGENEOUS MINIMUM-BOILING AZEOTROPES

This is another case in which azeotropes can be separated without the addition of a separating agent. The process applies to heterogeneous minimum-boiling azeotropes: those that form two liquid phases in certain regions of temperature, pressure, and composition. Figure 10.2 shows a schematic T–X diagram of a typical heterogeneous binary minimum-boiling azeotrope. The compositions are expressed as mole fractions of the lighter component.

A fresh feed with a composition in region L1 (single liquid that is predominantly component 1) may be distilled in a conventional column to produce essentially pure component 1 in the bottoms and an overhead at or close to the azeotropic composition X_{az}. If the feed composition is in region L2 (single liquid that is predominantly component 2), it may be separated by conventional distillation into a bottoms product that is relatively pure component 2 and an overhead that is around the azeotropic composition X_{az}. The overhead from either column may be cooled

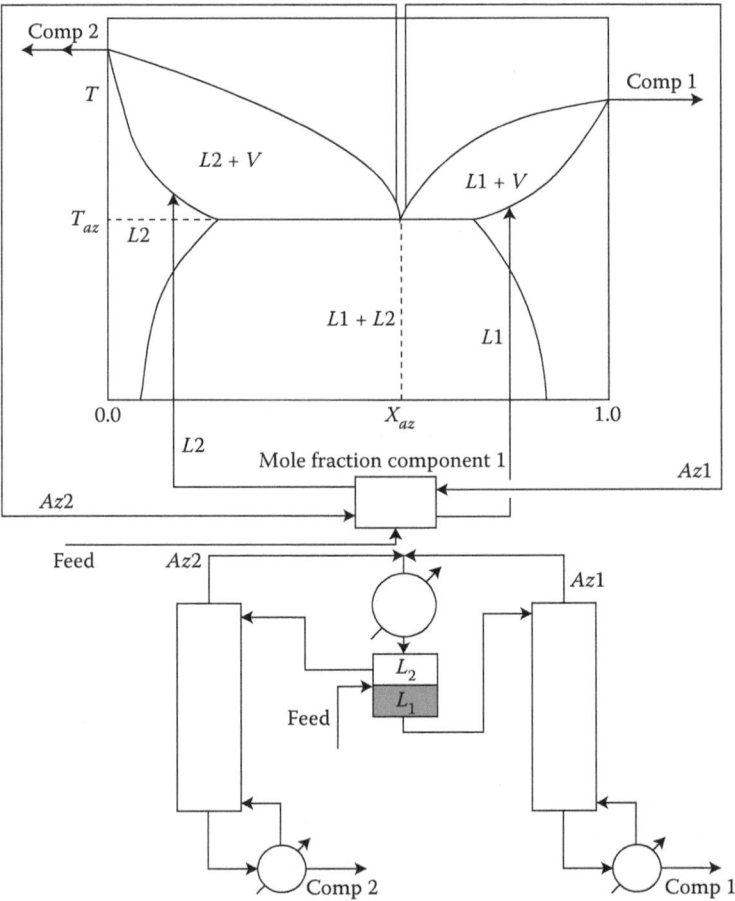

FIGURE 10.2 Separating heterogeneous minimum-boiling azeotropes.

below the azeotropic temperature, T_{az}, to the two-liquid phase region where the two phases, one rich in component 1 and the other rich in component 2, are separated by decantation.

Figure 10.2 shows a schematic of a separation process for a feed with a composition in the two-liquid phase region. The liquid phases are separated and each is distilled in a separate column as described above. The overhead streams from both columns (azeotropes $Az1$ and $Az2$) are mixed, condensed, and recycled to the liquid–liquid separator, where they are combined with the fresh feed.

Examples of the separation of heterogeneous azeotropes using such a process include n-butanol–water, ethyl acetate–water, and ethyl ether–water (Othmer, 1963).

10.1.2.1   Graphical Representation

Each of the distillation columns used in the heterogeneous azeotropic distillation is essentially a conventional column separating a binary: one column separates component 1 and the azeotrope, and the other separates component 2 and the azeotrope. As such, the corresponding Y–X diagrams can be used for preliminary evaluation of the process and the required number of stages using graphical methods as described in Chapters 5 and 6. On this foundation rigorous computer simulation provides accurate process computations for design and optimization.

EXAMPLE 10.2: SEPARATION OF BUTANOL-WATER BY HETEROGENEOUS AZEOTROPIC DISTILLATION

Normal butanol (1) and water (2) form a heterogeneous azeotrope at 100 kPa, 93°C, and mole fraction n-butanol $X_{az,1} = Y_{az,1} = 0.25$. At the azeotropic temperature the two liquid phases have compositions $X_1(L1) = 0.40$ and $X_1(L2) = 0.025$, where $L1$ is the butanol-rich liquid and $L2$ is the water-rich liquid. The equilibrium data at 100 kPa are given:

	VLE			LLE	
T (°C)	X_1 at BP	Y_1 at DP	T (°C)	X_1 in $L2$	X_1 in $L1$
100	0	0			
98	0.005	0.085			
96	0.015	0.155			
94	0.020	0.22			
92.8	0.025	0.25			
92.8	0.40	0.25	92.8	0.025	0.40
93	0.51	0.26	90.0	0.020	0.40
94	0.595	0.28	85.0	0.015	0.40
96	0.68	0.33			
100	0.80	0.42			
105	0.89	0.56			
110	0.91	0.70			
115	0.99	0.87			
117	1.00	1.00			

A 100 kmol/h feed stream, containing 35 kmol/h n-butanol and 65 kmol/h water, is to be separated into essentially pure n-butanol and pure water. It is required to plot the equilibrium data and calculate the flow rates and compositions of the products and all the connecting streams.

The operating data are as follows:

Feed, $F = 100$ kmol/h, $X_{F1} = 0.35$, $X_{F2} = 0.65$
Column 1 bottoms, B1, $X_{B1,1} = 1$, $X_{B1,2} = 0$
Column 2 bottoms, B2, $X_{B2,1} = 0$, $X_{B2,2} = 1$
Azeotrope overhead from Column 1, Az1, $Y_{Az1,1} = 0.25$
Azeotrope overhead from Column 2, Az2, $Y_{Az2,1} = 0.25$
Streams from the liquid–liquid separator, L1, $X_{L1,1} = 0.40$; L2, $X_{L2,1} = 0.025$

SOLUTION

Referring to Figure 10.2, overall material balances,

$F + Az1 + Az2 = L1 + L2$
$L1 = Az1 + B1$
$L2 = Az2 + B2$

n-Butanol material balance on the liquid-liquid separator,

$$FX_{F1} + Az1 \cdot Y_{Az1,1} + Az2 \cdot Y_{Az2,1} = L1 \cdot X_{L1,1} + L2 \cdot X_{L2,1}$$

n-Butanol material balance on column 1,

$$L1 \cdot X_{L1,1} = Az1 \cdot Y_{Az1,1} + B1 \cdot X_{B1,1}$$

Water material balance on column 2,

$$L2 \cdot X_{L2,2} = Az2 \cdot Y_{Az2,2} + B2 \cdot X_{B2,2}$$

Substituting the numerical values,

$L1 + L2 - Az1 - Az2 - 100 = 0$
$Az1 + B1 - L1 = 0$
$Az2 + B2 - L2 = 0$
$0.40L1 + 0.025L2 - 0.25Az1 - 0.25Az2 - 35 = 0$
$0.25Az1 + B1 - 0.40L1 = 0$
$0.75Az2 + B2 - 0.975L2 = 0$

These are six equations with six variables, L1, L2, Az1, Az2, B1, B2. The solution:

$B1 = 35$ kmol/h
$B2 = 65$ kmol/h
$L1 = 175$ kmol/h
$L2 = 72.22$ kmol/h
$Az1 = 140$ kmol/h
$Az2 = 7.22$ kmol/h

The compositions are as given in the operating data.

10.1.3 Separation by Forming an Azeotrope with One Component

The feed consists of two components to be separated, A and B, that are close boilers or that form an azeotrope with each other. An entrainer E that forms an azeotrope with one of the components is added to the feed. If the component forming the azeotrope is A, the resulting azeotrope AE is separated from B by distillation. The problem is shifted from separating azeotrope AB to separating azeotrope AE. For the process to be viable, there should be an economical means for separating AE. If AE has a pressure-sensitive azeotropic composition or is a heterogeneous azeotrope, the methods described in Section 10.1.1 or 10.1.2 could be used to round out the process.

The separation process depends on the nature of the vapor–liquid equilibrium relationships of the system, which can be represented on a ternary diagram. Figure 10.3a shows a ternary diagram at some fixed system pressure. Components A and B are close boilers, and A forms an azeotrope with the entrainer E. The curves in the triangle represent liquid isotherms. A corresponding vapor isotherm (not shown) could be drawn to represent the vapor at equilibrium with each liquid curve with tie lines joining vapor and liquid compositions at equilibrium. The temperature of the isotherms reaches a minimum at point Z that corresponds to the composition of the azeotrope formed between A and E.

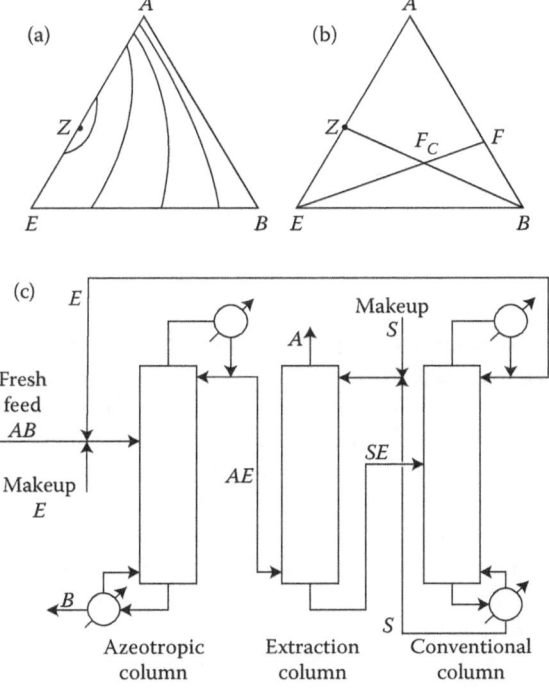

FIGURE 10.3 Separation by forming an azeotrope with one component, (a) ternary diagram of the A-B-E system, (b) composition points in the ternary system, and (c) schematic diagram of the separation process.

The feed composition is represented by point F in Figure 10.3b. An amount of entrainer is added to the feed, and the mixture is then separated into B and the azeotrope. Line EF represents compositions of mixtures formed by combining the feed with varying amounts of the entrainer. In order to separate the mixture into pure B and the azeotrope, the combined mixture must have a composition that lies on a straight line joining B and Z. Thus, the combined feed composition should correspond to point F_C, the intersection of BZ and EF.

The amount of entrainer required may be calculated by a material balance on component B. The total feeds (fresh feed + entrainer) equal the total products (azeotrope in the overhead + component B in the bottoms)

$$F_C = F + E = Z + B$$

where F_C is the combined feed rate, F is the fresh feed rate, E is the entrainer rate, Z is the azeotrope product rate, and B is the bottoms rate, assumed to be pure component B. A material balance on component A is written as

$$FX_{AF} = ZX_{AZ}$$

where X_{AF} is the fraction of component A in the fresh feed, and X_{AZ} is the fraction of component A in the azeotrope. It is assumed that all of component A leaves the column with the azeotrope and all of component B leaves in the bottoms. From a material balance on the entrainer

$$ZX_{EZ} = E$$

where X_{EZ} is the fraction of entrainer in the azeotrope. Here again it is assumed that all the entrainer leaves the column with the azeotrope. The above equations are combined to calculate the required ratio of entrainer to fresh feed:

$$\frac{E}{F} = \frac{X_{EZ} X_{AF}}{X_{AZ}} \tag{10.1}$$

Figure 10.3c is a schematic of a possible process for the separation of such a system. The fresh feed is mixed with the azeotrope-forming entrainer and is then fed to the azeotropic column, where pure component B is taken as column bottoms and the azeotrope AE is taken as overhead. The next step is to separate AE, an azeotrope that cannot be separated by simple distillation. A workable process involves liquid–liquid extraction where a solvent S is used to extract the entrainer, leaving a product that is virtually pure A. The solvent and entrainer are then separated by conventional distillation, the entrainer is recycled and mixed with fresh feed and makeup entrainer, and the solvent is recycled to the liquid–liquid extractor where some makeup solvent may be added.

This process can be used, for instance, to separate benzene and cyclohexane, which are very close boilers. Acetone, which forms an azeotrope with cyclohexane, is added as an entrainer. The cyclohexane is separated from the acetone by extraction with water, which dissolves the acetone. The cyclohexane is practically immiscible in water. The water–acetone solution is separated by simple distillation.

EXAMPLE 10.3: SEPARATING BENZENE AND
CYCLOHEXANE BY AZEOTROPIC DISTILLATION

The boiling points of benzene and cyclohexane are 80.1°C and 80.8°C, respectively, and they form a minimum boiling azeotrope at 100 kPa, 77°C, and 54 mole% benzene. It is proposed to separate them by adding acetone as an entrainer, which forms a minimum boiling azeotrope with cyclohexane at 100 kPa, 53°C, 73.9 mole% acetone and 26.1 mole% cyclohexane. The azeotrope is taken as the overhead stream in a distillation column, and the benzene is recovered as the bottoms product. Further processing will be used to separate the cyclohexane and acetone in the azeotrope distillate.

If the feed stream is 70 mole% benzene and 30 mole% cyclohexane, find the required acetone entrainer to feed ratio, assuming perfect separation between the azeotrope in the distillate and the benzene in the bottoms.

SOLUTION

The entrainer flow rate is determined by material balances. Cyclohexane material balance,

$$FX_{CF} = DX_{CZ}$$

Here F and D are the feed and distillate molar flow rates, and X_{CF} and X_{CZ} are the mole fractions of cyclohexane in the feed and azeotropic distillate. Another material balance results from the fact that all the cyclohexane in the feed and all of the entrainer make up the distillate rate:

$$FX_{CF} + E = D$$

Combining the two equations by eliminating D,

$$FX_{CF} = (FX_{CF} + E) X_{CZ}$$

The entrainer to feed ratio,

$$\frac{E}{F} = \frac{X_{CF}(1 - X_{CZ})}{X_{CZ}} = \frac{0.30(1 - 0.261)}{0.261} = 0.849$$

Alternatively, Equation 10.1 could be used directly.

Similar processes are used for separating other close boiling aromatics and paraffins such as the separation of toluene and n-heptane using as entrainer methyl ethyl ketone, which forms a minimum-boiling azeotrope with the heptane. The azeotrope is taken in the overhead and the toluene in the bottoms.

10.1.4   SEPARATION BY FORMING TWO BINARY AZEOTROPES

For certain close boilers or difficult-to-separate binaries, an entrainer may be selected that forms a binary azeotrope with each one of the components to be separated. Figure 10.4b shows a ternary diagram of such a system. Entrainer E forms azeotropes

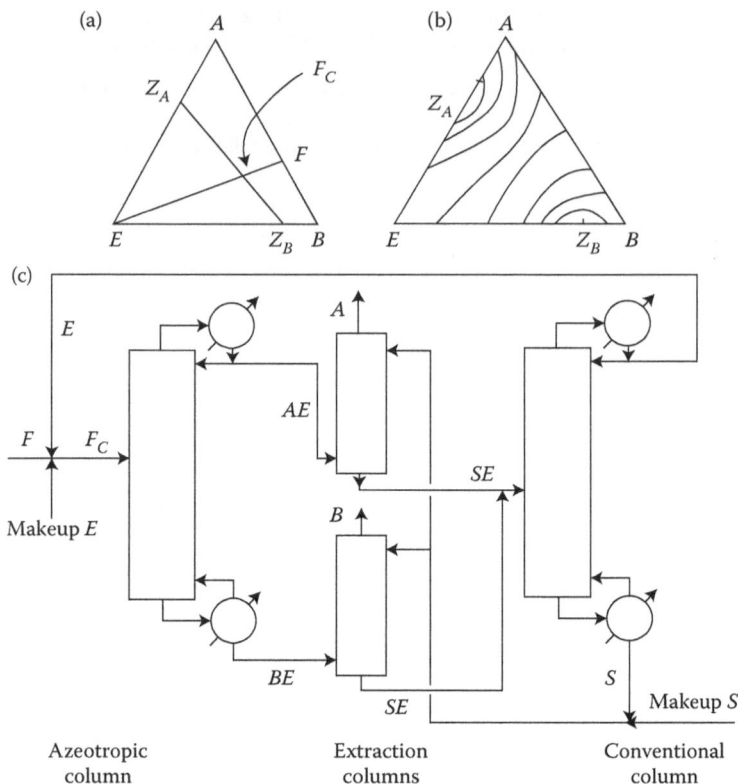

FIGURE 10.4 Separation by forming two binary azeotropes, (a) amount of entrainer by graphical material balances, (b) liquid composition isotherms, and (c) typical separation flowsheet.

Z_A with A, and Z_B with B. If the difference between the boiling points of the two azeotropes is large enough, they may be separated by fractionation. There should also be a monotonic equilibrium temperature variation with composition from one azeotrope to the other to allow the separation to take place. The curves in Figure 10.4b represent liquid composition isotherms. Equilibrium vapor composition isotherms could be drawn with tie lines connecting equilibrium phases, but are not shown. Temperature ridges in the ternary diagram could result in a maximum in the column temperature profile, which would cause column instability or make the separation impossible. Also, the feasibility of separating the azeotropes leaving the column should be ascertained.

Figure 10.4a represents graphically the material balance that determines the amount of entrainer that must be added to the feed to generate products Z_A and Z_B. The fresh feed composition is represented by point F. The combined composition of feed and entrainer must lie on EF. The intersection of EF and $Z_A Z_B$ at F_C represents the combined composition of the column feed.

The required amount of entrainer is calculated by material balance. From overall material balance,

$$F_C = E + F = Z_A + Z_B$$

where F, E, and F_C are the fresh feed rate, entrainer rate, and combined rate, and Z_A and Z_B are the rates of azeotropes Z_A and Z_B, respectively. The following equations represent material balances on components A, B, and E:

$$FX_{AF} = Z_A X_{AZA}$$
$$FX_{BF} = Z_B X_{BZB}$$
$$E = Z_A X_{EZA} + Z_B X_{EZB}$$

X_{AZA} = Mole fraction A in AE azeotrope
X_{EZA} = Mole fraction E in AE azeotrope
X_{BZB} = Mole fraction B in BE azeotrope
X_{EZB} = Mole fraction E in BE azeotrope

The mole fraction subscripts represent components A, B, and E in the feed and in the two azeotropes. It is assumed in these approximate calculations that components A and B and the entrainer are totally involved in forming the azeotropes and that the separation between the two azeotropes is perfect. Combining the above equations gives the required ratio of entrainer to feed:

$$\frac{E}{F} = \frac{X_{AF} X_{EZA}}{X_{AZA}} + \frac{X_{BF} X_{EZB}}{X_{BZB}} \qquad (10.2)$$

A typical flowsheet for the separation is shown in Figure 10.4c. The fresh feed is combined with the entrainer and then fed to the azeotropic column, where the two azeotropes are separated. Each azeotrope product is sent to a liquid–liquid extraction column to dissolve the entrainer in some suitable solvent. The combined entrainer–solvent solution from both extractors is sent to a distillation column to separate the entrainer and solvent. These are recycled to the azeotropic column and extractors with makeup added to each, as needed.

EXAMPLE 10.4: SEPARATING TOLUENE-2,5 DIMETHYLHEXANE (DMH) BY FORMING TWO BINARY AZEOTROPES WITH METHANOL

The boiling point of DMH at 100 kPa is 109.3°C, and that of toluene is 110.7°C, making it difficult to separate the two by conventional distillation. Each one of these components forms an azeotrope with methanol as an entrainer:

	B.P. (°C)	Azeotropic Composition (mole%)
DMH	109.3	15.7
Methanol	64.8	84.3
Azeotrope temp.	61.0	
Toluene	110.7	13.5
Methanol	64.8	86.5
Azeotrope temp.	63.8	

Although the azeotrope temperatures are still not too far apart, distillation can be done with adequate reflux ratio and number of stages.

For an equimolar DMH–toluene feed stream, it is required to find the required entrainer to feed ratio, assuming the distillate and bottoms are at the corresponding azeotropic compositions.

SOLUTION

From Equation 10.2,

$$\frac{E}{F} = \frac{X_{AF}X_{EZA}}{X_{AZA}} + \frac{X_{BF}X_{EZB}}{X_{BZB}} = \frac{0.5 \times 0.843}{0.157} + \frac{0.5 \times 0.865}{0.135} = 5.888$$

On a basis of feed rate, $F = 100$ kmol/h, the entrainer rate, $E = 588.8$ kmol/h The distillate rate, based on DMH material balance,

$$FX_{AF} = DX_{AZA}, \; D = FX_{AF}/X_{AZA} = 100 \times 0.50/0.157 = 318.47 \text{ kmol/h}$$

The bottoms rate, based on toluene bacterial balance,

$$FX_{BF} = BX_{BZB}, \; B = FX_{BF}/X_{BZB} = 100 \times 0.50/0.135 = 370.37 \text{ kmol/h}$$

Typical mixtures that can be separated by a process such as this include mixtures of close-boiling aromatics and paraffins using ethanol or methanol as entrainers. The aromatics such as toluene form the higher-boiling azeotrope while the alcohol and the paraffins form the lower-boiling azeotrope. Water is used to remove the alcohol from the azeotropes by extraction. Finally, the alcohol must be separated from the water and recycled. From this standpoint, methanol is preferred over ethanol because the latter forms an azeotrope with water, thus complicating the ethanol recovery.

10.1.5 SEPARATION BY FORMING A TERNARY AZEOTROPE

In this process, the entrainer E, added to the close-boiling or azeotrope-forming binary AB, forms a low-boiling ternary azeotrope, ABE. The ternary azeotrope boils at a temperature lower than A or B or their binary azeotrope, if it exists. The ABE azeotrope is removed in the column overhead, while pure component A or B is recovered in the bottoms.

The composition of the ternary azeotrope and that of the feed determine the relative rates of entrainer and feed required to produce a pure product in the bottoms and the azeotropic composition in the overhead. Figure 10.5 shows a typical ternary diagram for ABE at expected process temperature and pressure. Point F represents the fresh feed, which is a binary mixture of A and B. Point Z represents the ternary azeotropic composition. A straight line drawn through E and F represents compositions obtained by mixing fresh feed with variable amounts of entrainer. The combined feed composition should be such that it would separate into pure component A and the ternary azeotrope. Therefore, the combined feed composition should fall on a straight line joining A and Z. The amount of entrainer added to the fresh feed should yield the composition represented by the intersection point F_C.

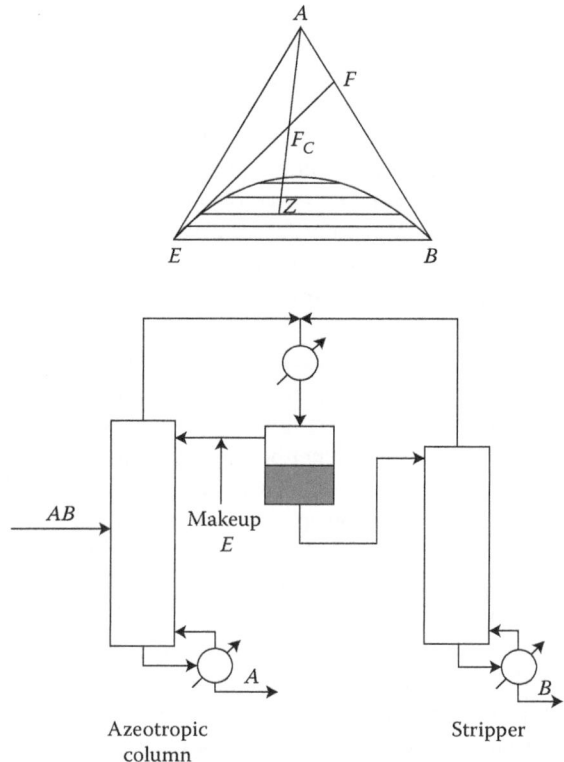

FIGURE 10.5 Separation by forming a ternary azeotrope.

The amount of entrainer is calculated by material balance. Using similar nomenclature for rates and compositions as in the previous section, the following equations are written for total and component material balances:

$$F_C = E + F = A + Z$$
$$FX_{AF} = A + ZX_{AZ}$$
$$FX_{BF} = ZX_{BZ}$$
$$E = ZX_{EZ}$$

The entrainer-to-feed ratio is calculated by combining the last two equations:

$$\frac{E}{F} = \frac{X_{EZ}X_{BF}}{X_{BZ}} \tag{10.3}$$

These derivations are based on the assumption of total azeotrope formation and perfect separation between component A and the ternary azeotrope.

A schematic of a flowsheet utilizing the ternary azeotrope phenomenon to separate components A and B is shown in Figure 10.5. The required amount of entrainer may be added either to the fresh feed or directly to the azeotropic column as reflux.

In this process the entrainer is added as reflux to the azeotropic column. The ternary azeotrope is taken as column overhead and condensed, whereupon it separates in the receiver into two liquid phases, one rich in the entrainer and the other rich in component B. Compositions below the curve in the ternary diagram (Figure 10.5) form two coexisting liquid phases, and those above the curve form one liquid phase. The tie lines connect compositions in the two liquid phases at equilibrium with each other.

Almost pure component A is taken as azeotropic column bottoms. The purity of A is determined by the relative amount of entrainer added. Too little entrainer will not remove sufficient amounts of B in the overhead, thus allowing part of it to flow down with the bottoms. Too much entrainer may result in part of it contaminating the bottoms.

The entrainer-rich liquid from the receiver is returned to the azeotropic column as reflux, and the liquid phase rich in component B is sent to a stripper column, where any amounts of A and E and some of B are removed in the overhead and combined with the azeotropic column overhead. The stripper bottoms is primarily component B.

At steady state, and aside from inevitable small losses, almost all of component A in the fresh feed is recovered as azeotropic column bottoms and B as stripper column bottoms. The entrainer, once introduced into the system, circulates within it indefinitely, and any losses are replaced by fresh makeup. The reflux rate and composition are controlled to maintain the ternary azeotrope composition in the overhead. The reflux composition is controlled by the inventory of the entrainer in the receiver and to some degree by the receiver temperature.

Typical of this process is the production of anhydrous ethanol from a concentrated ethanol solution in water, using benzene or cyclohexane or other components as the entrainer.

EXAMPLE 10.5: ETHANOL DEHYDRATION BY FORMING A TERNARY AZEOTROPE

Purifying ethanol from ethanol/water solution by simple distillation is limited by the formation of a minimum-boiling azeotrope containing 90.37 mole% ethanol at 78.14°C, 760 mmHg. Raising the ethanol concentration at 100 kPa above 90.37 mole%, or dehydrating it, can be accomplished by adding to the distillation column an entrainer to form a ternary azeotrope.

Using benzene as the entrainer to the ethanol/water solution forms a ternary azeotrope at 100 kPa, 64.86°C, containing 22.81 mole% ethanol, 23.32 mole% water, and 53.87 mole% benzene. By adding benzene to the distillation column, the ternary azeotrope, with its lower boiling point than that of ethanol at 78.4°C, leaves the column in the distillate, and purified ethanol is recovered in the bottoms.

Assuming perfect distillation, with pure ethanol in the bottoms product and only the ternary azeotrope in the distillate, find the entrainer to feed flow rate ratio if the ethanol solution feed stream contains 90 mole% ethanol and 10 mole% water.

SOLUTION

From the problem statement, the mole fraction of the entrainer in the ternary azeotrope,

$$X_{EZ} = 0.5387$$

The mole fraction of water in the feed, $X_{BF} = 0.10$.
The mole fraction of water in the ternary azeotrope, $X_{BZ} = 0.2332$.
From Equation 10.3,

$$\frac{E}{F} = \frac{0.5387 \times 0.10}{0.2332} = 0.231$$

10.1.6 SEPARATION BY EXTRACTIVE DISTILLATION

In extractive distillation a solvent is utilized that preferentially alters the relative volatilities of the components to be separated. In contrast to the entrainer in azeotropic distillation, the solvent in extractive distillation has a considerably higher boiling point than either component in the feed and is therefore always withdrawn in the column bottoms.

The solvent could depress the volatility of either the lighter or the heavier of the components to be separated. Thus, it is not unusual for the relative volatilities of the components in the feed to be reversed by the addition of the solvent. Since the solvent in extractive distillation is considerably less volatile than the feed components, no azeotropes are formed, and the separation of the solvent from the bottoms product is relatively easy.

The low volatility of the solvent also dictates that it be introduced in the column near the top, above the main feed, in order to maintain the required solvent concentration in as many trays as possible. The column thus contains three sections: the top section between the overhead product and the solvent feed, the middle section between the solvent feed and the main feed, and the bottom section between the main feed and the bottoms product. The top section serves to reflux any solvent that might have risen with the vapor above the solvent feed. The objective is to minimize the solvent concentration in the overhead product. The middle section serves to absorb the component whose volatility has been lowered by the solvent and to minimize its concentration in the overhead. The lower section serves to strip out the component with the higher volatility and minimize its concentration in the bottoms.

Figure 10.6 illustrates a typical extractive distillation process consisting of the extractive distillation column and the solvent recovery column. Fresh feed containing the binary AB is introduced around the middle of the extractive distillation column, and the solvent S is introduced near the top. Components A and B are close boilers and/or potentially azeotrope formers that are difficult or impossible to be separated by ordinary distillation. Whether individual component A is more volatile than B or vice versa, in the presence of the solvent, B becomes less volatile due to its higher affinity to the solvent. As a result, essentially pure A is distilled as the overhead of the extractive distillation column. Component B is entrained with the solvent in the bottoms stream, which is sent to the solvent recovery column. The solvent is substantially less volatile than component B, allowing easy separation by ordinary distillation. Practically pure B is recovered in the overhead, and pure solvent in the bottoms. The solvent is recycled to the extractive distillation column with makeup that might be required to compensate for losses.

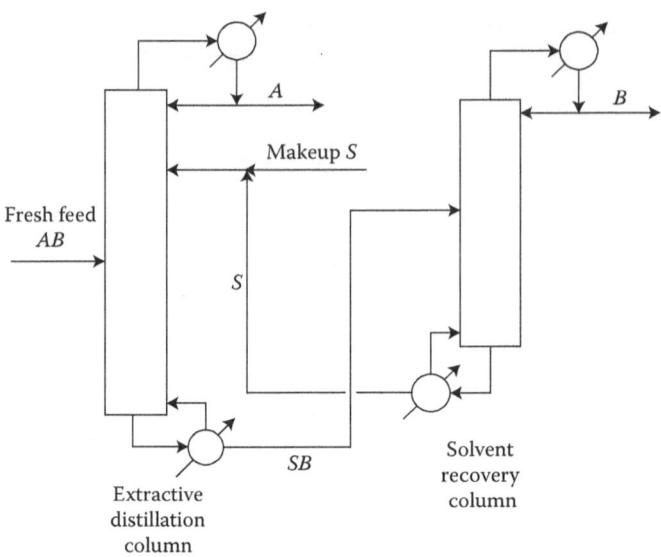

FIGURE 10.6 Separation by extractive distillation.

Typical mixtures that can be separated by extractive distillation in processes similar to the one described above include cyclohexane and benzene, and toluene and methylcyclohexane, both using phenol as the solvent. In another process, isobutane and 1-butene are separated using furfural as the solvent.

The relative rate of solvent to feed determines the sharpness of separation between the key components, much as the absorbent rate or reflux ratio influences separation in absorbers or conventional columns. An estimate of the required solvent rate may be obtained based on vapor–liquid equilibrium data at different solvent concentrations, as described later in this section.

The effect of furfural rate on the separation of isobutane and 1-butene was determined by simulation and is shown in Figure 10.7. The column has 23 theoretical stages, a total condenser, and a reboiler. The main feed, containing 40 mole percent isobutane and 60 mole percent 1-butene, is sent to the 11th tray from the top, and the solvent is introduced at the third tray. The column operates at 3580 kPa pressure and a temperature ranging from 40.5°C at the condenser to 42°C in the reboiler. The overhead rate is maintained at 40 mole% of the isobutane–1-butene feed rate, and the column reflux ratio is maintained at 3.0. The solvent-to-feed ratio is increased from 0 to 3, and the resulting separation is observed. Figure 10.7 shows a plot of the concentration of 1-butene in the overhead and isobutane in the bottoms as a function of the solvent-to-feed ratio. These concentrations decrease sharply as the solvent ratio is increased from low values but tend to level off asymptotically at a ratio of about 2 to 3.

10.1.6.1 Graphical Representation

The net effect of adding a solvent in extractive distillation is that new values emerge for the relative volatilities of the binary to be separated. If such data were available

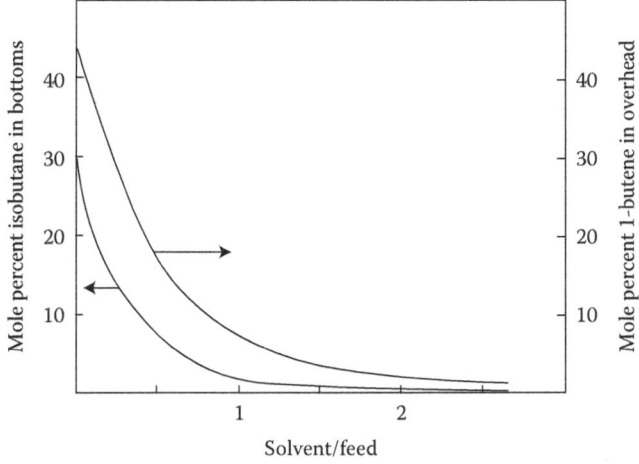

FIGURE 10.7   Effect of solvent-to-feed ratio on separation in extractive distillation.

for a given system, the vapor and liquid compositions could be plotted on a solvent-free basis to generate Y–X plots similar to ordinary binary Y–X diagrams. In the absence of direct relative volatility or Y–X data, liquid activity coefficient equations (Chapter 1) may be used to predict and plot the Y–X diagrams using computer simulation programs. The relative volatilities depend on the relative amount of solvent added to the binary. Therefore, different Y–X diagrams are obtained for different solvent-to-feed ratios.

The equilibrium curve, together with operating lines for the absorption and stripping sections, could then be used to estimate the number of stages required in each section at a given reflux ratio, as described in Chapters 5 and 6 for binary distillation. Since the Y–X diagram is specific to a given solvent-to-feed ratio, the graphical evaluation of the column is valid for that particular ratio. By trying diagrams corresponding to a number of ratios, the amount of solvent for a given feed rate may be estimated. As in binary distillation, the graphical representation could be invaluable for qualitative evaluation of the problem prior to rigorous column simulation. A thorough understanding of the effect of different parameters on the column performance can help in better defining the simulation model.

EXAMPLE 10.6: ACETONE/CHLOROFORM SEPARATION BY EXTRACTIVE DISTILLATION

A feed stream at the rate of 100 kmol/h contains 50% mole acetone and 50% mole chloroform. The two components form a maximum boiling azeotrope which prevents their separation by conventional distillation. It is proposed to separate them by extractive distillation using benzene as a solvent, at a rate of 800 kmol/h. Both the main feed and the solvent are at 75°C and 110 kPa, and the column pressure is assumed uniform, also at 110 kPa. A total condenser is used, with a reflux ratio of 4. The distillate composition is specified at 95% mole acetone and the bottoms at 5% mole acetone on a solvent-free basis. Using the pseudo-binary graphical method described in Section 10.1.6, estimate the required number of stages and the main feed stage and the solvent feed stage locations.

SOLUTION

The column configuration is similar to the extractive distillation column in Figure 10.6. The proposed solution method requires acetone/chloroform equilibrium data that are obtained when the solvent, benzene, is present in the solution.

Approximate calculations are initially made for the purpose of estimating the liquid compositions at different points in the column. These compositions are used to calculate activity coefficients, K-values and equilibrium vapor compositions as required for the pseudo-binary method. The initial calculations are based on the following assumptions:

1. The solvent-free-basis liquid composition on the solvent feed tray is about the same as the distillate composition. Recall that the trays above the solvent feed tray are mainly for the purpose of refluxing the solvent and have little effect on the separation of the two primary components.
2. The solvent concentration in the liquid between the solvent feed tray and the bottom tray is constant.
3. The main feed is saturated vapor, so that the combined liquid rate of the two primary components is constant throughout the column. This solvent-free liquid rate is determined by the reflux ratio.
4. The solvent feed is saturated liquid.

The distillate rate is determined by acetone material balance:

$$FX_{F1} = DX_{D1} + (F - D)X'_{B1}$$

where X'_{B1} is acetone mole fraction in the bottoms on a solvent-free basis.

$$(100) \times (0.50) = D(0.95) + (100 - D)(0.05)$$

$$D = 50 \text{ kmol/h}$$

With a reflux ratio $R = 4$, the solvent-free liquid rate, assumed constant in the column, is

$$L' = RD = 4 \times 50 = 200 \text{ kmol/h}$$

Designate the stage numbers, counting from the condenser, as follows:

N_S = solvent feed stage
N_F = main feed stage
N_B = bottom stage (reboiler)

Following these definitions and the above assumptions, the solvent-free liquid rate on stages 1 through N_B is L', and the solvent rate on stages N_S through N_B is F_S, the solvent feed rate.

As an approximation, the solvent-free liquid composition on the main feed stage is assumed to be about the same as the main feed composition. On this basis the estimated liquid compositions at different points in the column are as tabulated below. The estimated compositions on a total liquid basis (including the solvent) are calculated for components 1 and 2 as

$$X_i = \frac{L'X_i'}{L' + F_S} = \frac{200X_i'}{200 + 800} = 0.2X_i'$$

where X_i' is the solvent-free basis mole fraction. For component 3,

$$X_3 = \frac{F_S X_{S3}}{L' + F_S} = \frac{800 \times 1}{200 + 800} = 0.8$$

| | Feed | Solvent | Estimated Liquid Composition | | | | | |
| | | | Solvent-Free | | | Including Solvent | | |
	X_{Fi}	X_{Si}	On N_S	On N_F	On N_B	On N_S	On N_F	On N_B
1. Acetone	0.50	0.0	0.95	0.50	0.05	0.19	0.10	0.01
2. Chloroform	0.50	0.0	0.05	0.50	0.95	0.01	0.10	0.19
3. Benzene	0.00	1.0	0.00	0.00	0.00	0.80	0.80	0.80

Vapor pressures are calculated by the Antoine Equation 2.19, with the constants given below (p_i^0 in kPa, T in K). Also listed are normal boiling points and liquid molar volumes.

	A_i	B_i	C_i	NBP (°C)	Liq. mol. vol, V_i (cm³/gmol)
1. Acetone	14.636	2940.5	−35.9	56.16	77
2. Chloroform	16.055	3977.5	13.44	61.10	81
3. Benzene	13.886	2788.5	−52.4	80.24	94

The Wilson equation, with parameters given below in cal/gmol, will be used to calculate the activity coefficients.

$\lambda_{12} - \lambda_{11} = -72.20$	$\lambda_{13} - \lambda_{11} = 494.92$	$\lambda_{23} - \lambda_{22} = -204.22$
$\lambda_{12} - \lambda_{22} = -332.23$	$\lambda_{13} - \lambda_{33} = -167.91$	$\lambda_{23} - \lambda_{33} = 141.62$

Assuming a constant column pressure of 110 kPa, the vapor compositions and temperatures are determined by bubble point temperature calculations at liquid compositions on stages N_S, N_F, and N_B, with additional intermediate points. The bubble points are calculated as described in Example 2.8A, with the assumption of ideal gas behavior in the vapor phase. The results are tabulated below:

Stage	T (°C)	Liquid Composition X_1	X_2	X_3	Vapor Composition Y_1	Y_2	Y_3	Solvent-Free X_1'	Y_1'
N_S	72.90	0.190	0.010	0.800	0.3911	0.0096	0.5994	0.9500	0.9760
a	73.59	0.170	0.030	0.800	0.3573	0.0302	0.6125	0.8500	0.9221
a	74.98	0.130	0.070	0.800	0.2841	0.0774	0.6385	0.6500	0.7859
N_F	76.00	0.100	0.100	0.800	0.2245	0.1183	0.6572	0.5000	0.6549
a	76.98	0.070	0.130	0.800	0.1611	0.1641	0.6748	0.3500	0.4954
a	78.22	0.030	0.170	0.800	0.0710	0.2330	0.6960	0.1500	0.2336
N_B	78.79	0.010	0.190	0.800	0.0239	0.2707	0.7053	0.0500	0.0811

a Points at intermediate liquid compositions.

10.1.6.2   Sample Equilibrium Calculations

Bubble point temperatures and vapor compositions on a stage are calculated iteratively by assuming a temperature and computing the bubble point pressure. A solution is reached when the calculated pressure matches the specified value (110 kPa). The calculations below demonstrate the final iteration for the solvent feed stage N_S.

$$\text{Assumed temperature: } T = 72.90°C = 346.05 \text{ K}$$

Calculate the Wilson coefficients:

$$A_{12} = \frac{V_2}{V_1}\exp\left[-\frac{\lambda_{12} - \lambda_{11}}{RT}\right] = \frac{81}{77}\exp\left[-\frac{-72.20}{346.05 \times 1.987}\right] = 1.1684$$

$A_{21} = 1.5411$
$A_{13} = 0.5943$
$A_{31} = 1.0457$
$A_{23} = 1.5618$
$A_{32} = 0.7013$

The activity coefficients are calculated by Equation 1.36a (see Example 1.6A):

$$S_1 = \sum_j X_j A_{1j} = X_1 A_{11} + X_2 A_{12} + X_3 A_{13}$$

$$= 0.19 \times 1 + 0.01 \times 1.1684 + 0.80 \times 0.5943 = 0.6771$$

$S_2 = 1.5523$
$S_3 = 1.0057$

$$T_1 = \sum_k \left[\frac{X_k A_{k1}}{\sum_j X_j A_{kj}}\right] = \frac{X_1 A_{11}}{S_1} + \frac{X_2 A_{21}}{S_2} + \frac{X_3 A_{31}}{S_3}$$

$$= \frac{0.19 \times 1}{0.6771} + \frac{0.01 \times 1.5411}{1.5523} + \frac{0.80 \times 1.0457}{1.0057} = 1.1224$$

$$\ln \gamma_1 = 1 - \ln S_1 - T_1 = 1 - \ln 0.6771 - 1.1224 = 0.2675$$

$\gamma_1 = 1.3067$
$\gamma_2 = 0.7176$
$\gamma_3 = 1.0223$

Vapor pressures,

$$p_1^0 = \exp\left[14.636 - \frac{2940.5}{-35.9 + 346.05}\right] = 173.31\,\text{kPa}$$

$p_2^0 = 147.04\,\text{kPa}$
$p_3^0 = 80.64\,\text{kPa}$

The total pressure (Equations 1.29a and 2.16),

$$PY_i = \gamma_i p_i^0 X_i$$

$$P\sum Y_i = P = \sum \lambda_i p_i^0 X_i$$
$$= 1.3067 \times 173.31 \times 0.19 + 0.7176 \times 147.04 \times 0.01 + 1.0223 \times 80.64 \times 0.80$$
$$= 110.0 \ \text{kPa}$$

The vapor compositions,

$$Y_1 = \gamma_1 p_1^0 X_i / P = 1.3067 \times 173.31 \times 0.19/110.0 = 0.3911$$
$$Y_2 = 0.0096$$
$$Y_3 = 0.5993$$

10.1.6.3 Determining the Number of Stages

The number of stages between the solvent feed stage and the main feed stage, and between the main feed stage and the reboiler are estimated using the Y–X diagram of the acetone/chloroform solvent-free equilibrium curve Y_1' vs. X_1':

$X_1' = X_1/(X_1 + X_2)$
$Y_1' = Y_1/(Y_1 + Y_2)$

With a reflux ratio $R = 4$, the solvent-free L_r/V_r slope of the rectifying section operating line is

$$L_r/V_r = R/(R + 1) = 4/(4 + 1) = 0.8$$

The rectifying section operating line is drawn based on this slope and the distillate composition $X_1' = 0.95$, at the diagonal. The intersection of this operating line with the q-line (horizontal for a saturated vapor feed) is joined to the bottoms composition $X_1' = 0.05$, at the diagonal to form the stripping section operating line. The stages are stepped off between the operating lines and the equilibrium curve, giving $N_F = N_S + 8$, and $N_B = N_S + 18$.

10.1.6.4 Benzene Recovery Section

The benzene recovery section, between the condenser and the solvent feed stage, is represented on a Y–X pseudo-binary diagram where the acetone and chloroform are lumped as one component in solution with the benzene.

Bubble point temperature calculations are made, similar to the lower column sections, to determine vapor compositions at equilibrium with the liquid compositions. If a benzene mole fraction of 0.01 is allowed in the distillate, and with a benzene mole fraction of 0.8 on the solvent feed tray, the separation requirement is defined for this section of the column.

Assuming the relative amount of acetone to chloroform in this section remains essentially unchanged, the following liquid compositions are used for the equilibrium calculations. Also listed are the corresponding bubble point temperatures and vapor compositions.

Stage	X_1	X_2	X_3	T (°C)	Y_1	Y_2	Y_3
1(condenser)	0.95	0.04	0.01	59.48	0.9738	0.0195	0.0067
a	0.9116	0.0384	0.05	57.06	0.9481	0.0189	0.0330
a	0.8636	0.0364	0.10	57.60	0.9163	0.0182	0.0655
a	0.4798	0.0202	0.50	62.90	0.6625	0.0130	0.3245
N_S	0.1919	0.0081	0.80	69.84	0.3958	0.0077	0.5965

a Points at intermediate liquid compositions.

The equilibrium curve used for estimating the number of stages in this section is a plot of Y_{12} vs. X_{12}, where $Y_{12} = Y_1 + Y_2$ and $X_{12} = X_1 + X_2$.

X_{12}	0.99	0.95	0.90	0.50	0.20
Y_{12}	0.9933	0.9670	0.9346	0.6755	0.4035

The number of stages is determined based on the $Y_{12}X_{12}$ equilibrium curve. The operating line has a slope of 0.8, as determined from the reflux ratio of 4. It intersects the diagonal at $X_{12} = 0.99$. The lumped component mole fraction, X_{12}, changes from 0.99 at the condenser to 0.20 at the solvent feed stage, requiring 19 stages.

The total number of stages in the column is 38. With the total condenser designated as #1, the solvent feed stage is #20, the main feed stage is #28, and the reboiler is #38.

10.2 THREE-PHASE DISTILLATION

The phase behavior of a system determines the nature of a distillation problem. Depending on this behavior, column profiles can range from those characteristic of narrow-boiling to wide-boiling mixtures and from those typical of ideal to highly nonideal systems.

Nonideal behavior is manifested, for instance, in azeotropic and extractive distillation, discussed in this chapter. Another case of nonideality is the condition of vapor–liquid–liquid equilibrium (Sections 1.3.5 and 2.3.3). If the composition, temperature, and pressure on any column tray or number of trays put the mixture in the vapor–liquid–liquid region, three-phase distillation ensues. Sections 10.1.2 and 10.1.5 describe cases where the formation of two liquid phases makes up part of the separation process.

One of the more common cases of three-phase distillation is where the two liquid phases occur primarily in the condenser. Such occurrence is very common in crude oil and related distillation columns, where water separates in the condenser. When two liquid phases form in the column, one of them is usually taken out as a side draw at the tray where the liquid phase split occurs. This practice helps stabilize the column operation.

The computations in three-phase distillation involve two sets of vapor–liquid equilibrium coefficients, or K-values, derived from the activity coefficients of each component in each liquid phase (Section 2.3.3). The calculations may be simplified in hydrocarbon systems if the second liquid phase is mostly water. In these situations it is possible to assume the aqueous phase to be pure water and account only for water dissolved in the organic phase. The description of rigorous solution methods for three-phase distillation is deferred to Chapter 13. The objective at this point is to consider the effect a liquid phase split can have on distillation.

It has already been shown in Sections 10.1.2 and 10.1.5 how the formation of two liquid phases plays a part in the separation of azeotropes or close boilers. The two phases, having distinct compositions, are separated in a single stage by simple decantation, followed by further processing of each phase. These may be considered as special cases of three-phase distillation since the formation of two liquid phases is confined to a single stage: the condenser or the liquid–liquid separator.

Another example of potential three-phase distillation is provided by the methyl ethyl ketone (MEK)–water binary. This system forms two miscible regions and a two-liquid-phase region. Figure 10.8 is a Y–X plot of this binary at 100 kPa. Below 0.051 mole fraction MEK, a single, water-rich liquid phase exists, and above 0.652 mole fraction MEK, a single, MEK-rich liquid phase exists. Between these two concentrations, two liquid phases coexist.

Consider a column operating at 100 kPa to be used to produce essentially pure MEK from a mixture with water containing 75 mole% MEK. The purified MEK is taken as the column bottoms, and the overhead removes almost all of the water in the feed and some of the MEK. If the MEK concentration in the overhead is reduced below 65.2%, two liquid phases will form in the condenser. The water-rich phase is decanted, and the MEK-rich phase is refluxed. This makes for a much more efficient separation since the decanted water contains only about 5.1% MEK. If the overhead MEK concentration is not lowered to the two-phase region, the distillate would contain over 65.2% MEK.

This example involves two liquid phases only in the condenser and is therefore also a special case of three-phase distillation. It is possible to reflux a mixed-liquid

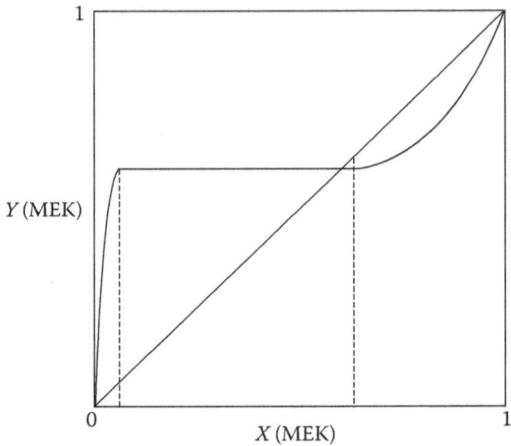

FIGURE 10.8   Methyl ethyl ketone–water binary.

phase to the column, creating three-phase conditions on a number of trays. At some tray down the column, the MEK concentration might become high enough to restore total miscibility. It is obviously advantageous, however, to decant the water phase and reflux only the MEK phase.

10.3   REACTIVE MULTISTAGE SEPARATION

When components entering a multistage vapor–liquid separation column are mutually reactive, chemical reactions and phase separation can occur simultaneously in what is generally described as reactive distillation. This phenomenon is found in several operations in the petroleum, chemical, and petrochemical industries.

Phase separation is controlled by phase equilibrium relations or rate-based mass and heat transfer mechanisms. Chemical reactions are controlled by chemical equilibrium relations or by reaction kinetics. For reactive distillation to have practical applications, both these operations must have favorable rates at the column conditions of temperature and pressure. If, for instance, the chemical reaction is irreversible, it may be advantageous to carry out the reaction and the separation of products in two distinct operations: a reactor followed by a distillation column. Situations in which reactive distillation is feasible can result in savings in energy and equipment cost. Examples of such processes include the separation of close-boilers, shifting of equilibrium reactions toward higher yields, and removal of impurities by reactive absorption or stripping.

In reactive distillation, chemical reactions are assumed to occur mainly in the liquid phase. Hence the liquid holdup on the trays, or the residence time, is an important design factor for these processes. Other column design considerations, such as number of trays, feed and product tray locations, can be of particular importance in reactive distillation columns. Moreover, since chemical reactions can be exothermic or endothermic, intercoolers or heaters may be required to maintain optimum stage temperatures. Column models of reactive distillation must include chemical reaction equilibrium or kinetic equations along with the material and energy balance equations and the phase equilibrium relations. These models and methods for solving them are discussed in Chapter 13.

The following are examples of reactive distillation applications.

10.3.1 SEPARATION OF CLOSE BOILERS

Reactive distillation is used to facilitate the separation of very close boilers, such as isomers, by the addition of a reactant that reacts at different rates with each of the components to be separated (Saito et al., 1971; Tierney and Riquelme, 1982). One example of such a process is the separation of xylenes. The normal boiling points of m-xylene and p-xylene are 139.3°C and 138.5°C, making it highly impractical to separate a mixture of these components by conventional distillation. If a reactive agent can be found that reacts preferentially with one of the components, the separation becomes much easier. One such agent is tertiary butyl benzene, which reacts with m-xylene in the presence of a catalyst to form benzene and tertiary butyl m-xylene in chemical equilibrium. Since chemical equilibrium is assumed, the tray liquid holdup is not critical as long as it is large enough for chemical equilibrium to be reached. Tertiary butyl m-xylene has considerably higher boiling point than the xylenes. The xylenes feed is introduced close to the column bottom and the reactive agent close to the top. With this configuration, and as a result of the combined effects of the chemical equilibrium and the phase equilibrium relationships, the components are split between the distillate and the bottoms such that the distillate contains mainly p-xylene and benzene, while the bottoms product contains mainly m-xylene, the balance of p-benzene, and tertiary butyl m-xylene.

10.3.2 ESTERIFICATION OF ACETIC ACID

Also in this process, the esterification reaction and the separation of the products take place in the same equipment: the distillation column. Acetic acid and ethanol react to produce ethyl acetate and water:

$$CH_3CO_2H + C_2H_5OH \leftrightarrow CH_3CO_2C_2H_5 + H_2O$$

The reaction occurs in the liquid phase and the conversion is kinetically determined. The liquid holdup on the trays should, therefore, be carefully considered. By separating the ethyl acetate as it is formed, the reaction can be driven toward completion. The feed to the column is a mixture of acetic acid, ethanol, and water. The distillate is predominantly ethyl acetate and most of the unreacted ethanol, plus small amounts of water and unreacted acetic acid. The bottoms product contains most of the water and unreacted acetic acid, plus small amounts of ethyl acetate and unreacted ethanol.

10.3.3 OTHER APPLICATIONS

The purification of gases is accomplished by removing undesirable components by reactive absorption as in amine treatment (De Leye et al., 1986).

In reactive strippers, hydrogen sulfide and/or ammonia are removed from water in a sour water stripper. This is a common process for lowering the level of sour gases in effluent water. One question is how much boilup or stripping steam is required to lower the sour gas concentration in the bottoms to acceptable levels. The electrolytic reactions taking place in the column must be considered in order to correctly predict the column performance (Chen et al., 1982).

In other applications, chemicals are produced by contacting gases and liquids in multistage processes where reaction products are concentrated in the same operation. Examples of such processes include the production of nitric acid (Koukolik and Marek, 1968) and chlorination.

PROBLEMS

10.1. The feed stream in the azeotropic distillation process described in Figure 10.1 is of variable composition. The column pressures are held constant. What ranges of feed composition can produce a product containing mostly component A and another containing mostly component B using the existing process configuration? With a variable feed composition, describe a control scheme to maintain constant product purities. What strategy may be followed if the feed composition moves outside the feasible range?

10.2. Assuming the process in Figure 10.2 has an external feed of fixed flow rate, composition, and thermal conditions, determine its degrees of freedom. How would you specify the performance of this process?

10.3. For the process shown in Figure 10.3, what control strategy alternatives may be used to maintain a specified purity for product B if the feed composition changes?

10.4. In the process shown in Figure 10.4 for separating a close boiling binary or azeotrope by forming two azeotropes through the addition of an entrainer, describe the sequence of events that would happen if the solvent makeup is cut off. What corrective measures should be taken to maintain product purities until the solvent supply is restored?

10.5. How many degrees of freedom exist for the process shown in Figure 10.5, assuming fixed feed AB? What variables control the purity of products A and B?

10.6. In the extractive distillation process in Figure 10.6, what effect would lowering the solvent rate have on the purity of product A? What changes in operating conditions could be made to maintain the specified purity of A? What potential problems should be considered?

10.7. A mixture of ethanol and benzene is to be separated into its constituents. The mixture forms a minimum-boiling azeotrope whose composition shifts with pressure. At 25 kPa the azeotropic composition is 36.0 mole% ethanol and at 100 kPa it is 44.8 mole% ethanol. The feed stream to be separated, F, contains 60.0 kmol/h ethanol and 30.0 kmol/h benzene. Stream F is mixed with a recycle stream D_2 and then sent to the low pressure column. The distillate from this column, D_1, is at the azeotropic composition, and the bottoms, B_1, may be assumed pure ethanol. Stream D_1 is sent to the higher pressure column where it separates into the distillate,

D_2, at the azeotropic composition, and the bottoms, B_2, which may be assumed pure benzene. The distillate D_2 is recycled and mixed with the feed F. Determine the flow rates of streams D_1, D_2, B_1, and B_2.

10.8.   Benzene and cyclohexane are separated by azeotropic distillation using acetone as an entrainer. At the column pressure of 1 atm, acetone forms a minimum boiling azeotrope with cyclohexane at 74.6 mole% acetone and 25.4 mole% cyclohexane. The feed contains 75 kmol/h benzene and 25 kmol/h cyclohexane. The entrainer, pure acetone, is mixed with the feed and sent to the column. The distillate is 99.5 mole% azeotrope and 0.5 mole% benzene. The bottoms is 99 mole% benzene and 1 mole% cyclohexane. Determine the entrainer flow rate.

10.9.   A stream of 100 kmol/h contains 70% mole benzene (1) and 30% mole cyclohexane (2). These components form a minimum-boiling azeotrope, which prevents their separation by conventional distillation. They can be separated by azeotropic distillation using acetone (3) as the entrainer. Acetone forms a minimum-boiling azeotrope with cyclohexane at the column pressure of 100 kPa, where the azeotropic composition is 73.9% mole acetone and 26.1% mole cyclohexane. The entrainer, at the rate of 75 kmol/h, is mixed with the feed and sent to the column. The distillate contains 99% mole acetone-cyclohexane azeotrope and 1% mole benzene, and the bottoms product is mostly benzene with a small percentage of cyclohexane. Use material balances to calculate the flow rates and compositions of the products.

10.10. A two-column process is to be designed for the separation of n-heptane and toluene by extractive distillation, using phenol as the solvent. One column is for the separation and the other for the recovery of toluene and the solvent. The feed stream is 100 kmol/h, containing 60 mole% n-heptane and 40 mole% toluene at 90°C and 130 kPa. The boiling points of the two components at this pressure differ by about 12°C, n-heptane being the more volatile. Although they do not form an azeotrope, their relative volatility approaches 1.0. In the presence of phenol, the relative volatility of n-heptane to toluene can be brought up to about 2.0. The solvent stream is pure phenol at 100°C, 130 kPa, and a flow rate two-to-three times the feed flow rate. Both columns operate at 130 kPa with negligible pressure drops, and at 100% efficiency.

The feed stream F and solvent stream S are sent to Column 1, where the distillate stream $D1$ is mainly n-heptane, and the bottoms $B1$ contains most of the toluene and the phenol. Stream $B1$ is sent to Column 2 for separating the toluene as distillate stream $D2$, and the solvent, phenol, as bottoms stream $B2$.

Based on material balances, determine the flow rates and compositions of streams $D1$, $B1$, $D2$, and $B2$ to satisfy the following preliminary specifications:

Column 1 distillate flow rate, $D1 = 60$ kmol/h
Purity of n-heptane in $D1$, $Y_{11} = 0.98$

Mole fraction of toluene in $D1$, $Y_{12} = 0.012$
Solvent flow rate, $S = 250$ kmol/h
Flow rate of phenol recovery stream $B2 = 0.99 S$

10.11. Iso-butane and 1-butene can be separated by extractive distillation, using furfural as the solvent. This agent alters the relative volatility of iso-butane to 1-butene according to the following equation:

$$\alpha_{12} = 1.1 + 0.1(F_S/F)$$

where F and F_S are the main feed and solvent rates, kmol/h. The following specifications in mole fractions are given on a solvent-free basis:

	Feed	Distillate	Bottoms
IC4	0.50	0.95	0.08
1-C4 =	0.50	0.05	0.92

The feed, at a flow rate of 100 kmol/h, is sent as saturated vapor to the distillation column. The column is equipped with a partial condenser with a vapor product, and a reboiler. For a solvent rate of 500 kmol/h, it is required to determine the required number of equilibrium stages and the optimum feed location for a reflux ratio of 1.5 times the minimum. The McCabe–Thiele method may be used on a solvent-free basis.

REFERENCES

Chen, C. C., H. I. Britt, J. F. Boston, and L. B. Evans, *AIChE J.*, 28, 588, 1982.
De Leye, L. and G. F. Froment, *Computers and Chemical Engineering*, 10(5), 493, 1986.
Koukolik, M. and J. Marek, Mathematical model of HNO_3 oxidation-absorption equipment, *Proc. Fourth European Symposium on Chemical Reaction Engineering*, 1968.
Othmer, D.F., Azeotropic separation, *Chem. Eng. Prog.*, 59, June 1963, 67.
Saito, S., T. Michishita, and S. Maeda, *J. Chem. Eng. Japan*, 4, 37, 1971.
Tierney, J. W. and G. D. Riquelme, *Chem. Eng. Commun.*, 16, 91, 1982.

11 Liquid–Liquid Extraction and Supercritical Extraction

Liquid–liquid extraction applies the phenomenon of liquid–liquid equilibria to carry out component separations. When the nonideality of a system causes two immiscible liquid phases to coexist at equilibrium, certain components may be more soluble in one phase than in the other.

Extraction involves the transfer of components between two liquid phases, much as absorption or stripping involves the transfer of components from liquid to vapor phase or vice versa. As in vapor–liquid multistage separation processes, the device employed to carry out liquid–liquid extraction is usually a counterflow column that performs the function of a number of equilibrium stages interconnected in counterflow configuration. In each stage, two inlet liquid streams mix, reach equilibrium, and separate into two outlet liquid streams. As in vapor–liquid columns, the lack of complete equilibrium in liquid–liquid extractors is accounted for by some form of tray efficiency. Liquid–liquid extraction may also be carried out in a cascade of mixing vessels connected in series in counterflow.

The column operation is gravity induced, with the heavier liquid flowing downward and the lighter liquid flowing upward by buoyancy. For the process to be workable, it is therefore required that one liquid phase be of a distinctly higher density than the other. The internal construction of an extractor column could be of several types, including trayed columns, packed columns, or spray columns with or without agitation.

Since liquid–liquid extraction merely accomplishes the transfer of components from one phase to another and not the separation of components, a complete separation process must include additional equipment such as distillation columns to recover the various components. The main parameters that should be considered in the design or performance evaluation of an extraction process include the choice of a solvent, the solvent-to-feed ratio, the number of stages, and any ancillary equipment that may be required to complete the separation.

In supercritical extraction, the components are transferred from a liquid phase to a supercritical dense phase at equilibrium with the liquid. A system would be suitable for supercritical extraction if the extracted component, or solute, is adequately soluble in the supercritical solvent, and its solubility is a strong function of the solvent density. This phenomenon facilitates the recovery of the solute from the solvent since lowering its pressure or raising its temperature will lower its density. The solubility of the solute drops at the lower solvent density allowing the solute to be recovered, possibly with no need for a distillation column following the extractor.

The extractor models, both liquid–liquid and supercritical, can be solved rigorously as multistage, multi-component counterflow equilibrium processes based on the same principles applied to distillation and similar processes, that is, material and energy balances and phase equilibrium relations. Appropriate thermodynamic methods must be used for predicting the phase equilibrium properties in any application. Rigorous solution methods are discussed in Chapter 13.

The modular shortcut column section method can also be applied to extractors, as described in Chapter 12.

This chapter is mainly concerned with the performance analysis of extraction processes and the estimation of design and operating parameters using graphical techniques.

11.1 EXTRACTION FUNDAMENTALS AND TERMINOLOGY

Liquid–liquid extraction may be represented by a three-component system forming two liquid phases. The solvent and the feed are two essentially immiscible liquids with possible partial mutual solubility. The extract component or solute, to be extracted from the feed by the solvent, is soluble in both phases.

11.1.1 SIMPLE EXTRACTORS

The basic extraction process is represented in Figure 11.1. The extract is the solvent phase that gets enriched in the extract component, and the raffinate is the feed phase that gets depleted in the extract component. If the solvent is lighter than the feed, it must be introduced at the column bottom.

Unlike absorption or stripping, extraction does not involve vaporization or condensation or the heat that accompanies these processes. For this reason extraction is nearly an isothermal process, although some temperature variation could occur as a result of heat of solution.

An extraction column is considered to be consisting of a number of equilibrium stages, each having two liquid phases at equilibrium. The conditions of temperature, pressure, and composition are such that no vapor phase exists. The number of independent variables required to define the column are the degrees of freedom available at the design phase. The complete set of column variables includes the variables needed to define the feed and the solvent, the number of stages, and the pressure and heat load on each stage. The variables are listed as follows for a feed and solvent with C components each and a column with N stages:

Component rates in the feed	C
Component rates in the solvent	C
Temperature and pressure of the feed	2
Temperature and pressure of the solvent	2
Number of stages	1
Stage pressures	N
Stage heat loads	N
Total	$2C + 2N + 5$

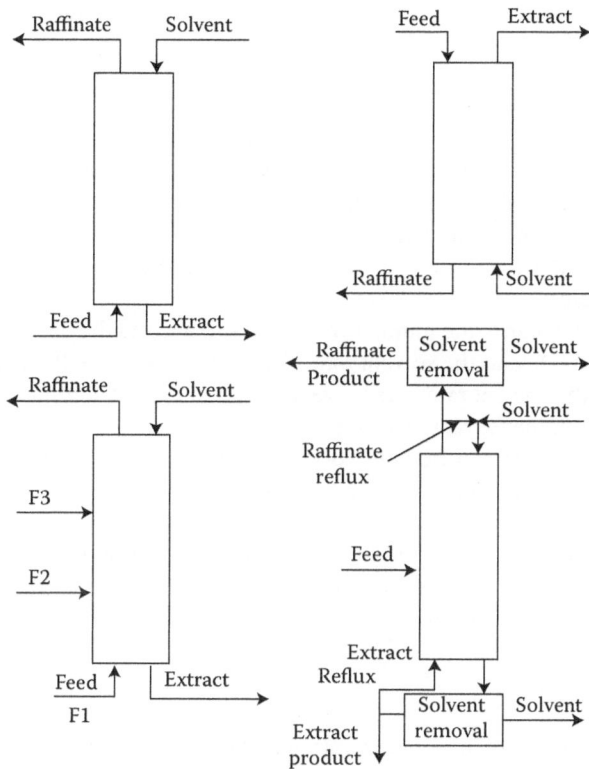

FIGURE 11.1 Schematics of liquid–liquid extractors.

This number of variables is matched by process constraints and performance specifications. The feed and solvent compositions and thermal conditions are constrained at values determined by the sources of these streams. Also, the stage pressures are not truly independent because once the pressure is set at one stage (by pumping capacity, etc.), the pressure profile is determined by the column hydraulics. The stage heat loads are constrained at values determined by heat losses if no controlled heat addition or removal exists at the stages. A listing of all these constraints follows:

Feed composition	$C-1$
Solvent composition	$C-1$
Temperature and pressure of the feed	2
Temperature and pressure of the solvent	2
Stage pressures	N
Stage heat loads	N
Total	$2C + 2N + 2$

At the design phase, three independent variables are available to satisfy performance specifications:

$$(2C + 2N + 5) - (2C + 2N + 2) = 3$$

The independent variables are the feed rate, the solvent rate, and the number of stages. For a given feed rate, the engineer must determine the solvent rate and the number of stages required to satisfy two specifications: the recovery of the extract component from the feed, and the concentration of the extract component in the extract or raffinate.

For an existing column the number of stages is fixed, and the only remaining variables are the feed rate and the solvent rate. For a given feed rate, the column operator can manipulate the solvent rate to achieve a specified extract component recovery, or its concentration in the extract or raffinate.

The description rule can be applied to confirm the above conclusions concerning the number of independent variables required to define an extraction process. The number of stages can be set by construction during the design phase. The feed and solvent rates can be controlled independently by external means during column operation. Thus, three variables are required to define the column during design and two variables during operation.

11.1.2 MULTIPLE FEEDS

For a given solvent stream, the concentration of the extract component in a feed stream determines the number of stages between the solvent and the feed that are required to achieve a specified extract component concentration in either product. The higher the extract component concentration in the feed, the more stages are required. If the extract component is to be removed from more than one feed, each with a different extract component concentration, the column could be designed to handle the feed stream with the highest extract component concentration. Alternatively, the feeds could be premixed and the column designed for the mixed feed composition. If a column is designed to handle the high extract concentration feed, the leaner feeds need not be introduced with the rich feed but could be introduced at intermediate stages. Such a setup is shown in Figure 11.1, with feed streams F1, F2, and F3.

Each additional feed adds $C + 3$ degrees of freedom to the column: one component rate for each of C components in the feed, and the feed temperature, pressure, and location. In designing a new column, the objective is to calculate the optimum number of stages, that is, the smallest number of stages required to achieve the specified separation for a given feed and solvent. For a multiple-feed column, the objective is to calculate the optimum total number of stages, as well as the optimum feed locations; that is, the optimum number of stages between adjacent feeds (including the top feed and solvent), such that the total number of stages is minimized.

The optimum number of stages and feed locations can be determined graphically as described in Section 11.2. For rigorous confirmation of graphical results, especially in most practical situations in which systems are not truly ternary but multicomponent, computer simulation becomes necessary. Although generally simulation algorithms are designed for handling fixed configuration columns (i.e., fixed number of stages and feed locations), it is often possible to determine the optimum through multiple solutions or by means of appropriate optimization routines.

11.1.3 REFLUXED EXTRACTORS

Extractors with reflux at one or both ends of the column (or series of mixing vessels) may be used to enhance the purity of the products. An extractor column with both extract and raffinate reflux is shown in Figure 11.1. In this configuration, the feed is sent to an intermediate stage, and the extractor performs as a distillation column, separating two components in the feed. In the column section above the feed, the raffinate phase is stripped of the extract component (the solute), and in the lower section the extract phase is enriched in the extract component.

Assume the feed contains two main components to be separated, designated as R, the raffinate component, and E, the extract component or solute. If it is not feasible to separate these components by distillation, an extraction process may be considered. A solvent stream containing mainly solvent component S enters at the top of the column, extracts a certain amount of component E from the rising raffinate phase to produce the raffinate product as column overhead, containing mostly component R. The solvent, forming the extract phase, flows down the column, extracting the solute E from the raffinate phase, and exits as the extract phase in the column bottoms, containing mostly components S and E. The solvent is chosen such that its removal from the raffinate and extract components can be readily accomplished, for example, by distillation.

With this configuration the separation process is accomplished with no refluxes: The overhead contains mostly the raffinate component, and the bottoms contains mostly the solute and solvent, from which the solvent can be easily removed. In the absence of refluxes, this arrangement is the same as a single-section extractor. In the section below the feed there is only one phase, the extract phase. It is flowing down the column, with no change in its composition.

A reflux arrangement is now added at the lower end of the column. The extract is sent to a solvent removal unit, and the solvent-free extract is split into an extract product and an extract reflux which is sent back to the bottom of the column. Without the solvent, the extract reflux is now composed of the raffinate component and the solute (components R and E), so that this reflux is actually on the raffinate side of the equilibrium curve (Section 11.2 and Figure 11.2) flowing countercurrent to the extract phase. The extract phase is thus interacting with a raffinate phase which is richer in the solute than the feed. As a result, the extract enrichment with the solute is greater than it would be if the extract were interacting directly with the external feed in the absence of the extractor section below the feed. A higher-purity extract product can therefore be expected.

Refluxing part of the raffinate, on the other hand, does not increase the removal of the extract component from the raffinate since maximum removal is achieved with pure solvent. The raffinate reflux, which amounts to recirculation of an equilibrium phase back to the same equilibrium stage, does not, in principle, change the product composition. In practice, the raffinate reflux may help in achieving a better approach

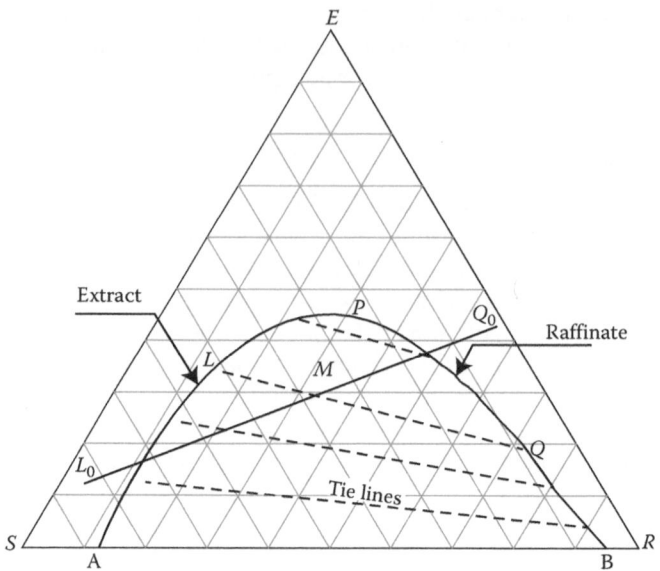

FIGURE 11.2   Liquid–liquid equilibrium for a ternary.

to phase equilibrium due to premixing with the solvent. The raffinate reflux could, that is, improve the stage efficiency.

The manner in which reflux influences an extractor performance and the extent of its effect depend on the equilibrium characteristics of the particular system. Also, the ranges of reflux and other operating conditions must be restricted to the two-phase region where separation is feasible, both thermodynamically and physically.

11.2   GRAPHICAL REPRESENTATION

Graphical representation of liquid–liquid equilibria provides a convenient graphical tool for studying extraction problems and doing preliminary extractor calculations. The results can be used as the basis for defining a computer simulation model for rigorous solution and optimization of the process.

Assuming the extraction system is a ternary, the liquid–liquid equilibrium relationships are best represented graphically on a triangular diagram (Section 1.3.5). The general shape of the equilibrium curve or curves depends on the mutual solubilities of the components in the ternary. In many extraction processes, there is total miscibility between the extract component (the solute) and each of the other components, and partial miscibility between the solvent and the raffinate components. A ternary having these characteristics is represented on an equilateral triangle in Figure 11.2, where E is the extract component, R the raffinate component, and S is the solvent. Each of the binaries ER and ES forms a single liquid phase at all compositions, while binary RS forms two liquid phases between compositions represented by points A and B and a single phase outside this composition range. The ternary mixture forms two liquid phases under curve APB and a single liquid phase outside the curve.

The tie lines shown in Figure 11.2 represent liquid compositions at equilibrium with each other. Thus, point M represents the combined composition of a mixture that separates into two liquid phases at equilibrium, with compositions represented by points L and Q. The compositions of the two phases approach each other as the tie lines become shorter. At point P, the plait point, the tie line vanishes, and the liquid compositions become identical, forming a single phase.

If more than one binary in the three-component mixture exhibits partial solubility (i.e., regions of immiscibility), additional two liquid phase regions would appear on the triangle, typically as domes on other sides of the triangle. These domes might merge in some systems, forming continuous regions of immiscibility. Extractors in practice operate in one particular region of immiscibility and, therefore, for the purpose of graphical calculation of extractors, only that region of immiscibility is of direct interest.

11.2.1 GENERATING EQUILIBRIUM DIAGRAMS

Before any graphical calculation of liquid–liquid extractors can be attempted, reliable liquid–liquid equilibrium data are necessary. These data should include the compositions defining the two-liquid phase envelope and tie lines at temperatures and pressures of interest. Note that a diagram such as Figure 11.2 is specific to one temperature and one pressure.

In the absence of experimental data, liquid–liquid equilibria may be predicted from liquid activity coefficient equations (Section 1.3). Some of these equations, such as NRTL and UNIQUAC, are well adapted for predicting liquid–liquid equilibrium compositions. Triangular diagrams may be generated based on computer-predicted activity coefficients much as they would be based on experimental data.

The computer-aided procedure, unless automated by the program, requires running a series of liquid–liquid equilibrium calculations (the equivalent of vapor–liquid flash calculations) at constant temperature and pressure. The composition is varied around the equilibrium curve, and the transition points from one phase to two, or vice versa, are noted. As many points as needed are obtained this way to generate the entire equilibrium curve. Also, each time an equilibrium calculation is done in the two-phase region, the compositions of the two phases are recorded. Each pair of data points thus obtained defines a tie line. The data obtained at one temperature and pressure generate one triangular diagram. If so desired, the procedure is repeated at other temperatures and pressures to determine the effect of these variables.

11.2.2 SINGLE-STAGE CALCULATIONS

According to the phase rule, a three-component, two-phase system has three degrees of freedom. Thus, by specifying the temperature, pressure, and concentration of one component in one phase, the state of the system is defined. The component concentration in one phase defines one point on the equilibrium curve, and this point marks one end of a tie line. The other end is determined thermodynamically either from experimental data or on the basis of liquid activity coefficient prediction methods.

Since all points on a tie line have the same temperature, pressure, component concentration in one phase, and component concentration in the other phase, one more variable must be specified to define the relative amounts of the phases at equilibrium. The specified variable could itself be either the fraction of one phase out of the total or the mixture composition. Note that only two component concentrations are independent in a ternary. In the single-phase region there are four degrees of freedom according to the phase rule. Therefore, if the fraction of one phase or whether one or two phases exist is unknown, four independent variables must be specified to completely define the system.

The temperature and pressure of the mixture represented by point M in Figure 11.2 are those for which this particular triangular diagram is constructed. The other two variables are the concentrations of two of the components E, R, and S, which fix the location of M. The lengths of the segments of the tie line passing through M determine the ratio of these phases according to the lever arm rule, which is a graphical statement of the material balance. With the extract phase designated by L and the raffinate phase by Q, the following component molar balances hold:

$$MX_{MS} = (L + Q)X_{MS} = LX_{LS} + QX_{QS}$$

$$MX_{MR} = (L + Q)X_{MR} = LX_{LR} + QX_{QR}$$

$$MX_{ME} = (L + Q)X_{ME} = LX_{LE} + QX_{QE}$$

where M, L, and Q are molar quantities, and X_{ij} are mole fractions. The first subscript refers to the mixture M, phase L, or phase Q, and the second subscript refers to the solvent S, the raffinate component R, or the extract component E. For each component, these equations may be written as

$$X_{Mi} = \frac{L}{M} X_{Li} + \left(1 - \frac{L}{M}\right) X_{Qi}$$

where i stands for component S, R, or E. This represents X_{Mi} as a linear function of L/M, so that M lies on a straight line between points Q and L. The same conclusion applies to the line joining Q_0 and L_0. The equations are rearranged to give the ratio of the phases:

$$\frac{L}{Q} = \frac{X_{MS} - X_{QS}}{X_{LS} - X_{MS}} = \frac{X_{MR} - X_{QR}}{X_{LR} - X_{MR}} = \frac{X_{ME} - X_{QE}}{X_{LE} - X_{ME}} = \frac{\underline{QM}}{\underline{LM}} \tag{11.1}$$

where $\underline{QM}$ and $\underline{LM}$ are the segment lengths.

In order to use the equilateral triangular phase diagram to calculate a single-stage extractor, the feed and solvent are first plotted on the diagram as points Q_0 and L_0, respectively (Figure 11.2). Note that, in general, L_0 may not be pure solvent and that the feed, Q_0, may contain some solvent component. The mixture is represented by point M on a straight line joining L_0 and Q_0. The lengths of segments $\underline{L_0M}$ and $\underline{Q_0M}$ are determined by the lever arm rule:

$$\frac{\underline{L_0M}}{\underline{Q_0M}} = \frac{Q_0}{L_0} \tag{11.2}$$

Here, L_0 and Q_0 are the molar rates of the solvent and feed.

The tie line through M together with an overall material balance determines the rates and compositions of the extract, represented by point L, and the raffinate, represented by point Q. The outcome of this process is that the extract component concentration is increased in the extract phase ($X_{LE} > X_{L0E}$) and decreased in the raffinate ($X_{QE} < X_{Q0E}$).

EXAMPLE 11.1: SINGLE-STAGE EXTRACTOR

A single-stage extractor is used to extract component E from a feed stream. The feed and solvent stream compositions and flow rates are as follows:

	Mole Fraction	
Component	Feed, Q_0	Solvent, L_0
E	0.43	0.13
S	0.01	0.83
R	0.56	0.04
Flow rate (kmol/h)	43.00	33.00

Using the phase equilibrium diagram in Figure 11.2, determine the compositions and flow rates of the extract and raffinate.

SOLUTION

The given feed and solvent compositions are used to locate points Q_0 and L_0 on the phase diagram. The mixture point M is then located based on the ratio of flow rates (the graphical solution is shown only schematically in Figure 11.2):

$$\frac{\underline{L_0M}}{\underline{Q_0M}} = \frac{Q_0}{L_0} = \frac{43}{33}$$

The tie line through M determines the raffinate and extract compositions, represented by points Q and L:

	Mole Fraction	
Component	Raffinate, Q	Extract, L
E	0.200	0.350
S	0.078	0.510
R	0.722	0.140

The ratio of raffinate to extract flow rates is calculated from the ratio of segment lengths as measured from the diagram:

$$\frac{Q}{L} = \frac{LM}{QM} = 0.5$$

By overall material balance,

$$Q + L = Q_0 + L_0 = 43 + 33 = 76 \text{ kmol/h}$$

Solving for Q and L, $Q = 25.33$ kmol/h, $L = 50.67$ kmol/h.

The ratio of product flow rates may also be calculated by component material balances. The mixture composition must first be determined either from the diagram or by component material balances:

$$X_{ME} = \frac{Q_0 X_{Q0E} + L_0 X_{L0E}}{Q_0 + L_0} = \frac{(43)(0.43) + (33)(0.13)}{43 + 33} = 0.300$$

$$X_{MS} = \frac{Q_0 X_{Q0S} + L_0 X_{L0S}}{Q_0 + L_0} = \frac{(43)(0.01) + (33)(0.83)}{43 + 33} = 0.366$$

$$X_{MR} = \frac{Q_0 X_{Q0R} + L_0 X_{L0R}}{Q_0 + L_0} = \frac{(43)(0.56) + (33)(0.04)}{43 + 33} = 0.334$$

The raffinate to extract flow rates ratio is calculated from component material balances (Equation 11.1):

$$\frac{Q}{L} = \frac{X_{LE} - X_{ME}}{X_{ME} - X_{QE}} = \frac{X_{LS} - X_{MS}}{X_{MS} - X_{QS}} = \frac{X_{LR} - X_{MR}}{X_{MR} - X_{QR}}$$

$$= \frac{0.350 - 0.300}{0.300 - 0.200} = \frac{0.510 - 0.366}{0.366 - 0.078} = \frac{0.140 - 0.334}{0.334 - 0.722} = 0.5$$

11.2.3 COUNTERCURRENT MULTISTAGE CALCULATIONS

A graphical solution of an extractor can be obtained by representing material balances with operating lines, and equilibrium relationships with tie lines on a triangular diagram. The solution is accurate to the extent of the accuracy of the equilibrium data used to generate the diagram. Since a given diagram is constructed for a fixed temperature and pressure, the graphical solution must assume isothermal and isobaric operation of the extractor.

The method assumes equilibrium stages; the stage efficiency data would be required to represent an actual column in terms of equilibrium stages. The stage efficiency depends on the physical properties of the liquids and the type of equipment used. It is usually estimated on the basis of experimental data or previous experience with similar columns.

An inherent limitation in using triangular diagrams for solving liquid–liquid extractors is the fact that these diagrams represent strictly ternary systems. Although three primary components are involved in a basic extraction process (the extract component (solute), the raffinate, and the solvent), actual systems are rarely truly ternary and are generally multi-component. Nevertheless, for many systems the graphical method provides at least a qualitative representation of the process and a visual means for studying the effect of design and operating parameters such as number of stages and flow rates. The final solution of an extractor column usually relies on computer simulation, which can be applied more effectively once a qualitative understanding of the problem is established.

As shown in Figure 11.3, an overall material balance on the column is written as

$$L_0 + Q_{N+1} = L_N + Q_1 = M \tag{11.3}$$

Thus, if the feed and solvent compositions and flow rates are known, the mixture point M can be determined (Figure 11.4). The solvent and feed points L_0 and Q_{N+1} are first plotted at their known compositions, and a straight line is drawn through them. The mixture point, M, is located on the basis of the lever arm rule:

$$\frac{L_0 M}{Q_{N+1} M} = \frac{Q_{N+1}}{L_0}$$

where $\underline{L_0 M}$ and $\underline{Q_{N+1} M}$ are segment lengths, and L_0 and Q_{M+1} are molar flow rates. The second part of Equation 11.3 is now used to determine points L_N and Q_1,

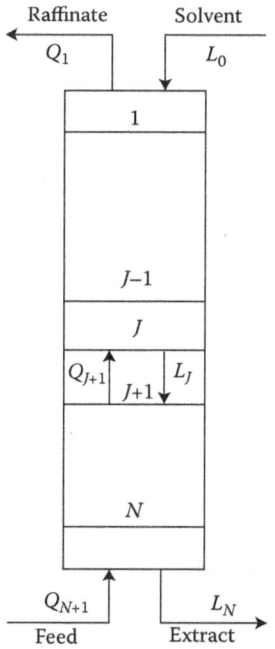

FIGURE 11.3 Equilibrium stages in a liquid–liquid extractor.

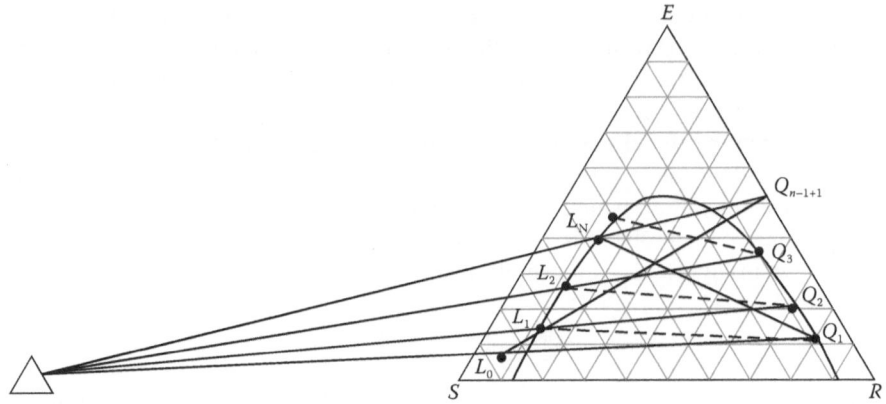

FIGURE 11.4 Graphical construction of extractor equilibrium stages.

corresponding to the raffinate and extract compositions. Both L_N and Q_1 must lie on the equilibrium curve since they are saturated liquids. The concentration of one component in one of these phases must be known, such as in a design situation. If, for instance, the concentration of the extract component in the extract phase is specified, point L_N could be plotted on the diagram. Next, Q_1 is located by drawing a straight line through L_N and M and extending it until it intersects the raffinate side of the equilibrium curve.

Rearranging Equation 11.3 defines the difference point Δ for streams Q_{N+1} and L_N and streams Q_1 and L_0:

$$Q_{N+1} - L_N = Q_1 - L_0 = \Delta \tag{11.4}$$

The difference point can be located on the basis of either side of this equation by passing a line through either pair and extending it to Δ to satisfy the lever arm rule:

$$\frac{Q_{N+1}\Delta}{L_N\Delta} = \frac{L_N}{Q_{N+1}} \quad \text{or} \quad \frac{Q_1\Delta}{L_0\Delta} = \frac{L_0}{Q_1} \tag{11.5}$$

The quantities $Q_{N+1}\Delta$, $L_N\Delta$, $Q_1\Delta$, and $L_0\Delta$ designate segment lengths and Q_{N+1}, L_N, Q_1, and L_0 designate molar flow rates. Since Δ is the same for both pairs of points, it is more easily located as the intersection between the straight lines passing through Q_{N+1} and L_N, and Q_1 and L_0.

The above procedure establishes the product rates and compositions based on overall material balance and the premise that streams leaving the extractor are saturated streams, falling on the equilibrium curve. To determine the number of equilibrium stages required to achieve the performance specification (such as the concentration of the extract component in the extract product) and to determine the compositions on each stage, a material balance is calculated around the top section of the column through stage j:

$$Q_{j+1} - L_j = Q_1 - L_0 = \Delta \tag{11.6}$$

Note that Δ in this equation is the same as Δ defined in Equation 11.4. It is constant for all passing streams L_j and Q_{j+1} between any two stages j and $j + 1$.

The graphical construction of equilibrium stages on the triangular ternary phase diagram can start either at the top or the bottom of the column. Starting at the top, the first operating line is the line drawn through Δ, L_0, and Q_1 (Figure 11.4). It relates the solvent feed, L_0, to the raffinate product, Q_1. Extract liquid L_1, leaving stage 1, is at equilibrium with Q_1 and is therefore represented by point L_1 at the other end of the tie line through Q_1. The operating line relating passing streams L_1 and Q_2 is obtained by drawing a straight line through Δ and L_1 and extending it to Q_2 on the raffinate side of the equilibrium curve. A tie line drawn through Q_2 locates L_2 on the extract side of the equilibrium curve. Another operating line is drawn through Δ and L_2 to locate Q_3, and so on. The procedure is continued until a tie line is reached that either meets L_N or crosses the operating line between L_N and Q_{N+1}. Thus, according to Figure 11.4, three theoretical stages would surpass the desired performance specifications although two theoretical stages would not be sufficient. The required number of stages could, therefore, include a fraction of a theoretical stage. The actual number of stages is obtained by dividing the number of theoretical stages by the efficiency. Thus, whether or not the theoretical stages include a fractional stage, the resulting calculated number of actual stages could include a fractional stage, which must be rounded up.

The procedure for graphical construction of extractor stages is modified if the set of specifications that define the column is different from the above. In most cases L_0 and Q_{N+1} (the external feeds) are known, but the third specification may not relate to the products. For an existing column, for instance, the number of stages is known, and it is required to calculate the products' compositions. A trial and error procedure must be used by assuming the Q_1 or L_N location on the equilibrium curve. Once one of these two points is assumed, the other one is located by passing a straight line through M. The stepping off of stages then proceeds in the usual manner to determine the number of stages. If this number does not match the number of theoretical stages corresponding to the existing column, the procedure is repeated with a new location for Q_1 until a match is obtained.

A ternary diagram constructed for a given system can be used as a tool to study the effect of parameters, such as the number of stages and solvent-to-feed ratio on the extractor performance. From Figure 11.4, it can be seen that, as Δ is moved farther away from the triangle, the operating lines become more closely spaced so that more stages are required for the same separation. Conversely, as Δ is moved closer to the triangle, fewer stages are required. According to Equation 11.5, as Δ approaches L_0, L_0/Q_1, the solvent-to-raffinate ratio, increases. Generally, this also implies an increase in the solvent-to-feed ratio. Thus, fewer stages are required with a higher solvent-to-feed ratio.

EXAMPLE 11.2: MULTISTAGE EXTRACTOR

A countercurrent extractor is used to lower the concentration of component E in a feed stream and produce a raffinate containing 12% mole component E. The compositions of the feed and of the solvent used for the extraction are given below:

Component	Mole Fraction	
	Feed, $Q_{N\pm1}$	Solvent, L_0
E	0.52	0.09
S	0.00	0.85
R	0.48	0.06

What are the flow rates and compositions of the products, and how many equilibrium extractor stages are required to meet the solute reduction specification if the feed rate is 90 kmol/h and the solvent rate is 60 kmol/h? The phase equilibrium diagram in Figure 11.4 may be used for this system.

SOLUTION

The feed, Q_{N+1}, and the solvent, L_0, are plotted based on their given compositions. The graphical solution is shown only schematically in Figure 11.4. Accurate results require a more detailed diagram. The mixture point M is plotted on the line joining Q_{N+1} and L_0 with segment lengths determined by the flow rates ratio:

$$\frac{L_0M}{Q_{N+1}M} = \frac{Q_{N+1}}{L_0} = \frac{90}{60} = 1.5$$

Next, Q_1, the raffinate product point is plotted on the raffinate equilibrium curve at $X_{Q1E} = 0.12$, as specified. The extract product point is located by joining Q_1 to M and extending the line to its intersection with the extract equilibrium curve at L_N. The product compositions are read from the diagram:

Component	Mole Fraction	
	Raffinate, Q_1	Extract, L_N
E	0.12	0.42
S	0.08	0.42
R	0.80	0.16

The ratio of product flow rates is calculated from the segment lengths as measured on the diagram:

$$\frac{Q_1}{L_N} = \frac{L_NM}{Q_1M} = 0.31$$

By overall material balance,

$$Q_1 + L_N = Q_{N+1} + L_0 = 90 + 60 = 150$$

Solving for the product rates,

$$Q_1 = 35.5 \text{ kmol/h}$$

$$L_N = 114.5 \text{ kmol/h}$$

The difference point Δ is located at the intersection point of straight lines through Q_{N+1} and L_N, and Q_1 and L_0. The number of equilibrium stages required for the specified extraction is found graphically. The operating lines are Q_1L_0, Q_2L_1, Q_3L_2, and so on. The tie lines are Q_1L_1, Q_2L_2, and Q_3L_3. The third tie line crosses the last operating line; therefore the required number of equilibrium stages is two and a fraction.

11.2.4 MULTIPLE FEED AND REFLUXED EXTRACTORS

The formulation of the operating line equations and equilibrium relationships and their graphical implementation is not restricted to the basic extractor configuration but can be applied to columns with multiple feeds and reflux streams. The resulting complexity is handled by breaking up the column into sections, each with its own difference point. This is demonstrated for a column with an intermediate feed, resulting in two sections.

Figure 11.5 depicts a two-section extractor with an intermediate feed going on stage f. The raffinate section consists of stages 1 through f–1 and the extract section consists of stages f through N. An overall material balance on the column is written as

$$Q_{N+1} + F + L_0 = Q_1 + L_N$$

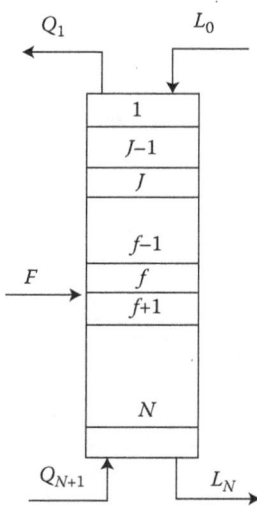

FIGURE 11.5 Extractor with two feeds.

which can be rearranged in two ways:

$$Q_1 - L_0 = Q_{N+1} + F - L_N = \Delta' \tag{11.7}$$

$$Q_{N+1} - L_N = Q_1 - F - L_0 = \Delta'' \tag{11.8}$$

A typical design and optimization problem would be to determine the required number of stages and optimum feed location when the feeds and product compositions are specified. The difference points Δ' and Δ'' are first located in accordance with Equations 11.7 and 11.8. In a design situation all the feed and product compositions are known and their corresponding points, Q_1, Q_{N+1}, L_0, L_N, and F, are plotted on the ternary diagram. Each difference point is determined by the intersection of lines corresponding to the first and second parts of Equations 11.7 and 11.8. Thus, to determine Δ', a straight line is first drawn through Q_1 and L_0. Next, the mixture point for Q_{N+1} and F is located, and a straight line is drawn through this point and L_N. The intersection of these two straight lines determines Δ'. Difference point Δ'' is determined in a similar manner.

A material balance on the top section of the column through any tray j above the feed gives

$$Q_{j+1} - L_j = Q_1 - L_0 = \Delta' \tag{11.9}$$

A material balance on the lower section of the column through any tray j below the feed gives

$$Q_j - L_j - 1 = Q_{N+1} - L_N = \Delta'' \tag{11.10}$$

The graphical construction of operating lines for the equilibrium stages proceeds in the usual manner, using difference point Δ' for the upper column section and Δ'' for the lower column section (Figure 11.6). The procedure is continued from both ends of the column until the external feed F point is reached on the diagram. This determines

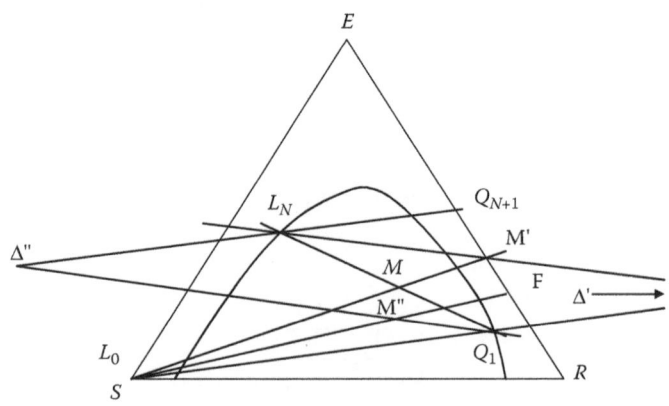

FIGURE 11.6   Graphical construction for extractor with two feeds.

the optimum feed location. The required number of stages is the total of the two sections. The relative stream flow rates are determined from the segment lengths.

11.2.5 LLE RECTILINEAR REPRESENTATION

An alternative to representing liquid–liquid equilibrium data on triangular diagrams is to use rectilinear coordinates. This method is particularly convenient for applying analytical methods for solving problems and considering options in liquid–liquid extraction applications.

The two rectilinear coordinates define the component fractions of two of the three main components in an extraction process. The third component is determined as the difference from unity. For example, the raffinate component fraction, X_r, may be defined on the horizontal coordinate and the extract component fraction, X_E on the vertical coordinate. The solvent component fraction is then determined as $X_s = 1 - X_r - X_e$. The resulting diagram for a given $R–E–S$ system is shown schematically in Figure 11.7.

A possible application using rectilinear coordinates is also shown in Figure 11.7. The diagram shows the phase envelope, which consists of the equilibrium composition curve of the extract phase (L) and the equilibrium composition curve of the raffinate phase (Q).

11.2.5.1 Analytical Approach

The graphical methods described in the sections above for analyzing and solving liquid–liquid extraction applications may be supplemented by the analytical approach, based on rectilinear coordinates. For investigation options and possible alternative approaches relating to an extraction process, it may be more efficient to change numbers in a program such as a spreadsheet than to redraw the graphics.

An example application of the analytical method could be to determine the number of stages required to achieve a specified raffinate composition, where the feed Q_0 and solvent stream L_{N+1} flow rates and compositions are given.

Based on available equilibrium data, equations would be fitted for the raffinate and equilibrium curves, as well as the tie lines. The equilibrium curves represent relevant

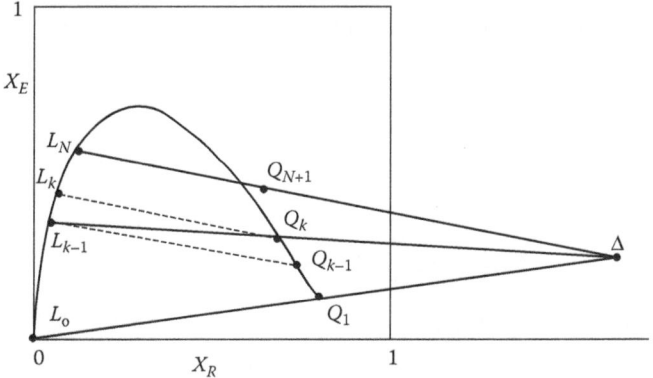

FIGURE 11.7 Liquid–liquid extractor using rectilinear coordinates.

portions of the complete equilibrium curve (Figure 11.7). For the specified raffinate composition, a simultaneous solution of material balances and the equilibrium equation defines flow rates L_1 and Q_N, and the composition of L_1 on the equilibrium curve.

Two straight line equations would then be derived, one through points Q_0 and L_1, and the other through Q_N and L_{N+1}. These are the top and bottom operating lines; their intersection defines the difference point Δ.

To find the required number of equilibrium stages, the calculations alternate between finding a raffinate composition at equilibrium with an extract composition using the tie line equation, then locating the extract composition on the next stage by passing a straight line (the operating line) through the current raffinate composition and Δ. This sequence progresses from one stage to the next as follows:

$$L_1 \rightarrow Q_1 \quad \text{Using a tie line equation}$$

$$Q_1 \rightarrow L_2 \quad \text{Using an operating line equation}$$

$$\cdot$$
$$\cdot$$
$$\cdot$$

This process is continued until point Q_N is either met or first time surpassed. The number of repetitions gives the required number of equilibrium stages.

EXAMPLE 11.3: ACETONE EXTRACTOR DESIGN BY THE NUMERICAL METHOD

In a chemical processing plant, the acetone concentration in a water–acetone solution stream must be lowered. Acetone extraction using vinyl trichloride (VTC) as the solvent is considered, and the required number of equilibrium stages in a countercurrent liquid–liquid extraction system must be determined. Equilibrium relations were developed based on available data at the expected operating conditions. The extract and raffinate phase and tie-line equations are expressed in terms of the extract component (acetone) and raffinate component (water) weight fractions in each phase.

Extract phase equilibrium curve,

$$X_{eL} = 1044.6(X_{rL})3 - 301.94(X_{rL})2 + 26.19(X_{rL}) - 0.1315$$

Raffinate phase equilibrium curve,

$$X_{rQ} = -0.6389(X_{eQ})2 - 0.7781(X_{eQ}) + 0.9758$$

Tie-lines,

$$X_{eQ} = 0.4336(X_{eL})2 - 0.5212(X_{eL}) + 0.0121$$

From the summation equations,

$$X_{sL} = 1 - X_{eL} - X_{rL}$$
$$X_{sQ} = 1 - X_{eQ} - X_{rQ}$$

The variables in the equations above are defined as follows:

X_{eL} = weight fraction of the extract component (acetone) in the extract phase L.
X_{eQ} = weight fraction of the extract component (acetone) in the raffinate phase Q.
X_{rL} = weight fraction of the raffinate component (water) in the extract phase L.
X_{rQ} = weight fraction of the raffinate component (water) in the raffinate phase Q.
X_{sL} = weight fraction of the solvent component (VTC) in the extract phase L.
X_{sQ} = weight fraction of the solvent component (VTC) in the raffinate phase Q.

The extraction process will operate at the given stream data:

Feed flow rate, $Q_0 = 1200$ kg/h; Composition, $X_{eQ0} = 0.45$, $X_{rQ0} = 0.55$, $X_{sQ0} = 0$
Solvent flow rate, $L_{N+1} = 420$ kg/h; Composition, $X_{sLN+1} = 1$
Specification, acetone fraction in the raffinate product, $X_{eQN} = 0.10$

In the above expressions, subscripts 0 and $N + 1$ refer to the external feeds to stages 1 and N, respectively.

SOLUTION

From the raffinate phase equilibrium equation and the specified acetone fraction in the raffinate product, and from the summation to 1,

$$X_{eQN} = 0.10$$

$$X_{rQN} = 0.8916$$

$$X_{sQN} = 0.0084$$

The following equations are solved simultaneously to determine the product rates L_1 and Q_N, and the extract product composition in terms of X_{eL1} and X_{rL1}:

Overall material balance,

$$Q_0 + L_{N+1} - Q_N - L_1 = 0$$

Acetone material balance,

$$Q_0 X_{eQ0} - Q_N X_{eQN} - L_1 X_{eL1} = 0$$

Water material balance,

$$Q_0 X_{rQ0} - Q_N X_{rQN} - L_1 X_{rL1} = 0$$

Extract equilibrium equation as given above.
    The solution to these four equations gives

$$L_1 = 921.56 \text{ kg/h}$$

$$Q_N = 698.44 \text{ kg/h}$$

$$X_{eL1} = 0.5102$$

$$X_{rL1} = 0.0404$$

The difference point, (X_{ed}, X_{rd}) is defined by the top and bottom operating lines:

$$(X_{eL1} - X_{eQ0})/(X_{rQ0} - X_{rL1}) - (X_{eL1} - X_{ed})/(X_{rd} - X_{rL1}) = 0$$

$$(X_{eQN} - X_{eLN+1})/(X_{rQN} - X_{rLN+1}) - (X_{ed} - X_{eLN+1})/(X_{rd} - X_{rLN+1}) = 0$$

Solution to these equations gives

$$X_{ed} = 0.2557$$

$$X_{rd} = 2.3723$$

From the tie line equation,

$$X_{eQ1} = 0.3909$$

From the raffinate equilibrium equation,

$$X_{rQ1} = 0.5741$$

The extract composition on stage 2 (L_2) is found from the intersection of the operating line through (X_{ed}, X_{rd}) and (X_{eQ1}, X_{rQ1}), and the extract equilibrium curve.

The raffinate composition on stage 2 (Q_2) is found from the tie line equation.

The following stage-by-stage results are obtained:

n	X_{eLn}	X_{eQn}	X_{rQn}	X_{rLn}
1.	0.5102	0.3909	0.5741	0.0404
2.	0.4250	0.3119	0.6710	0.0317
3.	0.3422	0.2412	0.7513	0.0243
4.	0.2419	0.1635	0.8316	0.0181
5.	0.1191	0.0804	0.9092	0.0104

Acetone in the raffinate, $X_{eQ5} = 0.0804$, more than satisfies the specified value of 0.10, therefore the required number of stages is five.

11.3   EXTRACTION EQUIPMENT

Extractor equipment considerations are discussed here in the context of their effect on the process performance and not for the purpose of describing detailed design. The main parameter of interest at this point is the number of equilibrium stages that represent the process. Liquid–liquid extraction requires thorough mixing of two liquid phases to achieve thermodynamic equilibrium, followed by complete separation of the phases. The particular equipment selected for a given process is determined, in part, by the mixing and separation characteristics of the phases.

In one arrangement, each stage of the extraction process consists of a mixing vessel followed by a settler where the phases are separated. The mixers could be mechanically agitated. A multistage process consists of a number of such vessels and settlers interconnected in countercurrent flow. In this process each stage could, in principle, be made to approach an equilibrium stage as desired by controlling the agitation and vessel volume (which determines the residence time).

The other type of extractor equipment falls under the general category of columns, with vertical, gravity-induced counterflow. Columns come with a variety of internal designs, including spray columns, packed columns, sieve or bubble cap trayed columns, or baffled columns. Some columns may contain internal mechanical agitators. The heavier phase is introduced at the top of the column and the lighter phase at the bottom. Usually one phase is continuous and the other discontinuous. In the case of packed columns, one of the phases preferentially wets the packing, and the wetting phase could be either the continuous or the discontinuous phase. Extractor columns usually include a distributor to disperse the discontinuous phase and an interface region to separate the phases.

Compared to mixer settler equipment, extractor columns have the advantage of being gravity-driven, thereby minimizing the need for pumping, piping, and so on. In any extraction arrangement the problem of mixing and separating the liquids is a design challenge. This is a more severe problem than in vapor–liquid separation processes where the properties of the two phases are substantially different. In liquid–liquid extraction, unless aided by mechanical agitation, low efficiency can be expected if the densities of the two liquids are not significantly different, their viscosities are high, and/or their interfacial surface tension is high. Extractor columns relying solely on gravitational forces would generally have low efficiencies. The benefits of their compactness must be supplemented by agitation to make them more efficient. Commercial extractor columns are equipped with a variety of agitation mechanisms.

The efficiency of extractors and the equivalent number of equilibrium stages depends on internal design and other factors. Detailed description of extractor equipment is provided by Najim (1989) and is outside the scope of this book.

11.4   SUPERCRITICAL EXTRACTION

In absorbers or strippers, components are transferred from the vapor phase to the liquid phase or vice versa. In liquid–liquid extraction, components are transferred from one liquid phase to the other. In supercritical extraction, components (the solute) are transferred from the liquid phase to the supercritical phase (the solvent) at equilibrium with the liquid.

One of the benefits of using supercritical fluids as the solvent is the strong dependence of the solubility of the solute on the solvent density. This is a property that could be exploited for facilitating the separation of the solute from the solvent as it leaves the column, by dropping its pressure or raising its temperature, thereby lowering its density and the solubility of the solute. As a result, the extract separates into a liquid solute and a vapor solvent. Another favorable property of supercritical fluids as solvents is the high diffusivity of the solute in these fluids compared to that in liquids. Supercritical fluids also have a substantially lower viscosity than liquids. Because of these properties the mass transfer rate of the solute in supercritical fluids is considerably higher than in liquids, resulting in a faster approach to equilibrium.

For a supercritical fluid to be suitable as a solvent in extraction, a high solubility of the solute is required. If the objective is to separate components, the solvent should also have selective dissolution properties. Moreover, the pressure effect on the solubility is a factor. High compression costs may be incurred if the conditions for desirable solubility require excessively elevated pressures. The critical temperature of a potential solvent is also important. If the solvent is to be around its critical point for optimal performance, it is preferred that its critical temperature not be too far from ambient temperature.

As in any multiphase system, the conditions for phase equilibrium in a supercritical extraction operation are the equality of the temperature, pressure, and component fugacities in the phases (Section 1.3). Supercritical extraction involves a liquid phase and a supercritical phase at or approaching equilibrium. The component fugacities in the liquid can be calculated from activity coefficient equations (Section 1.3.3). The supercritical phase may be represented by one of the cubic equations of state such as the Soave-Redlich-Kwong or Peng-Robinson equations (Section 1.2.4). For a better fit, the binary interaction coefficients used in the mixing rules for these equations should be regressed from experimental data around the extraction operating conditions of temperature, pressure, and composition. Once liquid-supercritical equilibrium data are developed, the supercritical extraction column could be modeled with the general rigorous multi-component multistage computational methods (Chapter 13). The equilibrium data could also be used to represent the column using simplified graphical methods. In the graphical methods, the supercritical extraction process is modeled as a ternary system consisting of a liquid component, a supercritical solvent component, and a solute component. As in the graphical methods for computing liquid–liquid extractors (Section 11.2), ternary liquid-supercritical fluid–solute diagrams may be employed to graphically estimate the number of stages required for a supercritical extraction process, and to calculate the product rates and compositions.

Among the practical applications of supercritical extraction is the use of supercritical carbon dioxide as the solvent in a number of processes. Carbon dioxide has several favorable properties as a supercritical solvent. It is nontoxic, low-cost, and noncorrosive. Its critical temperature is 304.2 K, which is near ambient. These properties are especially desirable in food processing, for the extraction of food components that must not be exposed to high temperatures. Examples are the removal of caffeine from coffee and the extraction of oil from beans and corn.

In hydrocarbon processing, residual streams and asphaltenes contain oils and resins mixed with high molecular weight hydrocarbons and undesirable metals and carbon residues. The oils and resins may be recovered more effectively by supercritical extraction than by other processes. One possible extraction fluid is supercritical pentane, which selectively dissolves the oils and resins. The extracted components are then conveniently separated from the pentane by raising the extract temperature. The extract separates into a liquid phase containing the oils and resins, and a vapor phase containing the pentane.

NOMENCLATURE

C Number of components
E Extract
F Feed stream
L Liquid phase: solvent and extract
M Mixture point
N Number of stages
Q Liquid phase: feed and raffinate
R Raffinate
S Solvent
X Mole fraction
Δ Difference point

SUBSCRIPTS

i Component designation
j Stage designation

PROBLEMS

11.1. A solvent is used to recover component E from a liquid feed containing components R and E by extraction in a single-stage extractor. The compositions and flow rates of the feed and solvent are given below:

	Mole Fraction	
Component	Feed, Q_0	Solvent, L_0
S	0.02	0.84
R	0.56	0.04
E	0.42	0.12
Flow rate (kmol/h)	100.00	75.00

Using liquid–liquid equilibrium data from Figure 11.2, calculate the flow rates and compositions of the raffinate and extract.

11.2. A pure solvent S is used to remove component E from a binary liquid solution containing 30% mole E and 70% mole R. In a single-stage extractor how many kmol solvent per 100 kmol feed are required to bring the concentration of E in the raffinate to 3% mole? What are the resulting rates and compositions of the raffinate and extract? Use liquid–liquid equilibrium data from Figure 11.2.

11.3. It is required to recover 93% of component E from a binary mixture of components E and R by treating with a solvent in a single-stage extractor. What is the required solvent rate per 100 kmol of feed, and what are the product rates and compositions? Assume the liquid–liquid equilibrium data in Figure 11.2 are applicable. The feed and solvent compositions are as follows:

	Mole fraction	
Component	Feed, Q_0	Solvent, L_0
S	0.02	0.95
R	0.59	0.00
E	0.39	0.05

11.4. In the extraction service of Problem 11.2, the solvent rate is limited to 85 kmol per 100 kmol of feed. How many extractor stages would be required to achieve the same objective of 3% mole concentration of component E in the raffinate?

11.5. Component E is to be removed by extraction with a pure solvent out of two feed streams containing components E and R. The compositions and flow rates of the feeds and solvent are given below. The concentration of E in the raffinate is specified at 5% mole. The main feed is sent to the bottom of the column and the side feed to an intermediate stage. Determine the number of stages required, the optimum location of the side feed, and the compositions of the products. Liquid–liquid equilibrium data may be obtained by modifying Figure 11.2 such that the concentration of E in both phases is the same. (The tie lines are parallel to the base of the triangle.)

	Mole Fraction		
Component	Main Feed	Side Feed	Solvent
S	0.00	0.00	1.00
R	0.60	0.80	0.00
E	0.40	0.20	0.00
Flow rate (kmol/h)	70.00	30.00	42.40

11.6. Feed stream Q_0, flowing at a rate of 1000 kg/h, contains 45 wt% acetone in solution with water. It is required to extract the acetone in an extractor column, using 1,1,2-trichloroethane as the solvent, L_{N+1}, at a rate of 350 kg/h. The raffinate, Q_N, should contain 10 wt% acetone.

Using rectilinear representation of the operating lines and equilibrium data, determine the required number of equilibrium stages. Use the horizontal coordinate for X_c and the vertical coordinate for X_a. The equilibrium data at the extractor temperature of 25°C are given below.

a = acetone (solute), c = water (carrier), s = 1,1,2-trichloroethane (solvent).
Q = Raffinate phase, L = Extract phase.

The raffinate-side equilibrium curve is based on the following data, weight fractions:

$(X_c)_Q$	0.35	0.43	0.57	0.68	0.79	0.895
$(X_a)_Q$	0.55	0.50	0.40	0.30	0.20	0.10

The extract-side equilibrium curve is based on the following data, weight fractions:

$(X_c)_L$	0.075	0.04	0.029	0.0215	0.015	0.01
$(X_a)_L$	0.575	0.50	0.40	0.30	0.20	0.10

The tie line weight fractions are as follows:

$(X_a)_L$	0.56	0.40	0.18
$(X_a)_Q$	0.44	0.29	0.12

The following are given data:

$Q_0 = 1000$ kg/h	$L_{N+1} = 350$ kg/h
$(X_a)_{Q0} = 0.45$	$(X_c)_{Q0} = 0.55$
$(X_a)_{QN} = 0.1$	$(X_c)_{QN} = 0.895$ (from the raffinate curve)
$(X_s)_{LN+1} = 1$	$(X_a)_{LN+1} = (X_c)_{LN+1} = 0$

11.7. Ethyl alcohol in water solution is to be treated with supercritical carbon dioxide solvent to lower by extraction the alcohol concentration in the water. The process is carried out in a six-stage countercurrent extractor. The stream data are as follows:

	Inlet Mole Fractions		
	Feed, X_{Q7}	Solvent, X_{L0}	K-values, X_{Qi}/X_{Li}
1. Ethanol	0.04	0	8.33
2. Water	0.96	0	166.7
3. Carbon dioxide	0	1	0.0286

Q_7: Feed flowrate = 2 kmol/h; Temperature = 300 K; Pressure = 10,000 kPa
L_0: Solvent flowrate = 5 kmol/h; Temperature = 300 K; Pressure = 10,000 kPa

The extractor is at constant temperature and pressure, consistent with the inlet streams conditions. It is required to determine the extractor products flow rates and compositions using the Kremser equation described in Chapter 12, Equations 12.43 through 12.46. Additionally, find the solvent flow rate that will bring the alcohol in the extract stream down to 0.02 kmol/h.

11.8. Solve Problem 11.6 graphically, based on the given equilibrium data.

REFERENCE

Najim, K., *Process Modeling and Control in Chemical Engineering*, New York, Marcel Dekker, 1989.

12 Shortcut Methods

In Chapters 7 through 11, the performance of multistage separation processes was analyzed qualitatively on the basis of fundamental principles developed in Chapters 3 through 6. The objective was to gain an understanding of the different types of separation processes and columns and the factors that affect their performance. Chapters 5 and 6 employed graphical and semi-quantitative methods to represent a limited set of separation processes, namely binary distillation. Chapters 10 and 11 also used combinations of graphical and analytical methods applied to binary or ternary systems to represent specific classes of nonideal separations.

This chapter treats multi-component separations using approximate or "shortcut" methods that rely on certain simplifying assumptions to solve the column equations. Some of these techniques consider limiting conditions of total reflux and minimum reflux with an infinite number of trays. The results can only approximate real processes where the reflux rate and the number of trays are finite. The other shortcut simplifying assumptions may or may not be valid for a given problem. Nevertheless, analyzing a distillation problem on the basis of these methods, where applicable, is useful for preliminary estimations and for determining practical column operating limits. Some shortcut methods are capable of calculating the required number of stages for a given separation problem, whereas rigorous methods usually assume a fixed number of stages. Further, the information derived from shortcut calculations may be used for evaluating alternative trade-offs of number of trays vs. reflux rate for achieving desired separation criteria.

In spite of the simplifications, these methods are still quite computation-intensive and in most cases must be solved using computer programs. However, due to the much shorter computing time, these methods are useful in situations where computing time is crucial, such as in online, real-time applications. Especially suited for these applications are the modular shortcut methods based on column sections— another topic discussed in this chapter.

The shortcut solution can often be used as the foundation for progressing toward a rigorous solution of the problem. In many applications the two methods are used in conjunction with each other to attain efficient and reliable simulation strategies.

12.1 COLUMNS AT TOTAL REFLUX

The concept of total reflux was discussed in Chapters 3 and 6. The conditions of total reflux may be approached either by operating the column at finite feed and product rates with a very large reflux rate or by cutting off the feed and products and maintaining internal boilup and reflux by adding and removing heat at the reboiler and condenser.

By using a simplified model of binary distillation with two stages, it was shown in Chapter 3 that maximum separation is achieved at total reflux. Also, the Y–X diagram was applied in Chapter 6 to verify that for a given number of stages, maximum separation of a binary mixture is obtained at total reflux. A corollary to this statement is that for a specified separation, the required number of stages is least at total reflux. The performance of multi-component columns at total reflux is discussed next.

12.1.1 MODEL DESCRIPTION

The model considered here is a single-feed, multi-component, multistage column with a partial condenser and reboiler, a vapor overhead product, and a liquid bottoms product. The feed is assumed to be of fixed rate, composition, and thermal conditions. The column pressure is also assumed fixed. The column has N stages, including the condenser and the reboiler, and the feed is introduced on stage f. A schematic of the model is shown in Figure 12.1. The model assumes that the column diameter, the condenser, and the reboiler are large enough to handle the internal reflux and boilup required to approach the conditions of total reflux. The column reflux or stripping fluid may not be supplied by external streams such as in absorbers or strippers, where the concept of total reflux is meaningless.

It will become evident in the mathematical derivation that the feed location in a total reflux column is immaterial. Hence, although the model is developed for

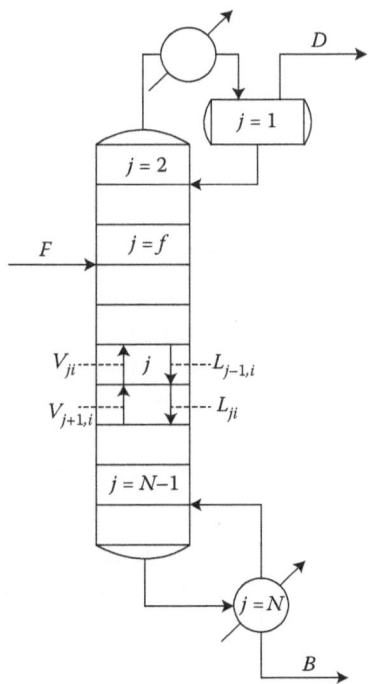

FIGURE 12.1 Column schematic.

a single feed, multiple feeds can be handled just as well, as long as they are not intended to serve as external reflux or stripping streams. The model may be extended to simulate multiple product columns as described in Section 12.1.5.

The other assumption in the model relates to the vapor–liquid equilibrium coefficients, or K-values. The K-values at a given pressure are assumed to be a function of temperature only, and not of composition. It is further assumed that the temperature dependence of the K-values for the different components is similar, that is, the ratio of the K-values of any pair of components is independent of temperature. Thus, the relative volatilities, defined as the ratios of K-values of any two components, are assumed constant throughout the column.

The feed, distillate, and bottoms flow rates are F, D, and B, respectively. The stages are numbered from the top down, the condenser being stage 1 and the reboiler stage N. The total vapor molar flow leaving stage j is designated as V_j and the total liquid molar flow leaving stage j as L_j. The vapor flow of component i leaving stage j is V_{ji}, and the liquid flow of component i leaving stage j is L_{ji}. Finally, the mole fractions of component i in the vapor and liquid leaving stage j are designated as Y_{ji} and X_{ji}, respectively. In general, the first subscript refers to the stage number and the second subscript refers to the component number.

12.1.2 MATHEMATICAL REPRESENTATION

From the above definitions, it follows that

$$V_{ji} = V_j Y_{ji}$$
$$L_{ji} = L_j X_{ji}$$

(12.1)

And, from the definition of the vapor–liquid distribution coefficient (Chapter 1),

$$K_{ji} = \frac{Y_{ji}}{X_{ji}}$$

(12.2)

Referring to Figure 12.1, an overall material balance around trays 1 through j for j between 1 and f (above the feed tray) is written as

$$V_{j+1} = L_j + D$$

(12.3)

For j between f and $N - 1$, the material balance must include the feed:

$$F + V_{j+1} = L_j + D$$

(12.4)

Dividing Equations 12.3 and 12.4 by V_{j+1} gives

$$\frac{L_j}{V_{j+1}} + \frac{D}{V_{j+1}} = 1$$

and

$$\frac{F}{V_{j+1}} + 1 = \frac{L_j}{V_{j+1}} + \frac{D}{V_{j+1}}$$

At total reflux, D and F approach zero compared to the internal vapor and liquid flows in the column. Hence, at total reflux,

$$\frac{D}{V_{j+1}} = 0, \quad \frac{F}{V_{j+1}} = 0, \quad \text{and} \quad \frac{L_j}{V_{j+1}} = 1$$

These relations apply throughout the column, that is, for j between 1 and N–1.

A material balance on component i around trays 1 through j takes the form

$$V_{j+1}Y_{j+1,i} = L_j X_{ji} + DY_{1i} \quad (1 \leq j < f) \tag{12.5}$$

and

$$FZ_i + V_{j+1}Y_{j+1,i} = L_j X_{ji} + DY_{1i} \quad (f \leq j \leq N - 1) \tag{12.6}$$

where Z_i is the mole fraction of component i in the feed. Division of either of Equations 12.5 or 12.6 by V_{j+1} and substitution of the total reflux expressions derived above results in the following:

$$Y_{j+1,i} = X_{ji} \quad (1 \leq j \leq N - 1) \tag{12.7}$$

This equation applies throughout the column at total reflux. Equation 12.2 is written for tray $j + 1$,

$$K_{j+1,i} = \frac{Y_{j+1,i}}{X_{j+1,i}}$$

and, combined with Equation 12.7 to eliminate $Y_{j+1,i}$,

$$X_{j+1,i} = \frac{X_{ji}}{K_{j+1,i}} \tag{12.8}$$

Equation 12.8 is next applied to trays $j = 1, 2, 3, \ldots, N - 1$ as follows:

$$X_{2i} = \frac{X_{1i}}{K_{2i}} \quad (j = 1)$$

$$X_{3i} = \frac{X_{2i}}{K_{3i}} = \frac{X_{1i}}{K_{2i}K_{3i}} \quad (j = 2)$$

$$X_{Ni} = \frac{X_{1i}}{K_{2i}K_{3i}...K_{Ni}} \quad (j = N - 1)$$

Substituting $X_{1i} = Y_{1i}/K_{1i}$ (Equation 12.2) in the last equation results in the following:

$$X_{Ni} = \frac{Y_{1i}}{K_{1i}K_{2i}...K_{Ni}} \tag{12.9}$$

Equation 12.9 relates the bottoms composition, X_{Ni}, to the distillate composition, Y_{1i}, at total reflux. The derivation of this equation on the basis of material balances and equilibrium relationships is equivalent to the graphical solution for binary mixtures described in Chapter 5. However, in the binary graphical solution, the material balance is represented by the operating curves, not necessarily at total reflux, while in the present derivation the material balance equation is obtained on the basis of total reflux.

If the molar flow rates of component i in the distillate and bottoms are designated as d_i and b_i respectively, the following relationships hold:

$$d_i = DY_{1i}$$
$$b_i = BX_{Ni} \tag{12.10}$$

Substituting the values of Y_{1i} and X_{Ni} from Equation 12.10 into Equation 12.9, the following is obtained:

$$\frac{b_i}{d_i} = \frac{B/D}{\prod_{j=1}^{N} K_{ji}} \tag{12.11}$$

The relative volatility, α_{ji}, for component i is defined relative to some *reference component, r*:

$$\alpha_{ji} = \frac{K_{ji}}{K_{jr}} \tag{12.12}$$

K_{ji} from this equation is now substituted in Equation 12.11 to give

$$\frac{b_i}{d_i} = \frac{B/D}{\prod_{j=1}^{N} \alpha_{ji} K_{jr}} \tag{12.13}$$

The factor $\prod K_{jr}$ may be replaced by $(K_r)^N$, where K_r is the K-value of reference component r at average column conditions. Also, since constant relative volatility is assumed, the product of the relative volatilities may be replaced by $(\alpha_i)^N$. Equation 12.13 may thus be rewritten as

$$\frac{b_i}{d_i} = \frac{B/D}{\alpha_i^N K_r^N} \tag{12.14}$$

A column overall material balance is written as

$$F = B + D \tag{12.15}$$

and a material balance on component i around the column is given by

$$f_i = d_i + b_i \tag{12.16}$$

where f_i is the flow rate of component i in the feed. The values of D and d_i from Equations 12.15 and 12.16 are now substituted in Equation 12.14 to give

$$b_i = \frac{f_i}{1 + (F - B/B)(\alpha_i^N K_r^N)} \tag{12.17}$$

This is a set of equations applicable to $i = 1, ..., C$, where C is the total number of components. Equation 12.17 is one form of the Fenske equation (Fenske, 1932) for a total reflux distillation column. The number of stages N, is the minimum required to achieve the separation corresponding to component flow rates f_i and b_i and bottoms rate B. In this equation, N, the number of stages, is also known as the fractionation index.

Another form of the Fenske equation can be derived for calculating the minimum number of stages required for a specified separation of two key components. Equation 12.9 is written for the light key component l and the heavy key component h:

$$\frac{Y_{1l}}{X_{Nl}} = \prod_{j=1}^{N} K_{jl}$$

$$\frac{Y_{1h}}{X_{Nh}} = \prod_{j=1}^{N} K_{jh}$$

The following is obtained by combining these two equations:

$$\left(\frac{Y_{1l}}{X_{Nl}}\right)\left(\frac{X_{Nh}}{Y_{1h}}\right) = \prod_{j=1}^{N}\left(\frac{K_{jl}}{K_{jh}}\right) = \prod_{j=1}^{N} \alpha_{jlh} = \alpha_{lh}^N$$

This equation is rearranged to give the minimum number of stages for a specified separation:

$$N_m = \frac{\log[(Y_{1l}/X_{Nl})(X_{Nh}/Y_{1h})]}{\log \alpha_{lh}} \qquad (12.17\text{a})$$

The separation of the key components may be specified by setting the values of any three of the mole fractions Y_{1l}, Y_{1h}, X_{Nl}, and X_{Nh}. The fourth mole fraction is determined by material balance. Thus, if for instance Y_{1l}, X_{Nl}, and X_{Nh} are given, the product rates are calculated from the following equations:

$$F = D + B$$

$$f_l = DY_{1l} + BX_{Nl}$$

The value of Y_{1h} is calculated by a material balance on the heavy key component:

$$f_h = DY_{1h} + BX_{Nh}$$

In these equations, f_l and f_h are the flow rates of the key components in the feed.

The relative volatility α_{lh} of the light key relative to the heavy key is an average value, and assumed constant on all the stages. Its value is determined independently from vapor–liquid equilibrium data or correlations. The temperature at which the relative volatility is calculated must be assumed. It is some average column temperature that could be verified by applying Equation 12.17 to check if $B = \Sigma b_i$ (see Section 12.1.4).

12.1.3 DEGREES OF FREEDOM

In equation set 12.17, F and f_i are known quantities. The relative volatilities α_i are assumed constant and known for each component. At a fixed pressure, K_r is a function of temperature only. The bottoms total flow rate, B, may be replaced by the sum of b_i, the component flow rates in the bottoms. There are therefore $C + 2$ variables in Equation 12.17: b_i ($i = 1, 2, ..., C$), N, and K_r (or an average column temperature). Equation 12.17 is a set of C independent equations for $i = 1, 2, ..., C$. With $C + 2$ variables and C equations, the column has two degrees of freedom, that is, two variables must be specified in order to define the column performance. The two specifications could be any two of the above variables or functions thereof. For instance, the specifications could be N and B, or b_l and b_h, the light and heavy key component flow rates in the bottoms, or Y_{Dh} and X_{Bl}, the mole fractions of the heavy key component in the distillate and the light key component in the bottoms, and so on. In each case, with two variables specified, all the other variables can be calculated from equation set 12.17.

According to Equation 12.17a, the column has three degrees of freedom: three variables must be specified. The additional degree of freedom is a consequence of the absence of a check on the temperature which is assumed in this equation. This is consistent with the fact that the condition $B = \Sigma b_i$ is not implied by Equation 12.17a.

12.1.4 SOLUTION METHODS

The set of nonlinear equations (Equation 12.17) can be solved by different techniques, depending on what pair of variables is specified. An iterative algorithm is outlined here for the case where the number of stages, N, and the bottoms rate, B, are specified.

Step 1. Assume an average column temperature, T_r, and calculate K_r, the reference component vapor–liquid equilibrium coefficient at average conditions, using some K-value correlation of the form $K_r = K_r(T_r)$. Also, at the same temperature, use a similar correlation to calculate all the other component K-values. Calculate the components' relative volatilities.
Step 2. Calculate b_i from Equation 12.17 for $i = 1, ..., C$.
Step 3. Calculate the sum of b_i.
Step 4. If the sum of b_i equals the specified B within a designated tolerance, a solution has been obtained. Otherwise, go to the next step.
Step 5. Update K_r using some convergence technique. A simple method might be to recalculate K_r as follows:

$$K_r(\text{new}) = K_r(\text{old}) \frac{\sum b_i}{B} \qquad (12.17b)$$

Calculate a new average temperature, T_r, corresponding to the new K_r. At the new temperature, calculate updated component K-values and relative volatilities.

The calculations are repeated at step 2 until a solution is reached. Also, based on the calculated distillate and bottoms compositions, the distillate and bottoms temperatures may be calculated as the dew point and bubble point temperatures, respectively (Chapter 2).

EXAMPLE 12.1: TOTAL REFLUX CALCULATIONS

It is required to calculate the maximum separation possible for a hydrocarbon mixture using a ten-theoretical-stage column. The column pressure is 500 kPa. The feed is a saturated liquid at 500 kPa. The feed component flow rates are as follows:

Component	f_i (kmol/h)
1. Ethane	12
2. Propane	48
3. n-Butane	25
4. n-Pentane	15

TABLE 12.1
Stream Component Flow Rates for Example 12.1

Component	Feed, f_i (kmol/h)	Bottoms, b_i (kmol/h)	Distillate, d_i (kmol/h)
Ethane	12.00	0.00	12.00
Propane	48.00	0.46	47.54
n-Butane	25.00	24.60	0.40
n-Pentane	15.00	15.00	0.00
Totals	100.00	40.06	59.94

The bottoms flow rate is specified at 40 kmol/h. The separation takes place between propane and n-butane, the light and heavy key components. Using n-pentane as the reference component, the following relative volatilities are given:

$$\alpha_1 = 13.304$$
$$\alpha_2 = 5.553$$
$$\alpha_3 = 2.315$$
$$\alpha_4 = 1.000$$

SOLUTION

Maximum separation is achieved at total reflux, where Equation 12.17 applies. For $N = 10$ and $B = 40$,

$$b_i = \frac{f_i}{1 + ((100 - 40)/40)(\alpha_i K_r)^{10}} = \frac{f_i}{1 + 1.5(\alpha_i K_r)^{10}}$$

An initial estimate for K_r is required, which could be either an arbitrary value or an estimate based on a K-value temperature function prediction method. Since the column temperature is not known, an arbitrary initial estimate is used for K_r. The reference component is pentane, the heaviest component in the mixture. Therefore, K_r should have a fairly small value, anywhere between zero and one. Using a starting value of $K_r = 0.1$, b_i is calculated from the above equation for $i = 1$, 2, 3, 4. The bottoms rate, calculated as the sum of b_i, is 88.244. Since the specified rate is 40, a second iteration is used, with an updated value of K_r:

$$K_r = (0.1) \times (88.244)/40 = 0.221$$

The iterations are continued until convergence is reached at $K_r = 0.275$, where the calculated bottoms rate is 40.06 kmol/h. The converged component flow rates in the bottoms are given in Table 12.1, along with the corresponding flows in the distillate.

EXAMPLE 12.1A: MINIMUM NUMBER OF STAGES

The feed stream in Example 12.1 is to be separated according to the following specifications for propane, the light key, and n-butane, the heavy key:

$$Y_{12} = 0.790, \quad Y_{13} = 0.0067, \quad X_{N2} = 0.0115$$

The relative volatility of propane to n-butane is $\alpha_{23} = 2.4$. What is the minimum number of stages required to meet the specifications?

SOLUTION

The product rates are calculated by overall and propane material balances:

$$100 = D + B$$

$$48 = 0.790D + 0.0115B$$

The calculated rates are $D = 60.2$ kmol/h, $B = 39.8$ kmol/h. The mole fraction of n-butane in the bottoms is calculated by a material balance on n-butane:

$$25 = (60.2) \times (0.0067) + 39.8X_{N3}$$

$$X_{N3} = 0.618$$

The minimum number of stages is calculated by Equation 12.17a:

$$N_m = \frac{\log\left[(0.790/0.0115)(0.618/0.0067)\right]}{\log(2.4)} = 10$$

12.1.4.1 General Specifications

The solution method described at the beginning of this section is workable only if the specifications are N and B. In general, the two specifications, designated as G_1 and G_2, could be any function of the variables in equation set 12.17. Thus, G_1 and G_2 may be expressed as

$$G_1(\overline{b}_i, N, K_r) = 0$$
$$G_2(\overline{b}_i, N, K_r) = 0 \tag{12.18}$$

Equations 12.17 and 12.18 comprise $C + 2$ equations which must be solved for the $C + 2$ variables. The entire set of equations could be solved simultaneously using the Newton–Raphson technique. In an alternative method which simplifies the calculations, equation set 12.18 is expressed as functions of N and B:

$$G_1(N, B) = 0$$
$$G_2(N, B) = 0 \tag{12.19}$$

Equation set 12.19 is equivalent to equation set 12.18 since N and B are implicit functions of b_i, N, and K_r. The independent variables are N and B, and the dependent variables are b_i and K_r. The calculations alternate between a Newton–Raphson solution of equation set 12.19 and an iterative solution of equation set 12.17 as described previously. The algorithm is outlined as follows:

Step 1. Assume initial estimates for N and B.

Step 2. Solve equation set 12.17 using the algorithm described previously for the case where N and B are specified.

Step 3. Check if the specifications, equation set 12.18, are satisfied within specified tolerances. If they are, the current values of b_i, N, and K_r constitute a converged solution. If equation set 12.18 is not satisfied, proceed to the next step.

Step 4. The values N and B are updated by adding corrections ΔN and ΔB calculated from a Taylor series expansion of Equation 12.19:

$$\frac{\partial G_1}{\partial N} \Delta N + \frac{\partial G_1}{\partial B} \Delta B + G_1 = 0$$
$$\frac{\partial G_2}{\partial N} \Delta N + \frac{\partial G_2}{\partial B} \Delta B + G_2 = 0 \tag{12.20}$$

The partial derivatives and the functions G_1 and G_2 are calculated from Equations 12.17 and 12.18 at current values of the variables. Equation set 12.20 is then solved for ΔN and ΔB, which are used to update N and B. The calculations are repeated at step 2 until a solution is obtained.

EXAMPLE 12.2: SHORTCUT COLUMN—GENERAL SPECIFICATIONS

Using the feed stream and relative volatility data given in Example 12.1, calculate the minimum number of stages and the distillate and bottoms component flow rates to satisfy the following specifications: n-butane mole fraction of 0.10 in the distillate and propane mole fraction of 0.02 in the bottoms.

SOLUTION

The specifications can be written in the general form:

$$G_1 = Y_{13} - 0.10 = 0$$

$$G_2 = X_{N2} - 0.02 = 0$$

Following the algorithm outlined above, assume the initial values of N and B:

$N = 5$

$B = 30$ kmol/h

Calculate b_i by Equation 12.17:

$$b_i = \frac{f_i}{(1 + (100 - 30/30))(\alpha_i K_r)^5}$$

Iterate on K_r until $\Sigma b_i = B = 30$ kmol/h. The results at $K_r = 0.343$ are as follows:

	Component	f_i (kmol/h)	b_i (kmol/h)	d_i (kmol/h)
1.	Ethane	12.00	0.0026	11.9974
2.	Propane	48.00	0.8069	47.1931
3.	n-Butane	25.00	14.3964	10.6036
4.	n-Pentane	15.00	14.8357	0.1643
		100.00	30.0416	69.9584

The calculated specification functions are

$$G_1 = \frac{10.6036}{69.9584} - 0.10 = 0.05157$$

$$G_2 = \frac{0.8069}{30.0416} - 0.02 = 0.006859$$

The partial derivatives are calculated by perturbing N to 6 with $B = 30$, and B to 35 with $N = 5$. The results are

$$\frac{\partial G_1}{\partial N} = -0.0047 \quad \frac{\partial G_1}{\partial B} = -0.010$$

$$\frac{\partial G_2}{\partial N} = -0.0077 \quad \frac{\partial G_2}{\partial B} = 0.004$$

Equations 12.20, with the above values of G_1, G_2 and their partial derivatives, are solved for ΔN and ΔB. The results are $\Delta N = 2.9$ and $\Delta B = 3.8$. Equation 12.17 is solved with the new values of N and B, $N = 5 + 2.9 = 7.9$ and $B = 30 + 3.8 = 33.8$, to calculate new product component rates. The iterative procedure is repeated until the values of G_1 and G_2 are sufficiently small. At $N = 6$ and $B = 33.5$, the results are $G_1 = 0.006$ and $G_2 = -0.001$. The product streams with $K_r = 0.3301$ are as follows:

	Component	f_i (kmol/h)	b_i (kmol/h)	d_i (kmol/h)
1.	Ethane	12.00	0.0008	11.9992
2.	Propane	48.00	0.6291	47.3709
3.	n-Butane	25.00	17.9170	7.0830
4.	n-Pentane	15.00	14.9616	0.0384
		100.00	33.5085	66.4915

12.1.5 MULTIPLE PRODUCTS

As in two-product columns, total reflux in multiple product columns is a limiting condition where the column internal liquid and vapor flows are very large compared to each of the products and feed(s). Multi-product columns are considered to

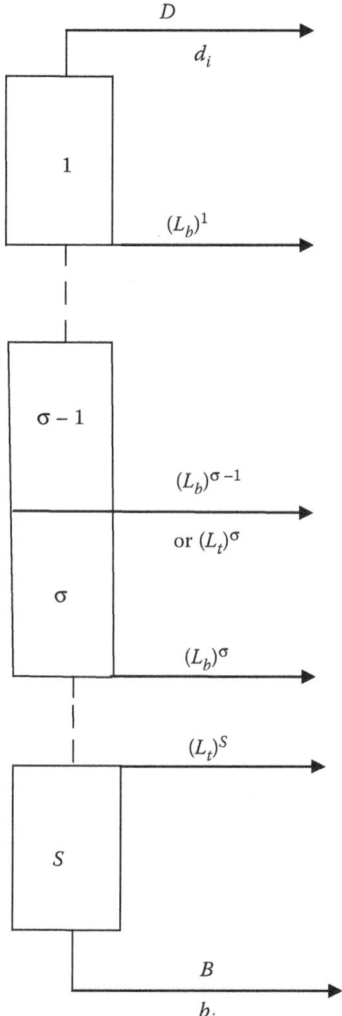

FIGURE 12.2 Multi-product column.

be consisting of column sections defined by the product locations. A multi-product column schematic is shown in Figure 12.2. Each section is bounded by two products, one at its top and the other at its bottom. Thus, a column with S sections has $S + 1$ products. As in two-product columns, multi-product columns operating at total reflux achieve the maximum separation possible with a given number of stages in each section and for a given set of product rates. Conversely, if the separation between the different products is specified, the minimum trays required in each section are evaluated using total reflux calculations.

Since the ratio of any feed or product rate to internal liquid or vapor flow is infinitesimal at total reflux, Equation 12.7, which is derived for two-product columns, applies to multi-product columns as well. This conclusion may be ascertained by carrying out component material balances over different column sections, similar to the material balances represented by Equations 12.5 and 12.6. The equilibrium relation, Equation 12.2, holds regardless of the number of products. The derivations that lead to Equation 12.9 can therefore be generalized to multi-product columns at total reflux, and that equation may be rewritten for component i in any column section as follows:

$$X_{bi} = \frac{Y_{ti}}{K_{ti}K_{t+1,i}\dots K_{bi}} \tag{12.21}$$

This equation relates the liquid composition at the lowest stage in a column section, X_{bi}, to the vapor composition at the highest stage in the section, Y_{ti}, in terms of the product of the K-values at each stage in the section.

Equation 12.21 is converted to a more usable form for a column section, as in the case of a two-product column. Constant relative volatilities are assumed for each column section or throughout the column. For a given section σ, a reference component equilibrium coefficient, K_r^σ, is defined at average section conditions. Referring to Figure 12.2 and substituting $K_{ji} = \alpha_i K_r^\sigma$ in Equation 12.21 and rearranging, the following equation is obtained:

$$\frac{Y_{ti}}{X_{bi}} = (\alpha_i K_r^\sigma)^{b-t+1} \tag{12.22}$$

Also, because $Y_{ti} = X_{ti}K_{ti} = X_{ti}\alpha_i K_r^\sigma$ Equation 12.22 can be expressed in terms of the component mole fractions in the liquid leaving the top and bottom trays in the section:

$$\frac{X_{ti}}{X_{bi}} = (\alpha_i K_r^\sigma)^{b-t} \tag{12.23}$$

If the liquid side draw from tray j has a molar flow rate of L_j^σ and the component flows in L_j^σ are L_{ji}^σ, then $L_{ji}^\sigma = L_j^\sigma X_{ji}$. Equation 12.23 is now rewritten in terms of product and component rates:

$$\frac{L_{ti}^\sigma}{L_{bi}^\sigma} = \frac{L_t^\sigma}{L_b^\sigma}(\alpha_i K_r^\sigma)^{b-t}$$

In the multi-product shortcut model, it is reasonable to consider the top product from section σ to be also the bottom product from section $\sigma-1$: $L_t^\sigma = L_b^{\sigma-1}, L_{ti}^\sigma = L_{bi}^{\sigma-1}$, and the above equation becomes

$$\frac{L_{bi}^{\sigma-1}}{L_{bi}^\sigma} = \frac{L_b^{\sigma-1}}{L_b^\sigma}(\alpha_i K_r^\sigma)^{b-t} \tag{12.24}$$

Equation 12.24 is applicable to all column sections that are bounded by side products on both sides. The uppermost and lowest sections are bounded by the overhead or bottoms on one side and a side draw on the other. If a partial condenser is used, Equation 12.22 should be applied to the uppermost section. Expressed in terms of stream and component flow rates, the equation for the uppermost section takes the form

$$\frac{d_i}{L_{bi}^1} = \frac{D}{L_b^1}(\alpha_i K_r^1)^b \tag{12.25}$$

Subscript b designates the stage number where the highest side product is drawn. Exponent b is the number of stages in the top section. For the lowest column section, Equation 12.24 is rewritten as

$$\frac{L_{ti}^S}{b_i} = \frac{L_t^S}{B}(\alpha_i K_r^S)^{N-t} \tag{12.26}$$

where S represents the bottom section, N is the total number of column stages, and t is the stage number where the lowest side product is drawn.

In the general case where the number of sections, S, is greater than two, the column has $S-2$ of Equations 12.24 plus Equations 12.25 and 12.26 for each component i. If there are two sections (one side product), only Equations 12.25 and 12.26 apply.

An overall column component material balance results in one more set of equations:

$$f_i = d_i + \sum_{\sigma=1}^{S-1} L_i^\sigma + b_i \tag{12.27}$$

The summation is over all the side products. The stage subscript of L_i^σ is dropped because side products are drawn only at the end stage of each section.

For a column with S sections, the total number of independent equations includes S sets of Equations 12.24 through 12.26 and one set of Equation 12.27. If C is the number of components, the total number of independent equations is $C(S+1)$. The variables are listed below:

d_i	$i = 1, ..., C$	C variables
L_i^σ	$i = 1, ..., C, \sigma = 1 ..., S-1$	$C(S-1)$ variables
b_i	$i = 1, ..., C$	C variables
N^σ	Number of stages in each section	S variables
K_r	Section reference component K-value	S variables

The quantities D, B, and L^σ are not considered independent variables since they may be replaced by the sums of d_i, b_i, and L_i^σ respectively. There is a total of

$C(S + 1) + 2S$ variables. Therefore, the column has $2S$ degrees of freedom, and in order to define its performance, $2S$ variables must be specified.

If N^σ in each section and any S of the product rates D, B, and L^σ are specified, the solution algorithm could proceed along the steps hereby outlined:

Step 1. Assume an average temperature for each section and calculate K_r^σ and relative volatilities in each column section.

Step 2. With the values for K_r^σ and the relative volatilities, Equations 12.24, 12.25, 12.26, and 12.27 are a set of linear equations. Solve them for d_i, b_i, and L_i^σ using a matrix inversion method.

Step 3. Check the equalities, $B = \Sigma b_i$ and $L^\sigma = \Sigma L_i^\sigma$. If they are satisfied within a specified tolerance, a solution has been reached. If the equalities are not satisfied, proceed to Step 4.

Step 4. Update the K_r^σ vector as described in Section 12.1.4. Calculate the average temperature in each column section from the temperature dependence of the updated K_r^σ values, or from bubble point or dew point calculations, and calculate new relative volatilities. Repeat the calculations beginning at Step 2.

The average temperatures of the sections calculated at the converged pass can be used for estimating the overall column temperature profile.

If the specifications are not N^σ, D, B, and L^σ, they may be expressed as functions thereof:

$$G_n(N^\sigma, B, D, L^\sigma) = 0 \quad \text{for } n = 1, 2, \dots, 2\sigma \qquad (12.28)$$

Equation set 12.28 is solved along with Equations 12.24, 12.25, 12.26, and 12.27. As in the method described for two-product columns, the calculations alternate between a Newton–Raphson solution of equation set 12.28 and an iterative solution of the rest of the equations as described above.

EXAMPLE 12.3: COLUMN WITH A SIDE DRAW

A distillation column is designed for the separation of a mixture of benzene, toluene, and biphenyl into a distillate (mostly benzene), a side draw (mostly toluene), and a bottoms product (mostly biphenyl). The column will operate at a pressure of about 175 kPa and a temperature ranging from about 90°C in the condenser to about 240°C in the reboiler. The three products define two column sections, and in each section a K-value for biphenyl, designated as the reference component, and relative volatilities for the other components are estimated based on the average temperature and pressure in the section. The feed stream component flow rates, relative volatilities, reference K-values, and product rates are given below:

Component	i	f_i (kmol/h)	α_{i3}^1 (Top Section)	α_{i3}^2 (Bottom Section)
Benzene	1	75.0	75.0	55.0
Toluene	2	23.0	30.0	28.0
Biphenyl	3	2.0	1.0	1.0

Reference K-values,

Top section	$K_r^1 = 0.02$
Bottom section	$K_r^2 = 0.03$

Product rates,

Distillate	$D = 74$ kmol/h
Side draw	$L^1 = 24$ kmol/h
Bottoms	$B = 2$ kmol/h

The minimum number of trays (at total reflux) for each section,

Top section	$N_m^1 = 6$
Bottom section	$N_m^2 = 3$

Calculate the expected component flow rates in each product.

SOLUTION

Equations 12.25, 12.26, and 12.27 are solved for each component:

$$d_i = (D/L^1)(\alpha_i^1 K_r^1)^{N_m^1} L_i^1$$

$$d_1 = (74/24)(75.0 \times 0.02)^6 L_1^1 = 35.12 L_1^1$$

$$d_2 = (74/24)(30.0 \times 0.02)^6 L_2^1 = 0.144 L_2^1$$

$$d_3 = (74/24)(1.0 \times 0.02)^6 L_3^1 = 1.97 \times 10^{-10} L_3^1$$

$$L_i^1 = (L^1/B)(\alpha_i^2 K_r^2)^{N_m^2} b_i$$

$$L_1^1 = (24/2)(55.0 \times 0.03)^3 b_1 = 53.91 b_1$$

$$L_2^1 = (24/2)(28.0 \times 0.03)^3 b_2 = 7.11 b_2$$

$$L_3^1 = (24/2)(1.0 \times 0.03)^3 b_3 = 0.000324 b_3$$

$$d_i + L_i^1 + b_i = f_i$$

$$d_1 + L_1^1 + b_1 = 75.0$$

$$d_2 + L_2^1 + b_2 = 23.0$$

$$d_3 + L_3^1 + b_3 = 2.0$$

These equations can be grouped by component and solved by substitution. The results are

i	d_i (kmol/h)	L_i^1 (kmol/h)	b_i (kmol/h)
1.	72.88	2.08	0.04
2.	2.58	17.90	2.52
3.	0.00	0.00	2.00
Totals	75.46	19.98	4.56

A discrepancy exists between the calculated product rates and the design product rates. A better match can be obtained by adjusting the reference component K-values, which would reflect an adjusted column temperature profile.

12.2 MINIMUM REFLUX RATIO

The operation of a distillation column at minimum reflux was discussed in Chapters 5 and 6 in the context of binary systems. Using the Y–X diagram, it was shown that the reflux ratio is directly related to the slope of the rectifying section operating line. As the operating line approaches the equilibrium curve, the reflux ratio becomes smaller. The operating lines may not cross the equilibrium curve; therefore, the minimum reflux corresponds to the smallest slope of the rectifying section operating line that will not result in either of the operating lines crossing the equilibrium curve. At minimum reflux, one of the operating lines touches the equilibrium curve or both of them touch it at their intersection (Figure 12.3a and b). The location of the point where contact occurs depends on the shape of the equilibrium curve and also on the feed and product compositions and the feed thermal conditions. The number of stages that can be stepped off goes up progressively as the operating line approaches the equilibrium curve. If the operating line touches the equilibrium curve the number of stages becomes infinite, and all these stages correspond to the same composition. Figure 12.3a shows an infinite number of trays in the rectifying section, and Figure 12.3b shows an infinite number of trays around the feed tray.

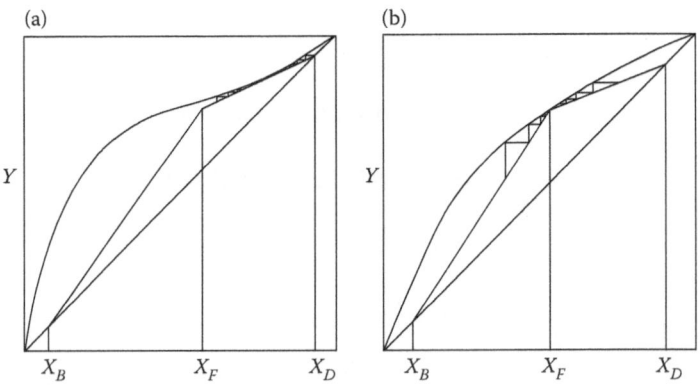

FIGURE 12.3   Minimum reflux in binary systems.

The concept of minimum reflux also applies to multi-component mixtures. It is defined as the reflux ratio below which a specified separation is infeasible, irrespective of the number of trays. At minimum reflux ratio, an infinite number of trays would be required to achieve the specified separation. With infinite trays there must exist in the column at least one section where the vapor and liquid compositions do not change from tray to tray.

The definition of minimum reflux is linked to a specification of separation between two components. In multi-component mixtures, the specified components are the light key and the heavy key components. Minimum reflux is meaningless if the column is specified in a manner that does not define a particular separation between two components, such as a specification of the number of trays and a product rate. Also, minimum reflux in this context is associated with single-feed, two-product columns. For a given feed composition, the minimum reflux depends on the key components, their separation specification, and the feed thermal conditions.

In rigorous column solution methods, the number of trays is usually assumed fixed and, therefore, the concept of minimum reflux does not apply. For binary mixtures, the minimum reflux is probably best determined by graphical methods such as the Y–X diagram method. For multi-component systems, the Underwood equation (Underwood, 1948), derived on semi-rigorous grounds, provides an estimate of the minimum reflux corresponding to a specified separation of two key components in a given stream. The assumptions used in the derivation are similar to the Fenske assumptions for calculating minimum trays, namely constant relative volatilities, with the additional assumption of constant liquid and vapor flows in the column. The Underwood equation should, therefore, be used only for columns and mixtures where these assumptions are valid.

The equation in its final form is presented here. The reader is referred to the original article (Underwood, 1948) for detailed derivation. The minimum reflux ratio R_m is given by the equation

$$R_m = \sum_{i=1}^{C} \frac{\alpha_i X_{Di}}{\alpha_i - \theta} - 1 \qquad (12.29)$$

where α_i is the relative volatility relative to any reference component, and X_{Di} are the mole fractions in the distillate at minimum reflux. This composition must be estimated independently, usually using the Fenske equation solved for given separation specs. The parameter θ has a value between the relative volatilities of the light and heavy key components. It is determined from the equation

$$\sum_{i=1}^{C} \frac{\alpha_i Z_i}{\alpha_i - \theta} = 1 - q \qquad (12.30)$$

where Z_i is the mole fraction of component i in the feed, and q is the feed thermal condition, as defined in Section 5.2.1.

The number of trays in an actual column is, of course, never infinite, and columns will never achieve the specified separation if operated at minimum reflux ratio. What the estimation of minimum reflux does provide is a limiting condition or a reference point. The operating reflux ratio is commonly expressed as a factor multiplied by the minimum reflux ratio. This factor must be greater than one in order to make the specified separation possible. The minimum reflux ratio is also used for correlating the required number of trays and reflux ratio as described in Section 12.3.

12.3 COLUMN DESIGN AND PERFORMANCE ANALYSIS

The minimum trays, N_m, and the minimum reflux ratio, R_m, estimated by the above shortcut methods, represent limiting conditions that require either infinite (total) reflux or infinite trays to achieve a specified separation. They also serve as correlating parameters for predicting the reflux ratio required to achieve the specified separation with a given number of theoretical trays, or the number of theoretical trays required with a given reflux ratio. The Gilliland correlation (Gilliland, 1940) is presented as a plot of $(N - N_m)/(N + 1)$ vs. $(R - R_m)/(R + 1)$ in Figure 12.4, where R is the reflux ratio required to achieve the separation with N theoretical trays. Thus, if the separation is given or specified, the minimum trays can be calculated by the Fenske method, the minimum reflux can be calculated by the Underwood method, and the reflux required with a given number of trays can be obtained from the Gilliland correlation.

An analytical equivalent to the Gilliland correlation is expressed in equation form (Molokanov et al., 1972):

Defining

$$X = \frac{R - R_m}{R + 1} \quad \text{and} \quad Y = \frac{N - N_m}{N + 1},$$

$$Y = 1 - \exp\left\{\left(\frac{1 + 54.4X}{11 + 117.2X}\right)\left(\frac{X - 1}{X^{0.5}}\right)\right\}$$

In a design situation the engineer can use the reflux vs. trays relationship to evaluate the economics of designing a column with more trays and lower reflux vs. fewer trays and higher reflux. The higher cost associated with a column with a large number of trays is offset by a reduced column diameter that would be required with a lower reflux ratio. The utility consumption for the reboiler and condenser is also reduced at the lower reflux. With fewer trays the column diameter and utility consumption are higher since a higher reflux ratio is required.

The calculations, aided by computer simulation programs, are started by making rough estimates of the overhead and bottoms compositions, based on the desired separation. The condenser pressure is determined so that it ensures condensation of the overhead. That is, the condenser pressure should be such that the estimated dew point temperature of the overhead stream is higher than the temperature of the cooling medium at hand. The column pressure should be higher

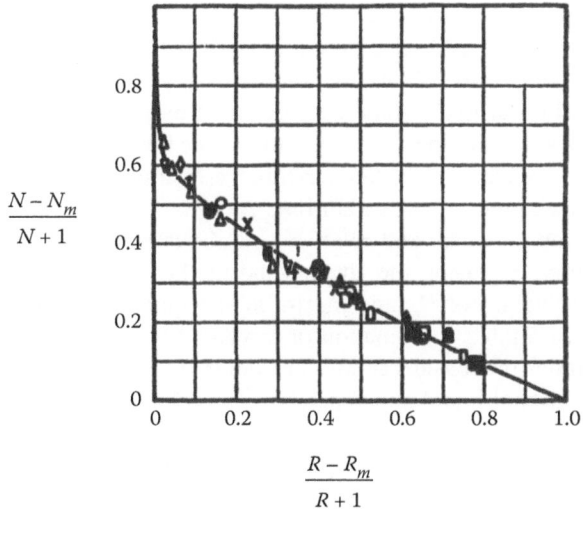

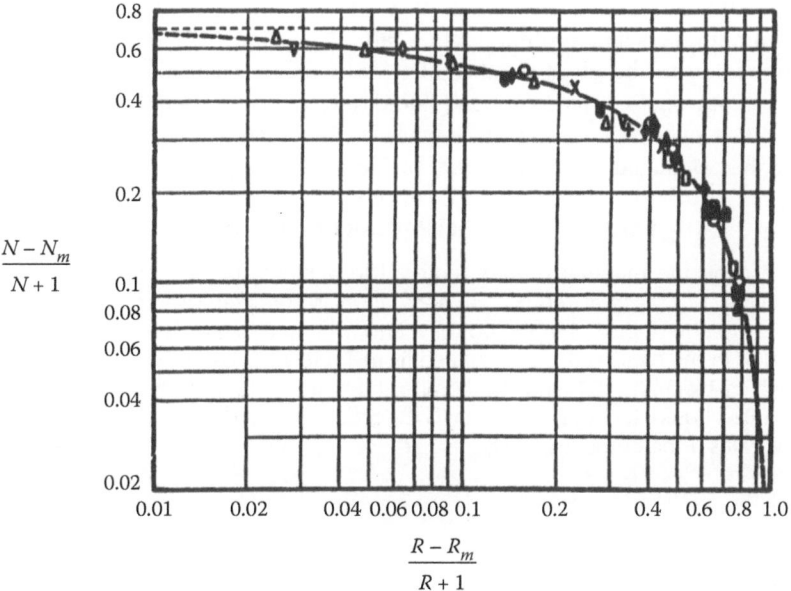

FIGURE 12.4 Gilliland correlation. The symbols on the curves represent data from three different systems. (Reprinted with permission from E. R. Gilliland, *Ind. Eng. Chem.*, 32, 1220, 1940. American Chemical Society.)

than the condenser pressure to allow for vapor-flow pressure drops across the column and condenser. Once the operating pressure is set, it is held constant in the first computation pass.

Next, the relative volatilities, α, are determined as averages of estimated α's obtained from K-value calculations at overhead and bottoms conditions. These conditions include the operating pressure, the estimated overhead and bottoms compositions, and the estimated dew point and bubble point temperatures. With a starting set of relative volatilities, the minimum trays, N_m, and the overhead and bottoms compositions at total reflux are calculated by the Fenske method (Equations 12.17 and 12.17a) for the specified separation. The new compositions can be used to recalculate more accurately the temperatures, pressures, and relative volatilities. The process is repeated until the α's stabilize.

The final compositions and relative volatilities are then used to calculate the minimum reflux, R_m, using the Underwood method (Equations 12.29 and 12.30). The calculated values of N_m and R_m are next applied in the Gilliland correlation to determine a suitable combination of trays and reflux, N and R, consistent with economic and design considerations. The column can then be approximately sized based on the internal liquid and vapor flows as calculated from the reflux ratio and product rates.

The shortcut methods can also be used for approximate analysis of the performance of an existing column. Here, the number of trays, N, is fixed, and the objective is to determine the reflux ratio required to meet a specified separation. The Fenske and Underwood methods (Equations 12.17, 12.29, and 12.30) are used to calculate the minimum trays and minimum reflux ratio, N_m and R_m. The operating reflux ratio corresponding to the given number of trays is then read from the Gilliland chart (Figure 12.4). The internal vapor and liquid rates are calculated from the reflux ratio and product rates. A check must be made to determine if the existing column can handle the calculated vapor and liquid traffic.

EXAMPLE 12.4: PRELIMINARY COLUMN DESIGN

Determine the minimum reflux required to achieve the separation obtained in Example 12.1 at total reflux. The separation is specified as propane (light key) mole fraction of 0.0115 in the bottoms and n-butane (heavy key) mole fraction of 0.0067 in the overhead, as calculated from Table 12.1. Use the relative volatilities given in Example 12.1 and the compositions in Table 12.1. Assume the feed is saturated liquid, that is, $q = 1$. What is the reflux ratio and internal liquid and vapor flow rates if the number of stages is 15?

SOLUTION

First calculate θ from Equation 12.30. Substituting the values of α_i, Z_i and q,

$$\frac{(13.304)(0.12)}{13.304 - \theta} + \frac{(5.553)(0.48)}{5.553 - \theta} + \frac{(2.315)(0.25)}{2.315 - \theta} + \frac{(1)(0.15)}{1 - \theta} = 0$$

The value of θ must be somewhere between the key components' relative volatilities. A trial and error solution gives $\theta = 2.86$.

The minimum reflux is computed directly from Equation 12.29, using the overhead composition at total reflux:

$$R_m = \frac{(13.304)(0.200)}{13.304 - 2.86} + \frac{(5.553)(0.793)}{5.553 - 2.86} + \frac{(2.315)(0.007)}{2.315 - 2.86} + \frac{(1)(0)}{1 - 2.86} - 1 = 0.86$$

The specified separation could thus be achieved with an infinite number of trays at a reflux ratio of 0.86. As shown in Example 12.1, the same separation could be achieved with ten trays at total reflux.

The reflux ratio required for a given number of trays may now be estimated from the Gilliland chart (Figure 12.4). If, for instance, $N = 15$, then

$$\frac{N - N_m}{N + 1} = \frac{15 - 10}{15 + 1} = 0.3125$$

From the chart, this corresponds to

$$\frac{R - R_m}{R + 1} = 0.39$$

The reflux ratio is calculated with $R_m = 0.86$, giving $R = 2$.

The column vapor and liquid flows at this reflux ratio are estimated from the relationships $V_r = L_r + D$ and $R = L_r/D$, where V_r and L_r are the vapor and liquid molar flows in the column rectifying section (assuming constant molar flows in the section). The distillate rate, D, is 60 kmol/h (Example 12.1). Therefore,

$$L_r = RD = (2)(60) = 120 \text{ kmol/h}$$

and

$$V_r = L_r + D = 120 + 60 = 180 \text{ kmol/h}$$

Since the feed is saturated liquid, the liquid flow in the stripping section is

$$L_s = L_r + F = 120 + 100 = 220 \text{ kmol/h}$$

By material balance, the stripping section vapor flow is the same as the rectifying section,

$$V_s = V_r = 180 \text{ kmol/h}$$

The column diameter must be checked for its ability to handle the vapor and liquid traffic (Chapter 14).

12.4 MODULAR SHORTCUT METHODS

The assumptions of total reflux or minimum reflux in the shortcut methods described above restrict their application to a limited class of columns. Another shortcut method is described in this section which is flexible enough to be generalized to a broad range of multistage, multi-component separation processes. The method is based on the column section, a module that could be used as a building block to model various column configurations. If such a module is to remain in the shortcut domain where computing time would not exceed about one-tenth of the rigorous computing time, it is obvious that simplifying assumptions would still have to be made.

The concepts of this generalized shortcut approach are outlined in this section along with methods for calculating the model parameters.

12.4.1 COLUMN SECTIONS

A column section is a group of contiguous adiabatic vapor–liquid (or liquid–liquid) equilibrium stages whose only feeds and products are a liquid feed to the top stage, a vapor feed to the bottom stage, a liquid product from the bottom stage, and a vapor product from the top stage (Figure 12.5). A simple absorber, stripper, or extractor can itself be considered a column section. Although the calculations developed here are focused on vapor–liquid column sections, they could be modified to apply to liquid–liquid extractors.

Based on Figure 12.5, a material balance for component i around tray j is written as

$$L_{j-1}X_{j-1,i} + V_{j+1}Y_{j+1,i} - L_jX_{ji} - V_jY_{ji} = 0 \qquad (12.31)$$

As in other shortcut methods, the liquid and vapor molar flow rates are assumed constant in the column section:

$$L_{j-1} = L_j = L$$

$$V_j = V_{j+1} = V$$

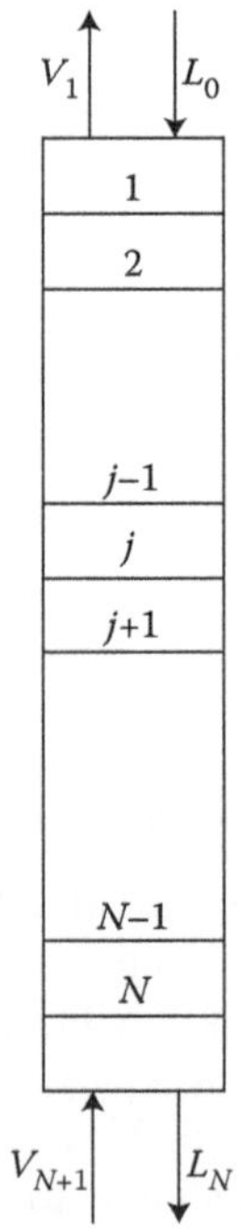

FIGURE 12.5   Schematic of a column section.

In addition, the method assumes a constant equilibrium coefficient for each component in the column section:

$$K_{ji} = K_i$$

This assumption is more restrictive than the assumption of constant relative volatilities, or relative K-values, that is used in the Fenske and Underwood methods. The payback for this assumption is the ability to generalize the model to different degrees of column complexity. The success of the method is dependent on proper evaluation of effective K-values or other model parameters that would represent actual behavior of the column section. The equilibrium coefficient is commonly lumped with the vapor and liquid molar flows in the column to define the stripping factor,

$$S_i = K_i \frac{V}{L} = \alpha_i \left(K_r \frac{V}{L} \right) \tag{12.32}$$

where K_r is the K-value of a reference component and α_i is the relative volatility of component i. Since K_i, V, and L are all assumed constant in the column section, the stripping factors are also constant.

The assumption of constant liquid and vapor molar flow may, in many situations, be grossly in error, especially in processes where mass transfer between the phases takes place mostly in one direction, such as in absorption and stripping. The method may, nevertheless, be used satisfactorily in such situations if appropriate stripping factors can be determined. This topic is addressed in more detail further along in this section.

By substituting $Y_{ji} = K_i X_{ji}$ in Equation 12.31 and replacing $K_i V/L$ with S_i, the following equation is obtained:

$$S_i X_{j+1,i} - (1 + S_i)X_{ji} + X_{j-1,i} = 0$$

This equation is now written for $j = 1, 2, \ldots, N - 1$:

$$S_i X_{2i} - (1 + S_i)X_{1i} + X_{0i} = 0$$

$$S_i X_{3i} - (1 + S_i)X_{2i} + X_{1i} = 0$$

$$\vdots$$

$$S_i X_{Ni} - (1 + S_i)X_{N-1,i} + X_{N-2,i} = 0$$

By successive substitution, and applying equation set 12.1, and rearranging, the following equation can be derived (Smith, 1963; Kremser, 1930):

$$(V_{N+1,i} + L_{0i})(1 - S_i^N) + L_{0i}(S_i^N - S_i) - L_{Ni}(1 - S_i^{N+1}) = 0$$

Rearranging,

$$L_{Ni} = \frac{V_{N+1,i}(1 - S_i^N) + L_{0i}(1 - S_i)}{1 - S_i^{N+1}}$$

(12.33)

This equation relates the rate of component i leaving the column section in the liquid product to the rates of component i entering the column section in the liquid and vapor feeds. For large values of S_i, as $S_i \to \infty$, L_{Ni} in Equation 12.33 goes to a limit:

$$L_{Ni} = V_{N+1,i}\left(\frac{S_i^N - 1}{S_i^{N+1} - 1}\right) + L_{0i}\left(\frac{S_i - 1}{S_i^{N+1} - 1}\right) \approx V_{N+1,i}\left(\frac{S_i^N}{S_i^{N+1}}\right) + L_{0i}\left(\frac{S_i}{S_i^{N+1}}\right)$$

$$\approx V_{N+1,i}\left(\frac{1}{S_i}\right) + L_{0i}\left(\frac{1}{S_i^N}\right) \approx \frac{V_{N+1,i}}{S_i}$$

As $S_i \to 1$, $L_{Ni} \to 0/0$. At this limit L_{Ni} may be expressed in terms of the derivatives with respect to S_i, of the numerator and denominator in Equation 12.33:

$$\lim_{S_i \to 1} L_{Ni} = \frac{-V_{N+1,i}NS_i^{N-1} - L_{0i}}{-(N+1)S_i^N} = \frac{V_{N+1,i}N - L_{0i}}{N+1}$$

The rate of component i leaving the column section in the vapor product is calculated by component material balance around the column section:

$$(V_{N+1,i} + L_{0i}) - (V_{1i} + L_{Ni}) = 0$$

(12.34)

An overall enthalpy balance is now written as follows:

$$(H_{N+1}V_{N+1} + h_0 L_0) - (H_1 V_1 + h_N L_N) = 0$$

(12.35)

where H and h are the molar enthalpies of the vapor streams and liquid streams, respectively. They are either rigorously calculated compositional enthalpies or simplified functions of temperature only (Chapter 1).

Each product is the sum of its constituent components:

$$V_1 = \sum_{i=1}^{C} V_{1i}, \quad L_N = \sum_{i=1}^{C} L_{Ni}$$

(12.36a)

And by overall material balance,

$$V_1 + L_N = V_{N+1} + L_0$$

(12.36b)

Two more relationships result from the fact that the vapor product is at its dew point and the liquid product is at its bubble point (Chapter 2):

$$\sum_{i=1}^{C} \frac{Y_{1i}}{K_{1i}} - 1 = 0 \qquad (12.37a)$$

$$\sum_{i=1}^{C} X_{Ni} K_{Ni} - 1 = 0 \qquad (12.37b)$$

where $Y_{1i} = V_{1i}/\sum_i V_{1i}$ and $Y_{1i} = L_{Ni}/\sum_i L_{Ni}$. The K-values could be either rigorous, composition-dependent, or simple functions of the temperature and pressure. The following is a complete list of the equations describing the column section:

Equation 12.32	C equations
Equation 12.33	C equations
Equation 12.34	C equations
Equation 12.35	1 equation
Equation 12.36a	1 equation
Equation 12.36b	1 equation
Equation 12.37a	1 equation
Equation 12.37b	1 equation
Total	$3C + 5$ equations

With a fixed number of stages in the column section, fixed stage pressures, no heat transferred across the column section boundaries, and with the feeds completely defined, the remaining variables associated with the column section are listed below:

S_i	Component stripping factors	C variables
V_{1i}	Component rates in the vapor product	C variables
L_{Ni}	Component rates in the liquid product	C variables
V_1	Vapor product rate	1 variable
L_N	Liquid product rate	1 variable
T_1	Top tray temperature	1 variable
T_N	Bottom tray temperature	1 variable
$K_r V/L$	Relative volatility multiplier	1 variable
Total		$3C + 5$ variables

The K-values, enthalpies, and relative volatilities are calculated (or otherwise determined) physical properties and are not considered as independent variables. The temperatures T_1 and T_N are implied variables since they determine the K-values K_{1i} and K_{Ni} in Equations 12.37a and 12.37b, and the enthalpies H_1 and h_N in Equation 12.35. The relative volatility multiplier $K_r V/L$ is a model variable that must be calculated to satisfy the set of equations. With equal number of variables and equations, the model is completely defined.

Equations 12.32 through 12.37b can be solved either simultaneously or iteratively using one or more iterative loops (Smith and Brinkley, 1960). In the procedure described here, Equations 12.32 through 12.34, 12.36a and 12.36b are first solved directly to calculate S_i, L_{Ni}, V_{1i}, L_N, and V_1 for a given value of $K_r V/L$. This requires some preliminary calculations: K_r and α_i are estimated based on K-values for the combined composition of the feed streams L_0 and V_{N+1} at their combined or average temperature. The relative volatilities are held constant at their calculated values through the entire procedure. The V/L ratio is estimated as the ratio of the feed streams, V_{N+1}/L_0.

Equations 12.37a and 12.37b determine the product temperatures T_1 and T_N through their effect on the K-values. The product temperatures determine the product molar enthalpies and hence their flow rates, V_1 and L_N, by energy balance, Equation 12.35, together with the overall material balance, Equation 12.36b. If these rates do not match those calculated by 12.36a, it may be necessary to adjust the relative volatility multiplier $K_r V/L$ and repeat the calculations, starting with Equation 12.32. This is equivalent to fine tuning the column temperature profile since K_r is based on the average column temperature.

The modular method as described above cannot directly calculate the column section to satisfy performance specifications such as product composition or quality. Performance specifications could be handled by superimposing an external iterative loop that would vary a fixed model parameter such as the feed rate or temperature to meet the specification.

EXAMPLE 12.5: ABSORBER CALCULATIONS

The feeds to a six-stage absorber are given below, along with the component relative volatilities at average column temperature and the column pressure of 2750 kPa.

		L_{0i} (kmol/h)	V_{7i} (kmol/h)	α_i
1.	C1	0	100.00	90.0
2.	C2	0	231.25	22.0
3.	C3	0	150.00	8.20
4.	nC4	0.03	15.62	2.68
5.	nC5	0.50	3.13	1.00
6.	nC10	99.47	0.00	0.002
Flow rate (kmol/h)		100.00	500.00	
Temperature (°C)		35.0	45.0	
Pressure (kPa)		2750.0	2750.0	

The K-value at 2750 kPa of nC5, selected as the reference component, is given by the equation

$$\ln K_5 = 8.3 - (3390/T) \qquad T \text{ in degrees Kelvin}$$

The parameters for estimating component enthalpies over a limited range of conditions are given below. Mixtures may be assumed to behave as ideal solutions for the purpose of calculating their enthalpies.

Vapor enthalpy: H_i (kJ/kmol) $= A_i + B_iT$ (°C)
Liquid enthalpy: h_i (kJ/kmol) $= C_i + D_iT$ (°C)

	A_i	B_i	C_i	D_i
C1	11669	42.63	8420	41.86
C2	20581	58.60	11174	92.12
C3	29103	66.99	12700	125.59
nC4	38240	83.73	15389	159.07
nC5	47209	117.21	18627	184.20
nC10	75083	167.46	31466	226.06

Using the modular column section method, calculate the absorber product rates and compositions. The relative volatilities and column pressure may be assumed constant for the entire computational cycle.

SOLUTION

An initial guess for $K, V/L$ is based on the temperatures and flow rates of the feed streams:

$$T_{avg} = (35.0 + 45.0)/2 + 273.15 = 313.15 \text{ K}$$

$$K_r = K_5 = \exp[8.3 - (3390/313.15)] = 0.080$$

$$K_r V/L = 0.080 \times 500/100 = 0.4$$

The stripping factors are calculated by Equation 12.32:

$$S_i = 0.4\alpha_i$$

The liquid product component flow rates are calculated by Equation 12.33:

$$L_{6i} = \frac{V_{7i}(1 - S_i^6) + L_{0i}(1 - S_i)}{1 - S_i^7}$$

The vapor product component flow rates are calculated by Equation 12.34,

$$V_{1i} = V_{7i} + L_{0i} - L_{6i}$$

The overhead and bottoms component mole fractions are calculated as

$$Y_{1i} = \frac{V_{1i}}{\sum V_{1i}}; \quad X_{6i} = \frac{L_{6i}}{\sum L_{6i}}$$

From Equations 12.37a and 12.37b,

$$\sum \frac{Y_{1i}}{K_{1i}} = \sum \frac{Y_{1i}}{\alpha_i K_{1r}} = 1; \quad K_{1r} = \sum \frac{Y_{1i}}{\alpha_i}$$

$$\sum X_{6i} K_{6i} = \sum X_{6i} \alpha_i K_{6r} = 1; \quad K_{6r} = \frac{1}{\sum X_{6i} \alpha_i}$$

The column top and bottom temperatures, T_1 and T_6, are calculated from K_{1r} and K_{6r} using the K_r temperature dependence equation,

$$T = \frac{3390}{8.3 - \ln K_r}$$

The component enthalpies are calculated by the equations provided, and the feed and product stream enthalpies are calculated based on the assumption of ideal solution behavior:

$$H_7 = \sum Y_{7i} H_{7i} \quad h_0 = \sum X_{0i} h_{0i}$$

$$H_1 = \sum Y_{1i} H_{1i} \quad h_6 = \sum X_{6i} h_{6i}$$

The enthalpy balance is checked by Equation 12.35. If it is not satisfied the computations are repeated with an updated value for $K_r V/L$ by trial and error. The converged solution is summarized below.

$$K_r V/L = 0.41468$$

	L_{0i} (kmol/h)	V_{7i} (kmol/h)	α_i	S_i	L_{6i} (kmol/h)	V_{1i} (kmol/h)
C1	0	100.00	90	37.3212	2.679	97.321
C2	0	231.25	22	9.1230	25.348	205.902
C3	0	150.00	8.20	3.4004	44.093	105.907
nC4	0.03	15.62	2.68	1.1113	12.627	3.023
nC5	0.50	3.13	1.00	0.4147	3.414	0.216
nC10	99.47	0.00	0.002	0.00083	99.388	0.082
	100	500			187.549	412.451

$$K_{1r} = 0.160, \quad T_1 = 334.54\ K = 61.39°C$$

$$K_{6r} = 0.157, \quad T_6 = 333.85\ K = 60.70°C$$

	h_{0i} (kJ/kmol)	H_{7i} (kJ/kmol)	H_{1i} (kJ/kmol)	h_{6i} (kJ/kmol)
C1	9885	13587	14286	10961
C2	14398	23218	24179	16766
C3	17096	32118	33216	20323
nC4	20956	42008	43380	25045
nC5	25074	52483	54405	29808
nC10	39378	82619	85364	45188

$$h_0 = \sum h_{0i} X_{0i} = 39301 \text{ kJ/kmol}$$

$$H_7 = \sum H_{7i} Y_{7i} = 24732 \text{ kJ/kmol}$$

$$H_1 = \sum H_{1i} Y_{1i} = 24334 \text{ kJ/kmol}$$

$$h_6 = \sum h_{6i} X_{6i} = 33376 \text{ kJ/kmol}$$

Checking the energy balance,

$$\text{Error} = (h_0 L_0 + H_7 V_7) - (h_6 L_6 + H_1 V_1) = -2.4 \text{ kJ/h}$$

This error is small enough relative to the enthalpy values.

12.4.2 REDUCED MODEL

The assumption of constant stripping factors is necessary for deriving the analytical solution of multistage separation in a column section, Equation 12.33. The simplicity of this approach and the resulting great savings in computing time are highly desirable.

The accuracy of the method hinges on proper determination of effective stripping factors. The method used in the previous section for calculating the stripping factors is based on solving the group of equations that describe the column section. The method can be enhanced by tuning the stripping factors and other model parameters to improve the model accuracy. The objective is to find model parameters that would make the shortcut results match actual performance data or rigorous solutions.

One approach involves periodic updating of the shortcut model parameters using the results of a rigorous model (Jirapongphan et al., 1980; Trevino-Lozano et al., 1984). The method is especially appropriate for online applications where repetitive, fast model solutions are required as the feed streams and other process conditions change, such as in process control or optimization. The rigorous solution provides a base case which serves to calculate the shortcut model parameters. The shortcut model is then used to solve the column section as conditions move away from the base case. A good shortcut model should be capable of predicting the column section performance at conditions reasonably distanced from the base case, using the same base case parameters. Additional periodic rigorous updates of the shortcut model parameters may be required as operating conditions move farther away from the base case.

The shortcut model is developed in terms of reduced parameters that are not strongly dependent on stream compositions, temperature, and pressure. The shortcut model, represented by Equations 12.33 through 12.37b, is solved in conjunction with reduced equations for calculating enthalpies, vapor–liquid equilibrium coefficients, and effective stripping factors based on the rigorous base case.

The enthalpies, which in the rigorous model are generally calculated by compositional methods appropriate to the mixture under consideration, are expressed in the shortcut model as a linear function of temperature only:

$$H = a + b(T - T_r) \tag{12.38}$$

where T_r is a reference temperature. The compositional effect on the enthalpy is small compared to the temperature effect, especially since moderate deviations from the base case are not expected to cause large changes in the composition. Parameters a and b are evaluated for each stream from rigorous model enthalpies calculated at T_r and one other temperature.

The vapor–liquid equilibrium coefficients are expressed in terms of the relative volatility, α, and a reference K-value at the top and bottom stages:

$$
\begin{aligned}
K_{1i} &= \alpha_{1i} K_{1r} \\
K_{Ni} &= \alpha_{Ni} K_{Nr}
\end{aligned}
\tag{12.39}
$$

The K_i values can be taken directly from the rigorous model, and the reference K-values can be calculated as weighted logarithmic averages:

$$\ln K_{1,r} = \sum Y_{1,i} \ln K_{1,i}$$

$$\ln K_{N,r} = \sum X_{N,i} \ln K_{N,i} \tag{12.40}$$

The relative volatilities are calculated from equation set 12.39, and are held constant for the same base case. Departures of the reference K-values from the base case are calculated from the relations

$$
\ln(K_{1r} P) = A_1 + B_1 \left(\frac{1}{T_1} - \frac{1}{T_1^*} \right)
$$

$$
\ln(K_{Nr} P) = A_N + B_N \left(\frac{1}{T_N} - \frac{1}{T_N^*} \right)
\tag{12.41}
$$

where T_1^* and T_N^* are the base case top and bottom temperatures, and P is the average column section pressure. The parameters A_1, B_1, A_N, and B_N are calculated for the top and bottom trays from rigorous K_r values at the base case temperature and one other temperature, with the composition held constant. With constant relative volatilities, equation set 12.39 is used once again to calculate the K-values away from the base case.

Reference component stripping factors, $S_{i,r}$, are also calculated from base case rigorous data using Equation 12.33. These are held constant for the same base case and deviations are calculated using a tuning parameter β:

$$S_i = \beta S_{i,r} \tag{12.42}$$

Once reference stripping factors, $S_{i,r}$, are calculated for a given base case, they are assumed constant until a new rigorous calculation is performed. Justification for this assumption is that, as conditions deviate from the base case, the component effective stripping factors will change, although their variation relative to each other will be much smaller.

The reduced model is completely defined by the column section mass balances, energy balances, and separation and equilibrium relations (Equations 12.33 through 12.37b) and the physical property models (Equations 12.38 through 12.42). The solution strategy starts by solving the rigorous model to calculate V_{1i}, V_1, L_{Ni}, L_N, T_1, and T_N for given feeds L_0 and V_{N+1}, number of stages N, and pressure profile (or constant average section pressure, P). Equation set 12.33 is then solved for $S_{i,r}$. The rigorous model results are also used to calculate K_{1i} and K_{Ni}, and Equations 12.40 and 12.39 are solved for K_{1r}, K_{Nr}, α_{1i}, and α_{Ni}. Next, the rigorous model temperature is perturbed to calculate parameters a and b in Equation 12.38 for streams V_1, V_{N+1}, L_0, and L_N and parameters A_1, B_1, A_N, and B_N in equation set 12.41.

The physical property parameters as derived here can be used with the other equations of the reduced model to calculate the column section. The reduced model should be able to reproduce the base case results exactly. The calculations can be carried out iteratively using a scheme similar to the one described in Section 12.4.1. The calculations start with the base case stripping factors $S_{i,r}$ and assumed value for β (initially $= 1$) to solve for the product stream component rates, Equations 12.33 and 12.34. Equations 12.37a and 12.37b are then solved with the reduced model K-value equations (Equations 12.39 through 12.41) to calculate the top and bottom tray temperatures, and the stream enthalpies are calculated using Equation 12.38. The computations are repeated by iterating on β until the energy balance, Equation 12.35, is satisfied.

With this method the column section products can be calculated if the feed streams are given. The calculated product rates, compositions, and thermal conditions can be used to predict physical properties, which may be used as control or optimization feedback data.

It is important to consider the ranges of departure of key variables from base conditions within which the reduced model predictions are accurate enough. These ranges should be determined prior to the actual application of the model. Whenever these ranges are exceeded, a signal should be sent to rerun the rigorous model and generate a new set of parameters for the reduced model.

EXAMPLE 12.6: ONLINE PROCESS MODEL

The bottoms product of a five-stage absorber column in a process plant is to be determined for control purposes. A reduced model is developed for continuous online monitoring of the column bottoms.

Base-case feed and absorbent streams, as well as rigorous calculated products of a five-stage absorber, are listed below. The Soave-Redlich-Kwong equation of state is used to calculate K-values and enthalpies.

Component	i	Feed, $V_{6,i}$	Absorbent, $L_{0,i}$	Ovhd, $V_{1,i}$	Btms, $L_{5,i}$	$S_{i,r}$
Methane	1	300.00	0.00	283.77	16.232	18.482
Ethane	2	152.00	0.00	121.74	30.255	5.0227
Propane	3	236.00	0.00	112.11	123.89	1.8648
n-Butane	4	167.00	0.00	7.2964	159.70	0.6491
n-Pentane	5	145.00	0.00	0.0952	144.90	0.2468
n-Octane	6	0.00	1000.0	8.7760	991.22	0.00878
Flow (kmol/h)		1000.0	1000.0	533.79	1466.2	
Temp (°C)		50.0	50.0	60.115	75.026	
Temp (K)		323.15	323.15	333.265	348.176	
Pres (kPa)		700.0	700.0	700.0	700.0	
Enth (MJ/h)		1223.0	−35190	852.5	−34819	
Enth (kJ/kmol)		1223.0	−35190	1597.1	−23748	

The reference stripping factors, $S_{i,r}$, are calculated from the base-case component flow rates above, using Equation 12.33 or its residual form representation (the equation preceding 12.33).

The rigorous model was perturbed by changing the absorbent temperature to 55°C, giving the following results:

Component	i	Feed, $V_{6,i}$	Absorbent, $L_{0,i}$	Ovhd, $V_{1,i}$	Btms $L_{5,i}$
Methane	1	300.00	0.00	284.21	15.790
Ethane	2	152.00	0.00	123.16	28.842
Propane	3	236.00	0.00	120.44	115.56
n-Butane	4	167.00	0.00	10.712	156.29
n-Pentane	5	145.00	0.00	0.1739	144.83
n-Octane	6	0.00	1000.0	11.129	988.87
Flow (kmol/h)		1000.0	1000.0	549.82	1450.2
Temp (°C)		50.0	55.0	65.127	77.597
Temp (K)		323.15	328.15	338.28	350.75
Pres (kPa)		700.0	700.0	700.0	700.0
Enth (MJ/h)		1223.0	−33891	1054.9	−33722
Enth (kJ/kmol)		1223.0	−33891	1918.63	−23253

The enthalpy reduced model constants for Equation 12.38 are calculated from the rigorous and perturbed cases. The resulting equations are as follows:

Feed	H (kJ/kmol) $= 1223 + 78.12$ $(T - 50)$
Absorbent	H (kJ/kmol) $= -35190 + 259.8$ $(T - 50)$
Overhead	H (kJ/kmol) $= 1597.126 + 64.17474$ $(T - 60.115)$
Bottoms	H (kJ/kmol) $= -22700 + 189.138$ $(T - 75.026)$

In the next step, the reduced model reference K-values, $K_{1,r}$ and $K_{5,r}$ are calculated from the rigorous base and perturbed results using equation set 12.40. The relative volatilities, $\alpha_{1,i}$ and $\alpha_{5,i}$, were calculated from equation set 12.39 using rigorous base case results.

Comp, i	1	2	3	4	5	6
Stage 1, column top, rigorous base case. T_1 (base) $= 60.115°C = 333.265$ K						
$X_{1,i}$	0.0175	0.03121	0.0815141	0.015023	0.00053	0.854219
$Y_{1,i}$	0.53126	0.22807	0.2100273	0.013669	0.00018	0.016441
$\alpha_{1,i}$	2.7535	0.6625	0.2336	0.08249	0.03063	0.001745
Stage 5, column bottom, rigorous base case. T_5 (base) $= 75.026°C = 348.176$ K						
$X_{5,i}$	0.01107	0.02064	0.0844975	0.108921	0.09883	0.676048
$Y_{5,i}$	0.34227	0.17375	0.2737189	0.135883	0.04972	0.024657
$\alpha_{5,i}$	252.736	68.8314	26.481138	10.19828	4.11302	0.298155
Stage 1, column top, rigorous perturbed case. T_1 (perturbed) $= 65.127°C = 338.277$ K						
$X_{1,i}$	0.01674	0.02904	0.0784393	0.019243	0.00082	0.855719
$Y_{1,i}$	0.51691	0.22400	0.2190516	0.019483	0.00032	0.020241
Stage 5, column bottom, rigorous pertutbed case. T_5 (perturbed) $= 77.597°C = 350.747$ K						
$X_{5,i}$	0.01089	0.01989	0.0796865	0.107773	0.09987	0.681894
$Y_{5,i}$	0.33893	0.17154	0.2676906	0.141127	0.05337	0.027339

Using equation set 12.40, the following values are obtained for the reference K-values:

At T_1 (base) $= 60.115°C = 333.265$ K	$K_{1,r}$ (base) $= 11.02987$
At T_5 (base) $= 75.026°C = 348.176$ K	$K_{5,r}$ (base) $= 0.12233$
At T_1 (perturbed) $= 65.127°C = 338.277$ K	$K_{1,r}$ (perturbed) $= 10.80299$
At T_5 (perturbed) $= 77.597°C = 350.747$ K	$K_{5,r}$ (perturbed) $= 0.12873$

The top and bottom stage relative volatilities for the rigorous base case, $\alpha_{1,i}$ and $\alpha_{5,i}$, are calculated using the base case reference K-values, and are tabulated above. These values are held constant as long as the same rigorous base case is used as a basis for the reduced model.

The temperature dependence of the reference K-values in the reduced model is expressed by equation set 12.41. Parameters A_1, A_5, B_1, B_5 are determined by solving these equations at the base and perturbed values above, giving the following values:

$A_1 = 8.95169$ $\qquad$ $B_1 = 467.5$ $\qquad$ $A_5 = 4.45003$ $\qquad$ $B_5 = -2424$

These constants are consistent with $P = 700$ kPa, $T_1^* = 333.265$ K, $T_5^* = 348.176$ K, and T_1 and T_5 being absolute temperatures (K) in equation set 12.41.

The reduced model can now be used to predict the performance of the absorber column at some conditions perturbed from the base case. In order to test the model, the same perturbation will be used as that done in the rigorous model, namely, changing the absorbent temperature from 50°C to 55°C.

A value is assumed for β, and the S_i values are calculated from base case $S_{i,r}$ using Equation 12.42. The column products' component flow rates are then calculated using Equations 12.33 and 12.34.

The overhead vapor component mole fractions, $Y_{1,i}$, and the bottoms liquid component mole fractions, $X_{5,i}$, are calculated from the products component flow rates. Then combining Equations 12.37a, 12.37b, and 12.39, the following equations are derived to calculate the reference K-values at the column top and bottom.

Top stage:

$$K_{1,r} = \frac{K_{1,i}}{\alpha_{1,i}} = \frac{Y_{1,i}}{X_{1,i}\,\alpha_{1,i}}$$

$$K_{1,r}X_{1,i} = \frac{Y_{1,i}}{\alpha_{1,i}}$$

$$K_{1,r}\sum X_{1,i} = K_{1,r} = \sum \frac{Y_{1,i}}{\alpha_{1,i}}$$

Bottom stage:

$$K_{5,r} = \frac{K_{5,i}}{\alpha_{5,i}} = \frac{Y_{5,i}}{X_{5,i}\,\alpha_{5,i}}$$

$$\frac{K_{5,r}}{Y_{5,i}} = \frac{1}{X_{5,i}\,\alpha_{5,i}}$$

$$\frac{K_{5,r}}{\sum Y_{5,i}} = K_{5,r} = \frac{1}{\sum X_{5,i}\,\alpha_{5,i}}$$

The relative volatilities, $\alpha_{1,i}$ and $\alpha_{5,i}$, are taken from the rigorous base case results.

The perturbed reduced model top stage and bottom stage temperatures are back-calculated from equation set 12.41, using the $K_{1,r}$ and $K_{5,r}$ values above.

Finally, the feeds and products enthalpies are calculated using the enthalpy expressions developed above, and an energy balance is carried out (Equation 12.35). Convergence is achieved iteratively by changing the value of β until Equation 12.35 is satisfied.

At convergence, $\beta = 0.97333$, and the following solution is obtained:

Comp	$S_{i,r}$	S_i	Feed, $V_{6,i}$	Absrbnt, $L_{0,i}$	Ovhd, $V_{1,i}$	Btms, $L_{5,i}$
CH_4	18.482	17.989	300	0	283.3231	16.6769
C_2H_6	5.0227	4.88872	152	0	120.9169	31.0831
C_3H_8	1.86479	1.81504	236	0	109.0249	126.975
C_4H_{10}	0.64914	0.63183	167	0	6.611587	160.388
C_5H_{12}	0.24679	0.24020	145	0	0.088115	144.912
C_8H_{18}	0.00878	0.00855	0	1000	8.545793	991.454
Flow (kmol/h)			1000	1000	528.510	1471.490
Temp (°C)			50	55	64.14	74.15
Temp (K)			323.15	328.15	337.29	347.30
Pressure (kPa)			700	700	700	700
Enthalpy (kJ/kmol)			1223.0	−33891.0	1855.43	−22865.68
Enthalpy (MJ/h)			1223.0	−33891.0	980.821	−33648.812

The enthalpy balance (Equation 12.35) is satisfied, giving a residual of ~8 kJ/h. Compared with the rigorous perturbed model results, the reduced model predicts the products' enthalpies and temperatures fairly well. The products' compositions can use some improvement. It is noted that Equation 12.38, used for calculating the stream enthalpies in the reduced model, does not take into account the effect of composition. This could be improved by using the ideal solution method as applied in Example 12.5. This method is still computationally simple enough to satisfy the reduced model conditions of reduced computing time.

12.4.3 Complex Configurations

By definition, the column section is limited in practical application to modeling simple absorbers, strippers, and liquid–liquid extractors—processes with one feed and one product at each end and with no heat addition or removal. Given a reliable method for solving the column section, however, it could be used as a module to build more complex configurations.

A distillation column, for instance, would be modeled with a column section for the stripping section, a column section for the rectifying section, and single equilibrium stages for the feed tray, the condenser, and the reboiler. In order to solve the distillation column separation equations as one unit, two sets of Equation 12.33, each with the appropriate stripping factors for the corresponding section, would have to be solved simultaneously along with a component balance around the feed tray, the condenser, and reboiler equations. Such a solution does exist for conventional distillation and for certain extraction problems (Smith and Brinkley, 1960).

With this approach it is possible to solve only a limited number of column configurations, with little flexibility regarding model parameter tuning. The alternative is to model the process using column sections, equilibrium stages (or general flash blocks), mixers, and splitters and to solve these modules sequentially. Depending on the sequence chosen, certain streams may have to be initialized to start the computations. Each time a module is to be solved, the streams feeding it must be either known or initialized. Another approach is to solve all the equations for all the column modules simultaneously. In principle, the modular approach could be applied to any complexity configuration. Coupled with the parameter updating technique described in Section 12.4.2, reliable modeling is possible for broad ranges of applications.

In building complex multistage separation models from column sections and other blocks, the question arises as to the number of variables that must be specified in a given situation to completely define the problem. That is, the number of degrees of freedom must be determined. Rather than tally the number of equations, variables, and so on, to determine the degrees of freedom, these can be inferred from the process configuration (see Section 9.2.2). Under conditions of fixed pressure, fixed number of stages, and fixed feeds, a column section has zero degrees of freedom. A single equilibrium stage at a fixed pressure and with fixed feeds and heat duty also has no degrees of freedom. A mixer falls in the same category. The degrees of freedom of a splitter (fixed feeds and pressure) equal the number of products minus one. Each time one of the fixed parameters is allowed to vary, one degree of freedom is created. The following are examples of different types of multistage separation processes modeled with column sections and other ancillary modules.

12.4.3.1 Reboiled Stripper

This process may be modeled with a column section and an equilibrium stage as shown in Figure 12.6. If the pressure profile, feed F, reboiler duty Q, and number

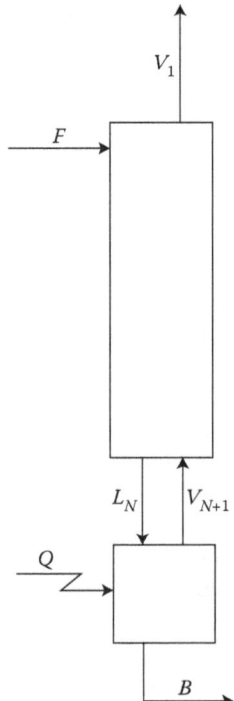

FIGURE 12.6 Reboiled stripper.

of stages N are all fixed, this process has zero degrees of freedom. An operator of such a column would have no control over the column products. If one variable, such as the reboiler duty, is allowed to vary, one other variable, such as the bottoms rate B, could be controlled and must be specified in order to completely define the problem.

The computations could start by initially solving the column section with an assumed vapor feed, V_{N+1}, to calculate V_1 and L_N. The reboiler is then solved with feed stream L_N, and one specification (reboiler duty, or the bottoms rate or temperature, etc.). The column section is solved again with the updated stream, V_{N+1}. The calculations are repeated until the product streams stabilize within acceptable tolerance.

EXAMPLE 12.7: SOLVING A REBOILED STRIPPER COLUMN

A liquid stream of hydrocarbons with a flow rate of 500 kmol/h is sent to the top of a reboiled stripper column. The column has the equivalent of seven equilibrium stages and a reboiler in a configuration similar to Figure 12.6. The column and reboiler pressure may be assumed uniform at 1000 kPa. The feed temperature to the column is controlled such that the average column temperature is 95°C. The heat duty to the reboiler is controlled such as to maintain the reboiler temperature at 120°C. The feed stream is given below, along with average component K-values in the column and reboiler, evaluated at the pressure and temperatures provided. Determine by the modular column section method the rates and compositions of the stripper column products.

		Feed, L_{0i} (kmol/h)	Column Avg K_i	Reboiler K_i
1.	Methane	24	21.82	23.30
2.	Ethane	61	7.08	8.69
3.	Propane	146	3.04	4.21
4.	n-Butane	176	1.25	1.79
5.	n-Pentane	59	0.53	0.87
6.	n-Hexane	34	0.21	0.43

SOLUTION

The vapor stream V_8 from the reboiler (stage 8) to the bottom of the column section must be estimated to start the calculations. A rough initial estimate can be obtained by assuming the liquid stream from the column section to the reboiler, L_7, to be the same as the liquid feed L_0. This stream is flashed at the reboiler conditions (using the reboiler K-values) to generate estimates for the vapor and liquid streams from the reboiler, V_8 and L_8.

The column section is solved with the liquid feed L_0 and the estimated vapor feed V_8, using Equations 12.32, 12.33, and 12.34. The flow rates L_0 and V_8 can be used for V and L in Equation 12.32 to calculate the stripping factors. The calculated liquid product L_7 is flashed at the reboiler conditions to update V_8, and the calculations are repeated until convergence, that is, until stream V_8 stabilizes. The results are as follows:

		Feed, L_{0i} (kmol/h)	Overhead, V_{1i} (kmol/h)	Bottoms, L_{8i} (kmol/h)
1.	Methane	24.0	24.00	0.00
2.	Ethane	61.0	61.00	0.00
3.	Propane	146.0	145.87	0.13
4.	n-Butane	176.0	149.52	26.48
5.	n-Pentane	59.0	21.10	37.90
6.	n-Hexane	34.0	4.78	29.21
	Totals	500.0	406.28	93.72

12.4.3.2 Distillation Column with a Partial Condenser

As shown in Figure 12.7, this process is modeled with six modules. The feed tray is calculated in module 1, which is actually a combination of a mixer and an equilibrium stage. The mixer serves to mix the external feed F, liquid from the rectifying section, and vapor from the stripping section. The equilibrium stage calculates the vapor going to Section 2 and the liquid going to Section 3. The condenser and reboiler are modeled with equilibrium stages 4 and 5. Since part of the condensed liquid is refluxed and the other part is taken as a distillate product, a splitter, unit 6, is used to model this part of the process.

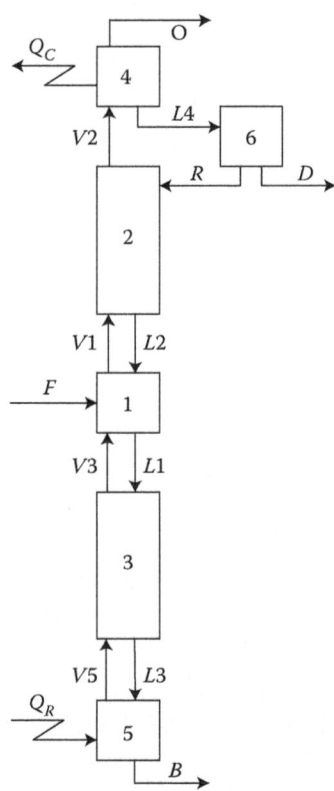

FIGURE 12.7 Distillation column with a partial condenser.

With fixed feeds, heat duties, pressure profiles, and number of stages, all the modules except unit 6 have zero degrees of freedom. Unit 6 is a splitter with two products and, therefore, has one degree of freedom. Hence, the entire column as defined has one degree of freedom and requires one more specification to be completely defined. If the parameters such as the condenser and reboiler duties are allowed to vary, the number of specifications must increase to match the number of variables.

In modular solution the computational sequence is usually arbitrary but may depend on the way the column is specified. In this model, if the only degree of freedom is that of the splitter, and the reflux rate R is specified, the modules could be solved according to the sequence 1, 2, 4, 6, 3, 5. With this sequence, streams $L2$, $V3$, R, and $V5$ would have to be initialized. Different streams would have to be initialized with a different computational sequence. Convergence is reached when all the streams have been stabilized within tolerance.

12.4.3.3 Multi-Column System

Figure 12.8 shows a model of a main column and a side-stripper. Vapor feed F goes to the bottom of the main column, which is steam-stripped with stream $S1$, using no reboiler. The column has a partial condenser with vapor distillate D. Side draw SD is taken from around the middle of the main column and steam-stripped in the side-stripper, with the overhead vapor, OH, returned to the draw tray in the main column. The side-stripper is steam-stripped with stream $S2$.

Module 1, a column section, represents the section in the main column below the draw tray. No separate module is needed at the bottom since no reboiler is used.

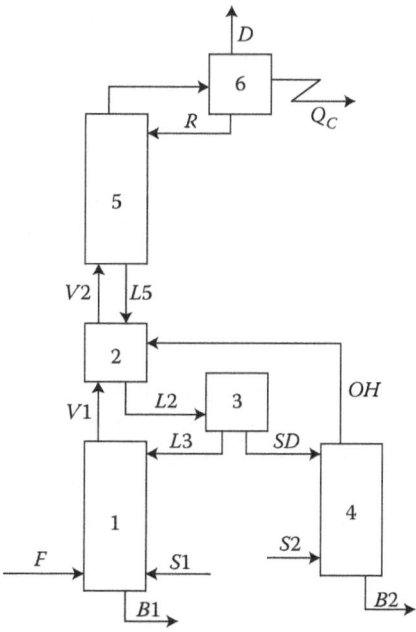

FIGURE 12.8 Multi-column system.

Module 2 models the draw tray and consists of a mixer and an equilibrium stage. Module 3 is a splitter that takes the liquid from the draw tray and splits it into side draw SD and the remaining liquid flowing down to the bottom column section. The side-stripper and the upper column section are modeled with column sections, modules 4 and 5, and the condenser is modeled with an equilibrium stage, module 6. Using computational sequence 1, 2, 3, 4, 5, 6 requires initialization of streams $L3$, $L5$, OH, and R.

12.4.4 LIQUID–LIQUID EXTRACTION BY THE SHORTCUT COLUMN SECTION METHOD

Another application of the shortcut column section method is for solving liquid–liquid extraction problems. The fundamentals of extraction were discussed in Chapter 11, and graphical solution methods were described based on ternary liquid–liquid equilibrium diagrams. These methods are limited to three-component extraction systems.

A multi-component countercurrent extractor model may be defined for the shortcut column section method in a manner very similar to the vapor–liquid model. The extractor model is developed for liquid–liquid equilibrium stages leading to component flow relationships comparable to the Kremser equation:

$$L_{Ni} = \frac{Q_{N+1,i}(1 - E_i^N) + L_{0i}(1 - E_i)}{1 - E_i^{N+1}} \tag{12.43}$$

$$Q_{1i} = L_{0i} + Q_{N+1,i} - L_{Ni} \tag{12.44}$$

The extraction model is shown in Figure 11.3. The equation variables L_{0i} and L_{Ni} are the solvent and extract component flows, $Q_{N+1,i}$ and Q_{1i} are the feed and raffinate component flows, and N is the number of equilibrium stages. Flow rates are on a mass or mole basis. The extraction factors E_i are defined as

$$E_i = K_i \frac{Q}{L} \tag{12.45}$$

$$K_i = \frac{X_i^Q}{X_i^L} \tag{12.46}$$

where Q and L are average raffinate and extract phase flow rates, K_i is the liquid–liquid equilibrium distribution coefficient, and X_i^Q and X_i^L are mole or mass fractions of component i in the raffinate and extract phases.

The two liquid phases are necessarily nonideal solutions and their component equilibrium coefficients are best calculated from the activity coefficients in each phase (Section 1.3.5):

$$K_i = \frac{X_i^Q}{X_i^L} = \frac{\gamma_i^L}{\gamma_i^Q} \tag{12.47}$$

where γ_i^L and γ_i^Q are the activity coefficients of component i in the extract phase and raffinate phase. Alternatively, the liquid–liquid K-values can be estimated directly from the available experimental data.

It can be seen from Equation 12.43 that for large values of N, the terms E_i^N and E_i^{N+1} can be very small or very large, depending on the value of E_i. Inspection of this equation shows that L_{Ni} approaches $Q_{N+1,i} + L_{0i}$ as E_i approaches zero, or it approaches zero as E_i approaches infinity.

At the start of the calculations the liquid flow rates in the column, Q and L in Equation 12.45, may be assumed equal to the inlet feed and solvent rates. The initial values for the activity coefficients may also be based on the inlet compositions and thermal conditions of the streams. The temperature and pressure variations in the extractor column are usually small, but the compositions will vary, and this may require recalculating the activity coefficients. The column calculations may be repeated with updated values of Q and L, taken as respective averages of each phase inlet and outlet stream flow rates calculated in the first trial. The activity coefficients can also be refined by recalculating them at the column top and bottom compositions for each phase. Averages of the top and bottom coefficients for each phase can be used in Equation 12.47 to calculate the new K-values. The extraction factors are then recalculated with the new values of Q, L, and K_i by Equation 12.45. The product component flow rates L_{Ni} and Q_{1i} are finally calculated by Equations 12.43 and 12.44. If large variations appear between the first and second trials, more trials may be considered.

For online applications, an alternative to the above procedure is to tune the extraction factors periodically to match plant data and laboratory analysis. The factors determined in this manner can be used to predict the column product compositions between parameter updates. The predictions should take into account the effect on the extraction factors of measured flow rates Q and L.

EXAMPLE 12.8: EXTRACTOR CALCULATIONS BY THE COLUMN SECTION METHOD

A four-stage extractor column is used to recover acetone from an acetone–chloroform solution using a water–acetic acid solution as the solvent. The feed and solvent are defined below, along with average K-values. Calculate the products' component flow rates.

	i	K_i	Feed, Q_{5i} (kmol/h)	Solvent, L_{0i} (kmol/h)
Acetone	1	2.2	20.0	0.0
Chloroform	2	9.8	80.0	0.0
Water	3	0.1	0.0	90.0
Acetic acid	4	0.3	0.0	50.0
Flow rate			100.0	140.0

SOLUTION

Use the inlet flow rates to estimate the extraction factors:

$$E_i = K_i \frac{Q}{L} = \frac{100.0}{140.0} K_i = 0.7143 \, K_i$$

From Equations 12.43 and 12.44,

$$L_{4i} = \frac{Q_{5i}(1 - E_i^4) + L_{0i}(1 - E_i)}{1 - E_i^5}$$

$$Q_{1i} = L_{0i} + Q_{5i} - L_{4i}$$

The results are as follows:

	i	E_i	Raffinate, Q_{1i} (kmol/h)	Extract, L_{4i} (kmol/h)
Acetone	1	1.571	8.12	11.88
Chloroform	2	7.000	68.58	11.42
Water	3	0.071	6.39	83.61
Acetic acid	4	0.214	10.68	39.32
Flow rate			93.77	146.23

The calculated product rates for each phase are not much different from the feed rates which were used to calculate the extraction factors. A recalculation of Q and L averages to update the extraction factors and to do another trial may, therefore, not be required. The K-values could be updated as averages of values based on the inlet and calculated outlet compositions.

NOMENCLATURE

B Bottoms stream designation or molar flow rate
b_i Molar flow rate of component i in the bottoms
C Number of components in column streams
D Distillate or overhead stream designation or molar flow rates
d_i Molar flow rate of component i in the distillate
E Extraction factor
F Feed stream designation or molar flow rate
f_i Molar flow rate of component i in the feed
G General column specification
H Vapor molar enthalpy
h Liquid molar enthalpy
K Vapor–liquid or liquid–liquid equilibrium distribution coefficient
L Extract phase flow rate
L_j Liquid molar flow rate leaving tray j
L_{ji} Component i liquid molar rate leaving tray j
N Number of theoretical stages
N_m Minimum number of stages
P Pressure
Q Heat duty or raffinate phase flow rate
q Feed thermal condition

R	Reflux ratio
R_m	Minimum reflux ratio
S	Stripping factor; number of column sections
T	Temperature
V_j	Vapor molar flow rate leaving tray j
V_{ji}	Component i vapor molar rate leaving tray j
X_{ji}	Mole fraction of component i in liquid leaving tray j
Y_{ji}	Mole fraction of component i in vapor leaving tray j
Z_i	Mole fraction of component i in the feed
α	Relative volatility
β	Stripping factor parameter
γ	Activity coefficient
θ	Parameter intermediate between relative volatilities of light key and heavy key

SUBSCRIPTS

B	Bottoms designation
b	Lowest stage in a column section
D	Distillate designation
f	Feed tray designation
h	Heavy key component designation
i	Component designation
j	Stage designation
l	Light key component designation
r	Reference component or rectifying section
s	Side product or stripping section
t	Highest stage in a column section

SUPERSCRIPT

σ	Column section designation

PROBLEMS

12.1. For the separation problem described in Examples 12.1 and 12.4, estimate the number of trays required if the condenser can handle a maximum vapor rate of 150 kmol/h and the overall tray efficiency is 65%.

12.2. A packed column is to be used for the separation of propane and butane as described in Examples 12.1 and 12.4. The column uses a partial condenser and reboiler, and is 12 m high. Assuming 1 m of packing is equivalent to an equilibrium stage, estimate the reflux rate required to achieve the specified separation.

12.3. Determine the optimum column size for the separation problem described in Examples 12.1 and 12.4. The following information is given:

Capital cost of the column (expected life of ten years): $C_C(\$) = 1000 N_a \sqrt{d}$
N_a = Actual number of trays

d = Column diameter, m
Overall tray efficiency = 65%
Energy cost = $2/10^6$ kJ
Cooling water cost = $2/10^7$ kJ
Average heat of condensation/vaporization = 17,500 kJ/kmol
Average vapor molar volume in the column = 7.8 m³/kmol
Target vapor velocity in the column = 1 m/s

Assume the feed to the column is saturated liquid.

12.4. a. Based on the Fenske equation, derive an expression for the minimum number of stages for a binary mixture in terms of the separation specifications. Assume constant relative volatility.

b. A saturated liquid binary mixture of components 1 and 2 is to be separated in a distillation column into a distillate product (mostly component 1) and a bottoms product (mostly component 2). The feed composition and separation specifications are as follows:

		Mole Fraction	
Component	Feed	Distillate	Bottoms
1.	0.40	0.90	0.05
2.	0.60	0.10	0.95

Assuming a constant relative volatility of 1.8, calculate the minimum number of theoretical stages and the minimum reflux ratio.

12.5. A distillation column is used to separate a three-component mixture into three products: the distillate, a side draw, and the bottoms. It is proposed to monitor the product compositions by measuring the product rates, using total reflux calculations. The minimum number of stages equivalent to the actual column are four stages above the sidedraw and five below it. Calculate the product compositions corresponding to the following feed composition and product rates. The relative volatilities, assumed constant, are also given.

		Relative Volatility	
Component	Feed Mole Fraction	Top Section	Bottom Section
1.	0.25	1.60	1.50
2.	0.40	1.30	1.25
3.	0.35	1.00	1.00
Feed flow rate	= 100.0 kmol/h		
Distillate flow rate	= 25.0 kmol/h		
Side draw rate	= 40.0 kmol/h		
Bottoms flow rate	= 35.0 kmol/h		

12.6. A 1000 kmol/h feed stream containing a mixture of hydrocarbons is separated in a distillation column between propane and n-butane. Based on total reflux calculations with 14 minimum equilibrium stages, calculate the distillate and bottoms compositions and temperatures when the distillate rate is 215 kmol/h. The column pressure is 1700 kPa and a partial condenser is used. The feed composition is given below, along with component relative volatilities referred to n-butane and assumed constant.

i	1	2	3	4	5
Component	Ethane	Propane	n-Butane	n-Pentane	n-Hexane
Mole%	3	19	38	36	4
α_{i3}	6.00	2.40	1.00	0.42	0.18

The K-value of n-butane can be calculated by Raoult's law. Its vapor pressure is given by the following equation, with the pressure in kPa and the temperature in K:

$$p_3^0 = \exp\left(13.692 - \frac{2140.8}{T - 36.504}\right)$$

12.7. A hydrocarbon gas stream contains some heavy components which must be removed by absorption. The absorber has six theoretical stages and operates at 2750 kPa. The overhead product should contain 0.01 kmol/h nC_5. Using the modular column section method, calculate the required absorbent flow rate and the product streams flow rates and compositions. The feed streams are defined below, along with the component relative volatilities which are referred to pentane and assumed constant. The K-value of pentane at the column pressure of 2750 kPa is given by

$$\ln K_5 = 8.3 - (3390/T) \text{ where } T \text{ is in } K$$

	C_1	C_2	C_3	nC_4	nC_5	Oil
Absorbent inlet, X_{0i}	0	0	0	0	0	1
Gas inlet, V_{7i} (kmol/h)	150	345	225	23	7	0
α_i	93.27	23.00	8.19	2.73	1	0.0014

The absorbent inlet is at 35°C and 2750 kPa, and the gas inlet is at 45°C and 2750 kPa. Enthalpies may be calculated as in Example 12.5 with the given parameters for the hydrocarbon components. The oil parameters are $A_6 = 78,000$, $B_6 = 175$, $C_6 = 32,500$, and $D_6 = 235$.

12.8. A debutanizer distillation column is to be designed to separate butanes and lighter components from the hydrocarbon stream defined below. The separation should result in 0.05 mole fraction isopentane in the distillate and 0.03 mole fraction n-butane in the bottoms. The column will operate at 800 kPa with a partial condenser.

Saturated liquid feed at 800 kPa:

C_3	iC_4	nC_4	iC_5	nC_5
6	14	26	19	35 kmol/h

a. Calculate N_m, the minimum number of stages at total reflux, by the second Fenske Equation 12.17a. It is estimated that the mole fraction of n-butane in the distillate is 0.50, and the average column temperature is 90°C. Vapor–liquid equilibrium may be assumed to follow Raoult's law, with vapor pressures calculated by the Antoine Equation 2.19 using constants given below to represent the expected conditions (p^0 in kPa and T in K):

	C_3	iC_4	nC_4	iC_5	nC_5
A_i	13.4410	13.9044	13.6981	13.8834	14.2697
B_i	1729.04	2211.04	2143.66	2436.35	2728.28
C_i	−34.8637	−23.3413	−36.3467	−37.7276	−26.5939

b. Use N_m and B (the bottoms rate) and the first Fenske Equation 12.17 to calculate the distillate and bottoms compositions and an average column temperature by iterating on K_r. (Select nC_4 as the reference component).

c. Calculate R_m, the minimum reflux, by the Underwood equations (12.29 and 12.30).

d. Determine N and R by the Gilliland correlation such that the liquid flow in any part of the column does not exceed 200 kmol/h.

12.9. In the absorber column of Example 12.5, it is required to determine the absorbent rate necessary to bring the mole fraction of n-butane in the overhead stream down to 0.001. Use data from Example 12.5, and solve the problem by the modular column section method to close all the model equations.

12.10. It is proposed to modify the design of the absorber column in Problem 12.9 with the objective of lowering the absorbent rate and raising the propane concentration and recovery in the overhead, while keeping the n-butane concentration in the overhead at 0.001 mole fraction. This is to be achieved by increasing the number of stages in the column from 6 to 12. Determine the required absorbent rate and the product rates, compositions and temperatures using the column modular section method to close all the model equations.

12.11. A distillation column is to be used to separate a hydrocarbon mixture between propane and butane. The feed stream is defined below, along with the K-values:

		Feed (kmol/h)	K-value
1.	Ethane	20.0	3.66
2.	Propane	55.0	1.52
3.	Butane	25.0	0.64

The separation specifications are mole fraction of propane in the distillate, $Y_{D2} = 0.70$; mole fraction of propane in the bottoms, $X_{B2} = 0.18$; mole fraction of butane in the distillate, $Y_{D3} = 0.04$. Calculate the distillate and bottoms compositions.

What is the minimum number of stages required to achieve the specified separation? Using a value of 1.18 for the parameter θ, calculate the minimum reflux ratio. If the column is operated at twice the minimum reflux ratio, what is the required number of stages?

12.12. Stream L_0 is an absorbent used to absorb heavier components in a light hydrocarbon vapor stream V_6 in a five-stage absorber. The absorbent and inlet vapor stream compositions, and the component relative volatilities are given below:

	i	L_{0i} (kmol/h)	V_{6i} (kmol/h)	α_{i3}
1.	C2	0.0	75.0	8.42
2.	C3	0.0	15.0	3.25
3.	C4	1.0	10.0	1.00
4.	C10	99.0	0.0	0.01
T (°C)		24.0	38.0	

The K-value of C4 is given by the equation $\ln K_3 = 11.02 - (3000/T)$, where T is in Kelvin. It is assumed that the stripping factors can be estimated based on the average of the inlet stream temperatures, and the inlet liquid and vapor flow rates. Calculate the C4 flow rate in the liquid outlet. What would be the value if the absorbent rate is doubled?

12.13. The stream defined below is sent to a column at a fixed pressure. The column has a partial condenser with a vapor distillate product and a partial reboiler with a liquid bottoms product. Assuming total reflux and constant relative volatilities, calculate by the Fenske equations the minimum number of trays and the product rates and compositions to meet the following specifications:

$d_3 + d_4 = 13.85$ kmol/h	$G_1 = d_3 + d_4 - 13.85 = 0$
$b_1 + b_2 = 6.03$ kmol/h	$G_2 = b_1 + b_2 - 6.03 = 0$

Feed stream:

i	Component	f_i (kmol/h)	α_{i3}
1.	iC4	12	2.12
2.	nC4	448	1.73
3.	iC5	36	1.00
4.	nC5	15	0.777
5.	C6	23	0.389
6.	C7	39	0.185
7.	C8	272	0.087
8.	C9	31	0.046
	Total	876	

12.14. A distillation column separates the feed stream shown below into butanes in the distillate and pentanes in the bottoms. The relative volatilities (assumed constant) and the separation specifications are also given. Use the first Fenske equation to calculate the minimum number of stages that can meet the specs at total reflux. With an average K-value of isopentane $K_3 = 0.8$, use the second Fenske equation to calculate all the component flow rates in both products. Does this value of K_3 produce component flow rates consistent with total flow rates?

	i	f_i (kmol/h)	α_{i3}	Y_i (distillate)	X_i (bottoms)
iC4	1	14	1.9		
nC4	2	26	1.6	0.5570	0.0730
iC5	3	33	1.0	0.0859	
nC5	4	27	0.8		

12.15. The effluent gas of a fermentation process contains carbon dioxide and ethanol. It is required to recover 97.5% of the ethanol by absorption with water in a countercurrent absorber. Using the Kremser equation, determine the smallest number of stages required to meet the specification. Use trial calculations with number of stages ranging from 6 to 9. The K-value of ethanol at average column temperature and pressure may be assumed constant at 0.57. The stripping factor can be estimated based on the inlet vapor and liquid flows.

	Water	Effluent Gas
Water (kmol/h)	150	0
Carbon dioxide (kmol/h)	0	176.4
Ethanol (kmol/h)	0	3.6
Temperature (°C)	30	30
Pressure (kPa)	110	110

12.16. The feed stream given below with the indicated component relative volatilities is sent as saturated liquid to a distillation column.

	f_i (kmol/h)	α_{i4}
1.	150	2.375
2.	250	1.812
3.	320	1.527
4.	280	1.000

The performance specifications are as follows:

Mole fraction component 1 in the bottoms	$X_{B1} = 0.0846$
Mole fraction component 4 in the distillate	$Y_{D4} = 0.0164$
Recovery of component 4 in the bottoms	99.06%

Calculate the following:
a. The distillate and bottoms flow rates.
b. Component 1 mole fraction in the distillate and component 4 mole fraction in the bottoms.
c. The minimum number of stages (at total reflux) required to make the specified separation.
d. The distillate and bottoms component flow rates at total reflux for all the components, if the K-value for component 4 is $K_4 = 0.58$.
e. The minimum reflux ratio if $\theta = 1.673$.
f. The required reflux ratio if the number of stages is 10.

12.17. A debutanizer column is to be designed to separate a hydrocarbon mixture. The following data is given:

Component	Feed (kmol/h)	A_i	B_i	C_i
1. Propane	39	13.439	1728.4	−34.898
2. n-Butane	28	13.692	2140.8	−36.504
3. n-Pentane	19	14.246	2714.8	−27.294
4. n-Hexane	14	12.812	2154.3	−78.966
Total (kmol/h)	100			
Temperature (°C)	85			
Pressure (kPa)	1750			

Constants A_i, B_i, C_i are for the Antoine vapor pressure equation, $p_i^0 = \exp[A_i - B_i/(T + C_i)]$, where p_i^0 is in kPa and T is in K.

The average column pressure is 1700 kPa. The separation specifications are $X_{B2} = 0.02$, $Y_{D3} = 0.03$. Estimated X_{B3} is 0.54.
1. Estimate the minimum number of stages using one form of the Fenske equation. Use Raoult's law for the K-values at average column pressure and an average column temperature assumed equal to the feed temperature. Select pentane as the reference component.

2. Calculate component rates in the distillate and bottoms using the other form of the Fenske equation. Adjust the column average temperature to match the bottoms flow rate. If the adjusted temperature is considerably different from the temperature in Part 1, repeat that part with the new temperature then repeat Part 2.
3. Calculate the minimum reflux ratio by the Underwood equation, and determine the number of theoretical stages by the Gilliland correlation at a reflux ratio that is 1.5 times the minimum. Assume the feed is saturated liquid.

12.18. An absorber is used for rough separation of a hydrocarbon gas between propane and lighter components and n-butane. Most of the butane will be absorbed and most of the propane and lighter components will leave with the overhead vapor. The absorber has 10 actual trays with an overall efficiency of 50%. The feed gas and absorbent and the component relative volatilities are as follows:

i		α_{i4}	Feed Gas (kmol/h)	Absorbent, X_i
1.	C_1	48.0	1990	0
2.	C_2	10.5	210	0
3.	C_3	3.2	150	0.01
4.	nC_4	1.0	85	0.04
5.	Oil	0.0021	0	0.95
Flow rate (kmol/h)			2435	???
Temperature (°C)			15	20
Pressure (kPa)			400	400

It is required to determine the range of absorbent flow rates that would achieve the specified separation. Using the Kremser equation calculate the propane and butane flows in the products at different absorbent rates and recommend appropriate rates. Absorbent rates to be checked are 500, 600, and 700 kmol/h.

The stripping factors may be based on the inlet liquid and vapor flow rates and their average temperature. The K-value of n-butane can be calculated by Raoult's law, with the vapor pressure calculated by the Antoine equation:

$$p_4^0 \text{ (kPa)} = \exp[13.692 - 2140.8/(T(K) - 36.504)]$$

12.19. The hydrocarbon stream defined below is to be separated in a distillation column into distillate and bottoms products. For a column with seven stages calculate the distillate and bottoms compositions at total reflux if the distillate rate is 70 kmol/h.

What is the required minimum reflux ratio for achieving the same separation as above? The feed stream enters the column as saturated liquid and the key components are propane and n-butane.

What is the required number of stages if the reflux ratio is twice the minimum? Use the provided relative volatilities and assume a starting value of $K_r = K_3 = 0.5$.

	Feed, f_i (kmol/h)	α_{i3}
1. Ethane	14	5.75
2. Propane	56	2.40
3. n-Butane	30	1.00

12.20. A column is to be designed with a number of trays equivalent to 8 theoretical stages for an absorption operation to remove at least 60% of the butane in a hydrocarbon stream.

Absorbent rates of 200, 250, or 300 kmol/h are possible. It is required to determine the lowest among these that would meet the butane specification, and to calculate the product rates and compositions when the column is operated with the selected absorbent rate. Use the modular column section method with stripping factors based on inlet liquid and vapor flow rates. The gas and absorbent streams are defined below.

Component	K-Value	Feed Gas, Y_i	Absorbent, X_i
1. Ethane	7.05	0.49	0
2. Propane	1.97	0.30	0
3. n-Butane	0.64	0.21	0.05
4. n-Decane	0.0015	0.00	0.95
Flow rate (kmol/h)		500	

12.21. It is required to design a distillation column to separate a propylene-propane stream according to the following specifications:

	Feed (kmol/h)	Distillate (kmol/h)	Bottoms (kmol/h)
1. Propylene	60.0	58.0	
2. Propane	40.0		39.5

The relative volatility of propylene/propane at column pressure and average column temperature is $\alpha_{12} = 1.11$ ($\alpha_{22} = 1.00$). The feed stream enters the column as saturated liquid. In order to optimize the column design, a number of options will be considered, each using a different number of stages. Determine the reflux ratio required to meet the above specifications for $N = 100$, $N = 125$, and $N = 150$ stages.

12.22. A countercurrent absorber is to be designed for recovering 98% of the ethanol from a gas mixture of ethanol and carbon dioxide with water as the absorbent. Using the modular shortcut method, determine the expected ethanol recovery (ethanol in the liquid product/ethanol in the gas feed) as a function of the absorbent rate in a five-stage column. Absorbent rates of 80, 90, and 100 kmol/h should be considered. At the column conditions, the K-value of ethanol may be assumed constant at 0.6. The stripping factor may be assumed constant, based on the absorbent and gas feed flow rates.

	Gas Feed (kmol/h)	Absorbent (kmol/h)
1. Carbon dioxide	97.8	0
2. Ethanol	2.1	0
3. Water	0.1	Variable

If the highest absorbent rate considered is the maximum available but does not meet the required ethanol recovery, what design option(s) would you recommend to meet the specification? Support your recommendation with additional calculations.

12.23. The feed stream defined below is to be separated in a distillation column between propane and n-butane. Based on the total-reflux column model, calculate the number of stages and products' rates and compositions if it is desired to have in the bottoms 10 kmol/h propane and 360 kmol/h n-butane. The relative volatility values provided may be assumed constant. (Ref: Example 12.2)

	f_i (kmol/h)	α_{i3}
1. Ethane	30	6.00
2. Propane	190	2.40
3. n-Butane	380	1.00
4. n-Pentane	360	0.42
5. n-Hexane	40	0.18

REFERENCES

Fenske, M. R., Fractionation of straight-run Pennsylvania gasoline, *Ind. Eng. Chem.*, 24, 482, 1932.

Gilliland, E. R., *Ind. Eng. Chem.*, 32, 1220, 1940.

Jirapongphan, S., J. F. Boston, H. I. Britt, and L. B. Evans, "A nonlinear simultaneous modular algorithm for process flowsheet optimization, Paper presented at *AIChE 73rd Annual Meeting*, Chicago, November 1980.

Kremser, A., *Nat. Petrol. News*, 22, May 21, 1930.

Molokanov, Y. K., T. P. Korablina, N. I. Mazurina, and G. A. Nikiforov, *Int. Chem. Eng.*, 12(2), 209–212, 1972.

Smith, B. D., *Design of Equilibrium Stage Processes*, New York, McGraw-Hill, 1963.

Smith, B. D. and W. K. Brinkley, *AIChE J.*, 6(3), 446, 1960.

Trevino-Lozano, R. A., T. P. Kisala, and J. F. Boston, *Comput. Chem. Eng.*, 8(2), 105, 1984.

Underwood, A. J. V., *Chem. Eng. Progr.*, 44, 603, 1948.

13 Rigorous Equilibrium Methods

The rigorous column solution methods described in this chapter are applicable to multistage vapor–liquid and liquid–liquid separation processes in general, including complex distillation and absorption/stripping, as well as liquid–liquid extraction, reactive distillation, and three-phase distillation. The methods involve the formulation and solution of a large number of linear and nonlinear equations that represent material and energy balances and phase equilibrium relations. Consideration is given initially to steady-state processes. Relaxation methods, which take a dynamic route toward convergence, are also described, as well as detailed column dynamics.

In a rigorous solution no assumptions are made to simplify the model equations as in shortcut methods. It must be emphasized, however, that the model is only as good as the physical properties prediction methods used for estimating the system properties such as phase equilibrium distribution coefficients and enthalpies (Chapter 1). Moreover, in the rigorous equilibrium stage model, a tray is considered an equilibrium stage. This leads to the concept of tray efficiency, a parameter that relates actual trays to equilibrium stages for the purpose of design and performance evaluation. Tray efficiency and the rate-based approach to solving columns are discussed in Chapter 14.

Because of the complexity of the equations, the only practical way for solving them rigorously is by means of numerical solution algorithms implemented on computer programs. These programs have become indispensable for the design and performance evaluation of multistage separation processes. An understanding of the solution methods is important for efficient and intelligent use of the programs.

The model for which the solution methods are developed represents a fixed configuration column. Thus, parameters such as number of trays, feed and draw tray locations, and heater and cooler locations are all fixed. The column is completely specified by defining a number of additional performance specifications equal to the degrees of freedom of the column. The model is then solved to determine all the other variables.

13.1 MODEL DESCRIPTION

The model is shown in Figure 13.1 for a generalized multistage vapor–liquid counterflow column. Each stage j may have an external feed, F_j, a liquid side product, L_j^s, a vapor side product, V_j^s, and a heat duty, Q_j. Also, the net liquid flowing from stage j to stage $j + 1$ is designated as L_j, and the net vapor flowing from stage j to stage $j-1$ is designated as V_j. The stage temperature is T_j, and its pressure is P_j. Stages are numbered from the top down: from $j = 1$ at the condenser or top stage to $j = N$ at the

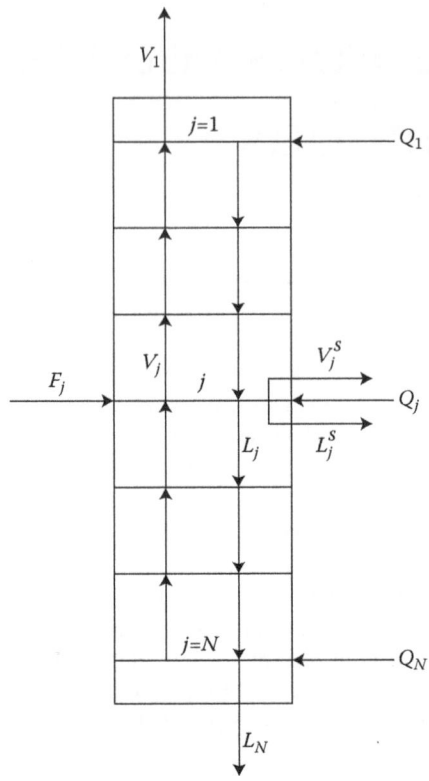

FIGURE 13.1   Generalized column model.

reboiler or bottom stage. Since in the generalized model each stage can have liquid and/or vapor products and a heat duty, there is no need for any special designation for the condenser or reboiler.

The component molar rates and fractions and total rates are interrelated as follows:

$$L_{ji} = L_j X_{ji}$$
$$V_{ji} = V_j Y_{ji}$$
$$F_{ji} = F_j Z_{ji} \tag{13.1}$$
$$L_{ji}^s = L_j^s X_{ji}$$
$$V_{ji}^s = V_j^s Y_{ji}$$

where X_{ji} and Y_{ji} are mole fractions in the liquid and vapor of component i on stage j. The initial thermal conditions of external feed streams are designated by superscript f, thus, the initial temperature, pressure, and enthalpy of a stream going to stage j are designated as T_j^f, P_j^f, and H_j^f, respectively. Regardless of its phase, single or mixed, a stream going to stage j is assumed to mix with the contents of that stage, reach phase equilibrium, then separate into vapor and liquid. A column pumparound is modeled by a pair of streams consisting of a product and a feed with the appropriate constraining equations that tie them together.

13.1.1 MODEL EQUATIONS

Since all stages in the model have a general designation, the equations are written for any stage ($j = 1, 2, ..., N$). The equations required to completely define a stage include material balance, energy balance, and phase equilibrium equations. The component material balance around stage j is written in terms of total flows and mole fractions:

$$L_{j-1}X_{j-1,i} - (L_j + L_j^s)X_{ji} + V_{j+1}Y_{j+1,i} - (V_j + V_j^s)Y_{ji} + F_jZ_{ji} = 0 \qquad (13.2)$$

The mole fractions must add up to unity:

$$\sum_i X_{ji} - 1 = 0$$
$$\sum_i Y_{ji} - 1 = 0 \qquad (13.3)$$

Note that Equation 13.2 can be summed up over all the components, and then combined with equation set 13.3 in order to yield the overall material balances on each tray. The condition that the sum of the feed mole fractions, Z_{ji}, must be equal to 1 follows from these balances. Further, the overall tray material balance, summed over trays 1 through N, results in a column overall material balance.

The energy balance around stage j is given by

$$L_{j-1}h_{j-1} - (L_j + L_j^s)h_j + V_{j+1}H_{j+1} - (V_j + V_j^s)H_j + F_jH_j^f + Q_j = 0 \qquad (13.4)$$

Liquid and vapor enthalpies are designated by h and H, respectively, and a feed enthalpy is expressed as H_f.

The phase equilibrium relations are written as

$$Y_{ji} - K_{ji}X_{ji} = 0 \qquad (13.5)$$

The fluid properties, K_{ji}, H_j, and h_j, are either supplied data or predicted by some thermodynamic methods that correlate them as functions of temperature, pressure, and composition (Chapter 1). Expressed mathematically, the data or correlations can be written as follows:

$$K_{ji} = K_{ji}(T_j, P_j, X_{ji}, Y_{ji}) \qquad (13.6)$$

$$H_j = H_j(T_j, P_j, Y_{ji}) \qquad (13.7)$$

$$h_j = h_j(T_j, P_j, X_{ji}) \tag{13.8}$$

In all the above equations, $j = 1, 2, ..., N$, the number of stages, and $i = 1, 2, ..., C$, the number of components. The total number of independent equations describing the model is arrived at as follows:

Equation set (13.2)	CN equations
Equation set (13.3)	$2N$ equations
Equation set (13.4)	N equations
Equation set (13.5)	CN equations
Equation set (13.6)	CN equations
Equation set (13.7)	N equations
Equation set (13.8)	N equations
Total	$(3C + 5)N$ equations

The model variables include all the variables in Equations 13.2 through 13.8. A quick inspection indicates that there are more variables than the total number of equations. The number of variables minus the number of equations equals the degrees of freedom of the model. In order to define the column performance, a number of variables equal to the degrees of freedom must be specified. In general, these specifications are expressed as functions of the column variables:

$$G(w) = 0 \tag{13.9}$$

where G represents the specifications expressed in terms of the variable vector, w. The model is mathematically defined if the total number of equations, Equations 13.2 through 13.9, is equal to the total number of variables w. A column solution is reached when a set of values for w is found that satisfies Equations 13.2 through 13.9.

The specification functions, equation set 13.9, are often simply a constant. For instance, the mole fraction of component i in the vapor from the top stage (the distillate) can be specified as $Y_{1i} = 0.95$. Certain other parameters such as tray pressures and feed compositions are considered in most solution algorithms as fixed quantities, thus reducing the number of variables.

Examine the number of equations and variables in the generalized model as applied to a specific column. Consider a column with a single feed, a condenser and a reboiler, a vapor distillate product, and a bottoms product. There are no side products and the only heaters or coolers are the reboiler and condenser. The feed is assumed to be of fixed rate, composition, and thermal conditions. Under these conditions the variables in Equations 13.2 through 13.8 are the following:

L_j	N variables
V_j	N variables
X_{ji}	CN variables
Y_{ji}	CN variables
h_j	N variables

H_j	N variables
Q_j $(j = 1, j = N)$	2 variables
K_{ji}	CN variables
T_j	N variables
P_j	N variables
Total	$2 + (3C + 6)N$ variables

The number of variables exceeds the number of equations by $N + 2$. If the N tray pressures are fixed, the column has 2 degrees of freedom and requires two specifications to define its performance, as concluded in Chapters 6 through 12 for this type of column.

13.2 STEADY-STATE SOLUTION METHODS

The equations that make up the generalized column model, Equations 13.2 through 13.9, include many that are nonlinear, and their number can run in the tens of thousands. (In a 10-component, 30-tray column with two specifications, the number of equations would be $(3 \times 10 + 5) \times 30 + 2 = 1052$.) It is clear that analytical solution of these equations is impossible, the alternative being numerical, iterative techniques using computer programs. Many algorithms have been proposed for solving the equations, a good review of which is presented by Wang et al. (1980). Due to the complexity of the equations, most algorithms are prone to convergence difficulties in certain column situations. Contributing to these difficulties is the large variation in the relative magnitudes of the variables, round-off errors, and the sparse matrices that result from the equations.

Certain algorithms may be better suited for certain types of columns. Convergence characteristics depend on whether the column is of a predominantly distillative type or absorption/stripping type, or both (Chapters 7, 8, and 9). In distillation columns, where mixtures are relatively narrow-boiling, the separation is brought about primarily as a result of the variation in equilibrium coefficients with temperature. In absorbers and strippers, the mixtures are wide-boiling, and the separation is more influenced by the heat of absorption associated with the transfer of components from one phase to the other. It may thus be seen that different equations in the column model have varying weights for each particular type of column. Algorithms that are designed for one type may not function very effectively for another. Complex columns could contain sections where the equilibrium equations are predominant and others where the energy balances are predominant. Thus, algorithms designed for solving the generalized column model should be capable of handling this variety of situations.

The differences among the solution methods result from several considerations, including the choice of independent variables, the grouping and arrangement of the equations, and the convergence path. Some methods are based on equation decoupling, while others solve the entire system of equations simultaneously.

In the equation decoupling methods, the model equations 13.2 through 13.9 are grouped into smaller sets of equations, which are solved one set at a time. The solution of one set of equations is substituted in the other sets to form iterative loops. Each algorithm has its unique way of sequencing the calculations. Within this group of methods, some algorithms are better suited for specific types of columns. Some of these methods are adapted for handling only limited types of column performance specifications.

Following is an overview of some of the methods, considered representative of the main algorithmic categories, and presented mostly in historical order.

13.2.1 METHOD OF THIELE AND GEDDES

Although not widely used anymore, this method is presented because it was the forerunner for many of the more recent and more efficient methods. It is one of the earlier methods for rigorous numerical solution of distillation columns. It was first described by Thiele and Geddes (1933) and later detailed by Lyster et al. (1959).

The method is basically designed for solving conventional distillation columns with one feed, F, a distillate, V_1, a bottoms product, L_N, a condenser with duty Q_1, and a reboiler with duty Q_N. In the generalized model of Figure 13.1, all the feeds except one, all the products except V_1 and L_N, and all the duties except Q_1 and Q_N are set to zero. The method becomes numerically unstable for columns other than conventional distillation. Another drawback of the method is that it can handle limited types of performance specifications. In the outline that follows, the specifications are the distillate rate, V_1, and the reflux rate, L_1. It is assumed that the number of stages and the column pressure profile are fixed at given values.

The stage temperatures are considered the independent variables. The calculations are started by assuming a set of stage temperatures T_j, vapor flows V_j, and liquid flows L_j. The equilibrium coefficients K_{ji} are estimated based on the assumed tray temperatures. Equations 13.1 and 13.5 are combined to give

$$L_{ji} = A_{ji} V_{ji}$$

where

$$A_{ji} = \frac{L_j}{V_j K_{ji}}$$

In the rectifying section, the above equation is divided by V_{1i}:

$$\frac{L_{ji}}{V_{1i}} = A_{ji} \frac{V_{ji}}{V_{1i}} \tag{13.10}$$

Next, a component material balance is written around the top of the column, including any tray j above the feed. The resulting equation is divided by V_{1i}:

$$\frac{V_{j+1,i}}{V_{1i}} = \frac{L_{ji}}{V_{1i}} + 1 \tag{13.11}$$

Replacing j by $j+1$ in Equation 13.10 and eliminating $V_{j+1,i}/V_{1i}$ from Equation 13.11, the following is obtained:

$$\frac{L_{j+1,i}}{V_{1i}} = A_{j+1,i}\left(\frac{L_{ji}}{V_{1i}} + 1\right) \tag{13.12}$$

This equation is written for $j = 1$, and L_{1i}/V_{1i} is replaced by R/K_{1i}, where R is the reflux ratio, L_1/V_1:

$$\frac{L_{2i}}{V_{1i}} = A_{2i}\left(\frac{R}{K_{1i}} + 1\right)$$

The assumed liquid and vapor rates are used to evaluate A_{2i}, and since the reflux ratio is specified, L_{2i}/V_{1i} may be computed. Its value is then substituted in Equation 13.12 written for $j = 2$, from which L_{3i}/V_{1i} is calculated. The procedure is continued through the stage just above the feed.

The equivalent of Equation 13.12 for the stripping section is derived in a similar manner by combining the equilibrium and component material balance equations. The result is

$$\frac{L_{j-1,i}}{L_{Ni}} = \frac{1}{A_{ji}}\frac{L_{ji}}{L_{Ni}} + 1 \tag{13.13}$$

Written for stage $j = N$, the reboiler, Equation 13.13 becomes

$$\frac{L_{N-1,i}}{L_{Ni}} = \frac{1}{A_{Ni}} + 1$$

After $L_{N-1,i}/L_{Ni}$ has been calculated, Equation 13.13 is applied in succession to calculate the remaining liquid component flow ratios, L_{ji}/L_{Ni}, from the bottom up through the feed stage. As in the rectifying section, the equilibrium coefficients K_{ji} are calculated from the assumed stage temperatures, then the K_{ji}'s and the assumed stage vapor and liquid flows are used to evaluate the factors A_{ji}. The ratio of component i in the products, L_{Ni}/V_{1i}, is determined by combining Equations 13.12 and 13.13 and matching the downward and upward stage calculations at the feed stage. For a bubble point feed, the result is as follows (Lyster et al., 1959):

$$\frac{L_{Ni}}{V_{1i}} = \frac{L_{f-1,i}/V_{1i} + 1}{L_{f-1,i}/V_{Ni} - 1} \tag{13.14}$$

where subscript f designates the feed tray.

The distillate component flow rates may now be calculated from an overall column component material balance:

$$V_{1i} = \frac{FZ_i}{1 + L_{Ni}/V_{1i}} \tag{13.15}$$

At convergence, the summation of V_{1i} over all components should equal the specified distillate rate, V_1. If this test is not satisfied, then new, corrected values of V_{1i} must be found. A factor θ is defined such that

$$V_1 - \sum_{i=1}^{C} \frac{FZ_i}{1 + \theta(L_{Ni}/V_{1i})} = 0 \tag{13.16}$$

This equation is solved iteratively to determine θ. Once θ has been evaluated, the corrected distillate and bottoms component rates are calculated as follows:

$$V'_{1i} = \frac{FZ_i}{1 + \theta(L_{Ni}/V_{1i})} \tag{13.17}$$

and

$$L'_{Ni} = \theta \frac{L_{Ni}}{V_{1i}} V'_{1i} \tag{13.18}$$

where the prime denotes corrected flow rates.

The stage liquid compositions are calculated from the corrected component liquid flow rates. In the rectifying section,

$$X_{ji} = \frac{(L_{ji}/V_{1i})V'_{1i}}{\sum_{i=1}^{C}(L_{ji}/V_{1i})V'_{1i}} \tag{13.19}$$

and in the stripping section,

$$X_{ji} = \frac{(L_{ji}/L_{Ni})L'_{Ni}}{\sum_{i=1}^{C}(L_{ji}/L_{Ni})L'_{Ni}} \tag{13.20}$$

The updated stage temperatures are obtained by calculating the bubble point temperature on each stage on the basis of the new liquid compositions given by Equations 13.19 and 13.20. The updated compositions and temperatures are used to calculate updated total liquid and vapor flows on each stage using enthalpy balances as described in the modified Thiele–Geddes method in the following section.

The entire procedure is repeated until convergence is achieved. The use of θ to correct the component flow rates tends to guide the computations toward faster convergence by forcing an overall material balance in each iteration. Convergence is considered reached when the difference between the specified and calculated distillate rates is within a preset tolerance.

13.2.2 MODIFIED THIELE–GEDDES METHOD

This method also considers the stage temperatures as the independent variables. The algorithm is applied to a single-feed, two-product column with a partial condenser and reboiler. As in the original Thiele–Geddes method, the problem definition is such that the feed component flow rates, f_{ji}, are known and fixed. The column pressure profile is also fixed, as well as its configuration, which defines the number of stages and feed location. In addition, one product rate (the distillate) and one internal flow (such as the reflux rate, L_1) are specified. The solution method, outlined below, is described in detail by Holland (1975).

1. Assume a set of stage temperatures and vapor rates.
2. Calculate stage liquid rates by applying total material balance on each stage.
3. Estimate equilibrium coefficients on the basis of a composition independent correlation.
4. Combine component material balance and phase equilibrium relations to calculate vapor rates of each component on each stage.
5. Calculate liquid rates of each component on each stage.
6. Calculate mole fractions of each component in the liquid and vapor on each stage.
7. Calculate a set of corrected mole fractions to satisfy overall material balance.
8. Calculate a set of improved stage temperatures based on an updated set of equilibrium coefficients.
9. Using the latest computed sets of stage temperatures and compositions, apply energy balances to update the vapor flow rates and compute the condenser and reboiler duties.
10. If the new vapor rates and temperatures agree with the previous set within a preset tolerance, the problem is converged. Otherwise, repeat the calculations beginning at step 4.

The first step is to assume a set of temperatures T_j and vapor flows V_j. The temperatures can be obtained by linear interpolation between the condenser and reboiler temperatures, determined by dew point and bubble point calculations of products estimated by shortcut methods or on the basis of past experience with similar columns. The vapor rates are estimated from the specified distillate and reflux rates. Constant vapor rates are assumed above and below the feed. The change in the vapor profile at the feed tray is estimated on the basis of the feed vapor fraction.

In step 2 the stage liquid rates L_j are calculated by total material balance on each stage, beginning at the condenser. Since L_1, the reflux rate and V_1, the distillate rate are both specified, V_2 is calculated from a material balance on stage 1, the condenser:

$$V_2 = V_1 + L_1$$

Next, a material balance on stage 2 yields

$$L_2 = L_1 - V_2 + V_3$$

In this equation V_3 is taken from the estimates in step 1. The calculations are continued through the bottom stage. At the feed tray the feed must be taken into account in the material balance.

The K-values in step 3 can be estimated using an equation based on Raoult's law or Henry's law (Chapter 1) such as

$$K_{ji} = \frac{P_i^0(T_j)}{P_j}$$

where P_j is the tray pressure, and $P_i^0(T_j)$ is the vapor pressure or Henry's law constant of component i at tray temperature T_j.

With a given set of V_j, L_j, and K_{ji}, step 4 applies component material balances and phase equilibrium relations to calculate component vapor rates using the tridiagonal matrix method. The formulation of equations at this point is general, in that each stage can have an external feed and vapor and liquid side products. The results may be applied to the special case of the modified Thiele–Geddes model by setting to zero all the nonexistent feeds and products. Equations 13.1, 13.2, and 13.5 are combined and written for $j = 1, 2, ..., N$. The resulting equation for component i on tray j is as follows:

$$A_{j-1,i}V_{j-1,i} - B_{ji}V_{ji} + V_{j+1,i} = -F_{ji} \tag{13.21}$$

where

$$A_{ji} = \frac{L_j}{K_{ji}V_j} \tag{13.22}$$

and

$$B_{ji} = A_{ji}\left(1 + \frac{L_j^s}{L_j}\right) + \frac{V_j^s}{V_j} + 1 \tag{13.23}$$

When Equation 13.21 is written for stages $j = 1, 2, ..., N$, the following set of equations results:

$$
\begin{array}{llllll}
-B_{1i}V_{1i} & +V_{2i} & & & & = -F_{1i} \\
A_{1i}V_{1i} & -B_{2i}V_{2i} & +V_{3i} & & & = -F_{2i} \\
& A_{2i}V_{2i} & -B_{3i}V_{3i} & +V_{4i} & & = -F_{3i} \\
\vdots & \vdots & \vdots & \vdots & \vdots & \vdots \\
& & A_{N-2,i}V_{N-2,i} & -B_{N-1,i}V_{N-1,i} & +V_{Ni} & = -F_{N-1,i} \\
& & & A_{N-1,i}V_{N-1,i} & -B_{Ni}V_{Ni} & = -F_{Ni}
\end{array}
\tag{13.24}
$$

The equations are solved for the vector $V_{1i}, V_{2i}, ..., V_{Ni}$. Inspection of this set of equations indicates that their coefficient matrix is made up of zero elements everywhere except along the three central diagonals, hence the name tridiagonal matrix. The matrix may be solved by recurrence formulas using the Thomas algorithm (Carnahan et al., 1964) or its modified version (Boston et al., 1972). The algorithm is discussed in more detail in the later part of this section.

The solution of equation set 13.24 gives the vapor flows of component i on all stages. The equation set must be solved repeatedly for each component to determine the vapor flows of all the components on all the trays.

The component liquid rates (step 5) are calculated from the component vapor rates using Equation 13.1 substituted into equation set 13.5:

$$
L_{ji} = \frac{L_j}{V_j K_{ji}} V_{ji}
$$

or

$$
L_{ji} = A_{ji} V_{ji}
$$

The mole fractions (step 6) are calculated directly from equation set 13.1. The mole fractions are adjusted in step 7 to satisfy both the overall material balance for each component and the distillate rate specification. The overall component material balance requires that

$$
FZ_i = V'_{1i} + L'_{Ni}
\tag{13.25}
$$

The prime indicates adjusted values. The feed stage subscript is dropped because there is only one feed in this model. The distillate rate specification requires that

$$
\sum_{i=1}^{c} V'_{1i} = V_1
\tag{13.26}
$$

where V_1 is the specified distillate rate. A correction factor θ is defined, relating corrected to calculated flow rates as follows:

$$\frac{L'_{Ni}}{V'_{1i}} = \theta \frac{L_{Ni}}{V_{1i}} \tag{13.27}$$

The multiplier θ must be determined such that both Equations 13.25 and 13.26 are satisfied. The following equation is obtained by eliminating L'_{Ni}/V'_{1i} between Equations 13.27 and 13.25, and rearranging:

$$V'_{1i} = \frac{FZ_i}{1 + \theta(L_{Ni}/V_{1i})} \tag{13.28}$$

Summation over components i results in

$$\sum_{i=1}^{C} V'_{1i} = V_1 = \sum_{i=1}^{C} \frac{FZ_i}{1 + \theta(L_{Ni}/V_{1i})} \tag{13.29}$$

The one unknown in this equation is θ, which is calculated by an iterative technique such as Newton–Raphson's technique. The liquid and vapor mole fractions are then updated using the following equations:

$$X_{ji} = \frac{(L_{ji}/V_{1i})V'_{1i}}{\sum\limits_{i=1}^{C}(L_{ji}/V_{1i})V'_{1i}} \tag{13.30}$$

$$Y_{ji} = \frac{(V_{ji}/V_{1i})V'_{1i}}{\sum\limits_{i=1}^{C}(V_{ji}/V_{1i})V'_{1i}} \tag{13.31}$$

The phase equilibrium coefficients are calculated next rigorously based on the latest mole fractions and stage temperatures and pressures (Chapter 1). These K-values are used to update the stage temperatures (step 8). A reference (hypothetical) component r is used to define the relative volatility:

$$\alpha_i = \frac{K_i}{K_r} = \frac{Y_i}{X_i K_r}$$

Rearranged and written for stage j, this expression becomes

$$Y_{ji} = K_{jr}\alpha_{ji}X_{ji}$$

Both sides are summed up over all the components on stage j, resulting in the following expression for K_{jr}:

$$K_{jr} = \frac{1}{\sum_{i=1}^{C} \alpha_{ji} X_{ji}}$$

(13.32)

Equation 13.32 is used to calculate an updated K_{jr} for each stage based on current α_{ji}'s and X_{ji}'s. The stage temperatures are then calculated from some function of temperature correlation of K_r (Holland, 1975).

Energy balances (Equation 13.4) are used in step 9 to update the vapor flows and to calculate the condenser and reboiler duties. First apply the energy balance to stage 1 (the condenser):

$$V_2 H_2 = V_1 H_1 + L_1 h_1 + Q_1$$

(13.33)

where V_1 is the distillate vapor rate, V_2 is the vapor rate from stage 2, and L_1 is the reflux rate. The vapor molar enthalpy on stage j is designated as H_j, and the liquid molar enthalpy as h_j. The condenser duty is Q_1. The term $V_2 H_2$ is evaluated from pure component enthalpies:

$$V_2 H_2 = \sum_i V_{2i} H_{2i}$$

(13.34)

where in ideal solutions, H_{2i} is the vapor molar enthalpy of pure component i on stage 2, and in nonideal solutions, it is the partial molar enthalpy (Chapter 1). Using a component material balance, V_{2i} is replaced by $V_{1i} + L_{1i}$:

$$V_2 H_2 = \sum_i V_{1i} H_{2i} + \sum_i L_{1i} H_{2i} = V_1 \sum_i Y_{1i} H_{2i} + L_1 \sum_i X_{1i} H_{2i}$$

(13.35)

Elimination of $V_2 H_2$ between Equations 13.33 and 13.35 yields an expression for the condenser duty:

$$Q_1 = V_1 \sum_i Y_{1i} H_{2i} + L_1 \sum_i X_{1i} H_{2i} - V_1 H_1 - L_1 h_1$$

(13.36)

The values of V_1 and L_1 are known, and the enthalpies in Equation 13.36 are calculated from the current values of stage compositions and temperatures.

Once the condenser duty is calculated, L_j is calculated in the rectifying section in a similar manner as above, by eliminating $V_{j+1} H_{j+1}$. The vapor rate V_{j+1} is then calculated from a total material balance, $V_{j+1} = V_1 + L_j$. The procedure is extended to the stripping section by including the feed as the feed tray is crossed. Finally, the reboiler duty is calculated by an overall energy balance.

The final step is to check for convergence by one of two criteria: either the variation of the V_j's and T_j's from the previous iteration is within a given tolerance, or Equations 13.2 through 13.5 are satisfied within tolerance.

The modified Thiele–Geddes method was mainly designed for conventional distillation columns although it has been generalized to handle complex columns (Holland, 1963). It is limited in the types of performance specifications it can handle and could be numerically unstable, especially for wide-boiling or nonideal mixtures (Wang et al., 1980).

13.2.3   Method of Wang and Henke

This method (Wang et al., 1966) is similar to the modified Thiele–Geddes method. Its authors introduced the tridiagonal matrix method described in the previous section for solving the component balance and equilibrium equations to determine the stage compositions. No "forcing technique" is employed to correct the compositions comparable to using the factor θ in the modified Thiele–Geddes method.

The stage temperatures are determined from the calculated stage compositions by bubble point calculations. The calculations are carried out iteratively using Muller's algorithm (Wang et al., 1966). The authors have determined that convergence by this method is more reliable than the Newton–Raphson technique.

In the next step the enthalpies of the stage vapor and liquid streams are calculated by some generalized method (Chapter 1). The energy balance relations (Equation 13.4) are then solved to compute updated vapor flows. The results are used in the next iteration to solve the tridiagonal matrix, and the remaining steps are repeated until convergence is reached. The problem is considered converged when the maximum stage temperature variation from iteration to iteration is within a given tolerance.

The method has been successfully employed for many column types, although convergence problems can occur for nonideal mixtures where the K-values are strongly composition dependent or for highly complex columns. The method is also limited in the types of performance specifications it can handle.

13.2.4   Method of Tomich

This method (Tomich, 1970) differs from the foregoing methods mainly in that the summation statements and the energy balances (Equations 13.3 and 13.4) are solved simultaneously. The benefits of the simultaneous solution are twofold. First, distillation columns and absorbers and columns that are hybrids of both types of processes can all be solved with the same method. Second, different types of column performance specifications can be incorporated in the simultaneous solution of the equations. The method is also computationally stable and efficient because it uses Broyden's modification of the Newton–Raphson technique for solving the equations (Broyden, 1965). A brief description of the method follows:

1. Assume initial temperature and vapor profiles, T_j and V_j (the independent variables).
2. Compute the liquid profile, L_j, from a total material balance on each stage.

3. Solve for the vapor and liquid compositions on each stage by setting up a tridiagonal matrix as described in the modified Thiele–Geddes method (Equation 13.24). The terms A_{ji} and B_{ji} in the tridiagonal solution require knowledge of the equilibrium coefficients, K_{ji}. In the first iteration K_{ji} are estimated from a composition-independent correlation of temperature and pressure. In subsequent iterations the current compositions are used in rigorous calculations of K_{ji}.

4. Solve Equations 13.3 and 13.4 simultaneously for a new set of T_j and V_j. At this point, any performance specifications that the column may be required to meet (Equation 13.9) are expressed as functions of T_j and V_j and solved simultaneously with Equations 13.3 and 13.4.

5. If the convergence criteria are met, a solution has been reached and the calculations are stopped. Otherwise, the calculations are repeated beginning at step 2, using the updated values of T_j and V_j.

Step 4 requires some further elaboration. Equation set 13.3 is combined and written as follows:

$$S_j = \sum_{i=1}^{C} Y_{ji} - \sum_{i=1}^{C} X_{ji} = 0 \tag{13.37}$$

If the left-hand side of Equation 13.4 is designated as E_j, it may be written as

$$E_j = 0 \tag{13.38}$$

For a given set of variables T_j and V_j, it is possible to calculate the corresponding values of L_j, X_{ji}, and Y_{ji} as described in steps 2 and 3. With these quantities known, the enthalpies can be calculated. Thus, S_j and E_j are implicit functions of T_j and V_j (the independent variables). The specification functions, Equation 13.9, can also be expressed in terms of L_j, X_{ji}, Y_{ji}, H_j, h_j. The specification functions are therefore also implicit functions of T_j and V_j.

In this method (Tomich, 1970) the summations and energy balances, Equations 13.37 and 13.38, are rewritten as functions of T_j and V_j.

$$S_j(T_j, V_j) = 0 \tag{13.39}$$

$$E_j(T_j, V_j) = 0 \tag{13.40}$$

If the only column specifications are the duties, Q_j, and the product rates, Equations 13.39 and 13.40 completely define the column. At convergence, S_j and E_j are equal to zero within tolerance. If other specifications are required, they must be expressed as functions of T_j and V_j and solved along with Equations 13.39 and 13.40. The following steps are described for the case where the only specifications are the duties and product rates. Equations 13.39 and 13.40 are expanded in a Taylor series truncated after the first derivative to calculate the functions at iteration $k + 1$ in terms of values at iteration k. The new functions are set to zero:

$$S_j^{k+1} = S_j^k + \left(\frac{\partial S_j^k}{\partial T_j}\right)\Delta T_j + \left(\frac{\partial S_j^k}{\partial V_j}\right)\Delta V_j = 0 \tag{13.41}$$

$$E_j^{k+1} = E_j^k + \left(\frac{\partial E_j^k}{\partial T_j}\right)\Delta T_j + \left(\frac{\partial E_j^k}{\partial V_j}\right)\Delta V_j = 0 \tag{13.42}$$

The values of S_j^k, E_j^k are calculated from Equations 13.37 and 13.38 (combined with Equation 13.4) at current values of the variables, and the derivatives are calculated by finite difference methods. Equations 13.41 and 13.42 are solved for ΔT_j and ΔV_j by inverting the Jacobian matrix, the matrix of partial derivatives of S_j and E_j with respect to T_j and V_j. The independent variables T_j and V_j are then updated for the next iteration:

$$T_j^{k+1} = T_j^k + \alpha\Delta T_j \tag{13.43}$$

$$V_j^{k+1} = V_j^k + \alpha\Delta V_j \tag{13.44}$$

where α is an adjustment factor. The value of α can range from -1 to $+1$ and is determined such that the sum of error squares at iteration $k + 1$ is less than that at the previous iteration:

$$\sum_{j=1}^{N}\left[(S_j^{k+1})^2 + (E_j^{k+1})^2\right] < \sum_{j=1}^{N}\left[(S_j^k)^2 + (E_j^k)^2\right] \tag{13.45}$$

Once the new T_js and V_js are calculated, the iterative procedure is repeated at step 2.

In the classical Newton–Raphson technique, the Jacobian matrix is inverted every iteration in order to compute the corrections ΔT_j and ΔV_j. The method of Tomich, however, uses the Broyden procedure (Broyden, 1965) in subsequent iterations for updating the inverted Jacobian matrix.

13.2.5 METHOD OF NAPHTALI AND SANDHOLM

This method was aimed at overcoming some of the weaknesses of other methods. While certain methods may be better suited for wide-boiling or narrow-boiling mixtures, this method (Naphtali et al., 1971), which solves all the model equations simultaneously, is an attempt to handle all types of columns using the same algorithm. The method is also capable of directly solving various types of performance specifications by including them in the solution algorithm. The original article (Naphtali et al., 1971) also describes how the Murphree tray efficiencies can be incorporated directly into the calculations.

The simultaneous solution uses the Newton–Raphson method, which is based on linearizing the model equations. Two characteristics are inherent in this method. Since the equations are highly nonlinear, the success of linearization usually requires good starting values. On the other hand, as the solution is approached, the linearized equations become progressively more accurate and convergence is accelerated.

The Newton–Raphson simultaneous solution procedure is formulated in a generalized terminology by representing the model equations by a function vector g, and all the variables by a variable vector w. The system of equations is written as

$$g(w) = 0 \qquad (13.46)$$

Naphtali and Sandholm (1971) grouped the equations by stages, writing Equation 13.46 in its expanded form as follows:

$$g_{11}(w_{11}, \ldots, w_{1i}, \ldots, w_{1,2C+1}, \ldots, w_{ji}, \ldots, w_{N,2C+1}) = 0$$
$$\vdots$$
$$g_{1i}(\ldots) = 0$$
$$\vdots$$
$$g_{1,2C+1}(\ldots) = 0 \qquad (13.47)$$
$$\vdots$$
$$g_{ji}(\ldots) = 0$$
$$\vdots$$
$$g_{N,2C+1}(\ldots) = 0$$

It is recalled that the first subscript designates the stage and the second one the component. Consider the following formulation of the system equations and variables: Variables w_{j1} through w_{jC} represent V_{ji}, the component vapor rates leaving stage j. Variable $w_{j,C+1}$ is the stage temperature T_j and variables $w_{j,C+2}$ through $w_{j,2C+1}$ represent L_{ji}, the component liquid rates leaving stage j. Function g_{j1} is the enthalpy balance on stage j, functions g_{j2} through $g_{j,C+1}$ are the component material balances on stage j, and functions $g_{j,C+2}$ through $g_{j,2C+1}$ are the phase equilibrium relationships on stage j. These are therefore $N(2C + 1)$ equations with $N(2C + 1)$ unknowns. The stage heat duties, pressures, and side draws are considered fixed in this formulation, and therefore no other specifications are required.

The equations are specifically ordered in the manner shown in equation set 13.47, namely the equations for each stage are grouped together, beginning at the top stage and going sequentially stage by stage down the column. The Jacobian matrix resulting from this grouping has a block tridiagonal structure, which lends itself to simple solution using a Gaussian elimination technique.

Equation set 13.47 is first linearized using a Taylor series expansion truncated after the first derivative. The resulting equations are used to calculate the corrections Δw to the variables w at each iteration:

$$\frac{\partial g_{11}}{\partial w_{11}} \Delta w_{11} + \cdots + \frac{\partial g_{11}}{\partial w_{N,2C+1}} \Delta w_{N,2C+1} = -g_{11}$$

$$\vdots$$

$$\frac{\partial g_{N,2C+1}}{\partial w_{11}} \Delta w_{11} + \cdots + \frac{\partial g_{N,2C+1}}{\partial w_{N,2C+1}} \Delta w_{N,2C+1} = -g_{N,2C+1}$$

(13.48)

The functions on the right-hand side and their partial derivatives are evaluated from the values of the variables at the current iteration. The partial derivatives are generally evaluated numerically by finite differences. When convergence is reached, the functions g_{ji} are equal to zero within tolerance. The matrix of derivatives, or Jacobian, may be written in the form shown below:

$$
\begin{array}{ccc|ccc|ccc}
\dfrac{\partial g_{11}}{\partial w_{11}} & \cdots & \dfrac{\partial g_{11}}{\partial w_{1,2C+1}} & \dfrac{\partial g_{11}}{\partial w_{21}} & \cdots & \dfrac{\partial g_{11}}{\partial w_{2,2C+1}} & \dfrac{\partial g_{11}}{\partial w_{31}} & \cdots & \dfrac{\partial g_{11}}{\partial w_{3,2C+1}} \cdots \\
\vdots & & \vdots & \vdots & & \vdots & \vdots & & \vdots \\
\dfrac{\partial g_{1,2C+1}}{\partial w_{11}} & \cdots & \dfrac{\partial g_{1,2C+1}}{\partial w_{1,2C+1}} & \dfrac{\partial g_{1,2C+1}}{\partial w_{21}} & \cdots & \dfrac{\partial g_{1,2C+1}}{\partial w_{2,2C+1}} & \dfrac{\partial g_{1,2C+1}}{\partial w_{31}} & \cdots & \dfrac{\partial g_{1,2C+1}}{\partial w_{3,2C+1}} \cdots \\
\hline
\dfrac{\partial g_{21}}{\partial w_{11}} & \cdots & \dfrac{\partial g_{21}}{\partial w_{1,2C+1}} & \dfrac{\partial g_{21}}{\partial w_{21}} & \cdots & \dfrac{\partial g_{21}}{\partial w_{2,2C+1}} & \dfrac{\partial g_{21}}{\partial w_{31}} & \cdots & \dfrac{\partial g_{21}}{\partial w_{3,2C+1}} \cdots \\
\vdots & & \vdots & \vdots & & \vdots & \vdots & & \vdots \\
\dfrac{\partial g_{2,2C+1}}{\partial w_{11}} & \cdots & \dfrac{\partial g_{2,2C+1}}{\partial w_{1,2C+1}} & \dfrac{\partial g_{2,2C+1}}{\partial w_{21}} & \cdots & \dfrac{\partial g_{2,2C+1}}{\partial w_{2,2C+1}} & \dfrac{\partial g_{2,2C+1}}{\partial w_{31}} & \cdots & \dfrac{\partial g_{2,2C+1}}{\partial w_{3,2C+1}} \cdots \\
\hline
\dfrac{\partial g_{31}}{\partial w_{11}} & \cdots & \dfrac{\partial g_{31}}{\partial w_{1,2C+1}} & \dfrac{\partial g_{31}}{\partial w_{21}} & \cdots & \dfrac{\partial g_{31}}{\partial w_{2,2C+1}} & \dfrac{\partial g_{31}}{\partial w_{31}} & \cdots & \dfrac{\partial g_{31}}{\partial w_{3,2C+1}} \cdots \\
\vdots & & \vdots & \vdots & & \vdots & \vdots & & \vdots \\
\dfrac{\partial g_{3,2C+1}}{\partial w_{11}} & \cdots & \dfrac{\partial g_{3,2C+1}}{\partial w_{1,2C+1}} & \dfrac{\partial g_{3,2C+1}}{\partial w_{21}} & \cdots & \dfrac{\partial g_{3,2C+1}}{\partial w_{2,2C+1}} & \dfrac{\partial g_{3,2C+1}}{\partial w_{31}} & \cdots & \dfrac{\partial g_{3,2C+1}}{\partial w_{3,2C+1}} \cdots \\
\vdots & & \vdots & \vdots & & \vdots & \vdots & & \vdots
\end{array}
$$

The Jacobian matrix is made up of matrix blocks, shown enclosed in boxes. A matrix block is a submatrix that includes the derivatives of the functions of one stage with respect to the variables of that stage or any other one stage. The matrix above shows only the first three block rows and first three block columns. The matrix blocks at the top right and bottom left corners contain derivatives of functions of one stage with respect to variables on stages other than that stage or the adjacent ones. Since the functions of a given stage involve only variables of that stage and the adjacent ones, all the derivatives in the top right and bottom left corners are zero.

As a consequence, the Jacobian matrix has a block tridiagonal structure. All matrix blocks other than the three central diagonals are zero. With this structure the matrix can be readily solved by a Gaussian elimination scheme (Naphtali et al., 1971).

Once the Jacobian matrix is solved for the corrections Δw, the straight Newton–Raphson method could be applied to update the variables for the next iteration:

$$w^{k+1} = w^k + \Delta w$$

If the current values of the variables are far from solution, applying the entire corrections may result in divergence. For this reason, the corrections are multiplied by an adjustment factor which can have a value between 0 and 1:

$$w^{k+1} = w^k + \alpha \Delta w$$

Two methods are suggested for determining α: either start with 1 and keep halving it until a smaller sum of error squares is obtained, or initiate a search for the value of α that minimizes the sum of error squares. This modification to the Newton–Raphson procedure helps stabilize the calculations, thereby highly enhancing the likelihood of convergence.

13.2.6 METHOD OF WANG AND OLESON

The method of Wang and Oleson (Wang et al., 1980) also solves the model equations (Equations 13.2 through 13.9) simultaneously by linearization. Since the resulting Jacobian is a large and sparse matrix, the matrix inversion procedure is crucial in these simultaneous methods. Whereas Naphtali and Sandholm grouped the equations by stages, resulting in a block tridiagonal matrix, Wang and Oleson partitioned the matrix by equation types.

The Newton–Raphson technique is also modified in this method. A damping factor α, between zero and one, is applied to the corrections as above. The way α is calculated, however, is different from the Naphtali–Sandholm method.

This technique, combined with Fenske shortcut calculations for generating initial estimates of temperature profiles and stage liquid or vapor flow rates, is a robust method that can solve a large percentage of different types of separation processes. The algorithm also has provision for handling inequality specifications (Brannock et al., 1977). For each inequality specification, an alternate equality specification is required to ensure a unique solution. In this manner, the so-called over-constrained problems may be solved since inequality specifications are not subject to degrees of freedom restrictions.

13.2.7 TWO-TIER METHODS

A more recent approach to the rigorous solution of multistage separation processes has been developed with three primary objectives: to be flexible in handling performance specifications, to require little information for generating initial estimates, and to have high computational speed.

The techniques employed for achieving these goals are embodied in a method described by Russell (1983). The method, based on the original article by Boston and Sullivan (1972), consists of an inner loop, where column tray compositions and flows and temperature profiles are calculated on the basis of simplified K-value and enthalpy calculations, and an outer loop, where these properties are calculated rigorously. The outer loop (and hence, the entire column) is considered converged when the K-values and enthalpies used in the inner loop are the same as those calculated rigorously in the outer loop. The method is fast because the rigorous thermodynamic calculations are relegated to the outer loop, where they are not performed as often. The term *inside-out* is commonly applied to this type of method because of the strategy of confining the time-consuming rigorous thermodynamic calculations to the outer loop. It should be noted that the overall solution is rigorous since the outer, rigorous loop must be converged before a solution is reached.

Another feature of the two-tier method is the use of special iteration variables for improved convergence stability. The equilibrium stage model is formulated as follows, beginning by writing the component material balances and equilibrium relations in terms of component molar flow rates instead of mole fractions:

$$L_{j-1,i} - (L_{ji} + L_{ji}^s) + V_{j+1,i} - (V_{ji} + V_{ji}^s) + F_{ji} = 0$$

$$V_{ji} = (K_{ji}V_j/L_j)L_{ji}^{\cdot}$$

The nomenclature is the same as before: Where double subscripts are used, the first designates the stage and the second the component. Superscript s designates a side product.

The withdrawal factors are a set of variables defined as

$$R_j^L = 1 + L_j^s/L_j \quad R_j^L \geq 1$$

$$R_j^V = 1 + V_j^s/V_j \quad R_j^V \geq 1$$

The withdrawal factor is equal to 1 when there is no side product from a given phase and stage. Since the internal flow and side product for each phase on a given stage are of identical composition, the withdrawal factors can also be written as

$$R_j^L = 1 + L_{ji}^s/L_{ji}$$

$$R_j^V = 1 + V_{ji}^s/V_{ji}$$

Incorporate these equations in the above component material balance:

$$L_{j-1,i} - R_j^L L_{ji} + V_{j+1,i} - R_j^V V_{ji} + F_{ji} = 0$$

The relative volatility is defined as

$$\alpha_{ji} = K_{ji}/K_{jb}$$

where K_{jb} is the K-value of a reference or base component. Another variable is the stripping factor of the base component:

$$S_{jb} = K_{jb}V_j/L_j$$

The relative volatility and stripping factor, together with the phase equilibrium relation, are combined to rewrite the component material balances:

$$L_{j-1,i} - (R_j^L + \alpha_{ji}S_{jb}R_j^V)L_{ji} + (\alpha_{j+1,i}S_{j+1,b})L_{j+1,i} + F_{ji} = 0 \qquad (13.49)$$

The phase equilibrium relations are also expressed in terms of relative volatilities and stripping factors:

$$V_{ji} = \alpha_{ji}S_{jb}L_{ji} \qquad (13.50)$$

The energy balances (Equation 13.4) are rewritten by incorporating the withdrawal factors:

$$E_j = L_{j-1}h_{j-1} - R_j^L L_j h_j + V_{j+1}H_{j+1} - R_j^V V_j H_j + F_j H_j^F + Q_j = 0 \qquad (13.51)$$

where Q_j is the rate of heat transfer to or from stage j.

Column performance specifications, such as product rates, compositions, and so on, are, in general, functions of R_j^L, R_j^V, and Q_j. If the general specifications are designated as G_n, where n is the number of specifications, the specification equations can be written as

$$G_n(R_j^L, R_j^V, Q_j) = 0 \qquad (13.52)$$

Equations 13.51 and 13.52 are written in residual form, such that $E_j = 0$ and $G_n = 0$ at solution.

Equations 13.49 through 13.52, together with the physical property equations, completely define the multistage, multi-component vapor–liquid separation model. In each of these equations $i = 1, ..., C$, the number of components, and $j = 1, ..., N$, the number of stages. There are, therefore, $2NC + N$ equations, plus the specification equations.

The model contains $2NC + 4N$ independent variables:

L_{ji}	NC variables
V_{ji}	NC variables
R_j^L	N variables
R_j^V	N variables
Q_j	N variables
T_j	N variables

The physical properties K_{ji}, h_j, H_j, α_{ji}, and S_{jb} are determined by property models and are not considered as main model variables. The stage pressures P_j are determined by tray hydraulics, and so on, and are also not considered as main model variables. The external feed rates and compositions, F_j, F_{ji}, Z_{ji} are determined by upstream units and are considered fixed in the column model. The liquid and vapor flows, L_j and V_j, are not independent variables since they can be replaced by the summations of L_{ji} and V_{ji}. The stage temperatures T_j do not appear explicitly in the model equations but are implied in the property computations and are, therefore, main model variables.

The degrees of freedom, variables minus equations, are $(2NC + 4N) - (2NC + N) = 3N$. If the duties Q_j and the liquid and vapor withdrawal factors R_j^L and R_j^V (or the side products L_j^S and V_j^S) are fixed, the column is completely defined. If any of these variables are freed, an equal number of specifications would be required to define the column operation.

13.2.7.1 Inner Loop Property Models

These are methods for calculating K-values and enthalpies in the inner loop. They are simple functions that use parameters derived from the rigorous property models. The parameters are updated at each outer loop iteration.

The following equations are used for the K-values:

$$K_{jb} = \exp(A_j - B_j/T_j)$$
$$\alpha_{ji} = K_{ji}/K_{jb}$$

Each time the simple model parameters are updated, K_{ji} and K_{jb} are calculated on each stage by the rigorous property model. The base component b may either be a component in the feed or a hypothetical reference component whose K-value is calculated as some weighted average of the component K-values on each stage. The relative volatilities are calculated as K_{ji}/K_{jb} and held constant until the next parameter update. The reference component K_{jb} is calculated at two temperatures, such as T_{j-1} and T_{j+1}, for calculating A_j and B_j:

$$B_j = \frac{\ln(K_{j-1,b}/K_{j+1,b})}{1/T_{j+1} - 1/T_{j-1}}$$

$$A_j = \ln K_{j-1,b} + \frac{B_j}{T_{j-1}}$$

The enthalpies in the simple model are expressed as linear functions of the temperature:

$$h_j = h_j^0 + c_j + d_j(T_j - T_r)$$

$$H_j = H_j^0 + e_j + f_j(T_j - T_r)$$

where h_j^0 and H_j^0 are ideal gas molar enthalpies for the liquid and vapor compositions on stage j at stage j temperature, with T_r as a reference temperature. The parameters c_j, d_j, e_j and f_j are calculated for each stage from the rigorous enthalpy model.

13.2.7.2 Outer Loop Property Models

These are rigorous, composition-dependent methods, generally based on equations of state and liquid activity coefficients (Chapter 1). The K-values and enthalpies are ultimately functions of temperature, pressure, and composition:

$$K_{ji} = K_{ji}(T_j, P_j, X_{ji}, Y_{ji})$$

$$h_j = h_j(T_j, P_j, X_j)$$

$$H_j = H_j(T_j, P_j, Y_j)$$

13.2.7.3 Two-Tier Algorithm

A summary of the main computational steps is presented here, noting that variations in the details are possible for different applications.

To begin the calculations the column variables must be first initialized to some estimated values. Simple methods can be used for this purpose, based on the column specifications and possibly supplemented by shortcut methods. The column temperature profile may be assumed linear, interpolated between estimated condenser and reboiler temperatures. The values for L_j and V_j may be based on estimated reflux ratio and product rates, assisted by the assumption of constant internal flows within each column section. The compositions X_{ji} and Y_{ji} may be assumed uniform throughout the column, set equal to the compositions of the liquid and vapor obtained by flashing the combined feeds at average column temperature and pressure. The other variables to be initialized are R_j^L, R_j^V, and S_{jb}, which are calculated from their defining equations. The values for Q_j may either be fixed at given values (zero on most stages) or estimated.

1. Using the current values of T_j, P_j, X_{ji}, and Y_{ji}, determine the values for A_j, B_j, c_j, d_j, e_j, f_j, K_{jb}, and α_{ji} from property values computed by the rigorous thermodynamic property model. This step marks the beginning of the outer loop.
2. The inner loop starts at this step. In this loop the variables S_{jb}, R_j^L, R_j^V, and Q_j are calculated to satisfy Equations 13.49 through 13.52, using the simple thermodynamic models. Begin by computing L_{ji} from equation set 13.49. For each component i, there are N linear equations represented by a tridiagonal matrix, which can be solved by a special sparse matrix method, the Thomas algorithm, described further along in this section. Next, V_{ji} are calculated from equation set 13.50, and L_j, V_j, and X_{ji} are calculated directly:

$$L_j = \sum L_{ji}$$

$$V_j = \sum V_{ji}$$

$$X_{ji} = L_{ji}/L_j$$

3. On each stage compute a new value for K_{jb} to satisfy the bubble point condition,

$$\sum K_{ji} X_{ji} = \sum \alpha_{ji} K_{jb} X_{ji} = 1$$

$$K_{jb} = \frac{1}{\sum \alpha_{ji} X_{ji}}$$

The new values for T_j are then calculated from the equation defining the temperature dependence of K_{jb}:

$$T_j = \frac{B_j}{A_j - \ln K_{jb}}$$

The next objective is to update S_{jb}, R_j^L, R_j^V, and Q_j to satisfy Equations 13.51 and 13.52. In the most general case all these parameters are variable, bringing the total number of variables to $4N$. The equations to be solved are N energy balances (Equation 13.51) and $3N$ specifications (Equation 13.52). The Newton method is used by numerically calculating the Jacobian matrix then inverting it to determine the corrections to the variables. The Jacobian elements are the partial derivatives of each of the residuals of Equations 13.51 and 13.52 with respect to each of the variables:

$$\frac{\partial E_j}{\partial S_{mb}} \quad \frac{\partial E_j}{\partial R_m^L} \quad \frac{\partial E_j}{\partial R_m^V} \quad \frac{\partial E_j}{\partial Q_m}, \quad i = 1,\dots,N; m = 1,\dots,N$$

$$\frac{\partial G_n}{\partial S_{mb}} \quad \frac{\partial G_n}{\partial R_m^L} \quad \frac{\partial G_n}{\partial R_m^V} \quad \frac{\partial G_n}{\partial Q_m}, \quad n = 1,\dots,3N; m = 1,\dots,N$$

4. Compute each of the residuals E_j and G_n from Equations 13.51 and 13.52.
5. Compute the Jacobian elements by perturbing each of S_{jb}, R_j^L, R_j^V, and Q_j and recalculating the residuals E_j and G_n, starting at step 2.
6. Compute corrections to the variables S_{jb}, R_j^L, R_j^V, and Q_j, then update these variables by adding the corrections:

$$(S_{jb})^{k+1} = (S_{jb})^k + (\Delta S_{jb})^k$$

$$(R_j^L)^{k+1} = (R_j^L)^k + (\Delta R_j^L)^k$$

$$(R_j^V)^{k+1} = (R_j^V)^k + (\Delta R_j^V)^k$$

$$(Q_j)^{k+1} = (Q_j)^k + (\Delta Q_j)^k$$

where k refers to the iteration number. The corrections may be multiplied by a damping factor if the magnitude of the residuals is not reduced.

7. If the residuals E_j and G_n are within tolerance, the inner loop is converged, and the computations continue at step 8. Otherwise, repeat the inner loop computations with the new values of S_{jb}, R_j^L, R_j^V, and Q_j, beginning at step 2.

8. Using the results of the converged inner loop, calculate the new K-values and enthalpies by the rigorous thermodynamic property model. If these property values match the latest values used in the inner loop, the problem is solved. Otherwise, determine new values for A_j, B_j, c_j, d_j, e_j, f_j, K_{jb}, and α_{ji} using the rigorous thermodynamic property model, calculate new values for R_j^L, R_j^V, and S_{jb}, and repeat the inner loop calculations starting at step 2.

13.2.7.4 Tridiagonal Matrix Algorithm

Equation set 13.49 can be written as

$$L_{j-1,i} + B_{ji}L_{ji} + C_{ji}L_{j+1,i} = D_{ji}$$

where

$$B_{ji} = -(R_j^L + \alpha_{ji}S_{jb}R_j^V)$$

$$C_{ji} = \alpha_{j+1,i}S_{j+1,b}$$

$$D_{ji} = -F_{ji}$$

From the model definition, it can be seen that $L_{0i} = 0$ and $V_{N+1,i} = 0$. The above equation is expanded for any component i on stages 1 to N:

$$
\begin{array}{llllll}
B_{1i}L_{1i} & + C_{1i}L_{2i} & & & & = D_{1i} \\
L_{1i} & + B_{2i}L_{2i} & + C_{2i}L_{3i} & & & = D_{2i} \\
& L_{2i} & + B_{3i}L_{3i} & + C_{3i}L_{4i} & & = D_{3i} \\
& \vdots & \vdots & \vdots & \vdots & \vdots \\
& L_{N-1,i} & + B_{N-2,i}L_{N-2,i} & + C_{N-2,i}L_{N-1,i} & & = D_{N-2,i} \\
& & L_{N-2,i} & + B_{N-1,i}L_{N-1,i} & + C_{N-1,i}L_{Ni} & = D_{N-1,i} \\
& & & L_{N-1,i} & + B_{Ni}L_{Ni} & = D_{Ni}
\end{array}
$$

In matrix form,

$$
\begin{bmatrix}
B_{1i} & C_{1i} & 0 & 0 & 0 & 0 & 0 \\
1 & B_{2i} & C_{2i} & 0 & 0 & 0 & 0 \\
0 & 1 & B_{3i} & C_{3i} & 0 & 0 & 0 \\
\vdots & \vdots & \vdots & \vdots & \vdots & \vdots & \vdots \\
0 & 0 & 0 & 1 & B_{N-2,i} & C_{N-2,i} & 0 \\
0 & 0 & 0 & 0 & 1 & B_{N-1,i} & C_{N-1,i} \\
0 & 0 & 0 & 0 & 0 & 1 & B_{Ni}
\end{bmatrix}
\cdot
\begin{bmatrix}
L_{1i} \\
L_{2i} \\
L_{3i} \\
\vdots \\
L_{N-2,i} \\
L_{N-1,i} \\
L_{Ni}
\end{bmatrix}
=
\begin{bmatrix}
D_{1i} \\
D_{2i} \\
D_{3i} \\
\vdots \\
D_{N-2,i} \\
D_{N-1,i} \\
D_{Ni}
\end{bmatrix}
$$

The equation matrix is tridiagonal, that is, all its elements are zero except the three middle diagonals. This matrix lends itself to a direct solution algorithm consisting of forward elimination followed by backward substitution. For simplicity the component subscript i is dropped since the matrix is solved for one component at a time.

To start the forward elimination, write the stage 1 equation and divided it by B_1:

$$B_1 L_1 + C_1 L_2 = D_1$$

$$L_1 + p_1 L_2 = q_1$$

where

$$p_1 = \frac{C_1}{B_1}$$

$$q_1 = \frac{D_1}{B_1}$$

For stage 2,

$$L_1 + B_2 L_2 + C_2 L_3 = D_2$$

Substituting for L_1 from the stage 1 equation and rearranging,

$$L_2 + p_2 L_3 = q_2$$

where

$$p_2 = \frac{C_2}{B_2 - p_1}$$

$$q_2 = \frac{D_2 - q_1}{B_2 - p_1}$$

Similarly for stage j,

$$L_j + p_j L_{j+1} = q_j$$

where

$$p_j = \frac{C_j}{B_j - p_{j-1}} \quad (1 < j < N)$$

$$q_j = \frac{D_j - q_{j-1}}{B_j - p_{j-1}} \quad (1 < j \le N)$$

For stage N, $L_N + p_N L_{N+1} = L_N = q_N$, since $L_{N+1} = 0$.
The new set of equations can be written in matrix form:

$$\begin{bmatrix} 1 & p_1 & 0 & 0 & 0 & 0 & 0 & 0 \\ 0 & 1 & p_2 & 0 & 0 & 0 & 0 & 0 \\ \vdots & \vdots & \vdots & \vdots & \vdots & \vdots & \vdots & \vdots \\ 0 & 0 & 0 & 1 & p_j & 0 & 0 & 0 \\ \vdots & \vdots & \vdots & \vdots & \vdots & \vdots & \vdots & \vdots \\ 0 & 0 & 0 & 0 & 0 & 1 & p_{N-2} & 0 \\ 0 & 0 & 0 & 0 & 0 & 0 & 1 & p_{N-1} \\ 0 & 0 & 0 & 0 & 0 & 0 & 0 & 1 \end{bmatrix} \cdot \begin{bmatrix} L_1 \\ L_2 \\ \vdots \\ L_j \\ \vdots \\ L_{N-2} \\ L_{N-1} \\ L_N \end{bmatrix} = \begin{bmatrix} q_1 \\ q_2 \\ \vdots \\ q_j \\ \vdots \\ q_{N-2} \\ q_{N-1} \\ q_N \end{bmatrix}$$

In backward substitution, the computations are started at the bottom:

$$L_N = q_N = r_N$$
$$\vdots$$
$$L_j = q_j - p_j L_{j+1} = r_j$$
$$\vdots$$
$$L_1 = q_1 - p_1 L_2 = r_1$$

In matrix form,

$$\begin{bmatrix} 1 & 0 & 0 & 0 & 0 \\ 0 & 1 & 0 & 0 & 0 \\ \vdots & \vdots & \vdots & \vdots & \vdots \\ 0 & 0 & 0 & 1 & 0 \\ 0 & 0 & 0 & 0 & 1 \end{bmatrix} \cdot \begin{bmatrix} L_1 \\ L_2 \\ \vdots \\ L_{N-1} \\ L_N \end{bmatrix} = \begin{bmatrix} r_1 \\ r_2 \\ \vdots \\ r_{N-1} \\ r_N \end{bmatrix}$$

EXAMPLE 13.1: COMPONENT FLOW RATES BY THE TRIDIAGONAL MATRIX METHOD

A distillation column consists of four equilibrium stages including a partial condenser and a reboiler. The feed stream defined below enters the column as saturated liquid on the second stage from the top. Also given below are the relative volatilities and K_{jb} values at the column pressure of 680 kPa and the indicated stage temperature estimates. The reference component is n-butane. The distillate rate is specified at 48 kmol/h, and the reflux ratio at 2. Using the provided data, calculate the liquid and vapor component rates of propane on each stage, using the tridiagonal matrix method. The reference component is n-butane ($\alpha_{j2} = 1.0$).

		Feed (kmol/h)		Stage, j			
				1	2	3	4
1.	Propane	32	α_{j1}	3.32	3.10	2.90	2.75
2.	n-Butane	29	α_{j2}	1.00	1.00	1.00	1.00
3.	n-Pentane	39	α_{j3}	0.340	0.364	0.387	0.405
	Flow rate	100	K_{jb}	0.49	0.70	0.93	1.21
			T_j (°C)	30	45	60	75

SOLUTION

The liquid rate from stage 1 is calculated from the distillate rate and reflux ratio:

$$L_1 = 2V_1 = (2)(48) = 96 \text{ kmol/h}$$

Since the feed is entering stage 2 as saturated liquid, the liquid rate from stage 2 is calculated as

$$L_2 = L_1 + F = 96 + 100 = 196 \text{ kmol/h}$$

Assuming constant liquid molar overflow,

$$L_3 = L_2 = 196 \text{ kmol/h}$$

By column overall material balance,

$$L_4 = F - V_1 = 100 - 48 = 52 \text{ kmol/h}$$

By total material balance on the condenser,

$$V_2 = V_1 + L_1 = 48 + 96 = 144 \text{ kmol/h}$$

By total material balance on the other stages,

$$V_4 = V_3 = V_2 = 144 \text{ kmol/h}$$

The stripping factors are calculated from the equation

$$S_{jb} = K_{jb} V_j / L_j$$

$S_{1b} = (0.49) \times (48)/96 = 0.245$
$S_{2b} = (0.70) \times (144)/196 = 0.514$
$S_{3b} = (0.93) \times (144)/196 = 0.683$
$S_{4b} = (1.21) \times (144)/52 = 3.351$

These values are used in Equation 13.49 for propane on stages 1 through 4 with R_j^l and R_j^v set to 1, since no side products exist.

$$- [1 + (3.32) \times (0.245)]L_{11} + (3.10) \times (0.514)L_{21} = 0$$

$$L_{11} - [1 + (3.10) \times (0.514)]L_{21} + (2.90) \times (0.683)L_{31} = -32$$

$$L_{21} - [1 + (2.90) \times (0.683)]L_{31} + (2.75) \times (3.351)L_{41} = 0$$

$$L_{31} - [1 + (2.75) \times (3.351)]L_{41} = 0$$

The equations reduce to

$$- 1.8134L_{11} + 1.5934L_{21} = 0$$

$$L_{11} - 2.5934L_{21} + 1.9807L_{31} = -32$$

$$L_{21} - 2.9807L_{31} + 9.2152L_{41} = 0$$

$$L_{31} - 10.2152L_{41} = 0$$

Written in matrix form,

$$\begin{bmatrix} -1.8134 & 1.5934 & 0.0 & 0.0 \\ 1.0 & -2.5934 & 1.9807 & 0.0 \\ 0.0 & 1.0 & -2.9807 & 9.2152 \\ 0.0 & 0.0 & 1.0 & -10.2152 \end{bmatrix} \cdot \begin{bmatrix} L_{11} \\ L_{21} \\ L_{31} \\ L_{41} \end{bmatrix} = \begin{bmatrix} 0.0 \\ -32 \\ 0.0 \\ 0.0 \end{bmatrix}$$

The constants are

$B_{11} = -1.8134$	$C_{11} = 1.5934$	$D_{11} = 0.0$
$B_{21} = -2.5934$	$C_{21} = 1.9807$	$D_{21} = -32$
$B_{31} = -2.9807$	$C_{31} = 9.2152$	$D_{31} = 0.0$
$B_{41} = -10.2152$	$C_{41} = 0.0$	$D_{41} = 0.0$

By forward elimination,

$$p_1 = \frac{C_{11}}{B_{11}} = \frac{1.5934}{-1.8134} = -0.8787$$

$$q_1 = \frac{D_{11}}{B_{11}} = \frac{0}{-1.8134} = 0$$

$$p_2 = \frac{C_{21}}{B_{21} - p_1} = \frac{1.9807}{-2.5934 + 0.8787} = -1.1551$$

$$q_2 = \frac{D_{21} - q_1}{B_{21} - p_1} = \frac{-32 - 0}{-2.5934 + 0.8787} = 18.6622$$

$$p_3 = \frac{C_{31}}{B_{31} - p_2} = \frac{9.2152}{-2.9807 + 1.1551} = -5.0478$$

$$q_3 = \frac{D_{31} - q_2}{B_{31} - p_2} = \frac{0.0 - 18.6622}{-2.9807 + 1.1551} = 10.2225$$

$$q_4 = \frac{D_{41} - q_3}{B_{41} - p_3} = \frac{0.0 - 10.2225}{-10.2152 + 5.0478} = 1.9783$$

The new equations in matrix form,

$$\begin{bmatrix} 1 & -0.8787 & 0 & 0 \\ 0 & 1 & -1.1551 & 0 \\ 0 & 0 & 1 & -5.0478 \\ 0 & 0 & 0 & 1 \end{bmatrix} \cdot \begin{bmatrix} L_{11} \\ L_{21} \\ L_{31} \\ L_{41} \end{bmatrix} = \begin{bmatrix} 0 \\ 18.6622 \\ 10.2225 \\ 1.9783 \end{bmatrix}$$

By backward substitution,

$L_{41} = q_4 = r_4 = 1.9783$ kmol/h
$L_{31} = q_3 - p_3 L_{41} = r_3 = 10.2225 + (5.0478) \times (1.9783) = 20.2086$ kmol/h
$L_{21} = q_2 - p_2 L_{31} = r_2 = 18.6622 + (1.1551) \times (20.2086) = 42.0052$ kmol/h
$L_{11} = q_1 - p_1 L_{21} = r_1 = 0.0 + (0.8787) \times (42.0052) = 36.9100$ kmol/h

The vapor propane rates are calculated by Equation 13.50:

$V_{11} = (3.32) \times (0.245) \times (36.9100) = 30.0226$ kmol/h
$V_{21} = (3.10) \times (0.514) \times (42.0052) = 66.9311$ kmol/h
$V_{31} = (2.90) \times (0.683) \times (20.2086) = 40.0272$ kmol/h
$V_{41} = (2.75) \times (3.351) \times (1.9783) = 18.2305$ kmol/h

13.2.8 STAGE EFFICIENCIES

The column solution methods described in this chapter are based on an equilibrium stage model. Normally, the vapor and liquid leaving a tray are not at equilibrium due to imperfect mixing or insufficient residence time on the tray. Tray efficiencies, discussed in Chapter 14, are parameters that relate actual performance to equilibrium stage performance. By replacing the equilibrium relationship in the model with an equation based on the Murphree tray efficiency, the actual column performance can be calculated. The accuracy of the model obviously depends on the reliability of the tray efficiencies used in the calculations.

The vapor and liquid streams leaving an equilibrium stage are both saturated at the same temperature, the stage temperature. A nonequilibrium stage implies that the saturation temperatures of the vapor and the liquid leaving the stage are not the same. In the model they are assumed to be saturated and must therefore be at different temperatures.

The vapor phase Murphree tray efficiency for tray j is defined as

$$E_{MVj} = \frac{Y_{ji} - Y_{j+1,i}}{K_{ji}X_{ji} - Y_{j+1,i}}$$

where Y_{ji} is the mole fraction of component i in the vapor leaving tray j. This is the actual vapor composition and is not necessarily equal to $K_{ji}X_{ji}$. The efficiency equation, rearranged as follows, is used instead of Equation 13.5 in the column calculations:

$$E_{MVj}K_{ji}X_{ji} - Y_{ji} + (1 - E_{MVj})Y_{j+1,i} = 0$$

In terms of L_{ji}, V_{ji}, and T_j, usually considered the independent variables, this equation becomes (Naphtali and Sandholm, 1971)

$$\frac{E_{MVj}K_{ji}L_{ji}}{L_j} - \frac{V_{ji}}{V_j} + \frac{(1 - E_{MVj})V_{j+1,i}}{V_{j+1}} = 0$$

The energy balance calculations for nonequilibrium stages must be modified since the temperatures of the vapor and liquid leaving a stage are different. Each phase enthalpy must be calculated based on its own temperature.

13.3 CHEMICAL REACTIONS IN MULTISTAGE SEPARATION

Chemical reactions and phase separation can occur simultaneously in multistage separation processes as in reactive distillation and absorption. This phenomenon is found in several operations in the petroleum, chemical, and petrochemical industries (Section 10.3).

One or more reactions could take place in one or more stages in the column, and different reactions could take place in different stages. They could either be equilibrium or kinetically controlled reactions and could occur in either phase. The trays are assumed to act as continuous, well-mixed reactors as well as phase separation devices. For catalytic reactions, a catalyst is placed on the trays where reactions are intended to take place.

To simulate multistage separation processes with reactions, the reaction equations must be included in the model equations. The component balances, Equation 13.2 or 13.49, will include an additional term that represents the rate of generation or disappearance of components by kinetic reactions. Another equation is included in the model to represent composition changes due to equilibrium reactions. The kinetic reaction rate is calculated by a power law expression. A holdup term is computed from tray geometry and fluid hydraulics. The equilibrium conversion is calculated from the equilibrium constant, which may be determined from experimental data or calculated from Gibbs free energies.

For a general reversible reaction, the reaction rate or the rate of disappearance of a reference reactant is expressed as

$$r = k_1 C_{R1}^{r1} C_{R2}^{r2} \cdots - k_2 C_{P1}^{p1} C_{P2}^{p2} \cdots$$

where $C_{R1},\ldots$ and $C_{P1},\ldots$ are the molar concentrations of reactants $R1,\ldots$ and products $P1,\ldots$, and $r_1,\ldots, p_1, \ldots$ are the exponents on the concentrations. These exponents could be the stoichiometric coefficients, but in general they could have other values as determined by the power law function of the reaction rate equation. Parameters k_1 and k_2 are the reaction rate constants for the forward and backward reactions. These constants are temperature-dependent, calculated by the Arrhenius function:

$$k_1 = k_{10} \exp(-E_R/RT)$$
$$k_2 = k_{20} \exp(-E_P/RT)$$

where k_{10} and k_{20} are pre-exponential constants and E_R and E_P are activation energies.

At chemical equilibrium $r = 0$, and the equilibrium constant is

$$K_e = \frac{k_1}{k_2} = \frac{C_{P1e}^{p1} C_{P2e}^{p2} \cdots}{C_{R1e}^{r1} C_{R2e}^{r2} \cdots}$$

where subscripts $P1e$, $P2e$, $\ldots R1e$, $R2e,\ldots$ refer to concentrations at chemical equilibrium. The reaction rate may be written in terms of the forward reaction rate constant and the equilibrium constant:

$$r = k_1 \left[C_{R1}^{r1} C_{R2}^{r2} \cdots - (1/K_e) C_{P1}^{p1} C_{P2}^{p2} \cdots \right] \tag{13.53}$$

A model for multistage separation accompanied by chemical reactions must include a term in the component material balance, Equation 13.2 or 13.49, to represent the change in the number of moles of component i on stage j due to chemical reaction. This term is expressed as

$$M_{Rji} = U_j \nu_i r_j \tag{13.54}$$

where U_j is the volumetric holdup on stage j (liquid holdup if the reaction is in the liquid phase or vapor holdup if it is in the vapor phase), ν_i is the stoichiometric coefficient of component i, and r_j is the rate of a reaction on stage j. The inclusion of M_{Rji} with the reaction rate expression (Equation 13.53) in the component material balance equation completes the reactive separation stage model. The model would obviously still include the standard separation stage model equations of phase equilibrium and energy balance. The reaction equations should be written for any number of reactions that can occur on any stage.

Equation 13.53 can be made to represent reversible reactions, irreversible reactions, and equilibrium reactions. For reversible reactions, k_1 and k_2 are nonzero positive, and for irreversible reactions k_1 is nonzero positive and k_2 is zero $(1/K_e = 0)$. For equilibrium reactions $r = 0$ (and $M_R = 0$), and Equation 13.53 simplifies to the equilibrium form:

$$K_e = \frac{C_{P1}^{p1} C_{P2}^{p2} \cdots}{C_{R1}^{r1} C_{R2}^{r2} \cdots}$$

If the elemental reference state is used to calculate stream enthalpies, no heat of reaction calculation is necessary, and the same energy balance, Equation 13.4 or 13.51, applies.

Several algorithms have been proposed to solve the model equations with chemical reactions (Holland, 1981; Saito et al., 1971; Tierney et al., 1982). Venkataraman et al. (1990) applied the two-tier method for this purpose.

13.4 THREE-PHASE DISTILLATION

Nonideal systems that can result in three-phase distillation are discussed in Section 10.3. This phenomenon, where the liquid splits into two phases, could occur on a number of trays or in the entire column. From a practical standpoint and for better control of the column operation, one of the liquid phases is usually withdrawn when it forms.

The basic three-phase distillation model is similar to the general model described in Section 13.1 and Figure 13.1. The only difference is the potential for the existence of two liquid phases flowing from each tray j to tray $j + 1$, designated as L_j' and L_j'', and two liquid draws from each tray j, designated as $L_j^{s'}$ and $L_j^{s''}$.

The prediction of the performance of three-phase multistage separation processes is dependent on the ability to describe the thermodynamics of three-phase behavior. The mathematical solution of three-phase distillation columns is similar to two-phase vapor–liquid columns, the difference being in the model used to calculate the K-values. If the K-value model predicts two liquid phases, two liquid profiles must be considered in the column instead of one.

By defining a mixed K-value model, programs developed for solving vapor–liquid distillation columns have been successfully modified and used for simulating three-phase distillation (Schuil and Bool, 1985). In this method a mixed K-value is defined as the ratio of the mole fraction of a component in the vapor to its mole fraction in the mixed liquid phase (Section 2.3.3). The column is solved using the mixed K-values instead of the usual vapor–liquid K-values to determine the temperatures, compositions, and flow rates of the vapor and total liquid on all the trays. The liquid phase split is then calculated on the basis of K-values for each liquid phase to determine the compositions and flow rates of the two liquid phases.

13.4.1 HYDROCARBON–WATER SYSTEMS

A special case of three-phase distillation exists in hydrocarbon–water systems when both an organic liquid phase and a water phase are formed. If the hydrocarbon solubility in the water phase is neglected, it is possible to calculate a simplified mixed K-value that, except for water, is not based on liquid activity coefficients. The computation of the mixed K-values for this application is discussed in detail in Section 2.3.3.

13.5 LIQUID–LIQUID EXTRACTION

Liquid–liquid extraction problems constitute a special case of two-phase multistage separation processes and can therefore be solved by many of the general rigorous methods described in Section 13.2. The algorithms can be specifically adapted to extractors to take advantage of the special characteristics of these processes.

Of particular significance is the fact that in liquid–liquid extraction the transfer of components between the two liquid phases does not involve vaporization or condensation or the heats associated with these processes. The heat associated with the transfer of components from one liquid phase to another, as well as the heat of mixing, may be neglected. Thus, if the feeds enter the extractor at the same temperature and if the process is adiabatic, it is practically also isothermal. The column temperature profile may optionally be controlled by adding or removing heat at different stages. In either case, the column temperature profile is known, and an energy balance is not required as an integral part of the algorithm. The heat duties, if required, may be calculated separately once a column solution has been obtained. These assumptions are not necessary for the solution of liquid–liquid extractors but are used in the discussion that follows to illustrate how a column algorithm can be adapted to a special situation.

Another point that should be observed in extraction calculations is the non-ideal nature of the system, which is responsible for the occurrence of two liquid phases in equilibrium. The liquid–liquid equilibrium distribution coefficients, or K-values, are highly composition-dependent and must be calculated by appropriate methods, namely those based on liquid activity coefficients. The NRTL and UNIQUAC liquid activity equations (Chapter 1) are among the more accurate ones for predicting liquid–liquid equilibria. The K-value is defined as the ratio of the mole fraction of a component in one liquid phase to its mole fraction in the other, and is calculated as

$$K_i = \frac{X_i''}{X_i'} = \frac{\gamma_i'}{\gamma_i'}$$

(13.55)

where $X_i', X_i'', \gamma_i', \gamma_i''$ are mole fractions and activity coefficients of component i in the two liquid phases designated as L' and L''.

The modified Thiele–Geddes method described in Section 13.2 is adapted here to liquid–liquid extraction. Reference is made to the steps described in that method.

In step 1 the temperatures are known if the process is considered isothermal. The flow rates of one of the liquid phases are assumed on each stage, and the flow rates of the other liquid phase are calculated by material balance (step 2). In step 3 the K-values may have to be estimated based on assumed compositions of the liquid phases on each tray rather than from a simplified correlation. Steps 4, 5, 6, and 7 are similar to the modified Thiele–Geddes method with the stipulation that liquid activity coefficients be used to calculate K-values by Equation 13.55. Steps 8 and 9 may be skipped if the special extractor model assumptions are used. Step 10 is the convergence check which needs to check only that the material balance is satisfied and that the K-values have stabilized.

One more variable that may require special consideration in extractor calculations is the column pressure. In the rigorous multistage methods the stage pressures are considered fixed. In extractor calculations the pressure profile is also considered fixed, and although its effect on liquid–liquid equilibrium may be neglected, it should be checked to ascertain that it is above the bubble point pressure, and that no vapor phase exists.

13.6　CONVERGENCE BY DYNAMIC ITERATION

In this method, the component material balance, enthalpy balance, and phase equilibrium equations are written for component i on stage j at unsteady state:

$$L_{j-1,i} + V_{j+1,i} - L_{ji} - V_{ji} + F_{ji} - L_{ji}^s - V_{ji}^s = \frac{dm_{ji}}{dt} \tag{13.56}$$

$$L_{j-1}h_{j-1} - (L_j + L_j^s)h_j + V_{j+1}H_{j+1} - (V_j + V_j^s)H_j + F_j H_j^f + Q_j = \frac{d(m_j C_{Pj} T_j)}{dt} \tag{13.57}$$

$$\frac{V_{ji}/V_j}{L_{ji}/L_j} = K_{ji} \tag{13.58}$$

Equation 13.56 is the unsteady-state counterpart of Equations 13.1 and 13.2 combined. The term on the right-hand side of Equation 13.56 is the rate of change of the moles of component i on stage j. In this equation, m_{ji} is the molar holdup of component i in both phases on stage j at time t. Equation 13.57 is the unsteady state form of Equation 13.4. In Equation 13.57, m_j, the sum of m_{ji}, is the total moles in both phases on stage j, and C_{Pj} is an average heat capacity for both phases. Thus, the term on the right-hand side of Equation 13.57 represents the rate of change of the total energy on stage j. The equilibrium relationship is written in its usual, steady-state from, Equation 13.58, since phase equilibrium is assumed to exist at all times.

Although the fundamental equations are written for unsteady-state conditions, the computational method presented here is not concerned with a quantitative prediction of the transient performance of the column. The unsteady-state analysis is merely used to define a convergence path that corresponds to the transition from unsteady- to steady-state conditions. As the column moves toward steady state, the terms on the right-hand side of Equations 13.56 and 13.57 approach zero and the equations reduce to steady-state relationships. Thus, reaching steady state is equivalent to reaching a converged solution. This is commonly referred to as the relaxation method. It is rigorous, in that no simplifying assumptions are used.

The method has been applied to absorbers (Khoury, 1980), where all heat duties, side draws, and all feeds except the liquid feed at the top of the column and the vapor feed at the bottom are set to zero. The rates of change of the component molar holdup and total energy on a stage may be broken down into liquid and vapor contributions (subscripts j and i are dropped in the rates of change terms for simplicity):

$$-\frac{dm_{ji}}{dt} = \dot{M}_L + \dot{M}_V$$

and

$$-\frac{d(m_j C_{Pj} T_j)}{dt} = \dot{h} + \dot{H}$$

Equations 13.56 and 13.57, applied to the absorber model, are rearranged as follows:

$$L_{ji} + V_{ji} = L_{j-1,i} + \dot{M}_L + V_{j+1,i} + \dot{M}_V \tag{13.59}$$

$$L_j h_j + V_j H_j = L_{j-1} h_{j-1} + \dot{h} + V_{j+1} H_{j+1} + \dot{H} \tag{13.60}$$

If two successive iterations are made to simulate two successive points in time in the approach to steady state and if quantities with superscript k represent iteration k, the following expressions will hold:

$$(L_{j-1,i} + \dot{M}_L)^k = (L_{j-1,i})^{k-1}$$

$$(V_{j+1,i} + \dot{M}_V)^k = (V_{j+1,i})^{k-1}$$

$$(L_{j-1} h_{j-1} + \dot{h})^k = (L_{j-1} h_{j-1})^{k-1}$$

$$(V_{j+1} H_{j+1} + \dot{H})^k = (V_{j+1} H_{j+1})^{k-1}$$

Substitution of these expressions in Equations 13.59 and 13.60 yields

$$(L_{ji})^k + (V_{ji})^k = (L_{j-1,i})^{k-1} + (V_{j+1,i})^{k-1} \tag{13.61}$$

$$(L_j h_j)^k + (V_j H_j)^k = (L_{j-1} h_{j-1})^{k-1} + (V_{j+1} H_{j+1})^{k-1} \tag{13.62}$$

As in other methods, the calculations are started with an assumed temperature and vapor profile, T_j and V_j, from which the liquid profile L_j and the K_{ji}s are estimated. The tridiagonal matrix, Equation 13.24, is then solved only at the start to generate initial estimates for the component liquid and vapor flows, L_{ji} and V_{ji}. Following this, liquid and vapor enthalpies, h_j and H_j, are calculated for all the stages.

The calculated values are used as initial estimates corresponding to iteration $k-1 = 0$. With known quantities on the right-hand side, Equations 13.61 and 13.62 are written for the first iteration ($k = 1$) as follows:

$$L_{ji} + V_{ji} = C_j \tag{13.63}$$

$$L_j h_j + V_j H_j = E_j \tag{13.64}$$

where C_j and E_j are evaluated from initial estimates. The equations are grouped by stage: Equations 13.63, 13.64, and the equilibrium relationship, Equation 13.58, are solved simultaneously for each stage. This grouping of equations is equivalent to single equilibrium stage calculations, which may be solved by methods described in Chapter 2. The solutions generate new values for C_j and E_j in Equations 13.63 and 13.64 for the next iteration. At this point the component flow rates may be adjusted, if necessary, to meet overall column material balance. The procedure is repeated until the maximum variation of C_j and E_j from iteration to iteration is within an acceptable tolerance.

The method may be generalized to other column configurations. Various computational techniques may be incorporated in the algorithm to reduce computing time, such as skipping rigorous compositional K-value calculations on iterations where the change in composition is not large. The method is stable, although convergence may tend to slow down as the solution is approached. For this reason, once the column is stable and close to solution, the computations may be switched over to one of the other methods for final convergence.

EXAMPLE 13.2: ABSORBER CONVERGENCE BY DYNAMIC ITERATION

A three-stage absorber recovers isobutane from an ethane/isobutane mixture using n-hexane as the absorbent. The column operates at constant temperature and pressure, 80°C and 125 kPa. The feed and absorbent streams are defined below, along with the K-values, which may be assumed constant. Solve the column by the dynamic iteration method to determine the component flow rates in the vapor and liquid on all three stages.

	Feed, V_4 (kmol/h)	Absorbent, L_0 (kmol/h)	K-values
Ethane	95.0	0.0	18.0
Isobutane	5.0	0.0	1.5
n-Hexane	0.0	50.0	0.1

SOLUTION

The iterative path is shown in Figure 13.2. The liquid and vapor feeds to all stages on iteration 0 (initial estimates) are assumed equal to the external liquid and vapor feeds. Thus on iteration 0, $V_2 = V_3 = V_4$, and $L_2 = L_1 = L_0$. The combined feeds on each stage are flashed, using the given K-values, to generate vapor and liquid streams for the next iteration. The computations may be represented symbolically as follows, with superscripts designating the iteration at which a stream was generated:

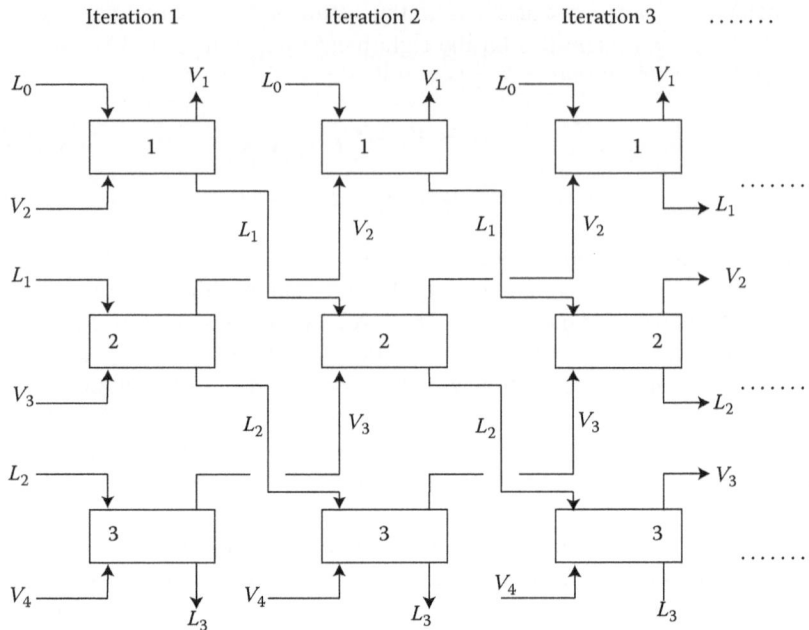

FIGURE 13.2 Iterative path for Example 13.2.

Iteration 0:

$$V_2^0 = V_3^0 = V_4$$

$$L_2^0 = L_1^0 = L_0$$

Iteration 1:

$$L_0 + V_2^0 \rightarrow V_1^1 + L_1^1$$

$$L_1^0 + V_3^0 \rightarrow V_2^1 + L_2^1$$

$$L_2^0 + V_4 \rightarrow V_3^1 + L_3^1$$

Iteration 2:

$$L_0 + V_2^1 \rightarrow V_1^2 + L_1^2$$

$$L_1^1 + V_3^1 \rightarrow V_2^2 + L_2^2$$

$$L_2^1 + V_4 \rightarrow V_3^2 + L_3^2$$

$$\vdots$$

The computations stabilized after six iterations. The results for selected iterations are summarized in the following table:

	Stage 1				Stage 2				Stage 3			
	Inlets		Outlets		Inlets		Outlets		Inlets		Outlets	
	L_{0i}	V_{2i}	L_{1i}	V_{1i}	L_{1i}	V_{3i}	L_{2i}	V_{2i}	L_{2i}	V_{4i}	L_{3i}	V_{3i}
Iteration 1												
C2	0	95	2.092	92.908	0	95	2.092	92.908	0	95	2.092	92.908
iC4	0	5	1.064	3.936	0	5	1.064	3.936	0	5	1.064	3.936
nC6	50	0	40.104	9.895	50	0	40.104	9.895	50	0	40.104	9.895
Iteration 2												
C2	0	92.908	2.633	90.274	2.092	92.908	2.092	92.908	2.092	95	1.553	95.538
iC4	0	3.936	1.021	2.916	1.064	3.936	1.064	3.936	1.064	5	0.990	5.074
nC6	50	9.895	50.314	9.582	40.104	9.895	40.104	9.895	40.104	0	29.891	10.214
...												
Iteration 6												
C2	0	95.548	2.629	92.919	2.630	95.536	2.619	95.548	2.619	95	2.084	95.535
iC4	0	4.757	1.206	3.551	1.162	5.159	1.564	4.757	1.564	5	1.362	5.203
nC6	50	10.203	50.321	9.882	50.318	10.216	50.332	10.203	50.332	0	40.114	10.218

13.7 COLUMN DYNAMICS

Multistage separation columns will operate at unsteady-state conditions during startup or shutdown, or when any of the operating variables change. While the condition of steady-state operation is a basic model assumption for most of the solution methods, it is an assumption that represents an operation that in reality may apply only to limited periods of time, in which steady-state conditions actually prevail. As column conditions change with time, a new steady-state solution will be required. Whereas steady-state models can simulate the column performance at a point in time, dynamic models can simulate the column performance on a continuous time basis.

It is important to model column dynamics, the variation of column conditions with time, for a variety of reasons. In the design phase, the simulation of column dynamics provides information about transient conditions. The column must be designed for anticipated column operating conditions, including those that can occur during transitions resulting from changes in operating conditions, and not only for steady-state conditions. Changes in operating conditions include changes in feed flow rate, composition, or properties, changes in product specifications, and so on. In operations, simulation of column dynamics provides a tool for predicting transient conditions, allowing operators to take necessary measures to prevent undesirable conditions during the transition, such as flooding. It also provides information for developing control strategies.

Dynamic simulation may be used for off-line or on-line applications. An off-line dynamic model runs independent of the plant. This is a predictive mode in which a column or an entire process is simulated to predict transient behavior with no input from the plant. Such a model is typically used for the design of equipment and control strategies and as a training simulator. An on-line dynamic model may be used to monitor the column performance and to provide vital information for the control system. It reads current plant conditions and, in real time, computes properties that are not measured on-line, such as product compositions. This makes it possible to control such properties directly. The on-line dynamic simulator can also predict future column trends, thereby allowing the control system to take corrective action in advance.

Steady-state simulators may also be used on-line, but they are accurate only when the actual column is running at steady state. Consider a situation where measured feed and product rates (along with other variables such as reflux rate, temperatures, and pressures) of a distillation column are sent regularly at small time intervals to a steady-state simulator. Based on the information it receives, the model predicts instantaneous product compositions. If a plant variable such as the reflux rate changes, the model predicts an instant change in the column compositions, while in reality a time lag exists between the two events, and the compositions would change gradually. Moreover, an on-line steady-state model may not reach a feasible solution during transient conditions: The plant data the model receives may not be steady-state data and may not be in material or energy balance, while the steady-state model must satisfy these conditions.

13.7.1   DYNAMIC MODEL DEFINITION

The basic equations of column dynamics were introduced in Section 13.6 in the context of developing a convergence strategy for a steady-state solution. For a full representation of column dynamics, these equations are rewritten here in a slightly modified form and additional equations are introduced to complete the model. The equations refer specifically to a column tray with liquid and vapor holdups. The vapor holdup is negligible compared to the liquid holdup and is usually omitted from the equations. The liquid and vapor on the tray are each assumed of uniform composition.

A component material balance is written for component i on tray j as follows:

$$F_{ji} + L_{j-1,i} + V_{j+1,i} - (L_{ji} + L_{ji}^s) - (V_{ji} + V_{ji}^s) = \frac{dM_{ji}}{dt} \qquad (13.65)$$

where M_{ji} is the molar liquid holdup of component i on tray j and t is the time. The total material balance on tray j is written as

$$F_j + L_{j-1} + V_{j+1} - (L_j + L_j^s) - (V_j + V_j^s) = \frac{dM_j}{dt} \qquad (13.66)$$

where M_j is the total molar liquid holdup on tray j. The summation equation may be written either for the vapor or the liquid:

$$\sum_i V_{ji} = V_j \quad \text{or} \quad \sum_i L_{ji} = L_j \qquad (13.67)$$

By energy balance on tray j,

$$F_j H_j^F + L_{j-1} h_{j-1} + V_{j+1} H_{j+1} - (L_j + L_j^s) h_j - (V_j + V_j^s) H_j + Q_j = \frac{d(M_j u_j)}{dt}$$

$$(13.68)$$

where u_j is the molar internal energy of the liquid on the tray.

The phase equilibrium equation is combined with the vaporization efficiency to define the relationship between phase compositions:

$$Y_{ji} = E_{ji} K_{ji} X_{ji}$$

or

$$L_{ji} = (L_j / E_{ji} K_{ji} V_j) V_{ji} \qquad (13.69)$$

The vaporization efficiency is defined as

$$E_{ji} = Y_{ji}/Y_{ji}^*$$

where $Y_{ji}^* = K_{ji}X_{ji}$, the equilibrium vapor composition.

The molar liquid holdups are related to the volumetric liquid holdup by the following equations:

$$M_j = U_j\rho_j^L$$

$$M_{ji} = U_j\rho_j^L X_{ji} = U_j\rho_j^L(L_{ji}/L_j) \tag{13.70}$$

$$M_j = \sum_i M_{ji} \tag{13.71}$$

where U_j is the total volumetric liquid holdup on tray j and ρ_j^L is the liquid molar density on tray j. The volumetric holdup is computed from the tray dimensions and tray hydraulics (Chapter 14) and is a function of the liquid and vapor flow rates within the column, and the tray temperature and pressure.

The time in the derivatives is approximated by a time increment set outside the model and is not considered a model variable. The volumetric liquid holdups are also computed or determined outside the model and are not considered as model variables. Quantities including external feed rates and compositions, K-values, vaporization efficiencies, pressures, molar enthalpies and densities are also determined outside the model. Accordingly, the model variables for N trays and C components are the following:

L_{ji}	NC variables
L_{ji}^s	NC variables
V_{ji}	NC variables
V_{ji}^s	NC variables
L_j	N variables
L_j^s	N variables
V_j	N variables
V_j^s	N variables
T_j	N variables
Q_j	N variables
M_{ji}	NC variables
M_j	N variables

The heat duties Q_j and the side draw rates, L_j^s and V_j^s, are assumed fixed. If they are allowed to vary, other parameters must be fixed to define the column operation. The component side draw rates, L_{ji}^s and V_{ji}^s, are determined by the side draw rates and tray compositions,

$$X_{ji} = L_{ji}/L_j$$
$$Y_{ji} = V_{ji}/V_j$$

The number of remaining independent variables is $3NC + 4N$.

The model equations include differential equations (13.65, 13.66, and 13.68) and algebraic equations (13.67, 13.69, 13.70, and 13.71). The numbers of these model equations are as follows:

Equation 13.65	NC equations
Equation 13.66	N equations
Equation 13.67	N equations
Equation 13.68	N equations
Equation 13.69	NC equations
Equation 13.70	NC equations
Equation 13.71	N equations
Total	$3NC + 4N$ equations

With an equal number of equations and independent variables, the model is completely defined.

Since any tray (or stage) can have a heat duty and side draws, there is no need for special consideration for condensers and reboilers, except with regards to hydraulics and liquid holdups, which are calculated outside the model.

13.7.2 SOLVING THE DYNAMIC MODEL EQUATIONS

The differential and algebraic equations that make up the dynamic distillation model are solved by numerical methods. If the differential equations cannot be solved analytically, they are first approximated in a number of ways as described below. For convenience, the left hand sides of Equations 13.65, 13.66 and 13.68 are abbreviated and represented simply as functions of the model variables v_j and v_{ji}:

$$f_c(v_{ji}) = dM_{ji}/dt \tag{13.72}$$

$$f_t(v_j) = dM_j/dt \tag{13.73}$$

$$f_e(v_j) = d(M_j u_j)/dt \tag{13.74}$$

where f_c, f_t, and f_e represent the component and total material, and energy balance equations. Following is a brief discussion of a number of differential equation solution methods.

In order to demonstrate the methods, a simple model is used, consisting of a tank with a constant-flow rate liquid stream feed. At the bottom of the tank the liquid exits in a stream whose flow rate is proportional to the liquid height in the tank. The tank being a vertical cylinder with constant cross-sectional area, the exit stream flow rate is also proportional to the liquid holdup in the tank. It is required to determine the holdup in the tank and the exit stream flow rate as a function of time.

At time $t = 0$, the holdup $M^0 = 100$ kmol. The feed flow rate is constant, $F = 10$ kmol/min. The exit stream flow rate, L, is proportional to the holdup, $L = cM$, where $c = 0.2$ min^{-1}. It follows that at $t = 0$, $L^0 = 0.2 \times 100 = 20$ kmol/min. Applying the equivalent of Equation 13.66 to this model, the material balance is expressed as

$$\frac{dM}{dt} = F - L = F - cM, \quad dt = \frac{dM}{F - cM}$$

This model being simple enough, its differential equation can be solved analytically:

$$t = \int_{M^0}^{M} \frac{dM}{F - cM} = \frac{1}{k} \ln\left(\frac{F - cM^0}{F - cM}\right)$$

Rearranging,

$$M = \frac{F}{c} + e^{-kt}\left(M^0 - \frac{F}{c}\right) = \frac{10}{0.2} + e^{-0.2t}\left(100 - \frac{10}{0.2}\right) = 50(1 + e^{-0.2t})$$

Thus M, and $L = cM$, are determined at any time t. The following are the results at time increments of 1 min, over a span of 5 minutes:

t (min)	0	1	2	3	4	5
M (kmol)	100	90.937	83.516	77.441	72.466	68.394
L (kmol/min)	20	18.187	16.703	15.488	14.493	13.679

Although this model can be solved analytically, the same model will be used here to demonstrate some numerical methods.

13.7.2.1 Euler's Method

The differential equations (Equations 13.72, 13.73, 13.74) are approximated by the following algebraic equations:

$$M_{ji} = M_{ji}^0 + \Delta t f_c(v_{ji}^0) \tag{13.75}$$

$$M_j = M_j^0 + \Delta t f_t(v_j^0) \tag{13.76}$$

$$M_j u_j = (M_j u_j)^0 + \Delta t f_e(v_j^0) \tag{13.77}$$

The superscript 0 refers to conditions at the beginning of a time increment, at which point the values of the variables are known or have been calculated at the end of the previous time increment. The above equations together with the other algebraic equations (13.67, 13.69, 13.70, 13.71) are solved to determine the values of the variables at the end of the current time interval. Methods similar to those described in Section 13.2 may be used to solve the equations.

Applied to the tank model described above,

$$\frac{dM}{dt} = f(M) = F - cM$$

The holdup at any point in time is approximated as

$$M = M^0 + \Delta t f(M^0) = M^0 + \Delta t(F - cM^0) = M^0(1 - c\Delta t) + F\Delta t$$

At $t = 0$ min, $M^0 = 100$ kmol
At $t = 1$ min, $M = 100(1 - 0.2) + 10 = 90$ kmol
At $t = 2$ min, $M = 90(1 - 0.2) + 10 = 82$ kmol
At $t = 3$ min, $M = 82(1 - 0.2) + 10 = 75.6$ kmol
At $t = 4$ min, $M = 75.6(1 - 0.2) + 10 = 70.48$ kmol
At $t = 5$ min, $M = 70.48(1 - 0.2) + 10 = 66.384$ kmol

The results compare fairly well with the analytical solution, and can be improved by using smaller time increments.

The Euler method is suitable for on-line applications where plant measurements can be sent to the model at the beginning of each time increment.

13.7.2.2 Two-Point Implicit Method

This method is predictive and may be used off-line or on-line. The differential equations are approximated as follows:

$$\left[\phi f_c(v_{ji}) + (1 - \phi)f_c(v_{ji}^0)\right]\Delta t = M_{ji} - M_{ji}^0$$

$$\left[\phi f_t(v_j) + (1 - \phi)f_t(v_j^0)\right]\Delta t = M_j - M_j^0$$

$$\left[\phi f_e(v_j) + (1 - \phi)f_e(v_j^0)\right]\Delta t = M_j u_j - (M_j u_j)^0$$

where ϕ is a weighting factor, $0 \leq \phi \leq 1$, whose value, typically around 0.6, can be tuned for best representation of the process dynamics. With $\phi = 0$, this method reduces to Euler's method. The above equations are solved with the remaining algebraic model equations to calculate the values of the variables at the end of a time increment based on known values at its beginning.

13.7.2.3   Runge-Kutta Method

In the fourth-order Runge-Kutta method, a set of formulas are used to calculate the model variables at the end of a time increment from their values at its beginning. Applied to the component material balance equations, the calculations proceed as follows:

$$M_{ji} = M_{ji}^0 + (k_{1ji} + 2k_{2ji} + 2k_{3ji} + k_{4ji})/6$$

The k factors in this equation are evaluated from the model variable values at the beginning of the time increment. For component material balances,

$$k_{1ji} = \Delta t f_c(v_{ji}^0)$$

$$k_{2ji} = \Delta t f_c(v_{ji}^0 + k_{1ji}/2)$$

$$k_{3ji} = \Delta t f_c(v_{ji}^0 + k_{2ji}/2)$$

$$k_{4ji} = \Delta t f_c(v_{ji}^0 + k_{3ji})$$

Similar equations are written for the total material and energy balance equations. These equations are solved with the model equations to compute the variables at the end of each time increment. The method can be used off-line or on-line in a predictive mode.

Using the simple tank model to illustrate the method, start by calculating the k factors:

$$k_1 = \Delta t \cdot f(M^0) = \Delta t(F - cM^0) = 1(10 - 0.2 \times 100) = -10$$

$$k_2 = \Delta t \cdot f(M^0 + k_1/2) = \Delta t \cdot f(100 - 10/2) = \Delta t \cdot f(95) = 1(10 - 0.2 \times 95) = -9$$

$$k_3 = \Delta t \cdot f(M^0 + k_2/2) = \Delta t \cdot f(100 - 9/2) = \Delta t \cdot f(95.5) = 1(10 - 0.2 \times 95.5) = -9.1$$

$$k_4 = \Delta t \cdot f(M^0 + k_3) = \Delta t \cdot f(100 - 9.1) = \Delta t \cdot f(90.9) = 1(10 - 0.2 \times 90.9) = -8.18$$

Next, the holdup at $t = 1$ min is calculated:

$$M = 100 - [10 + (2 \times 9) + (2 \times 9.1) + 8.18]/6 = 90.937 \text{ kmol}$$

This is followed by taking 90.937 kmol as the starting point for the next time increment, and completing the procedure as above. The results are tabulated herewith:

t (min)	0	1	2	3	4	5
M (kmol)	100	90.937	83.516	77.441	72.467	68.394

These results exactly match the analytical values. Such accuracy, however, may not be achievable with more complex models.

EXAMPLE 13.3: DYNAMICS OF AN EQUILIBRIUM STAGE

This example is intended to demonstrate the process dynamics methodology as implemented on a single equilibrium stage. A stream of light hydrocarbons is sent to a distillation column where the C_3's and lighter components are separated from the C_4's. Since the feed composition fluctuates substantially, it is sent to a flash drum located upstream of the column in order to attenuate the composition fluctuations and thereby improve the column controllability. The vapor and liquid products from the flash drum are then sent to different trays in the column.

It is required to evaluate the flash drum dynamics to help in the design of the column control strategy. In one particular test, the feed composition went through a step change while its flow rate and the drum temperature and pressure remained constant at 70°C and 1500 kPa. The steady-state conditions of the flash drum before and after the step change are given below. The K-values are also given and may be assumed composition-independent.

Assuming a completely mixed and constant liquid molar holdup of 50 kmol in the drum and neglecting the vapor holdup, determine the variation with time of the flash drum product rates and compositions between the initial and final states.

Initial Steady State:

	K_i	kmol/min		kmol/min		kmol/min	
		F_i	Z_i	L_i	X_i	V_i	Y_i
C_2H_6	4.01	1	0.01	0.0235	0.0027	0.9765	0.0107
C_3H_6	1.88	27	0.27	1.3181	0.1498	25.6819	0.2816
C_3H_8	1.60	14	0.14	0.7963	0.0905	13.2037	0.1448
iC_4H_{10}	0.88	19	0.19	1.8775	0.2133	17.1225	0.1877
nC_4H_{10}	0.69	39	0.39	4.7847	0.5437	34.2153	0.3752
Totals		100	1.00	8.8001	1.00	91.1999	1.00

Final Steady State:

	K_i	kmol/min		kmol/min		kmol/min	
		F_i	Z_i	L_i	X_i	V_i	Y_i
C_2H_6	4.01	6	0.06	3.2063	0.0390	2.7939	0.1565
C_3H_6	1.88	7	0.07	4.9698	0.0605	2.0302	0.1137
C_3H_8	1.60	8	0.08	5.9362	0.0723	2.0638	0.1156
iC_4H_{10}	0.88	22	0.22	18.4686	0.2248	3.5314	0.1978
nC_4H_{10}	0.69	57	0.57	49.5684	0.6034	7.4316	0.4164
Totals		100	1.00	82.1493	1.00	17.8507	1.00

SOLUTION

The stage dynamics are calculated based on the differential and algebraic equations described above. The method for solving these equations is adapted to the present problem with its stated assumptions. Since $X_i = M_i/M$, $Y_i = K_i M_i/M$, and $L = F - V$, Equation 13.65 may be reformulated as

$$\frac{dM_i}{dt} = F_i - LX_i - VY_i = F_i - (F/M)M_i + (V/M)(1 - K_i)M_i \quad (13.78)$$

Assuming $E_{ji} = 1$ (i.e., assuming phase equilibrium exists all the time), the equivalent of Equations 13.67 and 13.69 may be written as

$$\sum (1 - K_i)X_i = 0$$

or, by substituting $X_i = M_i/M$,

$$\sum (1 - K_i)M_i = 0$$

Since the K_is are constant (composition-independent) the following is also true:

$$\sum (1 - K_i)\frac{dM_i}{dt} = 0$$

Equation 13.78 is multiplied by $(1 - K_i)$ and summed over all the components to give

$$\sum (1 - K_i)\frac{dM_i}{dt} = \sum (1 - K_i)\left(F_i - \frac{FM_i}{M} \right) + \sum (1 - K_i)^2 \frac{VM_i}{M} = 0$$

Solving for V,

$$V = -\frac{M\sum (1 - K_i)(F_i - FM_i/M)}{\sum (1 - K_i)^2 M_i} \qquad (13.79)$$

Using the Euler method for solving the differential equations, the component holdups at the end of each time interval Δt are calculated as (Equation 13.75)

$$M_i = M_i^0 + \Delta t (dM_i/dt)^0 \qquad (13.80)$$

where superscript 0 refers to values at the beginning of the time interval which will be set at $\Delta t = 0.1$ min. The calculations start with the initial state conditions for X_i, Y_i, M_i, V, and L, and the new feed composition at $t = 0+$. The procedure is repeated for subsequent time intervals until the results stabilize at the final steady-state conditions. The first time interval calculations are as follows:

1. At $t = 0+$, $F = 100$ kmol/min

   $M = 50$ kmol
   $\Delta t = 0.1$ min

   Calculate V, the vapor flow rate at $t = 0+$

   $$V = -M\frac{\sum A_i}{\sum B_i}$$

   where

   $$A_i = (1 - K_i)(F_i - FM_i^0/M)$$

$$B_i = (1 - K_i)^2 M_i^0$$

The feed component flow rates F_i in these equations are the values at $t = 0+$, after the step change in the feed composition. In the first time interval calculations, M_i^0 is calculated as MX_i^0, where X_i^0 are the liquid component mole fractions at the initial steady-state conditions.

2. Calculate the derivatives $(dM_i/dt)^0$ from Equation 13.78, with $M_i = M_i^0$.
3. Calculate M_i from Equation 13.80.
4. Calculate the compositions and component flow rates of the vapor and liquid at the end of the time interval:

$$X_i = M_i/M, \quad Y_i = K_i X_i, \quad L_i = LX_i, \quad V_i = VY_i$$

Partial results of the first time interval are summarized below:

	K_i	F_i (kmol/min)	X_i^0	M_i^0 (kmol)	$(dM_i/dt)^0$	M_i (kmol)
C_2H_6	4.01	6	0.0027	0.135	5.410	0.6760
C_3H_6	1.88	7	0.1498	7.490	−12.965	6.1935
C_3H_8	1.60	8	0.0905	4.525	−3.130	4.2120
iC_4H_{10}	0.88	22	0.2133	10.665	1.670	10.8320
nC_4H_{10}	0.69	57	0.5437	27.185	9.080	28.0930

The product flow rates at $t = 0+$, $V = 38.1935$ kmol/min, $L = 100.0 - 38.1935 = 61.8065$ kmol/min.

Some of the dynamic trends are plotted in Figure 13.3.

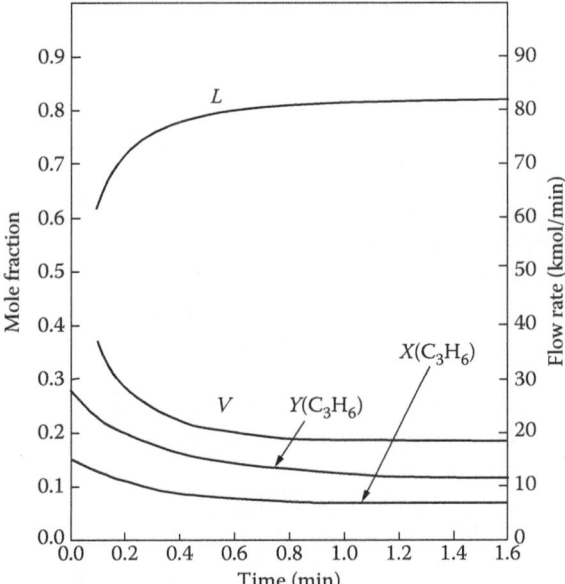

FIGURE 13.3 Equilibrium stage dynamics (Example 13.3).

NOMENCLATURE

A_{ji}	Parameter defined by Equation 13.22
B_{ji}	Parameter defined by Equation 13.23
C	Total number of components
C_j	Component balance function on tray j
C_{pj}	Average heat capacity on tray j
E_j	Energy balance function on tray j
E_{MVj}	Murphree tray efficiency in vapor terms
f	Fugacity
F_j	Molar rate of external feed to tray j
F_{ji}	Molar rate of component i in external feed to tray j
g	General function representing column equations
G	General column specification
h	Liquid molar enthalpy
H	Vapor molar enthalpy
K	Vapor–liquid equilibrium distribution coefficient
L_j	Liquid molar flow rate from tray j
L_{ji}	Liquid molar flow rate of component i from tray j
m_j, M_j	Total molar holdup on tray j
m_{ji}, M_{ji}	Holdup of component i on tray j
N	Number of theoretical stages
P_j	Pressure on tray j
Q_j	Heat duty on tray j
R	Reflux ratio
S	Stripping factor
S_j	Summation function on tray j
t	Time
T_j	Temperature on tray j
u_j	Liquid molar internal energy on tray j
U_j	Total volumetric liquid holdup on tray j
v	Model variable
V_j	Vapor molar flow rate from tray j
V_{ji}	Vapor molar flow rate of component i from tray j
w	General column variable
X_{ji}	Mole fraction of component i in the liquid on tray j
X_s	Solubility of water in hydrocarbon phase
X_w	Mole fraction water
Y_{ji}	Mole fraction of component i in the vapor on tray j
Z_{ji}	Mole fraction of component i in the feed to tray j
α	Adjustment factor
α	Ratio of one liquid phase to total liquid
α_{ji}	Relative volatility
γ	Activity coefficient
θ	Correction factor
ϕ	Fugacity coefficient

SUBSCRIPTS

b	Base component designation
f	Feed tray designation
i	Component designation
j	Tray designation
n	Specification number
r	Reference component designation
w	Designation of water phase or component water

SUPERSCRIPTS

F, f	Designation of initial thermal conditions of external feed
k	Iteration number designation
s	Side product designation

PROBLEMS

13.1. A three-stage absorber removes n-pentane from a hydrocarbon stream by absorption with n-decane. The column is maintained at a uniform temperature and pressure, 40°C and 2750 kPa. The feed streams are given below, along with estimated K-values at the column conditions. It is required to do a preliminary evaluation of the column performance.

	L_{0i} (kmol/h)	V_{4i} (kmol/h)	K_i
1. C2	0.10	350	1.84
2. nC5	1.00	50	0.08
3. nC10	98.90	0	0.00024

a. Calculate the rates and compositions of the products and the liquid and vapor on each stage using the dynamic iteration method.

b. As an alternative to the method above, calculate the product rates and compositions using the modular shortcut method (the Kremser equation). Initially use the liquid and vapor inlet stream rates to calculate the stripping factors. Compare the results with the dynamic iteration method. Explain any discrepancy and suggest strategy for improving the modular method.

c. Calculate the liquid and vapor component flow rates on each stage by solving the tridiagonal matrix. For the total liquid and vapor rates assume values equal to the liquid and vapor feed rates. Consider pentane as the base component.

13.2. A stream is sent to the third stage from the top of a five-stage distillation column with a partial condenser and reboiler. The stream is saturated liquid, and is defined as follows:

	i	f_i (kmol/h)	α_{i2}
nC_4	1	37	1.5
iC_5	2	42	1.0
nC_5	3	21	0.7

The distillate rate is 40 kmol/h and the reflux ratio is 2. The estimated K-values of iC_5 on each stage from the top are 0.68, 0.72, 0.77, 0.83, and 0.89. The given relative volatilities may be assumed constant on all the stages. In order to start a rigorous solution of the column, it is required to calculate initial estimates of the component flow rates in the liquid and vapor on each stage. Calculate these values for nC5 from the given information using the tridiagonal matrix method.

13.3. The following equations represent material balances on the liquid flow rates of one of the components at the five stages of a distillation column. Use the Thomas algorithm to solve for L_1, L_2, L_3, L_4, and L_5, the liquid flow rates of the component on trays 1, 2, 3, 4, and 5. These flow rates could constitute part of the initial estimates required for starting the rigorous column solution. The feed rate is 45 kmol/h, sent to stage 3.

$$-5L_1 + 4L_2 = 0$$

$$2L_1 - 3L_2 + 1.2L_3 = 0$$

$$L_2 - 3L_3 + 2L_4 = 45$$

$$3.5L_3 - 4.5L_4 + 5L_5 = 0$$

$$3L_4 - 7.6L_5 = 0$$

13.4. A situation existed where distillation column tray temperatures were to be evaluated based on energy balances on the trays. The following equations were developed to express the energy balances on two adjacent trays in a column equipped with side-heaters in terms of reduced tray temperatures, T_{r1} and T_{r2}:

$$E_1 = 3(T_{r1})^{2.5} - 1.5(T_{r2})^3 - 0.78 = 0$$

$$E_2 = 2(T_{r1})^{1.8} - 1.5(T_{r2})^{1.1} - 0.92 = 0$$

Solve these equations using the Newton–Raphson method.

REFERENCES

Boston, J. F. and S. L. Sullivan, Jr., An improved algorithm for solving mass balance in multistage separation processes, *Can. J. Chem. Eng.*, 50, 663, 1972.

Brannock, N. F., V. S. Verneuil, and Y. L. Wang, Rigorous distillation simulation, *Chem. Eng. Progress*, October, 83, 1977.

Broyden, C. G., *Mathematics of Computations*, 19, 577, 1965.

Carnahan, B., H. A. Luther, and J. O. Wilkes, *Applied Numerical Methods*, New York, John Wiley and Sons, 1964.

Chen, C. C., H. I. Britt, J. F. Boston, and L. B. Evans, *AIChE J.*, 28, 588, 1982.

De Leye, L. and G. F. Froment, *Computers and Chemical Engineering*, 10(5), 493, 1986.

Holland, C. D., *Fundamentals and Modeling of Separation Processes*, Englewood Cliffs, NJ, Prentice-Hall, 1975.

Holland, C. D., *Fundamentals of Multicomponent Distillation*, New York, McGraw-Hill, 1981.

Holland, C. D., *Multicomponent Distillation*, Englewood Cliffs, NJ, Prentice-Hall, 1963.

Khoury, F. M., Simulate absorbers by successive iteration, *Chemical Engineering*, December, 51, 1980.

Lyster, W. N., S. L. Sullivan, Jr., D. S. Billingsley, and C. D. Holland, *Petrol. Refiner*, 38(6), 221, 1959; 38(7), 151, 1959; 38(10), 139, 1959.

Naphtali, L. M. and D. P. Sandholm, Multicomponent separation calculations by linearization, *AICHE J.*, 17(1), 148, 1971.

Russell, R. A., A flexible and reliable method solves single-tower and crude-distillation-column problems, *Chem. Eng.*, October, 53–59, 1983.

Saito, S., T. Michishita, and S. Maeda, *J. Chem. Eng. Jpn*, 4, 37, 1971.

Schuil, J. A. and K. K. Bool, Three-phase flash and distillation, *Computers and Chemical Engineering*, 9(3), 295, 1985.

Thiele, E. W. and R. L. Geddes, *Ind. Eng. Chem.*, 25, 289, 1933.

Tierney, J. W. and G. D. Riquelme, *Chem. Eng. Comm*, 16, 91, 1982.

Tomich, J. F., A new simulation method for equilibrium stage processes, *AICHE J.*, 16(2), 229, 1970.

Venkataraman, S., W. K. Chan, and J. F. Boston, *Chem. Eng. Progress*, 86(8), 45, 1990.

Wang, J. C. and C. E. Henke, Tridiagonal matrix for distillation, *Hydrocarbon Processing*, 45(8), 155, 1966.

Wang, J. C. and Y. L. Wang, A review on the modeling and simulation of multi-stage separation processes, *Proceedings of the International Conference, Foundation of Computer-Aided Chemical Process Design*, July, 1980.

14 Tray Hydraulics, Rate-Based Analysis, and Tray Efficiency

In the column model discussed in Chapters 3 through 13, the internal liquid and vapor flows were considered only from the standpoint of their effect on the thermodynamic performance of the column. The performance parameters that were investigated included quantities such as compositions, temperatures, pressures, enthalpies, and K-values. The column was assumed capable of physically handling any liquid or vapor flow rate, regardless of hydraulic effects or pressure drops. The only flow limitations that were taken into account involved minimum reflux ratio and conditions under which the liquid or vapor "dried up," or approached zero flow in certain parts of the column.

The internal flow of liquid and vapor must be re-evaluated from the standpoint of column capacity, both in the design and performance studies of columns. The physical dimensions of a column can handle only limited ranges of vapor and liquid flow rates. The objective of this chapter is to evaluate the hydraulic aspects of fluid flow in trayed columns. The column performance is examined with regard to factors such as flooding, entrainment, pressure drop, mass transfer, and tray efficiency.

Details on tray hydraulics are often specific to the particular tray type, and information relating to their design may be manufacturer proprietary. Nonetheless, the fundamentals apply to trays in general, and understanding them is essential in evaluating the factors that affect the actual performance of multi-stage separation columns. The subject is discussed in this chapter on a generic level based on published material. References are provided for more detailed reading.

A major assumption made in the column models of Chapters 3 through 13 was the equilibrium stage. Tray hydraulics provides additional information essential for applying mass transfer theories to evaluate the column performance with a rate-based approach. This analysis provides a basis for calculating the tray efficiency associated with an equilibrium stage. The topics of rate-based analysis and tray efficiency are also discussed in this chapter.

A multi-stage tray column consists of a vertical cylinder ranging in diameter anywhere from a few inches to many feet and reaching heights of hundreds of feet. The number of trays in a column could be anywhere from a few trays to a few hundred.

Column trays come in many different types, but they all share certain features that may be described by means of a general, simplified model. Model simplification is essential in the study of tray hydraulics even when discussing one particular type of tray. The reason for this is the highly complex nature of countercurrent, that is, two-phase fluid flow in the column and across the trays. Certain assumptions must be made in the development of flow and pressure drop correlations.

Although the overall vapor and liquid flow in the column is countercurrent, what takes place on each individual tray for most types of trays is actually cross-flow, as shown in the model schematic, Figure 14.1. Counterflow tray types do exist, but their use is not as widespread as cross-flow trays. Therefore, counter-flow trays are only briefly mentioned here. Unlike cross-flow trays, counterflow trays occupy the entire cross section of the column. Flow through the tray openings is counterflow; that is, vapor flows up and liquid flows down through the same openings. Except for the spacing between trays, counterflow tray column hydraulics is to some degree similar to that of counterflow packed columns, discussed in Chapter 15.

A cross-flow tray consists of a horizontal metal plate that partially fills the column cross section. Figure 14.1 depicts a few single-pass trays in a column section. The other main components of each tray are the downcomer, the downcomer apron, and the weir. The top view of a tray is illustrated in Figure 14.2, where the downcomer cross section appears as the segment of a circle. Opposite the downcomer is another segment of the tray where liquid flows down from the upper tray. The tray area between the two segments, the tray active area, contains openings or valves of one type or another, through which vapor flows upward.

Although the general model focuses on single-pass trays, multi-pass trays are also used. In multi-pass trays the liquid splits and flows in opposite directions so that any liquid element travels only a certain fraction of the tray width. Liquid flows down from one tray to the tray below through more than one downcomer, except on alternating trays in two-pass trays where liquid flows down one central downcomer every other tray. Figure 14.3 illustrates tray arrangements and liquid paths in two-, three-, four-, and five-pass trays.

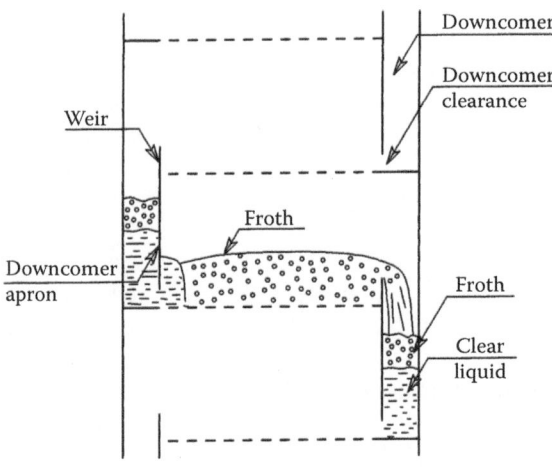

FIGURE 14.1   Tray schematic.

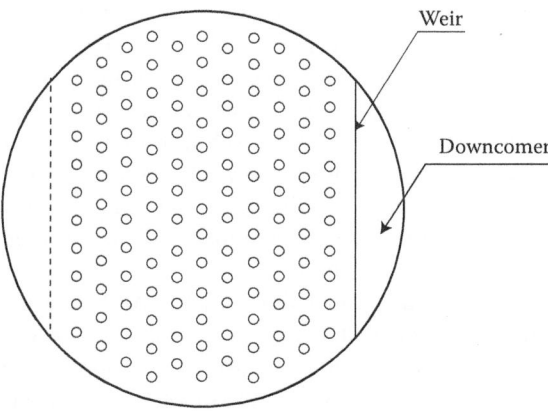

FIGURE 14.2 Single-pass tray.

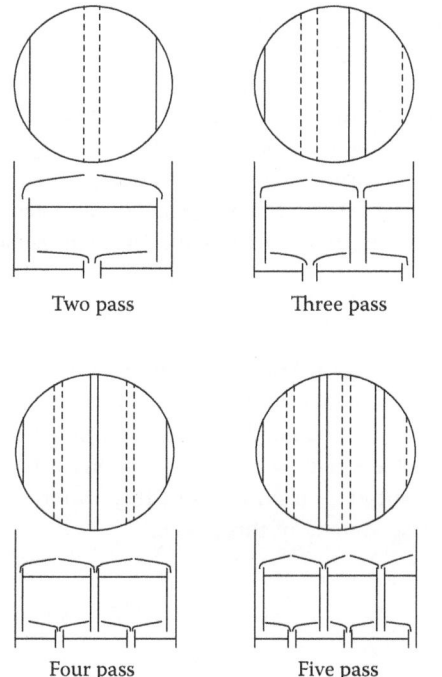

Two pass   Three pass

Four pass   Five pass

FIGURE 14.3 Multi-pass trays.

14.1 TRAY HYDRAULICS

From Figure 14.1, it is observed that liquid and froth flow down the downcomer from one tray to the tray below. As the fluid settles in the downcomer area, it separates into a layer of clear liquid and a layer of froth. The height of the froth depends on the foaming characteristics of the liquid. Clear liquid flows under the downcomer apron through the downcomer clearance and onto the active area of the tray between the downcomer apron and the weir. This part of the tray contains the openings through which vapor flows from the tray below. As a result of the vapor streaming upward through the openings and the liquid layer on the tray, froth is formed, starting a short distance from the downcomer apron. The froth flows across the tray, over the weir, and through the downcomer to the tray below. The weir height varies with tray type and in general is intended to ensure a minimum liquid height on the tray. As the froth moves from the downcomer to the weir, a liquid gradient develops across the tray to overcome the friction resistance to flow.

The hydraulics of liquid and vapor flow in and around the tray openings varies with tray type. One general characteristic of cross-flow trays is the fact that liquid is prevented from flowing down the openings by the upward flow of the vapor. Thus, while liquid flows horizontally across the tray, vapor flows vertically through the tray and between trays. This liquid–vapor cross-flow results in a composition gradient in the liquid across the tray. It is assumed that the vapor mixes thoroughly as it rises in the space between the trays and reaches a uniform composition before it penetrates the tray above. As the liquid travels across the tray, it interacts with the vapor and its composition changes progressively until it reaches the weir. Nevertheless, the simplified tray model assumes average liquid composition and equilibrium conditions on the tray.

The froth on the tray is a turbulent mass of usually liquid-continuous fluid with vapor dispersed in the form of small bubbles. The various designs of tray vapor openings attempt to maximize vapor dispersion by generating the smallest possible bubbles. The froth is where mass transfer takes place between the vapor and the liquid. Mass transfer and tray efficiency are enhanced by creating the largest possible interfacial area.

14.1.1 Types of Trays

Tray design varies with each particular manufacturer, and the design details are usually proprietary. Performance information is available directly from the manufacturers (Koch-Glitsch, The Nutter Engineering Company, 1976), some of whom provide detailed design for a given set of specifications as part of the purchase. A few features of the more commonly used types of tray vapor openings are summarized below. Figure 14.4 is a schematic of three types.

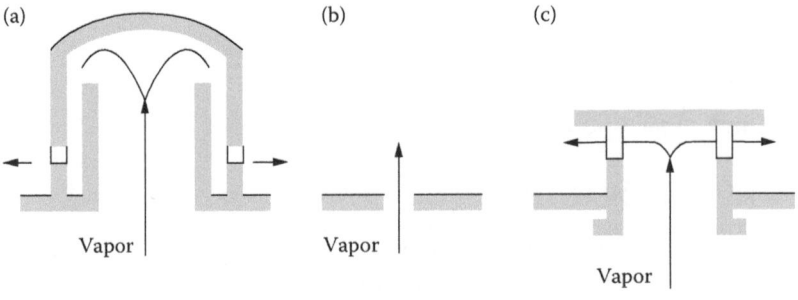

FIGURE 14.4   Types of tray openings: (a) bubble cap, (b) sieve, (c) valve.

14.1.1.1 Bubble Cap Trays

In this type of tray, each opening assembly consists of a cap, or inverted cup, with slots at its base, fixed above an opening. The opening consists of a hole and a riser through which vapor rises from the tray below. Its flow is reversed downward by the cap, after which the vapor flows down around the riser and bubbles out through the slots and into the liquid. Bubble cap diameters are usually about three to four inches. Because of the liquid seal created by the riser, bubble caps can operate at wide ranges of vapor and liquid flows with little loss in tray efficiency.

14.1.1.2 Sieve or Perforated Trays

Perhaps the simplest of cross-flow column tray designs is the sieve tray or perforated tray. The tray is a flat metal plate and the vapor openings are holes drilled in the plate. The holes are usually round, ranging from 1/8- to 1/2-inch diameter. Sieve trays have no liquid seals to prevent liquid from flowing down the holes. Liquid flow down the holes is prevented only by the upward flow of the vapor.

14.1.1.3 Valve Trays

This type of tray is designed to allow for wide variations in liquid and vapor flow. One typical design of a valve vapor opening consists of a one- to two-inch diameter orifice in the tray plate, an orifice cover, and a travel stop. At low vapor rates, the orifice cover is settled in its lower position. In this position, slots in the orifice cover allow small amounts of vapor to be distributed evenly. At higher vapor rates, the orifice cover is elevated to its upper position set by the travel stop. In this position, large amounts of vapor can flow through the valves.

14.1.2 Factors Affecting Tray Performance

The simplified tray model assumes idealized hydraulic conditions that enhance mass transfer (high tray efficiency) and maintain a low vapor pressure drop. Proper tray design aims at minimizing the effect of factors that tend to diminish "good" tray hydraulics. Inefficient performance may either be inherent in the tray type or design for a particular situation or may be the result of operating the column outside the design conditions. A look at some of these factors follows.

14.1.2.1 Foaming

The formation of froth on the trays is desirable for maximizing the interfacial area and mass transfer between liquid and vapor. Froth is formed through agitation and turbulence brought about by the vapor flow through the liquid. The amount of froth formed (foamability) and its tendency to linger (foam stability) are related to the physical properties of the liquid. High foamability can limit the allowable vapor flow, and high foam stability can limit the allowable liquid flow. It is clear that although the formation of foam is desirable, excessive foamability or foam stability should be avoided. The foaming properties must therefore be taken into account in the design and rating of column trays. In certain situations foamability may be adjusted by the addition of surfactants.

14.1.2.2   Vapor Entrainment

As liquid flows down through the downcomer, some froth flows with it and some more is formed in the downcomer by the turbulent flow. The downflowing froth should be allowed to break up in the downcomer before it reaches the tray below. Any froth flowing to the tray below carries with it some vapor, resulting in vapor entrainment. This lowers the tray efficiency on account of vapor flowing in the wrong direction. The downcomer volume must be large enough so that the residence time is sufficient to allow the froth to disintegrate.

14.1.2.3   Liquid Entrainment

When vapor disengages from the froth on a tray, liquid droplets may be entrained with it to the tray above. Liquid entrainment could be especially severe with highly foamable liquids where the froth occupies a large portion of the space between the trays. If the top of the froth is close to the tray above, the liquid droplets do not have enough time to fall back to the froth. Foamability therefore limits the vapor flow capacity of the trays. Handling high vapor flows with a foaming liquid requires larger tray spacing.

14.1.2.4   Liquid Gradient

As the liquid or froth moves across the tray, a liquid head develops to overcome resistance to flow caused by friction between the froth on the one hand and the tray and rising vapor on the other. The result is a liquid gradient with the liquid height at the bottom of the downcomer apron greater than that at the downflow weir.

The main problem that may arise from the liquid gradient, unless proper design and operating practices are followed, is its effect on the vapor distribution on the tray. Vapor flowing through tray openings where the liquid level is higher must overcome a greater liquid head than in places with lower liquid level. Hence, more vapor flows through the openings next to the weir than those next to the downcomer apron. Poor vapor distribution is detrimental to the overall performance of the column because it limits the extent of vapor–liquid interaction and consequently lowers the tray efficiency.

Liquid gradient problems are understandably more severe in larger diameter trays, where the flow paths are longer. For this reason, large diameter trays are usually designed with multiple passes (Figure 14.3) in order to reduce the liquid flow path and, hence, the liquid gradient.

14.1.2.5   Weeping

Weeping is another condition that could be aggravated by a liquid gradient. Weeping is the excessive flow of liquid down through the tray openings. As long as this flow does not cause an appreciable drop in tray efficiency, it is considered normal. If the vapor flow is reduced, liquid flow through the tray increases. Below a certain vapor flow, weeping ensues. With a liquid gradient, weeping is more severe at points on the tray where the liquid head is higher, that is, at the point of liquid entry, close to the downcomer apron.

14.1.2.6   Flooding

At certain vapor and liquid flow rates within the column, conditions could arise where liquid backs up between the trays and the trays cease to function as distinct stages. The column becomes "flooded" and completely inoperable because of a steep decline in tray efficiency and a sharp increase in pressure drop.

Flooding can result either from the froth rising on the tray and reaching the tray above or from liquid and froth filling the downcomer up to the liquid level at the downcomer overflow weir. The rise of the froth to the tray above is caused by excessive liquid entrainment resulting from too much vapor flow. The downcomer backup is caused by liquid rates exceeding the flow capacity of the downcomer. In either case the ultimate result is flooding, with liquid and froth filling the entire column.

It is therefore clear that flooding can be caused either by the vapor flow or the liquid flow exceeding certain limitations. Usually the vapor flow is limiting and the column preliminary design is based on it. The column is then checked for its liquid handling capacity.

The vapor velocity at which vapor-induced flooding occurs is the flooding vapor velocity. Since flooding is usually checked in terms of vapor flow, flooding vapor velocity is the most commonly used indicator. The vapor velocity is based on the net area available for vapor flow between the trays. Normally, the net area is the column cross-sectional area less the area occupied by downcomers, baffles, and so on. The actual column operation relative to its proximity to flooding is expressed as the ratio of the actual vapor velocity to the flooding vapor velocity. The ratio may be expressed as a percentage, commonly known as the percentage of flood. Columns are normally designed for operation at 75–85% of flood. If the correlation used for predicting the flooding vapor velocity does not take into account foaming characteristics, the design percentage of flood for foaming fluids should be about 70–75%.

The flooding vapor velocity correlations are empirical and specific to the type of tray. For a given tray type, the flooding vapor velocity is a function of the liquid flow, the vapor and liquid densities, and the tray spacing (Section 14.1.4). Corrections for surface tension are included in certain correlations.

14.1.2.7 Pressure Drop

Whereas liquid flow is caused by gravity in vapor–liquid countercurrent columns, a pressure gradient is necessary to induce vapor flow. Pressure drops exist from tray to tray, so the lower trays must be maintained at a higher pressure than the upper trays. One consideration in tray design is to attempt to keep the pressure drop at a minimum.

For a given tray type, the pressure drop from tray to tray is a function of vapor and liquid rates and properties as well as certain tray parameters, such as tray thickness, the geometry of the openings, and so on. Severe pressure drops are associated with certain operating conditions, such as too high downcomer backup, excessive liquid entrainment, and approach to flooding conditions. A minimum pressure drop is necessary to prevent weeping, or liquid flow down the openings.

High pressure drops can cause the downcomer backup to increase to a point where it fills the entire downcomer, causing flooding. For proper column operation, the downcomer backup should not exceed 40–60% of the tray spacing. The downcomer backup is balanced by the friction pressure drop of liquid flowing through the downcomer clearance, the liquid head on the tray on the other side of the downcomer apron, and the friction pressure drop of the vapor flowing through the tray openings and the liquid on the tray above (see Section 14.1.4).

14.1.2.8 Operable Ranges

Although a column may be designed for given values of vapor and liquid flows rates, actual operating conditions could vary considerably. It is therefore important to determine the ranges of flow rates over which the column can operate satisfactorily with only minor losses in tray efficiency. Different tray types have different operating flexibility, and the selection of a tray type should take into account its operable ranges. The tray should be checked for maximum anticipated vapor and liquid flows, then checked again for minimum anticipated flows. It is not unusual, for instance, for a tray designed for 75–85% of flood to operate at ranges extending down to 20% of flood without substantial loss of tray efficiency.

A qualitative diagram showing the constraints that limit operable vapor and liquid rates for a typical tray is presented in Figure 14.5. At low liquid flow rates, the operable vapor flow range is small: a quick transition from weeping conditions to excessive liquid entrainment takes place as the vapor flow rate is increased (area around C). Curve CB represents the lowest operable vapor rates at varying liquid rates. The curve corresponds to conditions where the liquid gradient could cause problems. At lower liquid rates, insufficient vapor could result in weeping and at higher liquid rates insufficient vapor could result in poor vapor distribution.

At high liquid rates, the tray can operate over a narrow range of vapor rates. As the vapor rates are increased, the tray hydraulics changes rapidly from poor vapor distribution to flooding (area around B). At lower liquid rates, a higher vapor rate can be maintained before flooding is incurred. Depending on the liquid rate, flooding can occur

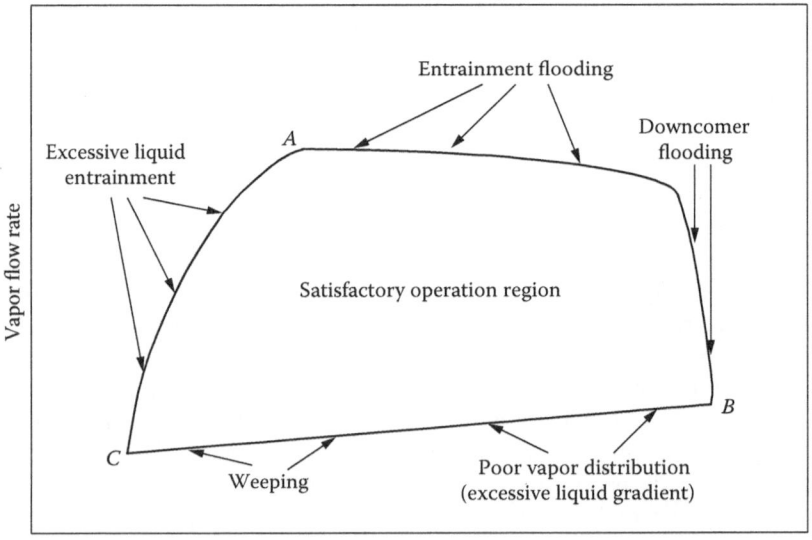

FIGURE 14.5   Column operable ranges.

either as entrainment flooding or as downcomer flooding (curve AB). Point A represents the upper limit of vapor flow. In that area the tray is susceptible either to excessive liquid entrainment at lower liquid rates or to flooding at higher liquid rates. Curve AC places constraints on the operable region because of excessive liquid entrainment. As the liquid rate is decreased, lower vapor rates can cause excessive liquid entrainment.

The diagram in Figure 14.5 is a qualitative one intended to demonstrate trends of tray hydraulics at varying liquid and vapor rates. The tray can be operated satisfactorily within the area defined by ABC. The diagram should be plotted quantitatively for a given tray type. The shape and characteristics of the operable region may vary from one tray type to another.

14.1.3 STEPS IN THE ANALYSIS OF TRAY HYDRAULICS

The factors influencing tray hydraulics are considered collectively in tray design or in checking tray performance under given operating conditions. Tray hydraulics is specific to tray type, and its characteristics are correlated on the basis of large amounts of operating data. The logical steps in designing a "general type" tray are described in principle here, relating the various hydraulics factors to tray performance.

Investigating column performance starts with heat and material balance calculations, which generate liquid and vapor flows and properties. The fluid foaming characteristics and corrosivity are also determined. Typically, the calculations are done on a computer simulator, which generates liquid and vapor flows as well as physical properties, such as densities, viscosities, and surface tension.

In the design mode, the next step is to make certain decisions and assumptions regarding the trays. Some of the assumptions may have to be refined at later stages of the design. The tray type and material of construction are selected. Tray spacing is assumed, as well as tray geometry: dimensions of openings, weir height and length, and downcomer type and size. Also, initially a single-pass crossflow tray is assumed for economy. Multiple-pass trays may be considered for large tray diameters to keep the liquid gradient in check.

The next step is to determine the flooding vapor velocity, which is a function of the vapor and liquid flow rates and their densities. The design vapor velocity is then calculated by multiplying the flooding vapor velocity by a flood factor such as 0.85. The vapor velocity is based on the vapor volumetric flow and the net flow area between the trays, that is, the column cross-sectional area minus the area blocked by the downcomers. With known vapor velocity, vapor flow, and fraction of the column cross section occupied by the downcomers, the net area can be calculated. The total area is then calculated, from which the tray diameter is determined. The actual column diameter is obtained by rounding off the tray diameter to the next larger standard size. The vapor velocity is then adjusted to the new diameter to calculate the expected flood factor.

At this point most of the column and tray parameters have been tentatively determined. Next, a performance check is carried out at various operating conditions to determine the operable ranges and whether any of the design parameters need to be adjusted. Performance checks include such items as pressure drop, liquid handling capacity, entrainment, and weeping.

If the calculated pressure drop at the vapor flow corresponding to the design flood factor is deemed excessive, consideration should be given to using larger tray openings. From another perspective, if the column is also required to operate at lower vapor rates, the weeping point should be determined, which sets the lower limit on the vapor rate. The operating range could be extended to lower vapor rates by using smaller tray openings, provided the pressure drop remains within acceptable ranges.

A liquid entrainment check is also usually made to ensure that it does not exceed acceptable limits. The fractional entrainment is a function of liquid and vapor flows and densities and of the flood factor. The column should not be operated at high fractional entrainment since this reduces the tray efficiency.

Finally, the trays must be checked for their liquid-handling capacity, which is controlled by the downcomer dimensions. The liquid level in the downcomer should not be allowed to exceed a certain height, normally about half the spacing between the trays. Liquid backup in the downcomer can cause flooding, although flooding is more often initially caused by excessive vapor flow. The liquid velocity in the downcomer is obtained by dividing the liquid flow by the downcomer cross-sectional area. The residence time is then evaluated from the ratio of the assumed height in the downcomer (half the tray spacing) to the liquid velocity. The residence time provides an empirical measure of the tray liquid-handling capacity. A minimum residence time is required to ensure froth separation and breakup to prevent excessive backup. More detailed calculations specific to the tray type are carried out to determine the actual liquid level in the downcomer.

In the following section many of these factors are discussed quantitatively based on general correlations.

14.1.4 General Tray Hydraulics Correlations

Many of the correlations presented here are of general applicability although some may be particularly suited for certain tray types as indicated. The main objectives are to determine the tray or column diameter for a given application, to calculate the tray liquid holdup and the pressure drop between the trays, and to check for flooding (entrainment and downcomer) and weeping.

14.1.4.1 Tray Diameter

The column cross-sectional area, and hence the tray diameter, is sized such as to prevent flooding. Tray sizing is initially based on preventing entrainment flooding, but is subsequently checked for downcomer flooding and other considerations. Above a certain vapor velocity, the vapor flood velocity, liquid droplets are entrained with the vapor, causing flooding. The tray should be sized such that the actual vapor velocity is below the vapor flood velocity.

At the vapor flood velocity U_f, the weight of the droplet, F_g, is balanced by the buoyant force F_b and the drag force F_d:

$$F_g = F_b + F_d$$

If the liquid droplet diameter is D_L, its density is ρ_L, and the surrounding vapor density is ρ_V, the following relations hold:

$$F_g = \left(\frac{\pi D_L^3}{6}\right)\rho_L g$$

$$F_b = \left(\frac{\pi D_L^3}{6}\right)\rho_V g$$

The drag force is given in terms of the droplet cross-sectional area, the vapor velocity and density, and the drag coefficient C_D. At the vapor flood velocity, the drag force is

$$F_d = C_D\left(\frac{\pi D_L^2}{4}\right)\rho_V\left(\frac{U_f^2}{2}\right)$$

The vapor flood velocity is obtained by substituting the above force expressions in the force balance equation and solving for U_f:

$$U_f = C\left(\frac{\rho_L - \rho_V}{\rho_V}\right)^{0.5} \tag{14.1}$$

where

$$C = \left(\frac{4D_L g}{3C_D}\right)^{0.5}$$

This parameter may either be calculated from the droplet diameter or from empirical correlations. Since the droplet diameter can be of many sizes and is usually unknown, C is estimated from correlations based on experimental data. In one such method by Fair (1961), developed specifically for sieve and bubble cap trays, C is correlated as a function of liquid and vapor flow rates and densities, tray spacing, surface tension, foaming properties, and the ratio of the combined hole area in the tray to its active area. In this correlation, U_f is based on the column cross-sectional area available for vapor flow, $A - A_d$, where A is the total tray area (or total inside column cross-sectional area) and A_d is the downcomer cross-sectional area. The parameter C is given as

$$C = F_{ST} F_F F_{HA} C_F \tag{14.2}$$

where F_{ST} is the surface tension factor, F_F is the foaming factor, F_{HA} is the hole area to active tray area factor, and C_F is the entrainment flooding capacity in ft/s. The flooding capacity is presented as a graphical function, Figure 14.6, of the tray spacing and F_{LV}, defined as

$$F_{LV} = \left(\frac{LM_L}{VM_V}\right)\left(\frac{\rho_V}{\rho_L}\right)^{0.5}$$

The flow rates L and V are the liquid and vapor molar rates in the column, and M_L and M_V are their molecular weights. The surface tension factor is calculated as

$$F_{ST} = \left(\frac{\sigma}{20}\right)^{0.2}$$

where σ is the surface tension in dyne/cm. The foaming factor F_F is a property of the fluid, equal to 1 for nonfoaming fluids and is less than 1 for foaming fluids. The hole area factor F_{HA} is a function of A_h/A_a, where A_h is the total hole area and A_a is the active tray area. This factor is estimated as follows:

$$\text{For } A_h/A_a \geq 0.10, \quad F_{HA} = 1.0$$

$$\text{For } 0.06 \leq A_h/A_a \leq 0.1, \quad F_{HA} = 5(A_h/A_a) + 0.5$$

The active tray area depends on the number of passes on the tray. For single-pass trays,

$$A_a = A - 2A_d$$

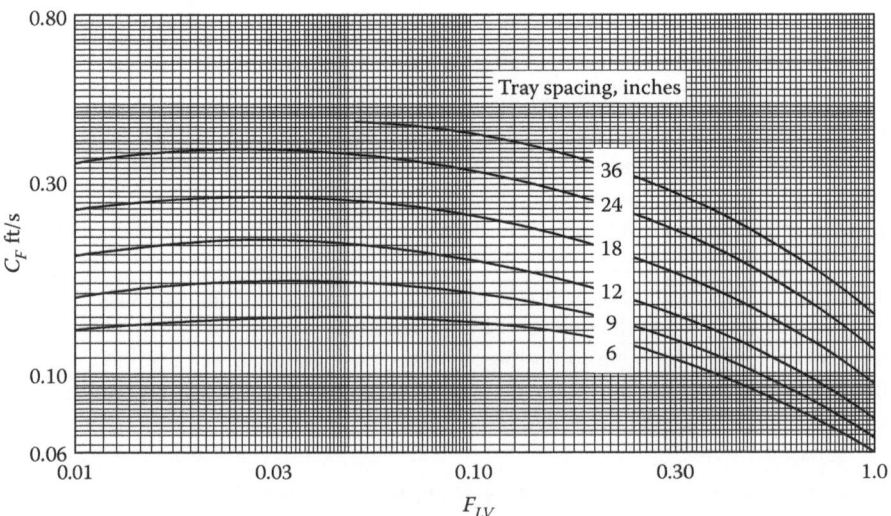

FIGURE 14.6 Flooding capacity for sieve and bubble cap trays. (From Fair, J. R., *Petro/ Chem Engineer*, 33(10), 211–218, Sept. 1961.)

The vapor flood velocity calculation is one step in the tray design, or rating of existing columns. The column should be designed or operated such that the actual vapor velocity is fU_f, where f is the fraction of flood velocity, and typically should be between 0.70 and 0.85. Since U_f is based on $A - A_d$, the actual vapor volumetric flow rate is

$$fU_f(A - A_d) = fU_f A\left(1 - \frac{A_d}{A}\right) = fU_f\left(\frac{\pi D^2}{4}\right)\left(1 - \frac{A_d}{A}\right)$$

where D is the tray diameter. The vapor molar flow rate is the volumetric flow rate multiplied by the molar density:

$$V = fU_f\left(\frac{\pi D^2}{4}\right)\left(1 - \frac{A_d}{A}\right)\left(\frac{\rho_V}{M_V}\right)$$

(14.3)

The area ratio A_d/A determines the downcomer cross-sectional area and the ability of the downcomer to handle the liquid flow. Since higher liquid to vapor flow ratios require higher values for A_d/A, this ratio is correlated with F_{LV} (Oliver, 1966):

$$\text{For } F_{LV} \leq 0.1, \quad A_d/A = 0.1$$

$$\text{For } 0.1 \leq F_{LV} \leq 1.0, \quad A_d/A = 0.1 + (F_{LV} - 0.1)/9$$

$$\text{For } F_{LV} \geq 1.0, \quad A_d/A = 0.2$$

Equation 14.3 is rearranged to calculate the tray diameter when designing a column, or to calculate the fraction of flood velocity to check for the possibility of flooding in an existing column:

$$D = 2\left[\frac{VM_V}{\pi f U_f \rho_V(1 - A_d/A)}\right]^{0.5}$$

(14.4)

$$f = \frac{4VM_V}{\pi D^2 U_f \rho_V(1 - A_d/A)}$$

(14.5)

14.1.4.2 Tray Pressure Drop

The flow of vapor upwards in the column from tray to tray requires a column pressure gradient to overcome the pressure drops through the trays. There is a friction pressure drop resulting from the flow of vapor through the tray openings. This pressure drop would occur even with no liquid on the tray, and is appropriately referred to as dry tray pressure drop. The vapor exiting the tray openings finds itself at the bottom of the liquid holdup on that tray, where the pressure is higher than that in the vapor space above, due to the liquid head. Additionally, the pressure inside the vapor bubbles coming out of the tray openings is slightly higher than the liquid pressure due to surface tension. All in all, the total pressure drop between two trays, or the difference between the pressure in the vapor space on tray $j + 1$ minus the pressure in the vapor space on tray j (Figure 14.1) is the sum of three pressure drops. In terms of clear liquid heads,

$$h_t = h_d + h_l + h_\sigma \tag{14.6}$$

where h_t, h_d, h_l, and h_σ represent total pressure drop, dry tray pressure drop, liquid head on the tray, and pressure drop due to surface tension, respectively.

The dry pressure drop calculation is based on the orifice equation for gases, in which the pressure drop is proportional to the square of the vapor velocity through the orifice, and the vapor density. In terms of liquid head, the equation for sieve trays is written as

$$h_d = 5.085 \times \frac{\rho_V}{\rho_L} \left(\frac{U_0^2}{C_0^2} \right) \tag{14.7}$$

In this equation h_d is in centimeters of clear liquid, U_0 is the hole velocity in m/s, and C_0 is the orifice coefficient. The hole velocity is calculated from the vapor volumetric flow and the total hole area A_h. The orifice coefficient for sieve trays is a function of the ratio of total hole area to active tray area, and of the ratio of tray thickness to hole diameter. This function may be approximated by the equation (Smith, 1963),

$$C_0 = 0.576 + 0.75(A_h/A_a) + 0.1826(l_t/d_h) \tag{14.8}$$

where l_t is the tray thickness and d_h is the hole diameter. Equation 14.8 is valid for values of A_h/A_a between 0.05 and 0.2, and l_t/d_h between 0.1 and 1.2.

To determine the clear liquid head, the froth height and the relative froth density must be estimated. The froth height is an average height that depends on the weir height and the liquid gradient. The following is an empirical equation for estimating h_l in centimeters of clear liquid (Bennett et al., 1983):

$$h_l = \phi_t \left[h_w + \{0.919 + 0.805 \exp(-1.378 h_w)\} \left(\frac{6.71 q_L}{L_w \phi_t} \right)^{2/3} \right] \tag{14.9}$$

The weir height, h_w, is in centimeters and the weir length, L_w, is in meters. The liquid flow rate across the tray, q_L, is in cubic meters per minute. The relative froth density on the tray, ϕ_t, defined as the ratio of clear liquid height to froth height, is given by another empirical equation:

$$\phi_t = \exp \left[-12.547 \left\{ U_a \left(\frac{\rho_V}{\rho_L - \rho_V} \right)^{0.5} \right\}^{0.91} \right] \tag{14.10}$$

where ρ_L and ρ_V are the liquid and vapor densities, and U_a, m/s, is the vapor velocity based on the active area, A_a. The weir length L_w is directly related to the downcomer area and tray diameter. It can be shown from the tray geometry that

$$\frac{A_d}{A} = \frac{\sin^{-1}(L_w/D)}{180} - \frac{1}{\pi}\left(\frac{L_w}{D}\right)\sqrt{1-\left(\frac{L_w}{D}\right)^2} \tag{14.11}$$

The arc-sin is in degrees. It is also recalled that $A_d = (A - A_a)/2$.

The pressure drop due to surface tension is calculated on the assumption that the diameter of the vapor bubbles is equal to the hole diameter d_h. The excess pressure inside the bubble, $h_\sigma\rho_L g$, acts on the bubble cross-sectional area. The resulting force is balanced by the surface tension acting on the bubble circumference:

$$h_\sigma \rho_L g\left(\frac{\pi d_h^2}{4}\right) = \sigma\pi d_h$$

$$h_\sigma = \frac{4\sigma}{g\rho_L d_h} \tag{14.12}$$

Taking h_σ in centimeters of clear liquid, σ in dyne/cm, ρ_L in kg/m³, and d_h in centimeters, this equation may be written in the following form (Van Winkle, 1967):

$$h_\sigma = \frac{4.133\sigma}{\rho_L d_h} \tag{14.12a}$$

14.1.4.3 Downcomer Backup

The liquid and froth flowing over the weir partially fill the downcomer, creating a backup of height h_{df} (Figure 14.1). If the overall relative froth density in the downcomer is ϕ_d, the equivalent clear liquid height is

$$h_{dc} = \phi_d h_{df}$$

If the pressure in the vapor space of tray j is P_j, then the pressure in the vapor space of tray $j + 1$ may be calculated as

$$P_{j+1} = P_j + \rho_L g h_{dc} - \rho_L g h_{da} - \rho_L g h_l$$

where h_{da} is the liquid head loss resulting from flow through the clearance under the downcomer apron, and h_l is the clear liquid head on tray $j + 1$. Recalling that $P_{j+1} - P_j = \rho_L g h_t$ and combining with the above equation, an expression may be obtained for the downcomer clear liquid head:

$$h_{dc} = h_t + h_{da} + h_l \tag{14.13}$$

The liquid heads h_t and h_l are calculated by Equations 14.6, 14.7, 14.9 and 14.12. The head loss through the downcomer clearance is calculated by the orifice equation. The orifice is a rectangle of area $L_w h_a$, where h_a is the height of the downcomer clearance. For straight downcomers the width of the downcomer clearance is the same as the weir length L_w. (Some downcomers may be sloped, in which case both the weir length and the downcomer clearance length must be specified.) The coefficient of the orifice equation for this particular type of orifice is 0.6, which leads to the following equation for the head loss in centimeters of clear liquid:

$$h_{da} = 45.9 \left(\frac{q_L}{L_w h_a} \right)^2$$

(14.14)

where L_w is in meters and h_a is in centimeters and q_L is the liquid flow in cubic meters per minute.

The actual downcomer backup, or the height of the froth in the downcomer, is calculated as $h_{df} = h_{dc}/\phi_d$, where ϕ_d depends on the foamability of the fluid and must be determined experimentally. As a check on the possibility of downcomer flooding, h_{df} should be safely less than the tray spacing, not more than 70%.

14.1.4.4 Weeping

The excessive flow of liquid through the tray openings at low vapor rates and high liquid rates is a pressure-related phenomenon. If the liquid head on the tray exceeds the sum of the dry tray pressure drop and the surface tension pressure drop, that is, if $h_l > h_d + h_\sigma$, weeping can be expected. It can be prevented during the design phase and/or during operation by ensuring that the vapor and liquid flows satisfy the condition that $h_d + h_\sigma > h_l$.

14.1.4.5 Liquid Holdup

The volumetric tray liquid holdup can be considered consisting of two main parts: the liquid on the active tray area and the liquid in the downcomer. These holdups can be estimated from the liquid heights, h_l and h_{dc}, and the corresponding areas:

$$U_t = A_a h_l$$

(14.15)

$$U_d = A_d h_{dc}$$

(14.16)

where U_t is the active tray area holdup and U_d is the downcomer holdup.

EXAMPLE 14.1: COLUMN SIZING AND RATING

An absorber column is to be designed for lowering the concentration of acetone in a stream of air using water as the absorbent. The results of a simulation run are summarized as herewith:

		Mole Fraction		
	Absorbent	Gas Feed	Liquid Bottoms	Vapor Overhead
Nitrogen		0.771	1.0×10^{-5}	0.763
Oxygen		0.207	5.0×10^{-6}	0.205
Acetone		0.015	0.005	8.19×10^{-5}
Water	1.00	0.007	0.995	0.032
Flow rate (kmol/h)	2000.0	725.0	1992.5	732.5
Temperature (°C)	26.0	27.0	23.0	25.0
Pressure (kPa)	105.0	105.0	105.0	100.0
Molecular weight	18.01	29.22	18.21	28.51
Density (kg/m³)	997.0	1.23	998.0	1.15

The absorber will be a trayed column, using sieve trays with the following specifications:

Hole area	$A_h/A = 0.10$
Hole diameter	$d_h = 0.5$ cm
Weir height	$h_w = 5$ cm
Tray thickness	$l_t = 0.2$ cm
Height of downcomer clearance	$h_a = 4$ cm
Tray spacing	46 cm (18 inch standard)

Additional properties and specifications include the following:

Surface tension	$\sigma = 68$ dyne/cm
Foaming factor	$F_F = 0.80$
Froth density in the downcomer	$\phi_d = 0.5$
Fraction of flood velocity	$f = 0.75$

It is required to size and rate the column.

SOLUTION

Since the vapor rate is highest at the top, the column will be sized based on the top tray conditions. The liquid and vapor on that tray will be represented by the absorbent and the overhead vapor.

Miscellaneous Parameters

A number of parameters required at different computation steps are first calculated. The liquid–vapor ratio, F_{LV}, is calculated from its definition:

$$F_{LV} = \left(\frac{LM_L}{VM_V}\right)\left(\frac{\rho_V}{\rho_L}\right)^{0.5} = \left(\frac{2000.0 \times 18.01}{732.5 \times 28.51}\right)\left(\frac{1.15}{997.0}\right)^{0.5} = 0.0586$$

Since $F_{LV} < 0.1$, $A_d/A = 0.1$

Since $A_a = A - 2A_d$, $A_a/A = 1 - 2(A_d/A) = 1 - 2(0.1) = 0.8$

Since $A_h/A = 0.1$, $A_h/A_a = \dfrac{A_h}{A} \bigg/ \dfrac{A_a}{A} = 0.1/0.8 = 0.125$

The ratio of the weir length to the tray diameter is calculated from Equation 14.11 for $A_d/A = 0.1$ by trial and error. The result is $L_w/D = 0.73$.

Tray Sizing

The tray diameter is calculated by Equation 14.4, which requires the vapor flood velocity, Equation 14.1. Parameter C in this equation is calculated from Equation 14.2:

$$F_{ST} = \left(\frac{\sigma}{20}\right)^{0.2} = \left(\frac{68}{20}\right)^{0.2} = 1.277$$

$$F_F = 0.80$$

Since $A_h/A_a > 0.1$, $F_{HA} = 1.0$

The entrainment flooding capacity, C_F, is found from Figure 14.6 for $F_{LV} = 0.586$ and 18-inch tray spacing: $C_F = 8$ cm/s.

From Equation 14.2,

$$C = (1.277) \times (0.80) \times (1.0) \times (8) = 8.2 \text{ cm/s}$$

The vapor flood velocity, Equation 14.1,

$$U_f = (8.2)\left(\frac{997.0 - 1.15}{1.15}\right)^{0.5} = 2.41 \text{ m/s}$$

The tray diameter, Equation 14.4,

$$D = 2\left[\frac{(732.5) \times (28.51) \times (1/3600)}{\pi(0.75) \times (2.41) \times (1.15) \times (1 - 0.1)}\right]^{0.5} = 1.98 \text{ m}$$

Tray Rating

The column will be rated with the nearest larger standard tray size:

$$D = 2 \text{ m}$$

The calculations include fraction of flood velocity, pressure drop per tray, downcomer backup, tray liquid holdup, and check for weeping.

Fraction of Flood Velocity

From Equation 14.5,

$$f = \frac{(4) \times (732.5) \times (28.51) \times (1/3600)}{\pi(2)^2(2.41) \times (1.15) \times (1 - 0.1)} = 0.74$$

Pressure Drop per Tray

The orifice coefficient for sieve trays (Equation 14.8):

$$C_0 = 0.576 + (0.75) \times (0.125) + \frac{(0.1826) \times (0.078)}{(3/16)} = 0.746$$

Total tray area,

$$A = \frac{\pi D^2}{4} = \frac{\pi (2)^2}{4} = 3.14 \, m^2$$

Hole area,

$$A_h = 0.1A = 0.314 \, m^2$$

Active tray area,

$$A_a = 0.8A = (0.8) \times (3.14) = 2.512 \, m^2$$

Vapor velocity based on active tray area,

$$U_a = \frac{VM_V}{\rho_V A_a} = \frac{(732.5) \times (28.51)}{(1.15) \times (2.512) \times (3600)} = 2.0 \, m/s$$

Hole velocity,

$$U_0 = \frac{U_a}{(A_h/A_a)} = \frac{2.0}{0.125} = 16 \, m/s$$

Dry tray pressure drop (Equation 14.7),

$$h_d = 5.085 \left(\frac{1.15}{997.0} \right) \left(\frac{16}{0.746} \right)^2 = 2.7 \text{ centimeters clear liquid}$$

Relative froth density on the tray (Equation 14.10),

$$\phi_t = \exp \left[-12.547 \left\{ 2 \left(\frac{1.15}{997.0 - 1.15} \right)^{0.5} \right\}^{0.91} \right] = 0.34$$

Liquid volumetric flow rate,

$$q_L = \frac{(2000.0) \times (18.01)}{(997.0) \times (60)} = 0.602 \, m^3/min = 0.01 \, m^3/s$$

Weir length,

$$L_w = (L_w/D)D \times 2 = 1.46 \text{ m}$$

Clear liquid head (Equation 14.9),

$$h_l = (0.38)\left[5 + \{0.919 + 0.805\exp(-1.378 \times 5.0)\}\left(\frac{6.71 \times 0.602}{1.46 \times 0.34}\right)^{2/3}\right] = 3.31\text{cm}$$

Surface tension pressure drop (Equation 14.12a),

$$h_\sigma = \frac{4.133 \times 68}{(997.0) \times (0.5)} = 0.56 \text{ cm}$$

Total pressure drop per tray (Equation 14.6),
$h_t = 2.7 + 3.31 + 0.56 = 6.57$ centimeter clear liquid

Downcomer Backup
Liquid head loss through downcomer clearance (Equation 14.14),

$$h_{da} = 45.9\left(\frac{0.602}{1.46 \times 4}\right)^2 = 0.488 \text{ centimeter clear liquid}$$

Downcomer clear liquid head (Equation 14.13),

$$h_{dc} = 6.57 + 0.488 + 3.31 = 10.368 \text{ cm}$$

Downcomer backup height,

$$h_{df} = \frac{h_{dc}}{\phi_d} = \frac{10.368}{0.5} = 20.736 \text{ cm}$$

This height is safely below half the tray spacing.

Check for Weeping
Since $h_d + h_\sigma < h_l$ (2.7 + 0.56 < 3.31), weeping is possible. Consideration should be given to a smaller tray diameter or smaller hole diameter.

Liquid Holdup
Liquid holdup on the tray (Equation 14.15),

$$U_t = A_a h_l = (2.512) \times \left(\frac{3.31}{100}\right) = 0.083 \text{ m}^3$$

Downcomer cross-sectional area,

$$A_d = 0.1A = 0.1 \times 3.14 = 0.314 \, \text{m}^2$$

Liquid holdup in the downcomer,

$$U_d = A_d h_{dc} = (0.314) \times \left(\frac{10.368}{100} \right) = 0.033 \, \text{m}^3$$

14.2 RATE-BASED ANALYSIS

The tray hydraulics model may be extended to include mass and heat transfer rates for calculating the liquid and vapor flow rates and compositions in trayed columns on the basis of a rate-based model. The objective is to more realistically represent the actual performance of the column by providing a basis for estimating a tray. For this approach to be practical, methods should be available for reliably predicting the mass and heat transfer rates. General rate-based models are also discussed in Chapter 15 for solving packed columns.

Reference is again made to Figure 14.1 for defining the rate-based model. The froth on the active area of each tray is assumed to be consisting of a well-mixed liquid phase with uniform composition and a vapor phase moving upward through the froth in plug flow. The mole fraction of component i in the liquid entering tray j from tray $j - 1$ is $X_{j-1,i}$, and leaving tray j is X_{ji}. Since uniform liquid composition is assumed, the mole fraction in the liquid on tray j is also X_{ji}. The mole fraction of component i in the vapor entering tray j from tray $j + 1$ is $Y_{j+1,i}$ and leaving tray j is Y_{ji}. By the assumption of vapor plug flow through the froth, the mole fraction of component i, Y_i, is uniform at any froth height z on the tray, but varies vertically from $Y_{j+1,i}$ at $z = 0$ to Y_{ji} at $z = h_f = h_l/\phi_l$, the height of the froth holdup on the tray. A schematic of the model is shown in Figure 14.7.

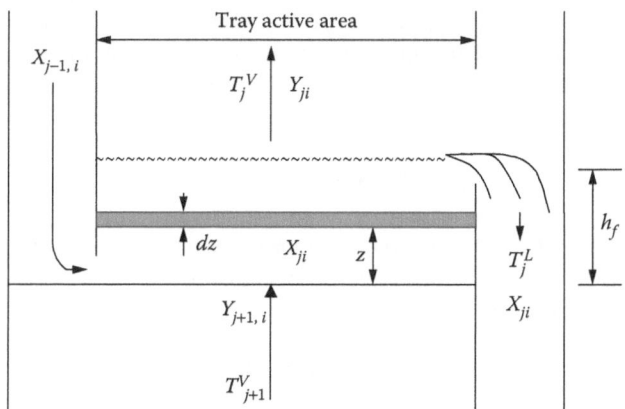

FIGURE 14.7 Tray rate-based analysis schematic.

The liquid phase on the tray being well mixed, its temperature is uniform, is equal to the temperature of the liquid leaving the tray, T_j^L. The vapor enters tray j at temperature T_{j+1}^V and leaves at T_j^V. An energy balance and heat transfer equation for tray j is written as

$$H_{j+1} - H_j = h_j - h_{j-1} = US\Delta T_m \qquad (14.17)$$

where U is the overall heat transfer coefficient, S is the interfacial area between the phases, and h and H are total liquid stream and vapor stream enthalpies. The mean temperature difference is defined as

$$\Delta T_m = \frac{(T_{j+1}^V - T_j^L) - (T_j^V - T_j^L)}{\ln \dfrac{T_{j+1}^V - T_j^L}{T_j^V - T_j^L}} \qquad (14.18)$$

The froth on the tray is where mass transfer takes place between the phases, assumed to exist at equilibrium at the vapor–liquid interface. The well mixed liquid phase is at its bubble point temperature T_j^L, at equilibrium with the vapor interfacial composition Y_{ji}^*. The vapor leaving tray j with composition Y_{ji} is at temperature T_j^V, determined by heat transfer rates according to Equations 14.17 and 14.18. This temperature may be above the dew point.

If the partial pressure of component i in the bulk vapor at froth height z is P_i and the partial pressure of component i at equilibrium with the mixed liquid is P_{ji}^*, then the driving force for mass transfer is $P_i - P_{ji}^*$. The rate of mass transfer of component i across the phase interface is proportional to the driving force and the interfacial area. The mass transfer in a differential area is expressed as follows:

$$dn_i = K_G(P_i - P_{ji}^*)dS \qquad (14.19)$$

where n_i is the rate of mass transfer of component i from the vapor to the liquid in moles per unit time, S is the interfacial area, and K_G is the overall vapor (or gas) mass transfer coefficient. This coefficient represents the resistances to mass transfer in both phases, and is based on the driving force in the vapor phase as defined above. The interfacial area for a differential height dz is calculated from the differential volume:

$$dS = aA_a dz$$

where A_a is the active tray area and a is the interfacial area per unit volume. Equation 14.19 may be written in terms of mole fractions instead of partial pressures by substituting the following relations:

$$P_i = P_j Y_i$$

$$P_{ji}^* = P_j Y_{ji}^*$$

where P_j is the total pressure on tray j, Y_i is the bulk vapor mole fraction of component i at froth height z, and Y_{ji}^* is the mole fraction of component i at equilibrium with the mixed liquid. Equation 14.19 becomes

$$dn_i = K_G a A_a P_j (Y_i - Y_{ji}^*) dz \qquad (14.20)$$

The rate of mass transfer taking place in the differential volume defined by dz may also be written in terms of V_j, the vapor flow rate, and dY_i, the change in the mole fraction of component i as the vapor flows through dz:

$$dn_i = -V_j dY_i \qquad (14.21)$$

Combining Equations 14.20 and 14.21 and rearranging, gives

$$A_a (K_G a P_j / V_j) dz = \frac{-dY_i}{(Y_i - Y_{ji}^*)} \qquad (14.22)$$

This equation is integrated over the froth on the tray from $z = 0$ and $Y_i = Y_{j+1,i}$ to $z = h_f$ and $Y_i = Y_{ji}$:

$$A_a \int_0^{h_f} \left(\frac{K_G a P_j}{V_j} \right) dz = \int_{Y_{j+1,i}}^{Y_{ji}} \frac{dY_i}{(Y_{ji}^* - Y_i)} \qquad (14.23)$$

The second integral defines N_{OG}, the number of overall gas phase mass transfer units. This is the ratio of the vapor composition change on the tray to the mean driving force. Applied to a process where only one component is transferred, such as in absorber columns, K_G, a, P_j, V_j, and Y_{ji}^* may be assumed constant, and both integrals can be evaluated analytically:

$$N_{OG} = \ln \left(\frac{Y_{j+1,i} - Y_{ji}^*}{Y_{ji} - Y_{ji}^*} \right) = \frac{K_G a P_j h_f A_a}{V_j} \qquad (14.24)$$

The overall gas mass transfer coefficient is related to the gas phase and liquid phase mass transfer coefficients, k_G and k_L. The concept of overall and individual phase mass transfer coefficients is discussed in Chapter 15. Equation 15.15 may be written in the following equivalent form by substituting $K_Y = P K_G$, $k_y = P k_G$, and $k_x = P k_X$, and dividing through by a:

$$\frac{1}{K_G a} = \frac{1}{k_G a} + \frac{K_{ji} P_j M_L}{\rho_L} \frac{1}{k_L a} \qquad (14.25)$$

where K_{ji} is the vapor–liquid equilibrium distribution coefficient, and M_L and ρ_L are the liquid molecular weight and mass density. The number of mass transfer units for the vapor phase and the liquid phase are defined in a manner analogous to the overall mass transfer units:

$$N_G = \frac{k_G a P_j h_f A_a}{V_j}$$

(14.26a)

$$N_L = \frac{k_L a \rho_L h_f A_a}{L_j M_L}$$

(14.26b)

By combining these definitions with Equation 14.25, the number of overall mass transfer units can be expressed in terms of the individual phase mass transfer units:

$$\frac{1}{N_{OG}} = \frac{1}{N_G} + \frac{K_{ji} V_j}{L_j} \frac{1}{N_L}$$

(14.27)

In order to calculate the number of mass transfer units, the mass transfer coefficients must be estimated, either based on experimental data or from empirical correlations. The following correlation (Chan and Fair, 1984) was developed for binary systems, and specifically applies to sieve trays. In this correlation N_G and N_L are expressed in terms of residence time and interfacial area instead of the standard forms of Equations 14.26:

$$N_G = k_G a t_G$$

(14.28a)

$$N_L = k_L a t_L$$

(14.28b)

The gas phase and liquid phase residence times, t_G and t_L, are calculated based on height-to-velocity or holdup-to-flow rate ratios.
Froth height = h_l/ϕ_t
Gas velocity = $U_a/(1 - \phi_t)$

$$t_G = \frac{(1 - \phi_t)h_l}{\phi_t U_a} \quad \text{(Froth height/Gas velocity)}$$

(14.29a)

$$t_L = \frac{h_l A_a}{q_L} \quad \text{(Holdup/Flow rate)}$$

(14.29b)

The clear liquid height on the tray, h_l, is calculated by Equation 14.9, and the relative froth density, ϕ, by Equation 14.10. The mass transfer coefficients are calculated from the following empirical correlations:

$$k_G a = \frac{1030 D_V^{0.5}(f - 0.842 f^2)}{h_l^{0.5}}$$

(14.30a)

$$k_L a = 78.8 D_L^{0.5} (U_a \rho_V^{0.5} + 0.425) \tag{14.30b}$$

where D_V and D_L are diffusion coefficients of the transferred component in the vapor and liquid, cm²/s, f is the fraction of flood velocity, U_a is the vapor velocity based on the active tray area, m/s, h_l is the clear liquid height, cm, and ρ_V is the vapor density, kg/m³.

This rate-based analysis of trayed columns gives an estimate of the number of overall gas phase mass transfer units corresponding to a given tray. The practical application of this analysis is to provide a means for estimating the tray Murphree vapor efficiency discussed in Section 14.3.

EXAMPLE 14.2: RATE-BASED CALCULATIONS IN A TRAYED COLUMN

Calculate the number of overall gas phase mass transfer units for acetone, equivalent to one tray in the absorber column of Example 14.1. The K-value of acetone and its diffusion coefficients in the liquid and vapor are given:

$$K = 1.875 \quad D_V = 0.125 \text{ cm}^2/\text{s} \quad D_L = 1.0 \times 10^{-5} \text{ cm}^2/\text{s}$$

SOLUTION

The gas phase and liquid phase residence times are calculated from Equations 14.29a and 14.29b, using tray and hydraulic data from Example 14.1:

$\phi_t = 0.38$
$h_l = 3.31$ cm
$U_a = 2$ m/s
$A_a = 2.512$ m²
$q_L = 0.01$ m³/s
$f = 0.74$
$\rho_V = 1.15$ kg/m³

$$t_G = \frac{(1 - \phi_t) h_l}{\phi_t U_a} = \frac{(1 - 0.38) \times (3.31)}{(0.38) \times (2.0) \times (100)} = 0.027 \text{ s}^{-1}$$

$$t_L = \frac{h_l A_a}{q_L} = \frac{3.31 \times 2.512}{0.01 \times 100} = 8.31 \text{ s}$$

The mass transfer coefficients are calculated from Equations 14.30a and 14.30b:

$$k_G a = \frac{1030 \times (0.125)^{0.5}(0.74 - 0.842 \times 0.74^2)}{3.31^{0.5}} = 55.8 \text{ s}^{-1}$$

$$k_L a = 78.8(1.0 \times 10^{-5})^{0.5}(2.0 \times 1.15^{0.5} + 0.425) = 0.64 \text{ s}^{-1}$$

Equations 14.28a and 14.28b are used to calculate the number of gas phase and liquid phase mass transfer units:

$$N_G = (55.8) \times (0.027) = 1.51$$

$$N_L = (0.64) \times (8.31) = 5.32$$

Finally, the number of overall gas phase mass transfer units is calculated by Equation 14.27:

$$\frac{1}{N_{OG}} = \frac{1}{N_G} + \frac{KV}{L}\frac{1}{N_L} = \frac{1}{1.51} + \frac{(1.875) \times (732.5)}{2000.0}\frac{1}{5.32} = 0.662 + 0.129 = 0.791$$

$$N_{OG} = 1.26$$

14.3 TRAY EFFICIENCY

Vapor–liquid multi-stage columns are commonly studied in terms of equilibrium stages, where the vapor and liquid phases are assumed to mix perfectly and achieve complete interaction with each other. A long-enough residence time is assumed on the stage to allow the phases to separate and leave the stage at the same temperature and with equilibrium compositions.

These assumptions must be re-evaluated when applying the model to real columns. A column tray can only approach, to varying degrees, the equilibrium stage. This section examines tray efficiency and its role in relating actual trays to equilibrium stages.

Although several different expressions are used for defining it, tray efficiency in general is a parameter that relates the actual performance of a column tray to a theoretical stage. Some of the reasons that actual trays do not achieve the separation of theoretical stages were discussed in Sections 14.1 and 14.2. They include factors such as incomplete contact between the phases, not enough residence time, vapor entrainment, liquid entrainment, weeping, leakage, and poor vapor distribution. Tray efficiency suffers a steep decline at flood conditions.

Tray efficiency is also a function of the basic tray hydraulics model. For instance, in a cross-flow tray the liquid is assumed to move across the tray in a plug flow manner with little back-mixing. The vapor is assumed to mix and reach a uniform composition in the tray below before it bubbles through the liquid. As the liquid flowing across the tray interacts with the rising vapor, its composition changes progressively along the width of the tray between the downcomer and the overflow weir. Since the rising vapor interacts with a changing liquid composition, its composition as it leaves the liquid also changes across the tray width, but is assumed to undergo complete mixing before it reaches the tray above. Even an ideal cross-flow tray does not perform as an equilibrium stage where the vapor leaving the stage is at equilibrium with the liquid leaving the stage. In a cross-flow tray all the vapor does not contact all the liquid.

14.3.1 MURPHREE EFFICIENCY

Since in general the liquid and vapor compositions vary across the vapor–liquid interface, a point efficiency is defined in terms of compositions at a given point on the tray. The Murphree point efficiency in vapor terms is defined as (Koch-Glitsch et al., 1925; Murphree, 1925)

$$E_V = \frac{Y - Y_{j+1}}{Y^* - Y_{j+1}} \tag{14.31}$$

where Y_{j+1} is the mole fraction of a given component in the mixed vapor coming from the tray below, designated as $j + 1$, and entering the tray in question, designated as j. The actual mole fraction of the same component in the vapor at a given point on tray j is Y, and Y^* is the vapor mole fraction of that component at equilibrium with X, the mole fraction of the component in the liquid at the same point on tray j. The point efficiency is thus defined as the ratio of the actual change in the vapor concentration of a component as it passes through a tray at a given point to the change that would occur if the vapor reaches equilibrium with the liquid. Note that the efficiency is defined for a particular component, implying that different components could have different efficiencies. This reflects the different rates of mass transfer for each component.

The Murphree point efficiency may be defined in liquid terms in analogy to Equation 14.31:

$$E_L = \frac{X_{j-1} - X}{X_{j-1} - X^*} \tag{14.32}$$

where X_{j-1} is an average mole fraction of a given component in the liquid flowing to tray j from the tray above, designated as $j-1$. The actual mole fraction of the same component in the liquid at a given point on tray j is X, and X^* is the mole fraction in the liquid of that component at equilibrium with Y, the vapor concentration at that point.

The Murphree tray efficiency relates to the tray as a whole and is defined as the ratio of the actual change in a component vapor concentration as it flows through the tray to the change that would occur at equilibrium conditions. The Murphree tray efficiency is expressed in vapor terms as

$$E_{MV} = \frac{Y_j - Y_{j+1}}{Y^* - Y_{j+1}} \tag{14.33}$$

where Y_{j+1} is the mole fraction of a component in the vapor coming from the tray below, designated as $j + 1$, and Y_j is the actual mole fraction of the same component in the vapor leaving tray j. In the denominator, Y^* is the vapor mole fraction of that component at equilibrium with X_j, the mole fraction of the component in the liquid leaving tray j. The Murphree tray efficiency may also be expressed in liquid terms as

$$E_{ML} = \frac{X_{j-1} - X_j}{X_{j-1} - X_j^*} \tag{14.34}$$

with similar definitions for the liquid mole fractions. As with point efficiencies, the Murphree tray efficiency is defined specifically for a given component and may have different values for different components. The Murphree tray efficiency is applicable under the assumption that the liquid phase and the vapor phase are completely mixed on each tray. That is, the exiting vapor and liquid phases, although not at equilibrium with each other, are each assumed to have a uniform composition.

The Murphree vapor efficiency may be estimated from N_{OG}, the number of overall gas phase mass transfer units (Section 14.2). For component i, the Murphree vapor efficiency is written as

$$E_{MV} = \frac{Y_{ji} - Y_{j+1,i}}{Y_{ji}^* - Y_{j+1,i}} = 1 - \frac{Y_{ji} - Y_{ji}^*}{Y_{j+1,i} - Y_{ji}^*}$$

From Equation 14.24,

$$\frac{Y_{ji} - Y_{ji}^*}{Y_{j+1,i} - Y_{ji}^*} = e^{-N_{OG}}$$

Combined with the above equation for E_{MV}, the Murphree vapor efficiency is expressed as a function of N_{OG}:

$$E_{MV} = 1 - e^{-N_{OG}} \tag{14.35}$$

Liquid entrainment can adversely affect the tray efficiency. As liquid is entrained from one tray to the tray above, the entrained liquid, e, joins the normally flowing liquid, L, resulting in a total liquid flow of $L + e$. The fractional entrainment ψ is defined as the ratio of entrained liquid to total liquid flow:

$$\psi = \frac{e}{L + e} \tag{14.36}$$

The tray efficiency is lowered by liquid entrainment because higher-boiling liquid is carried to a tray containing lower-boiling liquid, thereby countering the fractionation process. The lowering of the Murphree tray efficiency is expressed by the Colburn equation (Colburn, 1936):

$$\frac{E_e}{E_{MV}} = \frac{1}{1 + eE_{MV}/L} \tag{14.37}$$

where E_{MV} is the Murphree tray efficiency in the absence of entrainment, and E_e is the Murphree tray efficiency with entrainment.

EXAMPLE 14.3: ESTIMATING THE MURPHREE TRAY EFFICIENCY

Estimate the Murphree tray efficiency in the acetone absorber column of Examples 14.1 and 14.2.

The Murphree tray efficiency is calculated from Equation 14.35, with $N_{OG} = 1.42$ (Example 14.2):

$$E_{MV} = 1 - e^{-1.42} = 0.758$$

14.3.2 OVERALL COLUMN TRAY EFFICIENCY

Most commonly used in engineering practice is the overall column tray efficiency. It relates the number of theoretical stages in a column or column section to the number of actual trays that would be required to achieve a similar separation. The overall efficiency is used to translate actual trays to theoretical stages and vice versa in the analysis of the performance of existing columns and in the design of new columns. Since the degree of separation depends on the reflux ratio or liquid-to-vapor flows in the column as well as on the number of stages, it is important when applying overall efficiency to ensure that other factors are comparable. The overall column efficiency is defined as

$$E_O = \frac{N}{N_A} \tag{14.38}$$

where N is the number of theoretical stages, and N_A is the number of actual stages.

14.3.2.1 Theoretical Model

The theoretical derivation of the overall efficiency on the basis of fluid properties and tray dimensions consists of the following steps (A.I.Ch.E Tray Efficiency Research Program):

1. Predict Murphree point efficiencies from tray hydraulics and mass transfer rates in the vapor and liquid.
2. Assume a mathematical model to represent liquid mixing on the tray and use the model to predict Murphree tray efficiency from the point efficiency.
3. Correct the Murphree tray efficiency to account for entrainment.
4. Use the corrected Murphree tray efficiency to calculate overall column efficiency.

The model development is outside the scope of this book. The reader is referred to the original source for details. Due to the complexity and somewhat erratic nature of tray hydraulics, the accuracy of the resulting equations is not universal, and therefore their usefulness is limited to systems for which the equations were tested. The methods may be used for scaling up small column laboratory data to large columns.

The more common approach for estimating overall column efficiencies is to use empirical methods based on large amounts of experimental data.

14.3.2.2   Empirical Methods

The overall efficiency of an existing column is determined directly from experimental data obtained from the column. The data collected include the actual number of trays, rates and compositions of the feeds and products, the column pressure, and reflux ratio or internal liquid-to-vapor flow ratios. The theoretical number of stages required to achieve the measured separation of two key components at the column conditions is then calculated using a rigorous method (Chapter 13). Since most of these methods assume a known number of theoretical stages, this number must be determined by trial and error using multiple calculations, each with a different number of theoretical stages. The calculations can be conveniently carried out on a computer simulator where the number of stages and other parameters may be varied to generate the desired cases. Once the experimental data are matched with a particular number of theoretical stages, the overall efficiency E_O is calculated from Equation 14.38.

The efficiency may depend on the two components selected as the keys because, although the rigorous calculations can closely match measured key component compositions, other components may be off due to inaccuracies in the theoretical column model. Also, since the results of the column calculations depend on the thermodynamic properties prediction methods, the efficiency may also depend on the selection of these methods.

Column efficiency experimental data are usually compiled and saved and are used for estimating efficiencies of similar columns or for developing empirical correlations for predicting the efficiencies. When attempting to estimate a tray efficiency based on data available for another column, it is important to be aware of the factors that influence tray efficiency the most and take them into consideration. For instance, the tray dimensions should be comparable. Of particular importance are the tray diameter, tray spacing, number of passes, and liquid level on the tray.

The fluid properties should also be considered, among which are the viscosity, relative volatility, and surface tension. Lower viscosities are usually associated with higher efficiencies. It follows that columns operated at higher pressures can be expected to have higher efficiencies since higher pressures imply higher temperatures and lower viscosities. Also, columns for distilling narrow-boiling mixtures usually have higher efficiencies because most of the components are close to their boiling point and therefore have low viscosities. In contrast, columns used in absorption, stripping, and vacuum distillation have low efficiencies because many of the components in the liquid phase are far removed from their boiling points and therefore tend to have higher viscosities.

Relative volatilities also have a considerable effect on column efficiencies. A component with a high relative volatility has a low solubility in the liquid phase. This means high resistance to mass transfer in the liquid, which in general tends to lower the efficiency.

Surface tension influences efficiency through its effect on foaming. Foamability becomes a factor when it reaches levels that could increase entrainment, which in turn leads to a reduced efficiency.

Trends such as these have led to the development of empirical correlations for predicting column efficiencies. The Drickamer and Bradford (1943) correlation is based on data from numerous refinery fractionation columns. It correlates the overall efficiency E_O to a single parameter: the liquid average viscosity, μ_L, evaluated for the feed at average column temperature. The correlation is reproduced in Figure 14.8. Although it was developed primarily for hydrocarbon systems, it was found to predict non-hydrocarbon column efficiencies fairly well. However, care should be exercised when using the correlation for fluid types, operating conditions, and tray types or dimensions that are different from those for which the correlation was developed. Also, the correlation should be applied only to relatively narrow-boiling mixtures since it does not account for the effect of relative volatilities.

To include the effect of relative volatilities, O'Connell (1946) used the product of viscosity and relative volatility as the correlating parameter for predicting overall efficiencies. As in the Drickamer–Bradford correlation, the correlating viscosity, μ_L, is defined as the average liquid viscosity of the feed, calculated at the average column temperature. The correlating relative volatility, α, is that of the light key relative to the heavy key, calculated at the average column temperature. The overall efficiency E_O is correlated as a function of the product $\alpha\mu_L$ as shown in Figure 14.9, which is reproduced from the source. Although the O'Connell correlation somewhat extends the Drickamer–Bradford correlation to mixtures containing higher volatility components, it should still be used with care, especially outside the applications described in the source.

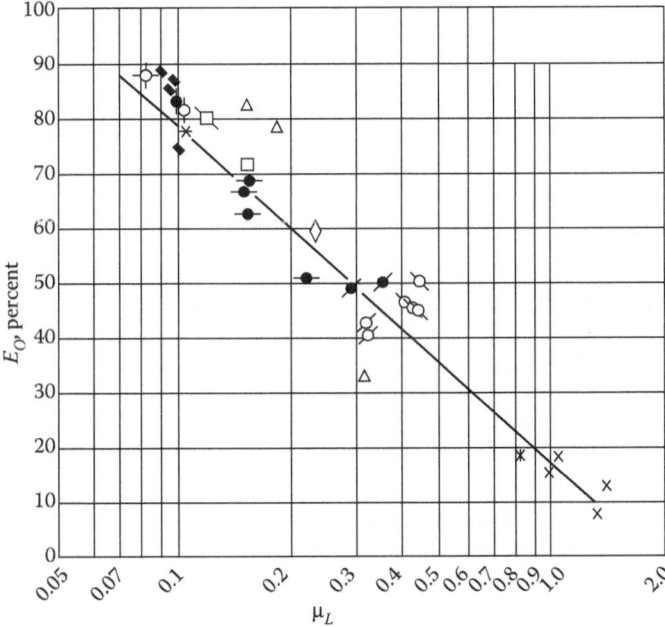

FIGURE 14.8 Drickamer–Bradford correlation. (Reprinted with permission from H. G. Drickamer and J. R. Bradford, Trans. A.I.Ch.E., 39, 319, 1943.)

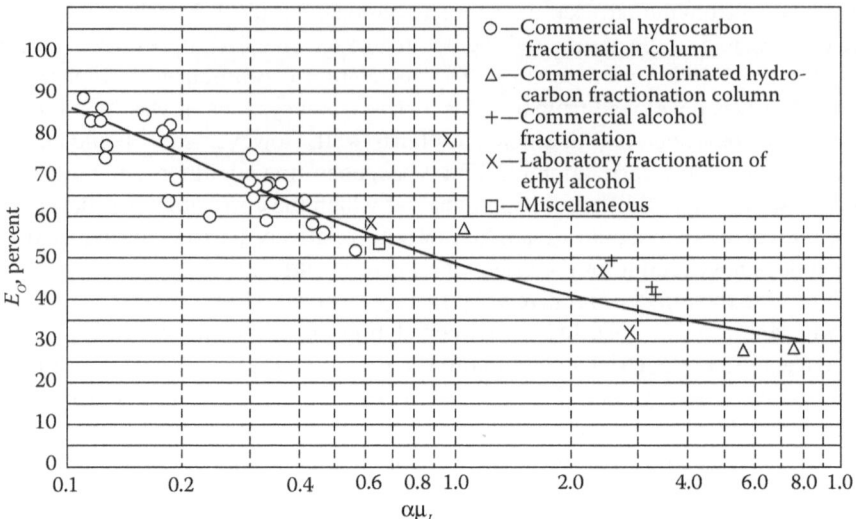

FIGURE 14.9 O'Connell correlation. (Reprinted with permission from H. E. O'Connell, Trans. A.I.Ch.E., 42, 741, 1946.)

NOMENCLATURE

a	Interfacial area per unit volume
A	Total tray area
A_a	Active tray area
A_d	Downcomer cross-sectional area
A_h	Total hole area
C	Parameter in Equation 14.1
C_D	Drag coefficient
C_F	Entrainment flooding capacity
C_0	Orifice coefficient
D	Tray diameter
d_h	Hole diameter
D_L	Diffusion coefficient in the liquid
D_L	Liquid droplet diameter
D_V	Diffusion coefficient in the vapor
e	Liquid entrainment
E_e	Murphree tray efficiency with entrainment
E_L	Murphree point efficiency in liquid terms
E_{ML}	Murphree tray efficiency in liquid terms
E_{MV}	Murphree tray efficiency in vapor terms
E_O	Overall tray efficiency
E_V	Murphree point efficiency in vapor terms
f	Fraction of flood velocity
F_b	Buoyant force

F_d	Drag force
F_F	Foaming factor
F_g	Droplet weight
F_{HA}	Hole area factor
F_{LV}	Correlating parameter, Figure 14.6
F_{ST}	Surface tension factor
H	Molar or total vapor enthalpy
h	Molar or total liquid enthalpy
h_a	Height of downcomer clearance
h_d	Dry tray pressure drop, clear liquid head
h_{da}	Liquid head loss through downcomer clearance
h_{dc}	Equivalent clear liquid downcomer backup height
h_{df}	Downcomer backup height
h_f	Froth height on the tray
h_l	Clear liquid head on the tray
h_t	Total pressure drop per tray, clear liquid head
h_w	Weir height
h_σ	Surface tension pressure drop, clear liquid head
K_{ji}	Vapor–liquid equilibrium distribution coefficient for component i on tray j
k_G	Gas phase mass transfer coefficient
K_G	Overall gas mass transfer coefficient
k_L	Liquid phase mass transfer coefficient
L	Liquid molar flow rate in the column
L_w	Weir length
l_t	Tray thickness
M_L	Liquid molecular weight
M_V	Vapor molecular weight
N	Number of theoretical stages
N_A	Number of actual trays
N_G	Number of gas phase mass transfer units
N_L	Number of liquid phase mass transfer units
N_{OG}	Number of overall gas phase mass transfer units
n_i	Rate of mass transfer of component i
P	Pressure
q_L	Liquid volumetric flow rate
S	Interfacial area between the liquid and vapor
T	Temperature
t_G	Gas phase residence time
t_L	Liquid phase residence time
U	Overall heat transfer coefficient
U_a	Vapor velocity based on active tray area
U_d	Volumetric liquid holdup in the downcomer
U_f	Vapor flood velocity
U_t	Volumetric liquid holdup on the tray
U_0	Hole velocity
V	Vapor molar flow rate in the column

X_{ji} Mole fraction of component i in the liquid on tray j
Y_{ji} Mole fraction of component i in the vapor on tray j
z Vertical distance
α Relative volatility
μ_L Liquid viscosity
ρ_L Liquid density
ρ_V Vapor density
σ Surface tension
ϕ_d Froth density in the downcomer
ϕ_t Relative froth density on the tray
ψ Fractional entrainment

SUBSCRIPTS

i Component designation
j Tray designation
m Logarithmic mean

SUPERSCRIPTS

L Liquid phase
V Vapor phase
$*$ Equilibrium composition

PROBLEMS

14.1. A sieve tray column is used as an absorber to remove acetone from a gas stream, using water as the absorbent. The process streams are as follows:

	Flow Rates (kmol/h)			
	Absorbent	Gas Feed	Liq Product	Gas Product
Argon	0	6.9	0	6.9
Oxygen	0	144.3	0.009	144.291
Nitrogen	0	536.0	0.017	535.983
Water	1943.0	5.0	1926.0	22.0
Acetone	0	10.3	10.25	0.05
Temp (°C)	25	25	22	25
Pressure (kPa)	101.3	101.3	101.3	90.0
Density (kg/m³)	1000.0			1.05
Mol. wt.	18.0			28.8
Diffusivity (m²/s)	0.0000112			0.127

Surface tension = 70 dynes/cm
K-value of acetone = 1.986

The sieve trays have 10% hole area, with 0.5 cm diameter holes, and 45 cm tray spacing. The weir height is 5 cm and the weir length is $0.73D_T$ (tray diameter).

Based on the top tray conditions, estimate the following:

a. The column diameter for a foaming factor of 0.85 and a fraction of flood of 0.75.
b. The vapor pressure drop per tray.
c. The number of transfer units, N_G and N_L.
d. The number of overall gas phase transfer units, N_{OG}.
e. The controlling resistance to mass transfer.
f. Murphree vapor efficiency for acetone, E_{MV}.

14.2. Based on preliminary studies, it was determined that a 10-tray column is appropriate for an absorption operation. The overall tray efficiency is estimated at 60%. For the indicated feed gas and absorbent, determine the expected flow rates and compositions of the products using the Kremser method.

Component	K-value	MW	Feed Gas (kmol/h)	Absorbent (kmol/h)
C1	29.1	16.04	1660	0
C2	6.5	30.07	173	0
C3	1.95	44.09	103	2
nC4	0.61	58.12	64	3
nC10	0.0011	142.28	0	495
Total (kmol/h)			2000	500
Temperature (°C)			30	30
Pressure (kPa)			250	250

The column is assumed to operate isothermally at a temperature of 30°C and an average pressure of 250 kPa. Stripping factors may be based on inlet liquid and vapor flow rates.

Sieve trays with 10% hole area and 0.5 cm diameter holes will be used. Trays are available in standard diameters of 0.25 m increments (0.25, 0.50, 0.75, 1, 1.25, 1.50, ..., m). Based on the top tray conditions, determine the required tray diameter rounded up to the nearest larger standard size. Assume a tray spacing of 0.5 m, a foaming factor of 0.80, and a fraction of flood of 0.80. The liquid density is given as 730 kg/m³, and the vapor density may be estimated based on the ideal gas equation. The liquid surface tension is 27 dynes/cm.

Calculate the vapor pressure drop per tray and the column total pressure drop. Check for the possibility of weeping. The weir height is 5 cm, and the weir length is 0.73 times the tray diameter.

Assuming a downcomer froth density of 0.5, estimate the downcomer backup. The downcomer clearance is 4 cm.

It is proposed, as an alternative to trays, to pack the same column with Koch-Sulzer structured packing, with a packing factor of 200 m²/m³. Using the same fluid flow rates and properties as in the trayed design, estimate the packed column pressure drop, and check for the possibility of flooding or channeling. The liquid viscosity is 0.35 centipoise.

The estimated HETP of the packing material is 0.7 m. Using the same stripping factors as in the trayed column, determine the expected flow rates and compositions of the products for the packed column.

14.3. A distillation column is designed based on conditions on the top tray, which are as follows:

Liquid flow: 500 kmol/h, MW = 141.38, density = 730 kg/m³, surface tension = 27 dyne/cm.

Vapor flow = 1946.9 kmol/h, MW = 19.585, density = 1.969 kg/m³.

a. Determine the required tray diameter rounded up to the nearest larger standard size. Sieve trays with 10% hole area and 0.5 cm diameter holes will be used. Trays are available in standard diameters of 0.25 m increments (0.25, 0.50, 0.75, 1, 1.25, 1.50, ..., m). Assume a tray spacing of 0.5 m, a foaming factor of 0.80, and a fraction of flood of 0.80.

b. Calculate the vapor pressure drop per tray and check for the possibility of weeping. The weir height is 5 cm and the weir length is 0.73 times the tray diameter.

c. Assuming a downcomer froth density of 0.5, estimate the downcomer backup. The downcomer clearance is 4 cm.

14.4. It is proposed to use an existing 3 m diameter distillation column with 17 actual trays, a total condenser, and a partial reboiler to separate the stream given below according to the indicated nC_4 mole fraction specifications in the distillate and bottoms. The estimated distillate composition is also given. It is required to determine if the column can perform this operation.

	Feed (kmol/h)	X_D	X_B
nC_4	200	0.950	0.010
nC_5	25	0.030	
nC_6	10	0.014	
nC_7	125	0.006	

a. Calculate by material balance the distillate and bottoms rates and compositions.

b. It was found by shortcut calculations that the specified separation could be achieved at total reflux with a minimum number of stages of 7. The same separation would require a minimum reflux ratio of 0.6. Existing pumping facilities can deliver a reflux rate of 275 kmol/h. Determine the required number of theoretical stages. What minimum overall tray efficiency is needed to make the specified separation with the existing column?

c. The feed will enter the column on the middle tray as saturated liquid. Estimate the liquid and vapor flows in the rectifying and stripping sections.

d. The average column pressure is 550 kPa, the top tray temperature is 50°C and the bottom tray temperature is 170°C. At these conditions the estimated liquid and vapor properties are as follows:

	Top Tray		Bottom Tray	
	Liquid	**Vapor**	**Liquid**	**Vapor**
Molecular weight	77	64	93	85
Density (kg/m³)	608	1.44	752	2.40
Surface tension (dyne/cm)	25		28	

The column has 3 m diameter sieve trays with 0.5 cm diameter holes and 10% hole area. The tray spacing is 45 cm. Assuming a foaming factor of 0.85, calculate the vapor flood velocity at the top tray. Check if the column diameter is acceptable. The fraction of flood velocity should be within a 60–85% range.

e. The weir height is 5 cm, and the weir length is 0.73 times the tray diameter. The downcomer clearance is 4 cm and the downcomer froth density is 0.6. For the bottom tray conditions, calculate the vapor pressure drop per tray and the downcomer backup.

14.5. An absorber column with 12 actual trays is used to lower the concentration of butane in a feed gas. The overall tray efficiency is estimated at 50%. The feed gas and absorbent are given below.

Component	K-value	MW	Feed Gas (kmol/h)	Absorbent (kmol/h)
C1	29.1	16.04	1675	0
C2	6.5	30.07	180	0
C3	1.95	44.09	90	1
nC4	0.61	58.12	60	4
nC10	0.0011	142.28	0	510
Total (kmol/h)			2005	515
Temperature (°C)			32	32
Pressure (kPa)			270	270

Calculate the expected flow rates and compositions of the products using the Kremser equation.

The column is assumed to operate at a constant temperature of 32°C and an average pressure of 270 kPa. Stripping factors may be based on inlet liquid and vapor flow rates.

Calculate the diameter of the column (rounded up to the nearest whole multiple of 0.25 m) based on the top tray conditions using the following data:

Liquid flow rate	515 kmol/h
Vapor flow rate	1950 kmol/h
Liquid density	725 kg/m^3
Vapor density	2 kg/m^3
Liquid mol. wt.	140
Vapor mol. wt.	20
Tray type	Sieve, with 10% hole area, 0.5 cm diameter holes
Weir height	5 cm
Weir length	0.73 times the tray diameter
Downcomer clearance	4 cm
Tray spacing	60 cm
Foaming factor	0.80
Fraction of flood	0.80
Surface tension	25 dynes
Liquid viscosity	0.36 centipoise

For the above column, calculate the vapor pressure drop per tray and the column total pressure drop. Check for the possibility of weeping.

Assuming a downcomer froth density of 0.6, estimate the downcomer backup.

It is proposed, as an alternative to trays, to pack the same column with Koch-Sulzer structured packing, with a packing factor of 200 m^2/m^3. Using the same fluid flow rates and properties as in the trayed design, estimate the packed column pressure drop, and check for the possibility of flooding or channeling.

14.6. A stream of hydrocarbon gas containing hydrogen is fed to an absorber for the purpose of raising the concentration of hydrogen in the gas. The feed gas stream and the absorbent, n-octane, are defined as follows:

	Mole Fraction in the Gas	Mole Fraction in the Absorbent
H_2	0.73	0.0
C_1	0.24	0.0
C_2	0.02	0.0
C_3	0.01	0.0
nC_8	0.00	1.0
Flow rate (kmol/h)	61,000	56,000
Temperature (°C)	35	40
Pressure (kPa)	2700	2675

The column has five equilibrium stages and will operate at 2650 kPa. It is required to size the column and rate its performance. The calculations may be based on the conditions at the bottom tray where the vapor rate is expected to be the highest. The flow rates and properties of the liquid and vapor around the bottom tray are as estimated below:

	Liquid	Vapor
Flow rate (kmol/h)	56,000	61,000
Molecular weight	114.0	6.34
Density (kg/m³)	703.5	6.60
Surface tension (dyne/cm)	20	

Sieve trays will be used with 60 cm spacing, 6 cm weir height, 0.6 cm hole diameter, 0.25 cm tray thickness, 5 cm downcomer clearance, and hole area 10% of the total tray area. The foaming factor is 0.80 and the froth density in the downcomer is 0.5. The target fraction of flood velocity is 0.70.

Determine by hand calculations the column diameter, then the following items with the diameter rounded up to the next standard tray diameter, available in 0.25 m increments. Recommend possible design modifications for situations where unacceptable column operation is indicated.

a. The fraction of flood velocity.
b. The vapor pressure drop per tray. A pressure drop greater than 1.5 kPa per tray is considered excessive.
c. The downcomer backup.
d. The possibility of weeping.
e. The holdups on the tray and in the downcomer.

14.7. A distillation column is to be designed based on the following top tray conditions:

	Liquid	Vapor
Flow (kmol/h)	600	2100
Temperature (°C)	27	27
Pressure (kPa)	2600	2600
Density (kg/m³)	725	2.1
Molecular weight	139	20.5
Surface tension (dyne/cm)	28	
Foaming factor	0.85	

It is proposed to use sieve trays with 10% hole area and 0.5 cm hole diameter, a weir height of 6 cm and a weir length that is 0.73 times the tray diameter. The tray thickness is 0.25 cm and the tray spacing is 50 cm.

a. Determine the required tray diameter for a fraction of flood of 0.75.
b. Calculate the expected pressure drop if the tray diameter is rounded up to the next standard diameter, available in whole feet.

14.8. The design of a column will be based on the data given below.

a. Determine the column diameter for a fraction of flood of 70%.
b. With the column diameter rounded up to the next standard diameter, available in 0.25 m increments, calculate the downcomer backup, assuming a froth density of 0.5.

Given data

Sieve trays, 10% hole area, 0.5 cm hole diameter.

Weir height, 5 cm; weir length, 0.73 times tray diameter.

Tray thickness, 0.25 cm; tray spacing, 50 cm; height of downcomer clearance, 5 cm.

The liquid and vapor properties on a typical tray are as follows:

Liquid: 200 kmol/h; density, 735 kg/m^3; molecular weight, 132; surface tension, 27 dyne/cm; foaming factor, 0.85.

Vapor: 500 kmol/h; density, 3.5 kg/m^3; molecular weight, 38.

Temperature, 35°C; pressure, 2400 kPa.

14.9. In a distillation column, the average pressure is 550 kPa, the top tray temperature is 50°C, and the bottom tray temperature is 170°C. At these conditions the estimated liquid and vapor properties at the top tray are as follows:

	Liquid	Vapor
Molecular weight	77	64
Density (kg/m^3)	610	1.4
Surface tension (dyne/cm)	25	
Flow rate (kmol/h)	280	490

The column has 2.75 m diameter sieve trays with 0.5 cm diameter holes and 10% hole area. The tray spacing is 45 cm. Assuming a foaming factor of 0.85, calculate the vapor flood velocity at the top tray. Check if the column diameter is acceptable. The fraction of flood velocity should be within a 60–85% range.

REFERENCES

Bennett, D. L., R. Agrawal, and P. J. Cook, *AIChE J.*, 29, 434–442, 1983.

Bubble Tray Design Manual, *Tray Efficiency Research Program,* New York, American Institute of Chemical Engineers, 1960.

Chan, H. and J. R. Fair, *Ind. Eng. Chem. Process Des. Dev.*, 23, 814–819, 1984.

Colburn, A. P., *Ind. Eng. Chem.*, 28, 526, 1936.

Drickamer, H. G. and J. R. Bradford, *Trans. A.I.Ch.E.*, 39, 319, 1943.

Fair, J. R., *Petro/Chem Engineer*, 33(10), 211–218, Sept. 1961.

Koch-Glitsch, L. P., Wichita, K. S., and Murphree, E. V., *Ind. Eng. Chem.*, 17, 747, 1925.

Murphree, E. V., *Ind. Eng. Chem.*, 17, 747, 1925.

Nutter Engineering Company, Float Valve Tray Design Manual, Tulsa, OK, 1976.

O'Connell, H. E., *Trans. A.I.Ch.E.*, 42, 741, 1946.

Oliver, E. D., *Diffusional Separation Processes, Theory, Design, and Evaluation*, New York, John Wiley and Sons, pp. 320–321, 1966.

Smith, B. D., *Design of Equilibrium Stage Processes*, New York, McGraw-Hill, 1963.

Van Winkle, M., *Distillation*, New York, McGraw-Hill, 1967.

15 Packed Columns

For the most part in this book, multistage columns were studied in terms of equilibrium stages, and a stage was assumed to represent a column tray. Multistage vapor–liquid countercurrent flow for the purpose of separation can also take place in packed columns. In these devices, instead of trays, the column is filled with an inert packing material designed for maximum mass transfer between the vapor and the liquid and for low-pressure drop. The vapor and liquid compositions vary continuously with packing height rather than discretely as in trayed columns.

In trayed columns the composition changes in a stagewise manner and each stage is treated as an equilibrium stage. Lack of equilibrium is accounted for by using some form of tray efficiency, which may be estimated on the basis of mass transfer analysis. In counterflow packed columns, the passing phases are never considered to be truly at equilibrium. In trayed columns, the operating line, such as in the McCabe-Thiele diagram for binary distillation, is not truly a continuous line but a series of points, each representing the compositions of passing phases between trays. The operating line for a packed column, on the other hand, is a continuous line representing the compositions of the passing phases over the height of the column.

In contrast to the cross-flow mechanism characterizing trayed columns, the flow in packed columns is vapor–liquid counterflow. Whereas in trayed columns the construction parameter that determines separation capacity is the number of trays, in packed columns it is the packing height. A certain packing height can accomplish the same separation as an equilibrium stage. This height is known as the height equivalent to a theoretical plate, or HETP. If the total column packing height and the HETP are known, the number of equilibrium stages can be calculated by dividing the packing height by the HETP. This is analogous to determining the number of equilibrium stages in a trayed column from the number of actual trays and the overall tray efficiency. Once the number of equilibrium stages is known, a packed column can be calculated by any of the methods described in Chapters 12 and 13.

In a design situation, the height of a packed column can be determined from the number of theoretical stages and the HETP. The number of theoretical stages is determined by analytical methods as described in Chapters 12 and 13, and the HETP is estimated independently. For a known packing height and equivalent number of equilibrium stages, the HETP may be calculated by dividing the former by the latter. The HETP thus obtained may be used as an estimate for the HETP of similar columns with the same packing type and under similar operating conditions.

An alternative approach to the HETP method for calculating the column packing height, also discussed in this chapter, is based on the height of a transfer unit (HTU) and the number of transfer units (NTU). In this approach the packing height is calculated by a theoretical analysis of mass transfer phenomena across the liquid and vapor phases.

Another parameter in the design of a packed column is the column diameter, which is determined on the basis of hydraulic considerations, such as pressure drop and flooding conditions. The hydraulic behavior of the column depends on the packing type and operating conditions. Correlations based on experimental data are available for calculating column packing hydraulics, and are discussed in this chapter.

The use of packing instead of trays in multistage separation columns is common for column diameters 3–4 ft or smaller. More recently, packing has been used for larger columns because of its low pressure drops, favorable efficiencies, and high vapor capacity. Packed columns are also the preferred choice where corrosion is a potential problem. The packing material used in these situations is ceramic or polymeric. Another characteristic of packed columns is their low liquid holdup, which could reduce the amounts of off-specification products at startup and shutdown.

15.1 CONTINUOUS DIFFERENTIAL MASS TRANSFER

In general, the performance of a packed column can be represented by an equilibrium curve, $Y^* = f(X)$, and an operating line. The latter relates crossing vapor and liquid compositions at any point in the column, and is based on material balance for the transferred component. For constant vapor and liquid flows, L and V, if at some reference point the concentrations of that component in the liquid and vapor are X_T and Y_T, and at another point in the column they are X and Y, then by material balance, $L(X - X_T) = V(Y_T - Y)$, or, $Y = Y_T + (L/V)(X_T - X)$.

In a simplified trayed column section, if the operating line and the equilibrium curve are assumed straight and parallel, the number of equilibrium stages required to change the vapor composition by a certain amount may be calculated simply by dividing the required composition change by the composition change per tray. Figure 15.1 represents such a process for a ternary system where one component, the solute, is absorbed from an inert gas by a counterflowing liquid. The concentration in the gas phase of the component being transferred is represented by Y_B at the inlet and by Y_T at the outlet. (No component subscripts are used since reference is always made to the component being transferred.) In the liquid, the concentrations of the transferred component are represented by X_T at the inlet and X_B at the outlet. Points (X_T, Y_T) and (X_B, Y_B) lie on the operating line because they both represent concentrations of passing phases, one at the top of the column and the other at its bottom.

If the concentrations of the solute in the liquid and vapor flowing between any two stages are X and Y, represented by point P on the operating line, the concentration in the vapor at equilibrium with the liquid is Y^*, and the concentration in the liquid at equilibrium with the vapor is X^* (Figure 15.1). The change per tray in the concentration in the vapor is $Y - Y^*$. Since this difference is constant when the two lines are parallel, the required number of equilibrium stages is calculated as

$$N = \frac{Y_B - Y_T}{Y - Y^*} \tag{15.1}$$

If instead of the discrete stages a packed column is used, an equivalent number of stages would be calculated by the same equation. In general, the operating line and

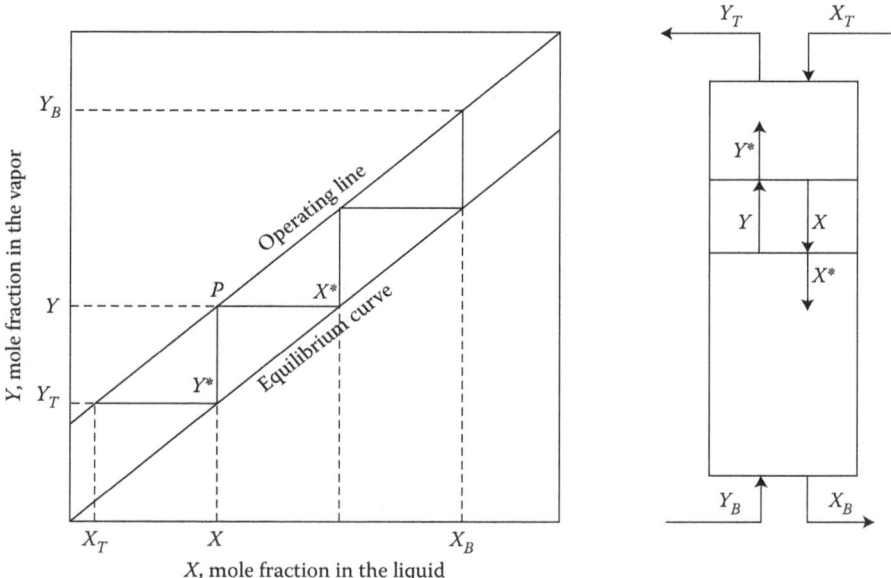

FIGURE 15.1   Simple absorber: straight, parallel, operating line and equilibrium curve.

the equilibrium curve are neither straight nor parallel, and Equation 15.1 would be written in differential form:

$$dN = \frac{dY}{Y - Y^*} \tag{15.2}$$

where dN is the differential NTU required to bring about a change dY in the vapor composition. The difference $Y - Y^*$ is the driving force causing the composition change. The NTU, which is the ratio of the composition change to the mean driving force, is obtained by integrating Equation 15.2 between the inlet and outlet vapor compositions:

$$NTU = \int_{Y_T}^{Y_B} \frac{dY}{Y - Y^*} \tag{15.3}$$

In the special case where the operating line and the equilibrium curve are straight and parallel, the difference $Y - Y^*$ is constant and the integration gives

$$NTU = \frac{Y_B - Y_T}{Y - Y^*} \tag{15.4}$$

which is the same as the number of equilibrium stages calculated by Equation 15.1. In general, the NTU is close to but not identical to the equivalent number of equilibrium stages. It will also be shown that the NTU may be defined in a number of different ways, giving potentially different NTU values for the same separation. The NTU could, for instance, be based on the change in the liquid composition instead of the vapor composition. Referring to Figure 15.1, the liquid-based NTU is calculated as:

$$NTU = \int_{X_T}^{X_B} \frac{dX}{X^* - X} \tag{15.5}$$

For straight and parallel operating line and equilibrium curve,

$$NTU = \frac{1}{X^* - X} \int_{X_T}^{X_B} dX \tag{15.5a}$$

or

$$NTU = \frac{X_B - X_T}{X^* - X} \tag{15.5b}$$

From the geometry of the $Y - X$ diagram in Figure 15.1, it can be seen that, for this simplified case, the liquid-based NTU equals the gas-based NTU (and also equals the equivalent number of equilibrium stages).

15.1.1 Nonparallel, Straight Operating Line, and Equilibrium Curve

Figure 15.2 is a Y–X diagram for a simple absorber column section, similar to the process described above, but with nonparallel, straight operating line, and equilibrium curve. The model being analyzed consists of three components: a liquid component, a vapor component, and a solute. The solute is at low concentrations, and it is the only component that can transfer across the phases. From the geometry of the diagram, the following relationships may be written as

$$\frac{\overline{PQ}}{\overline{BC}} = \frac{\overline{AP}}{\overline{AB}}$$

In terms of the compositions,

$$\overline{PQ} = (Y - Y^*) - (Y_B - Y_B^*)$$

$$\overline{BC} = (Y_T - Y_T^*) - (Y_B - Y_B^*)$$

$$\frac{\overline{AP}}{\overline{AB}} = \frac{Y_B - Y}{Y_B - Y_T}$$

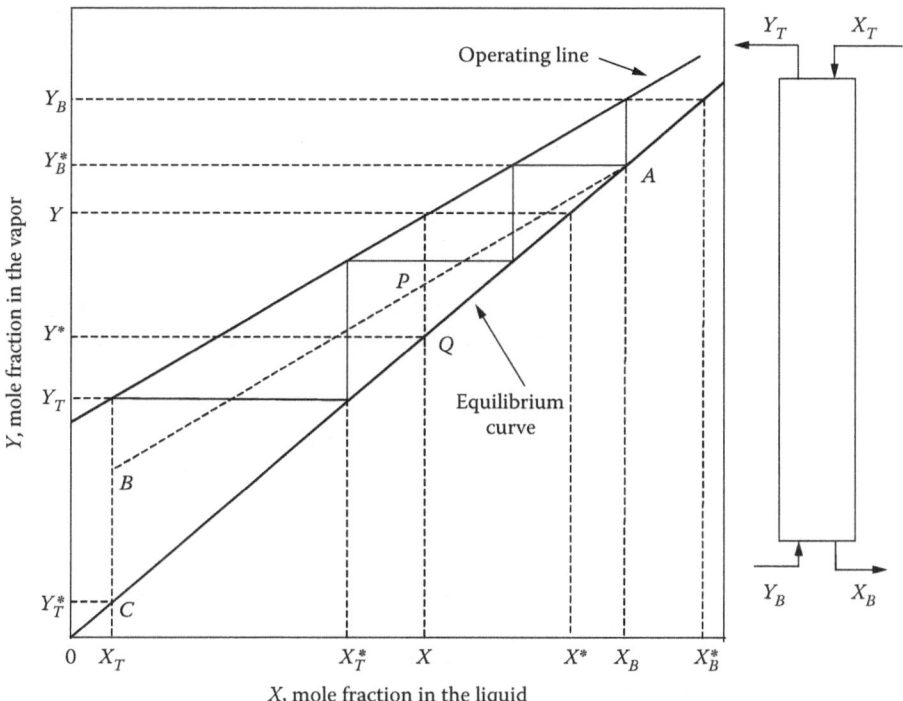

FIGURE 15.2 Dilute solution absorber.

This leads to the equation,

$$\frac{(Y - Y^*) - (Y_B - Y_B^*)}{(Y_T - Y_T^*) - (Y_B - Y_B^*)} = \frac{Y_B - Y}{Y_B - Y_T} \tag{15.6}$$

Rearranging,

$$Y - Y^* = (Y_B - Y_B^*) + E(Y_B - Y)$$

where

$$E = \frac{(Y_T - Y_T^*) - (Y_B - Y_B^*)}{Y_B - Y_T}$$

By substituting the value of $Y-Y^*$ in Equation 15.2 and integrating from $Y = Y_T$ to $Y = Y_B$, the following equation is obtained for the number of overall gas-phase transfer units:

$$
\begin{aligned}
N_{OG} &= \int_{Y_T}^{Y_B} \frac{dY}{(Y_B - Y_B^*) + E(Y_B - Y)} \\
&= \frac{1}{E} \ln \left[\frac{(Y_B - Y_B^*) + E(Y_B - Y_T)}{Y_B - Y_B^*} \right] \\
&= \frac{Y_B - Y_T}{(Y_B - Y_B^*) - (Y_T - Y_T^*)} \ln \left(\frac{Y_B - Y_B^*}{Y_T - Y_T^*} \right)
\end{aligned}
\tag{15.7}
$$

The term N_{OG} is one of several ways for defining NTU discussed in this chapter. A similar derivation for the number of overall liquid-phase transfer units gives

$$
N_{OL} = \frac{X_B - X_T}{(X_B^* - X_B) - (X_T^* - X_T)} \ln \frac{X_B^* - X_B}{X_T^* - X_T}
\tag{15.8}
$$

Although this analysis was made in reference to an absorber model, Equations 15.7 and 15.8 apply equally well to a stripper as long as it is a three component system where one of the components, the solute, is at dilute concentrations, and is the only one that can transfer across the phases.

The NTU are calculated numerically in Example 15.1 and compared to the equivalent number of equilibrium stages.

EXAMPLE 15.1: DILUTE-SOLUTION ABSORBER

A component is absorbed from a gas by a counterflow liquid in a process similar to that depicted in Figure 15.2. The following information is given, where mole fractions refer to the transferred component:

Inlet vapor composition:	$Y_B = 0.006$
Outlet liquid composition:	$X_B = 0.0002$
Equilibrium relationship:	$Y^* = 25X$
Operating line slope:	$m = 20$
Number of equilibrium stages:	$N = 3$

It is required to calculate the outlet vapor and the inlet liquid compositions and the NTU required for the separation.

Solution

The operating line, with the given slope of 20, is written as

$$
Y = 20X + b
$$

Since the operating line must pass through the point $X_B = 0.0002$, $Y_B = 0.006$, the intercept b is calculated as

$$
b = Y_B - 20X_B = 0.006 - (20)(0.0002) = 0.002
$$

The operating line equation is, therefore,

$$Y = 20X + 0.002$$

or

$$X = (Y - 0.002)/20$$

The equilibrium stage calculations are first performed by alternating between the operating line and the equilibrium curve. This is the analytical equivalent of the graphical construction of equilibrium stages described in Section 8.5. The numerical subscripts of the mole fractions refer to the equilibrium stage number from where the stream is leaving. Equilibrium stages in this example are numbered from the bottom up.

$Y_B = 0.006$
$X_1 = X_B = 0.0002$
$Y_1 = 25X_1 = (25)(0.0002) = 0.005$
$X_2 = (Y_1 - 0.002)/20 = (0.005 - 0.002)/20 = 0.00015$
$Y_2 = 25X_2 = (25)(0.00015) = 0.00375$
$X_3 = (Y_2 - 0.002)/20 = (0.00375 - 0.002)/20 = 0.0000875$
$Y_3 = Y_T = 25X_3 = (25)(0.0000875) = 0.0021875$
$X_T = (Y_T - 0.002)/20 = (0.0021875 - 0.002)/20 = 0.000009375$

The calculations are stopped after three equilibrium stages as specified in the problem definition. The equilibrium compositions at the inlet and outlet:

$$Y_B^* = 25X_B = (25)(0.0002) = 0.005$$

$$X_B^* = Y_B/25 = 0.006/25 = 0.00024$$

$$Y_T^* = 25X_T = (25)(0.000009375) = 0.000234375$$

$$X_T^* = Y_T/25 = 0.0021875/25 = 0.0000875$$

The NTU is now calculated from Equations 15.7 and 15.8:

$$N_{OG} = \frac{0.006 - 0.0021875}{(0.006 - 0.005) - (0.0021875 - 0.000234375)}$$
$$\times \ln\frac{0.006 - 0.005}{0.0021875 - 0.000234375} = 2.678$$

$$N_{OL} = \frac{0.0002 - 0.000009375}{(0.00024 - 0.0002) - (0.0000875 - 0.000009375)}$$
$$\times \ln\frac{0.00024 - 0.0002}{0.0000875 - 0.000009375} = 3.347$$

The liquid-based NTU is somewhat larger than the number of equilibrium stages, while the gas-based NTU is somewhat smaller. This is expected since the mean driving force for the vapor $(Y - Y^*)$ is greater than that for the liquid $(X^* - X)$.

In general, when the operating line and equilibrium curve are not straight, the NTU must be determined by numerical or graphical integration of Equation 15.3 or 15.5 or other equivalent form of these equations.

15.2   RATE OF MASS TRANSFER

When countercurrent vapor–liquid contacting for the purpose of inducing mass transfer between the phases is carried out in a continuous differential manner as in packed columns, phase equilibrium is not, in general, achieved at any point in the column. In a trayed column, the approach to equilibrium on each tray is determined by the tray efficiency. In a packed column, the separation is controlled by the rate of mass transfer between the phases. The rate of mass transfer determines the column packing height required to achieve a specified separation. Using the concept of transfer units, the column packing height is calculated as the product of the NTU and the HTU.

The relationship between the compositions of a given component in the liquid and vapor phases at a given column height is determined by the mass transfer rate of that component from one phase to the other. The mass transfer rate is a function of the resistance in the vicinity of the phase boundary and the composition driving force.

Several models have been proposed to represent the mechanics of mass transfer across the phases. In the two-film theory, the resistance to mass transfer is assumed to take place in a liquid film and a vapor film at the phase boundary interface, as shown schematically in Figure 15.3b. The coordinates X and Y in Figure 15.3a represent the mole fractions of a given component in the liquid and vapor, respectively. For simplicity, component subscripts are dropped and the discussions refer to one

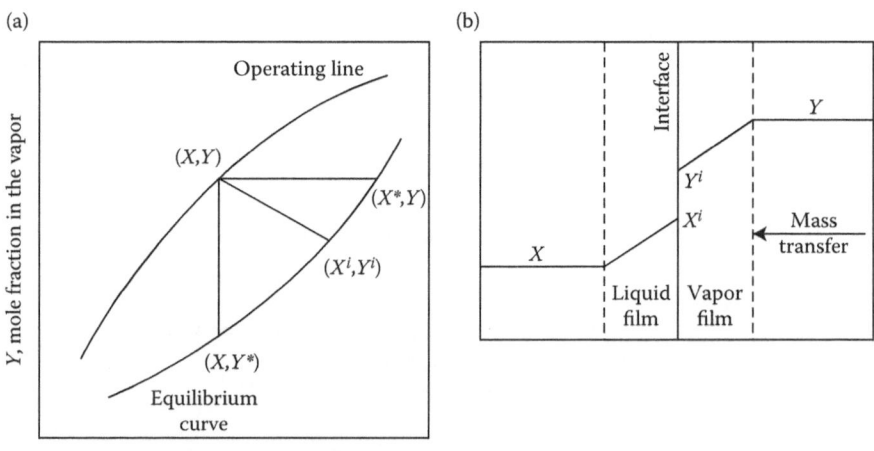

FIGURE 15.3   Two-film mass transfer model, (a) operating line and equilibrium curve, (b) mass transfer at the interface films.

typical component. Point (X,Y) on the operating line represents a point at a given column height where the mole fraction in the bulk liquid is X and the mole fraction in the bulk vapor is Y. The compositions in each phase and at the phase interface at a given column height are assumed constant across the column cross section since plug flow is assumed.

A liquid film and a vapor film are assumed to exist at the interface between the two phases, at which point the mole fractions in the liquid and vapor are X^i and Y^i. These compositions are, in general, different from the bulk liquid and vapor compositions and are assumed to correspond to equilibrium compositions. Thus, the point (X^i,Y^i) lies on the equilibrium curve (Figure 15.3a) at the same column height as point (X,Y). The equilibrium curve is obtained from vapor–liquid equilibrium data or correlations, as described in Chapter 1.

Two more points may be defined on the equilibrium curve corresponding to the same column height as (X,Y). Point (X,Y^*) represents the liquid bulk composition, X, and the vapor composition, Y^*, that would be at equilibrium with it. Point (X^*,Y) represents the vapor bulk composition, Y, and the liquid composition, X^*, that would be at equilibrium with it. Note that X^* and Y^* do not exist at the same column height as (X,Y).

Figure 15.3b is a schematic of the variation of the compositions in the vicinity of the phase interface according to the two-film theory. At a given column height, the liquid and vapor compositions, X and Y, are assumed constant, except in the mass transfer films where composition gradients are hypothesized. The compositions at the interface are the equilibrium compositions X^i and Y^i. The compositions shown in Figure 15.3b correspond to a component that is being transferred from the vapor to the liquid.

The rate of mass transfer across a film per unit of interface area is determined by the composition gradient, diffusivity, and film thickness. One way of expressing this relationship is through a mass transfer coefficient that includes the effects of both diffusivity and film thickness. If the compositional driving force is represented as the difference between the component mole fraction in the bulk and at the interface, the rate of mass transfer in the liquid film is given as

$$N = k_X(X^i - X) \tag{15.9}$$

and in the vapor film

$$N = k_Y(Y - Y^i) \tag{15.10}$$

where k_X and k_Y are the mass transfer film coefficients in the liquid and vapor, respectively, based on the mole fraction driving force. The mass transfer rate, N, is the number of moles of the transferred component crossing the film per unit area per unit time. The dimensions of k_X and k_Y are mole/(mole fraction × time × area). These coefficients are specific to the diffusing component and the mass transfer medium. Depending on the way the compositions are defined and on other considerations, the mass transfer rate equation may be expressed in different alternative forms, as presented later in this discussion. Appropriate mass transfer coefficients must be used with the particular form of the mass transfer rate equation.

Equations 15.9 and 15.10 are empirical with respect to the definition of the mass transfer coefficients, but the form of the equations is based on molecular diffusion theory. Applying the theory to a multi-component mixture where each component has a distinct diffusivity is impractically complex and must rely on diffusivity data for all the components in the mixture. To derive usable equations from the diffusion theory, certain simplifying assumptions must be made. The basis for the derivation of Equations 15.9 and 15.10 is to assume that mass transfer takes place either as equimolar counterdiffusion or as unimolar diffusion under dilute conditions.

In equimolar diffusion, a binary mixture is assumed where the two components diffuse in opposite directions at equal rates. These conditions exist in binary distillation where a mole of component 1 is vaporized for each mole of component 2 that is condensed. In unimolar diffusion, one component diffuses through a second, stagnant one. This is typical of an absorption process where one component diffuses through the gas phase to the interface boundary, is absorbed by the liquid, and then diffuses to the bulk of the liquid. The other gas components are assumed to remain in the gas and the liquid components to remain in the liquid.

Note that mass is transferred in the opposite direction of the composition gradient. For instance, for the case described in Figure 15.3 and Equations 15.9 and 15.10, the component in question flows from a higher X^i to a lower X and from a higher Y to a lower Y^i.

Since the rates of mass transfer across the liquid and vapor films must be equal at steady state (by material balance), Equations 15.9 and 15.10 may be combined to give the following:

$$N = k_X(X^i - X) = k_Y(Y - Y^i) \tag{15.11}$$

The values of X^i and Y^i are usually unavailable and difficult to measure. It is possible, however, to calculate from vapor–liquid equilibrium correlations the vapor composition Y^* that would be at equilibrium with X, or the liquid composition X^* that would be at equilibrium with Y. The rate of mass transfer could then be defined in terms of these compositions as follows:

$$N = K_Y(Y - Y^*) = K_X(X^* - X) \tag{15.12}$$

Since the mole fractions used in this equation do not represent film compositions but rather bulk compositions, the mass transfer coefficients, K_Y and K_X, are referred to as overall mass transfer coefficients for the vapor and liquid, respectively, based on the mole fraction driving force. The dimensions of K_Y and K_X are the same as those for k_Y and k_X.

Equations 15.11 and 15.12 may be written based on partial pressures in the vapor phase and concentrations in the liquid phase as the driving forces:

$$N = k_L(c^i - c) = k_G(p - p^i) \tag{15.11a}$$

$$N = K_G(p - p^*) = K_L(c^* - c) \tag{15.12a}$$

The concentrations are expressed as $c = \rho_L X/M_L$ where c is in moles/volume, and M_L and ρ_L are the liquid molecular weight and mass density. The partial pressures are $p = PY$ where P is the total pressure. By comparing Equations 15.11, 15.11a and 15.12, 15.12a, the different mass transfer coefficients are interrelated as follows:

$$k_X = k_L(\rho_L/M_L) \quad K_X = K_L(\rho_L/M_L)$$

$$k_Y = k_G P \qquad K_Y = K_G P$$

(15.12b)

The mass transfer film coefficients (k_X and k_Y) are more directly related to physical properties, such as diffusivities and hydrodynamic conditions, than the overall mass transfer coefficients (K_X and K_Y) and are, therefore, easier to predict and correlate. On the other hand, since the interfacial compositions are difficult to measure or predict, it is more convenient to use overall mass transfer coefficients for process design and analysis. The relationship between the two sets of mass transfer coefficients may be derived by equating different groups in Equations 15.11 and 15.12 as follows:

$$K_Y(Y - Y^*) = k_Y(Y - Y^i)$$

This can be rearranged as

$$\frac{1}{K_Y} = \frac{1}{k_Y}\frac{Y - Y^*}{Y - Y^i}$$

$$= \frac{1}{k_Y}\frac{(Y - Y^i) + (Y^i - Y^*)}{Y - Y^i}$$

$$= \frac{1}{k_Y} + \frac{1}{k_Y}\frac{Y^i - Y^*}{Y - Y^i}$$

The second term in the last equation is evaluated from Equation 15.11; the resulting equation is

$$\frac{1}{K_Y} = \frac{1}{k_Y} + \frac{1}{k_X}\frac{Y^i - Y^*}{X^i - X}$$

(15.13)

A similar expression is derived for K_X:

$$\frac{1}{K_X} = \frac{1}{k_X} + \frac{1}{k_Y}\frac{X^* - X^i}{Y - Y^i}$$

(15.14)

From Figure 15.3a, the ratio of mole fraction differences in Equation 15.13 and the reciprocal of the ratio of mole fraction differences in Equation 15.14 represent an average slope of the equilibrium curve. If this slope is H, Equations 15.13 and 15.14 may be written as

$$\frac{1}{K_Y} = \frac{1}{k_Y} + \frac{H}{k_X} \tag{15.15}$$

and

$$\frac{1}{K_X} = \frac{1}{k_X} + \frac{1}{Hk_Y} \tag{15.16}$$

If the equilibrium curve is a straight line passing through the origin, $H = K$, the K-value of the solute (the transferred component). Equations 15.15 and 15.16 may be written in terms of the other mass transfer coefficients (equation set 15.12b), with H replaced with K:

$$\frac{1}{K_G} = \frac{1}{k_G} + \frac{KPM_L}{\rho_L} \frac{1}{k_L} \tag{15.15a}$$

$$\frac{1}{K_L} = \frac{1}{k_L} + \frac{\rho_L}{KPM_L} \frac{1}{k_G} \tag{15.16a}$$

If the mass transfer coefficients were to be determined experimentally based on their defining equations (Equations 15.11 or 15.12), it is quite clear that the overall coefficients (K_X or K_Y) can be more conveniently determined than the film coefficients (k_X or k_Y). Whereas film coefficients involve almost impossible-to-measure interfacial compositions, overall coefficients can be determined based on bulk and equilibrium composition data.

The interfacial area is another quantity that is impractical to measure. Therefore, *capacity coefficients*, defined as $(k_X a)$, $(k_Y a)$, $(K_X a)$, and $(K_Y a)$, are used instead of the mass transfer coefficients. The quantity a is the interfacial area per unit of active column volume. Since both a and the transfer coefficients must be determined in one way or another (usually experimentally), there is no benefit gained from determining them separately. They are, therefore, most commonly reported as groups. In terms of these coefficients, Equations 15.11 and 15.12 are rewritten as

$$Na = (k_X a)(X^i - X) = (k_Y a)(Y - Y^i) \tag{15.17}$$

and

$$Na = (K_X a)(X^* - X) = (K_Y a)(Y - Y^*) \tag{15.18}$$

The quantity Na in Equations 15.17 and 15.18 represents the number of moles transferred between the phases per unit column or packing volume per unit time. Similar expressions may be written for the other mass transfer coefficients:

$$Na = (k_L a)(c^i - c) = (k_G a)(p - p^i) \tag{15.17a}$$

$$Na = (K_G a)(p - p^*) = (K_L a)(c^* - c) \tag{15.18a}$$

With Equation 15.18 as the defining equation, the mass transfer coefficients in a column are conceptually determined by analyzing samples of liquid and gas for the concentration of the transferred component. The samples are drawn at a given height of the column, operated with the fluids (absorbent, gas, and solute) and packing material under study at designated temperature and pressure conditions. The measured quantities represent X and Y in Equation 15.18 and Figure 15.3 (the bulk compositions). The compositions at equilibrium with the bulk compositions, X^* and Y^*, are obtained from the equilibrium curve (Figure 15.3). Additional samples across a small incremental height of packing are analyzed to determine, by material balance, the rate of mass transfer Na taking place in the packing volume contained in the incremental packing height. The mass transfer coefficients would then be calculated from Equation 15.18. In Example 15.2 the coefficients are calculated based on overall column data.

15.2.1 Mass Transfer Correlations

Mass transfer coefficients in packed columns are influenced by the packing type and size, by the liquid and vapor velocities in the column, and by the diffusivities of the solute in the liquid and vapor. Mass transfer coefficients are usually correlated in terms of these parameters by empirical correlations based on extensive experimental data (Seader and Henley, 1998).

15.3 MASS TRANSFER IN PACKED COLUMNS

The mass transfer equations discussed above are now combined with a material balance on the transferred component to calculate the column or packing height required for a given separation. The column cross-sectional area A is assumed known at this point although in a complete column design A must be determined based on pressure drop considerations. The column, which is in countercurrent flow with only liquid feed and vapor product at the top, and vapor feed and liquid product at the bottom (absorber, stripper, column section), is defined as follows:

Column top:
  Liquid in:
    L_T = liquid rate, kmol/h
    X_T = mole fraction of transferred component in L_T
  Gas out:
    V_T = vapor rate, kmol/h
    Y_T = mole fraction of transferred component in V_T
Column bottom:
  Liquid out:
    L_B = liquid rate, kmol/h
    X_B = mole fraction of transferred component in L_B

Gas in:

V_B = vapor rate, kmol/h

Y_B = mole fraction of transferred component in V_B

The column height h is measured from the column bottom. The total height at the top is h_T. At any column height h, the liquid and vapor rates and the mole fractions of the transferred component in the liquid and vapor are L, V, X, and Y. In general, these variables change continuously over the height of the column. A material balance on the transferred component is therefore carried out over a differential column height dh. The differential volume is Adh, and the differential mass transfer area is $aAdh$. The rate of mass transfer in this differential volume is $NaAdh$. This rate is equated to the rate of change of the transferred component in either phase:

$$NaAdh = d(LX) = -d(VY) \tag{15.19}$$

Or, in terms of concentrations and partial pressures,

$$NaAdh = d(LcM_L/\rho_L) = -d(Vp/P) \tag{15.19a}$$

The negative sign indicates that the component is transferred from the vapor to the liquid. Substituting the value of Na from Equations 15.17 and 15.18 into 15.19 results in the following:

$$d(LX) = -d(VY) = (k_X a)(X^i - X)Adh = (k_Y a)(Y - Y^i)Adh \tag{15.20}$$

$$d(LX) = -d(VY) = (K_X a)(X^* - X)Adh = (K_Y a)(Y - Y^*)Adh \tag{15.21}$$

Equivalent results may be obtained in terms of concentrations and partial pressures, using Equations 15.17a and 15.18a:

$$d(LcM_L/\rho_L) = -d(Vp/P) = (k_L a)(c^i - c)Adh = (k_G a)(p - p^i)Adh \tag{15.20a}$$

$$d(LcM_L/\rho_L) = -d(Vp/P) = (K_L a)(c^* - c)Adh = (K_G a)(p - p^*)Adh \tag{15.21a}$$

Any one of these equations may be used to calculate the required column height, h_T, by integrating over the specified composition change. In the following derivation, the mole fraction driving force is used with overall mass transfer. Equation 15.21 is rearranged and integration is carried out over the specified composition change, X_B to X_T or Y_B to Y_T

$$h_T = \int_0^{h_T} dh = \int_{X_B}^{X_T} \frac{d(LX)}{(K_X a)A(X^* - X)} \tag{15.22}$$

or

$$h_T = \int_0^{h_T} dh = -\int_{Y_B}^{Y_T} \frac{d(VY)}{(K_Y a)A(Y - Y^*)} \qquad (15.23)$$

In equimolar counterdiffusion or unimolar dilute diffusion, V and L are assumed constant, allowing Equations 15.22 and 15.23 to be written as

$$h_T = \frac{L}{(K_X a)A} \int_{X_B}^{X_T} \frac{dX}{X^* - X} \qquad (15.24)$$

and

$$h_T = \frac{V}{(K_Y a)A} \int_{Y_T}^{Y_B} \frac{dY}{Y - Y^*} \qquad (15.25)$$

The applicability of these equations could be expanded by keeping more terms inside the integral and carrying out the integration numerically. Another technique for the unimolar diffusion case is to express liquid and vapor flow rates on a solute-free basis, so they could be assumed constant over broader operating conditions. Compositions in this case would be expressed as mole ratios instead of mole fractions, and the phase equilibrium data would be converted accordingly.

The quantities before the integrals and the integrals themselves in Equations 15.24 and 15.25 are referred to as the height of a transfer unit (HTU) and the number of transfer units (NTU), respectively. Thus, the column height calculation is reduced to evaluating the product of two quantities, each of which has a certain physical significance. The HTU has the dimension of length and expresses the efficiency of the device in facilitating mass transfer. It is a function of the vapor and liquid flow rates and the mass transfer characteristics of the system. Shorter HTU's represent higher mass transfer efficiencies. The NTU, as indicated in Section 15.1, reflects the relationship between the operating line and the equilibrium curve. It is a dimensionless number, conceptually equivalent to the ratio of the required change in the vapor or liquid composition to the mean compositional driving force.

The HTU is determined from experimental data or empirical correlations, and the NTU is related to equilibrium and operating data. Since both HTU and NTU depend on the particular definition of mass transfer coefficients and compositions in the mass transfer equations, it is important to use compatible pairs of HTU and NTU. For instance, Equations 15.24 and 15.25 are rewritten as

$$h_T = H_{OL} N_{OL} \qquad (15.26)$$

$$h_T = H_{OG} N_{OG} \qquad (15.27)$$

where H_{OL} and H_{OG} are overall HTU's based on the liquid phase and the vapor phase, respectively, and N_{OL} and N_{OG} are overall NTU's based on the liquid phase and the vapor phase, respectively. It is obviously incorrect to calculate the column height as, for example, $H_{OG}N_{OL}$. If film coefficients were used in the column height derivation instead of overall coefficients, that is, if Equation 15.20 were used instead of Equation 15.21 as the starting point, the column height would be calculated as

$$h_T = H_L N_L \tag{15.28}$$

or

$$h_T = H_G N_G \tag{15.29}$$

where subscripts L or G refer to liquid film or gas film.

Example 15.2 is based on Example 15.1, where the NTU's were calculated from operating and equilibrium data. With additional experimental data, it is possible to determine HTU's, HETP, and mass transfer coefficients.

EXAMPLE 15.2: COLUMN PACKING CHARACTERISTICS FROM EXPERIMENTAL DATA

In reference to Example 15.1 and Figure 15.2, additional data are given, including a column diameter of 1 m, a column height of 3 m, and an inlet gas rate of 225 kmol/h. Calculate the HTU, HETP, and the mass transfer coefficients.

SOLUTION

The total liquid rate is calculated from a material balance on the transferred component:

$$V(Y_B - Y_T) = L(X_B - X_T)$$

or

$$L = V \frac{Y_B - Y_T}{X_B - X_T}$$

Using data from Example 15.1,

$$L = (225)\frac{0.006 - 0.0021875}{0.0002 - 0.000009375} = 4500 \text{ kmol/h}$$

The NTU calculated in Example 15.1 is based on overall concentration gradients, $(X^* - X)$ and $(Y - Y^*)$, and is designated as OL or OG:

$$N_{OL} = 3.347$$

$$N_{OG} = 2.678$$

The HTU's are calculated directly from Equations 15.26 and 15.27:

$$H_{OL} = 3/3.347 = 0.90 \text{ m}$$

$$H_{OG} = 3/2.678 = 1.12 \text{ m}$$

The HETP is calculated from the number of equilibrium stages:

$$HETP = 3/3 = 1 \text{ m}$$

From the definition of HTU in Equations 15.24 and 15.25:

$$H_{OL} = \frac{L}{(K_X a)A}$$

$$H_{OG} = \frac{V}{(K_Y a)A}$$

The column cross-sectional area is

$$A = \pi(1/2)^2 = 0.785 \text{ m}^2$$

The capacity coefficients are

$$(K_X a) = \frac{L}{H_{OL}A} = \frac{4500}{(0.9)(0.785)} = 6369 \text{ kmol/m}^3\text{h}$$

$$(K_Y a) = \frac{V}{H_{OG}A} = \frac{225}{(1.12)(0.785)} = 256 \text{ kmol/m}^3\text{h}$$

The mass transfer coefficients may also be calculated based on overall column performance, using Equation 15.18. The mole fraction differences are replaced with logarithmic means of the mole fraction differences at both ends of the column:

$$K_X a = \frac{Na}{(X^* - X)_{lm}}$$

$$K_Y a = \frac{Na}{(Y - Y^*)_{lm}}$$

where lm designates logarithmic mean.

$X_B = 0.0002$ $X_B^* = 0.00024$ $X_B^* - X_B = 0.00024 - 0.0002 = 0.00004$

$X_T = 0.000009375$ $X_T^* = 0.0000875$ $X_T^* - X_T = 0.0000875 - 0.000009375$
                                                                          $= 0.000078125$

$Y_B = 0.006$ $Y_B^* = 0.005$ $Y_B - Y_B^* = 0.006 - 0.005 = 0.001$

$Y_T = 0.0021875$ $Y_T^* = 0.000234375$ $Y_T - Y_T^* = 0.0021875 - 0.000234375$
                                                                          $= 0.001953125$

$$(X^* - X)_{lm} = \frac{0.000078125 - 0.00004}{\ln(0.000078125/0.00004)} = 0.00005695$$

$$(Y - Y^*)_{lm} = \frac{0.001953125 - 0.001}{\ln(0.001953125/0.001)} = 0.001424$$

The mass transfer rate per unit column volume, Na, is calculated from the mass transfer rate and the column volume. The mass transfer rate is calculated from the inlet and outlet compositions:

$$\text{Rate} = V(Y_B - Y_T) = L(X_B - X_T) = 225(0.006 - 0.0021875)$$
$$= 4500(0.0002 - 0.000009375) = 0.8578 \text{ kmol/h}$$

The column volume $= (0.785)(3) = 2.355 \text{ m}^3$

$$Na = 0.8578/2.355 = 0.3642 \text{ kmol/m}^3\text{h}$$

The mass transfer coefficients,

$$K_X a = 0.3642/0.00005695 = 6394 \text{ kmol/m}^3\text{h}$$
$$K_Y a = 0.3642/0.001424 = 255.8 \text{ kmol/m}^3\text{h}$$

These results agree within math errors with the results calculated from the HTU values because both the operating line and the equilibrium curve are straight lines.

The above mass transfer equations, although based on sound molecular diffusion principles, are limited in their applicability in a number of ways. A basic condition for their validity is the assumption of equimolar or dilute unimolar mass transfer. This limits the NTU and HTU approach to processes that are essentially either binary (distillation) or ternary (absorption or stripping) with only one component crossing the phase boundary. Another shortcoming of the transfer units technique is its exclusion of energy balances or temperature calculations.

15.3.1 GENERAL RATE-BASED MODEL

The rate-based analysis introduced in Section 14.2 can be developed as outlined below to define a general, rigorous, multi-component rate-based stage model. It takes into account mass and energy transfer between the phases within a stage, and can be connected to other stages to form multistage columns.

Rate-based models are not limited to packed columns but can be applied to trayed columns as well. A rate-based stage can represent a section of a packed column or a tray. It consists of a liquid phase and a vapor phase interacting based on the following principles:

- Separate liquid phase and vapor phase component material balances, coupled with
- Mass transfer rate equations for both phases across the interface,
- Energy balances for both phases, coupled with
- Heat transfer rate equations for both phases across the interface,
- Phase equilibrium equations relating the compositions in both phases at the interface, and
- Mole fraction summation equations for both phases at the interface.

The material balances are implemented based on Equation 13.2, written separately for each phase on a stage. Each phase equation is then expanded to include vapor–liquid interface mass transfer, Equations 15.11 or 15.12.

Similarly, the energy balances on the stage are expressed as in Equation 13.4, applied separately for each phase. The resulting equations are coupled with phase boundary heat transfer expressions in terms of interfacial area, heat transfer coefficients, and temperature differences between liquid or vapor inlet temperature and the interfacial temperature.

Finally, the interfacial component mole fractions in each phase must add up to unity and satisfy the equilibrium relations.

The stage pressures may either be fixed or interrelated based on column hydraulics (Chapter 14 and Section 15.4).

It is obvious from the conditions defined above that the rate-based model equations and variables are more numerous and complex than those in the equilibrium stage model described in Chapter 13. Other features of the rate-based model are that the exiting liquid and vapor from a stage can be at different temperatures since separate balance equations are written for each phase. Each phase on a stage can have a different externally transferred heat duty. The exiting phases in general are not at equilibrium: the liquid may be subcooled and the vapor may be superheated. In a rate-based model the phase interface must be defined. The variables defining the interface include the liquid and vapor compositions and the temperature at the interface, and the molar flux across the interface.

The advantage of the rate-based model for simulating multistage separation columns is that it is designed to model the actual performance of the column without the need for HETP or tray efficiencies. As in the case of the equilibrium stage model, the accuracy of the rate-based model depends on the quality of the physical data used in the model. The equilibrium stage model requires vapor–liquid equilibrium and enthalpy data, plus HETP or tray efficiencies to correct for inaccuracies resulting from the assumption of phase equilibrium. The rate-based model requires phase equilibrium and enthalpy data, and mass transfer data including binary mass transfer coefficients for all the binaries in the mixture. It is recognized that the rate-based model can be, in principle, more realistic as far as simulating the actual process. The ultimate quality of the model is determined by the level of accuracy of predicting the mass transfer coefficients, as well as the other physical data.

The rate-based model equations are solved by methods similar to those used with the equilibrium stage model, although the number of the rate-based model equations is considerably larger and more complex. Rate-based models are available in most commercial simulators.

15.4 PACKED COLUMN DESIGN

The alternative for calculating the column height from the HTU and NTU is based on the HETP. The equilibrium stage model for packed columns uses the HETP concept as described in the introduction to this chapter. With the HETP treated as an equilibrium stage, the same computational methods used for trayed columns could be used for packed columns. The problem of calculating the required column packing height, or of predicting the performance of a column with a given packing height, is thus reduced to estimating the HETP.

Having determined the column packing height on the basis of equilibrium calculations and whatever mass transfer correlations, the capacity of the column to handle the required vapor and liquid flows hydraulically must be checked. The specific items of concern in this regard are the pressure drop across the packing and the tendency of the column to flood. These are the considerations that determine the required column diameter.

15.4.1 Estimating the HETP

The HETP depends on the type and size of the packing material, the column dimensions, the fluid properties, the operating conditions, and so on. It also depends on whether the packing is dumped or ordered. The HETP is usually supplied by the packing vendors or estimated based on experimental data or empirical correlations.

Bolles and Fair (1979) estimate the HETP from vapor and liquid heights of transfer units (HTU). The vapor phase H_G and liquid phase H_L are first calculated separately from empirical correlations. The overall HTU is then calculated as follows:

$$H_{OG} = H_G + (mV/L)H_L \tag{15.30}$$

where m is the slope of the equilibrium curve (Y^*/X), and V and L are the vapor and liquid flow rates. The HETP is calculated from the overall HTU using the equation

$$\text{HETP} = H_{OG} \frac{\ln(mV/L)}{(mV/L) - 1} \tag{15.31}$$

Equations 15.30 and 15.31 are derived based on the assumption of constant operating line and equilibrium curve slopes (Vital et al., 1984).

The HETP correlations are described in detail by Vital et al. (1984), who also provide charts for estimating packing parameters to be used in these correlations. The same article contains HETP values for a variety of packing types and fluids as well as a number of rules of thumb for estimating the HETP. According to the simplest of these rules, the HETP is set equal to the column diameter. The authors recommend using the most conservative, that is, the largest, among the HETP values obtained from available data, correlations, or rules of thumb.

15.4.2 PACKED COLUMN CAPACITY

Multistage separation in packed columns is achieved by contacting counterflowing vapor and liquid through the packing. Under favorable conditions, the liquid wets the packing and forms the dispersed downward flowing phase, while the continuous vapor phase flows upward around the liquid-covered packing.

As the vapor flows upward, a pressure drop develops in the vapor phase in the direction of vapor flow as a result of friction against the packing and the falling liquid. The liquid flows down by gravity against this pressure drop, resulting in some steady-state liquid-to-vapor-flow ratio, L/V.

A packed column (or a trayed column) designed for a given service is usually expected to operate over a range of L/V ratios in anticipation of variations in separation requirements, feed flows and compositions, and other factors.

If the vapor rate is increased, the pressure drop increases, which raises the resistance to the liquid flow. At a certain vapor rate, the pressure drop rises to a point where liquid flow is practically blocked and liquid fills the column, resulting in a condition known as flooding. The column ceases to function as a separation device. A larger diameter column would be required to handle these flows.

Below certain vapor rates and pressure drops, channeling takes place, where the vapor and liquid pass each other with no appreciable contacting. To prevent this from happening, a smaller column diameter should be selected.

In general, columns should be designed such that pressure drop is maintained between 0.01 in. of water per foot of packing for vacuum columns and 1 in. of water per foot of packing for atmospheric fractionators. Columns should be operated above 50% and below 90% of the flooding vapor rate.

The pressure drop and the flooding vapor rate are functions of the packing type and size, the column diameter, and the vapor and liquid flow rates and properties. These rates and properties may change with the column height, resulting in different calculated diameters. The column should be designed with the most conservative (largest) diameter, although it should be checked for excessively low pressure drops that could cause channeling. The packing type and size are characterized by a packing factor, F, which has the dimension of area-per-unit volume of packing.

Figure 15.4 shows the Eckert (1975) pressure drop correlation, which may also be used to check the approach to flooding conditions. The variables in the coordinates are defined as follows: L and V are the liquid and vapor flow rates, lbmol/hr, G is the vapor mass velocity, lb/ft^{2}s, ρ_L and ρ_V are the liquid and vapor densities, lb/ft^3, F is the packing factor, ft^2/ft^3, and μ_L is the liquid viscosity, centipoise. The column diameter is implied since G is the vapor rate per unit column cross-sectional area. The pressure drop, in inches of water per foot of packing, is reported as a parameter in this correlation. Flooding can be expected to occur at any point above the pressure drop curve of 1.5 in. of water per foot of packing.

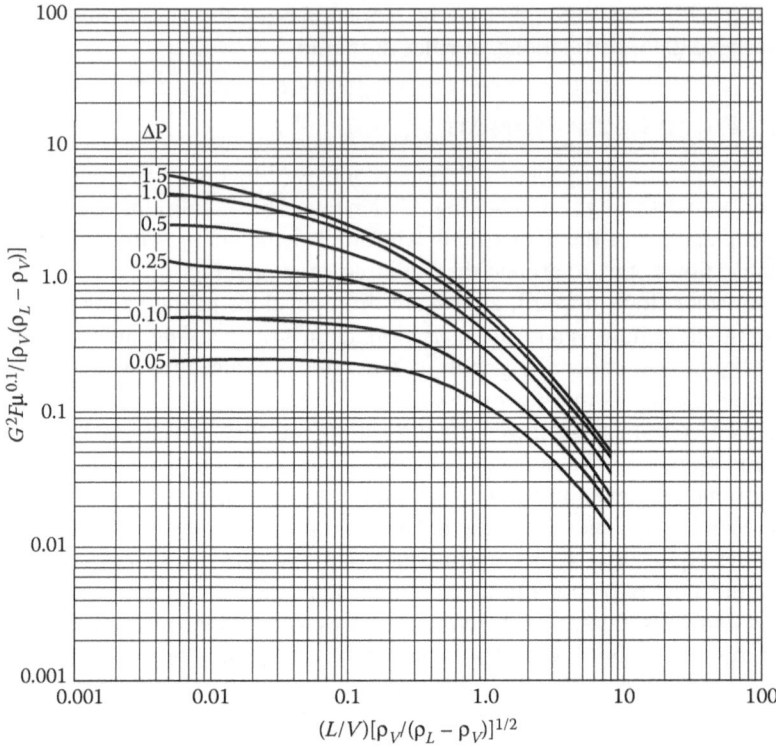

FIGURE 15.4 Packed column pressure drop correlation. (Reprinted with permission from J. S. Eckert, *Chem. Eng.*, 82(8), 70, 1975.)

15.4.3 PACKED COLUMN DESIGN OUTLINE

Designing a column usually starts with heat and material balance calculations, preferably using a computer simulation program. These calculations determine the liquid and vapor flow rates and the number of equilibrium stages required to meet the design performance specifications (separation, recovery, etc.). Fluid properties, such as densities and viscosities, may also be generated by the computer program.

Next, the packing material type and size are selected. Among the different types of packing material available, of which mesh, rings, and saddles are the most common, the type most suited for a particular application is best determined based on available data for the different types. Information on each type includes pressure drop per meter, HETP, corrosion resistance, and so on. Once a decision has been made regarding the packing type, its nominal size should be selected. Smaller sizes tend to give lower HETP's at the expense of higher pressure drops across the packing, while larger sizes can induce channeling. A good starting point is a packing size about one-tenth the column diameter or smaller. Since the column diameter is yet to be determined, an assumed value is used for selecting the initial packing size.

The pressure drop/flooding rate correlation, Figure 15.4, is next used to determine G, the vapor mass rate per unit column cross-sectional area. The information needed includes the packing factor of the selected packing material, the L/V ratio and the fluid properties (vapor and liquid densities and liquid viscosity), and a pressure drop within the acceptable range. The column cross-sectional area and diameter are then calculated from G and V. These calculations should be carried out at different column heights since the vapor and liquid rates and properties may vary.

If the resulting ratio of packing size to column diameter is reasonable, an appropriate column diameter has been determined. Otherwise, the calculations are repeated with a new packing size.

The next step is to determine the column packing height. The HETP is estimated using any of the methods described in Section 15.4.1. The column packing height is simply the number of equilibrium stages multiplied by the HETP.

A check on the column height may be carried out based on mass transfer rate. This requires the availability of mass transfer coefficient data. Depending on appropriate assumptions, Equation 15.22 or 15.24, or their equivalent counterparts may then be used to calculate the column height, as discussed in Section 15.3.

15.4.3.1 Packed Columns versus Trayed Columns

Many considerations influence the choice between a packed column and a trayed column for a particular application, foremost among which are economics and performance versatility. The decision involves many variables and is best handled on a case-by-case basis. Example 15.3 compares the performance of the same column structure, using either trays or packing.

EXAMPLE 15.3: DEBUTANIZER COLUMN—PACKING VERSUS TRAYS

An existing column is to be used for separating butanes from a mixture of butanes, pentanes, and other hydrocarbons. The column is required to recover 98% of the butanes in the distillate with a combined butanes molar purity of 94.5%. Table 15.1 shows stream compositions and conditions as obtained from simulation.

TABLE 15.1
Stream Data for Example 15.3

Rates (kmol/h)	Feed	Distillate	Bottoms
C3	2.3	2.3	0.0
IC4	16.0	15.7	0.3
NC4	38.6	37.7	0.9
IC5	11.3	0.7	10.6
NC5	11.4	0.2	11.2
NC6	11.4	0.0	11.4
Total	91.0	56.6	34.4
Temperature (°C)	90.0	78.0	136.0
Pressure (kPa)	1100	1050	1090

The column is 1 m diameter and can fit 22 trays with 0.60 m spacing between the trays. With an estimated column overall efficiency of 65%, the number of theoretical stages is

$$(22)(0.65) + 1 \text{ condenser} + 1 \text{ reboiler} = 16$$

The feed is introduced around the middle of the column. With 16 theoretical stages, and in order to meet the butanes recovery and purity specifications, the required reflux ratio is computed by simulation at 2.75.

The same column could be packed to give a packing height equal to the height between the top and bottom trays:

$$\text{Packing height} = (22 - 1)(0.6) = 12.6 \text{ m}$$

Spaces above and below the packing are reserved for the reflux and boilup distributors. The feed location is not changed from the trayed column design.

The packing material under consideration consists of 2.5 cm Pall rings. The estimated HETP is 45 cm and the packing factor, $F = 170$ m^2 per cubic meter of packing (Vital et al., 1984; Eckert, 1975). The equivalent number of equilibrium stages, including the condenser and reboiler, is

$$12.6/0.45 + 2 = 30$$

To meet the same butanes recovery and purity specifications, the required reflux ratio with 30 equilibrium stages is calculated by simulation at 1.95.

For the purpose of estimating column pressure drops, vapor and liquid flow rates are estimated from the reflux ratio. Calculations are performed at two points in the column: one in the rectifying section and the other in the stripping section.

TRAYED COLUMN

For the trayed column, the reflux ratio is 2.75 and the distillate rate is 56.6 kmol/h. The liquid flow in the rectifying section,

$$L_r = (2.75)(56.6) = 155.65 \text{ kmol/h}$$

The feed is introduced to the column as a liquid slightly below its bubble point; so the liquid flow in the stripping section is estimated as

$$L_s = L_r + F = 155.65 + 91.0 = 246.65 \text{ kmol/h}$$

The boilup rate is estimated as

$$V_s = L_s - \text{bottoms rate} = 246.65 - 34.4 = 212.25 \text{ kmol/h}$$

With the feed all liquid, the vapor rate in the rectifying section is assumed the same as in the stripping section:

$$V_r = V_s = 212.25 \text{ kmol/h}$$

Using the methods discussed in Chapter 14, the average pressure drop predicted by simulation is 0.1 kPa per tray, giving a total column pressure drop of (0.1) (22) = 2.2 kPa.

PACKED COLUMN

The packed column internal flows are estimated in a similar manner as the trayed column, based on a reflux ratio of 1.95:

$L_r = (1.95)(56.6) = 110.37$ kmol/h
$L_s = 110.37 + 91.0 = 201.37$ kmol/h
$V_s = 201.37 - 34.4 = 166.97$ kmol/h
$V_r = V_s = 166.97$ kmol/h

The following average fluid properties are given. For the rectifying section:

$M_V = 60$
$\rho_L = 482.1$ kg/m³
$\rho_V = 34.0$ kg/m³
$\mu_L = 0.24$ cp

The vapor mass velocity:

$$G = \frac{V_r M_V}{\pi D^2/4} = \frac{(166.97)(60)}{\pi(1)^2/4} = 12755.6 \text{ kg/m}^2\text{h}$$

For the stripping section:

$M_V = 76$
$\rho_L = 552.6$ kg/m³
$\rho_V = 38.4$ kg/m³
$\mu_L = 0.17$ cp

The vapor mass velocity:

$$G = \frac{V_s M_V}{\pi D^2/4} = \frac{(166.97)(76)}{\pi(1)^2/4} = 16157.0 \text{ kg/m}^2\text{h}$$

The pressure drop in the packed column is determined from Figure 15.4. In the rectifying section,

$$\text{Abscissa} = \frac{L}{V}\left(\frac{\rho_V}{\rho_L - \rho_V}\right)^{\frac{1}{2}} = \frac{110.37}{166.97}\left(\frac{34.0}{482.1 - 34.0}\right)^{\frac{1}{2}} = 0.182$$

$$\text{Ordinate} = \frac{G^2 F \mu_L^{0.1}}{\rho_V(\rho_L - \rho_V)} = \frac{(12755.6)^2(170)(0.24)^{0.1}}{(34.0)(482.1 - 34.0)(3600)^2} = 0.1214$$

In Figure 15.4 units, Ordinate = 0.1213×3.28 ft/m = 0.40.

The pressure drop corresponding to these coordinates is 0.22 in. of water per foot of packing, or 1.8 cm per meter of packing.

In the stripping section,

$$\text{Abscissa} = \frac{L}{V}\left(\frac{\rho_V}{\rho_L - \rho_V}\right)^{\frac{1}{2}} = \frac{201.37}{166.97}\left(\frac{38.4}{552.6 - 38.4}\right)^{\frac{1}{2}} = 0.330$$

$$\text{Ordinate} = \frac{G^2 F \mu_L^{0.1}}{\rho_V(\rho_L - \rho_V)} = \frac{(16157.0)^2(170)(0.17)^{0.1}}{(38.4)(552.6 - 38.4)(3600)^2} = 0.1452$$

In Figure 15.4 units, Ordinate = 0.1452×3.28 ft/m = 0.48.

The pressure drop corresponding to these coordinates is 0.35 in. of water per foot of packing, or 2.9 cm per meter of packing. These pressure drops are well below flood conditions, yet not so low as to cause channeling. With the feed at the column height middle, the total packing pressure drop is

$$(6.3 \text{ m})(1.8 \text{ cm H}_2\text{O/m}) + (6.3 \text{ m})(2.9 \text{ cm H}_2\text{O/m}) = 29.6 \text{ cm H}_2\text{O}$$

$$= (29.6)(0.098 \text{ kPa/cm H}_2\text{O}) = 2.9 \text{ kPa}$$

This is comparable to the trayed column pressure drop.

In conclusion, this case study indicates that replacing trays with Pall rings packing in this particular column reduces by about 30% the reflux ratio required to achieve the same separation. The corresponding energy savings would have to be weighed against the cost of the packing and its installation.

It is also important to check the column performance at off-design conditions to determine the operable ranges. The above calculations could be repeated for different liquid and vapor rates to check for excessive or inadequate pressure drops.

15.4.4 PACKED COLUMN DESIGN BY THE GROUP METHOD

An alternative method for the design of an absorption or stripping column is based on the modular group method (Kremser equation, 12.33), primarily for dilute mixtures. The required number of stages for a specified separation is first determined. Once this is known, the packing height can be estimated from the relation

$$h_t = N^* \text{HETP}$$

The HETP itself has to be estimated because no rigorous methods exist for its evaluation. Its value typically depends on the type and size of the packing material particles, and is usually supplied by the packing vendor.

To complete the column design, its diameter must be determined such that the pressure drop through the packing lies within practically acceptable ranges.

The design of a packed absorption column is illustrated in Example 15.4.

EXAMPLE 15.4: PACKED ABSORBER DESIGN

The mole fraction of heavier components in a hydrocarbon gas stream must be dropped to produce a lean gas stream. This is to be achieved by absorption in a packed column using normal decane as the absorbent. The gas feed stream and the absorbent are defined below. The column pressure is 500 kPa, and the packing is high-efficiency random packing with HETP = 0.5 m, and a packing factor given as $F = 40$ ft^{-1} (130 m^{-1}). The component K-values provided correspond to the column pressure of 500 kPa and a temperature that is taken as the average of the feed and absorbent temperatures, 25°C. The butane in the gas should be lowered to below 0.03 mole fraction. Determine the required packing height and column diameter. An acceptable pressure drop in the column can be considered as 0.2 kPa/m packing.

Component	Feed (kmol/h)	Absorbent (kmol/h)	K-Value
C_1	685	0	30
C_2	173	0	6
C_3	89	0	1.7
nC_4	53	0	0.5
nC_{10}	0	240	0.0006
Total	1000	240	
T (°C)	18	32	
P (kPa)	500	500	

SOLUTION

The modular group calculation results for four stages, $N = 4$, are as follows, for the column at an average temperature of 25°C, and 500 kPa pressure:

Component	C_1	C_2	C_3	nC_4	nC_{10}
$S_i = K_i V/L$	125	25	7.083	2.083	0.0025
L_{4i} (kmol/h)	5.48	6.92	12.56	24.72	239.4
Y_{1i}	0.715	0.175	0.080	0.030	0.0006

The lean gas flow rate is 950.92 kmol/h, and the bottoms rate is 289.08 kmol/h. The calculation results conclude that four equilibrium stages satisfy the specification of $Y_{1,4} \leq 0.03$. Fewer stages result in $Y_{1,4} > 0.03$.

The packing height, based on the given value of HETP,

$$h_t = 4 \times 0.5 = 2.0 \text{ m}$$

The column diameter calculations will target a pressure drop of 0.2 kPa/m packing, or 0.25 in. H$_2$O/ft of packing. Figure 15.4 will be used to determine the column diameter. The lean gas stream and the bottoms stream will be used for the calculations. Some additional values are required for applying the correlation represented in Figure 15.4.

The density of the lean gas stream is calculated based on the assumption of ideal gas and ideal solution at 25°C, giving the result $\rho_V = 4.68$ kg/m³ $= 0.292$ lb/ft³. The average molecular weight of the stream is based on the same assumptions, giving $(MW)_V = 23.19$ kg/kmol.

For the liquid stream, n-decane properties are used at 25°C: $\rho_L = 730$ kg/m³ $= 45$ lb/ft³, and viscosity, $\mu_L = 0.861$ cP.

The horizontal coordinate in Figure 15.4 is defined as

$$\text{Abscissa} = (L/V)[\rho_V/(\rho_L - \rho_V)]^{0.5} = (289.08/950.92)[4.68/(730 - 4.68)]^{0.5} = 0.0244$$

At 0.0244, and a pressure drop of 0.25 in. H_2O, the corresponding ordinate value is 1, so that

$$G^2 F(\mu_L)^{0.1}/[\rho_V(\rho_L - \rho_V)] = 1$$

Preserving consistency with units used in the packed column pressure drop correlation (Figure 15.4),

$$G^2 = \rho_V(\rho_L - \rho_V)/[F(\mu_L)^{0.1}] = 0.292(45 - 0.292)/(40 \times 0.861^{0.1}) = 0.331$$

$$G = 0.575 \text{ lb/ft}^2.\text{s}$$

From the definition of G,

$$G = V(MW)_V/(\pi D^2/4)$$

where D is the column diameter. Rearranging,

$$D^2 = 4V(MW)_V/(\pi G) = 4 \times 950.92 \times 23.19 \times 2.2/(3600 \times 3.1416 \times 0.575) = 29.8 \text{ ft}^2$$

Thus, $D = 5.46$ ft $= 1.66$ m, which could be rounded up to 2 m. With this diameter the column is expected to operate satisfactorily, away from flooding or channeling.

NOMENCLATURE

a	Interfacial area per unit volume of packing
A	Column cross-sectional area
c	Concentration, moles/volume
F	Packing factor
F	Feed stream
g_C	Acceleration of gravity
G	Vapor mass velocity in the column
h	Column height
HETP	Height equivalent to a theoretical plate
HTU	Height of a transfer unit
k	Mass transfer coefficient
K	Overall mass transfer coefficient

L	Liquid rate in the column
M	Molecular weight
N	Number of equilibrium stages
N	Number of moles transferred per unit area per unit time
NTU	Number of transfer units
p	Partial pressure
P	Total pressure
V	Vapor rate in the column
X	Component mole fraction in the liquid
Y	Component mole fraction in the vapor
μ	Viscosity
ρ	Mass density

SUBSCRIPTS

B	Column bottom
G	Gas or vapor phase
L	Liquid phase
lm	Logarithmic mean
OG	Overall gas
OL	Overall liquid
r	Rectifying section
s	Stripping section
T	Column top
V	Vapor phase
X	Liquid phase
Y	Vapor phase

SUPERSCRIPTS

i	Interfacial conditions
$*$	Equilibrium conditions

PROBLEMS

15.1. In the absorber column of Example 15.1, calculate the liquid rate per 100 kmol/h gas. For the same inlet gas and liquid compositions, calculate the outlet compositions if the liquid rate is halved. Determine the number of transfer units.

15.2. A 20 kmol/h effluent stream that is mostly air contains 0.02 mole fraction acetone. Prior to releasing it to the atmosphere the acetone content must be lowered to 100 ppm mole as a maximum. It is proposed to accomplish this by countercurrent water scrubbing in a 1 m-diameter packed column. The effluent scrubbing water stream is treated and recycled at the rate of 600 kmol/h with a residual acetone content of 10 ppm mole before it is sent back to the column. Calculate the required packing height if the mass transfer coefficient $K_ya = 28.1$ kmol/m³-h. The equilibrium acetone mole fraction is given as $Y^* = 1.67X$.

15.3. The sulfur dioxide content of a gas stream is lowered by 90% by scrubbing with water in a countercurrent packed column. The column is 1.25 m in diameter and 5 m in height. The gas rate is 500 kmol/h, and the inlet sulfur dioxide concentration is 2% mole. The column was designed such that the sulfur dioxide concentration in the bottoms does not exceed 0.04% mole. Determine the required water rate, the number of gas-phase transfer units, the height of the gas-phase transfer unit, and the mass transfer coefficient K_ya. The equilibrium relationship for sulfur dioxide is given as $Y^* = 38X$.

15.4. In order to check the operable range of the debutanizer packed column of Example 15.3, its performance is evaluated at various reflux ratios. Check for the possibility of flooding or channeling at reflux ratios of 1.0 and 3.0.

15.5. It is required to design a packed column at 100 kPa for the absorption of CO_2 from air using a dilute caustic solution in water as the absorbent, which will recover 97% of the CO_2. The following data are given:

Flow rate of inlet air, $V_{in} = 360$ kmol/h
Estimated flow rate of inlet caustic solution, $L_{in} = 830$ kmol/h
Mole fraction of CO_2 in inlet air, $Y_{in} = 0.03$
Mole fraction of CO_2 in inlet caustic solution, $X_{in} = 0$

The equilibrium concentration of CO_2 may be expressed as $Y^* = 1.75X$. The caustic solution and air are assumed to undergo no phase change.

It is proposed to use a 0.75 m diameter column and 5 cm Intalox saddles packing with $K_ya = 180$ kmol/m³.h

Based on the given data calculate the following:
a. The required number of theoretical stages using the Kremser equation.
b. The number of transfer units, N_{OG}, assuming a straight operating line.
c. The packing height.
d. The height equivalent of a theoretical plate, HETP.
e. The pressure drop per foot of packing, and check for the possibility of flooding or channeling, given the following:
    $M_V = 29$, $M_L = 18$, $\rho_V = 1.22$ kg/m³, $\rho_L = 998.2$ kg/m³, $\mu = 1$ cp, Packing factor, $F = 120$ m²/m³.

15.6. A packed absorber column is used to remove sulfur dioxide form an air stream, using pure water as the absorbent. The following data are given:

Total inlet gas flow rate, 0.062 kmol/s
Mole fraction of SO_2 in inlet gas, 0.016
Mole fraction of SO_2 in outlet gas, 0.004
Inlet water flow rate, 2.2 kmol/s

The SO_2 vapor–liquid equilibrium relation at the column temperature and pressure is expressed as $Y^* = 40X$. The column cross-sectional area is 1.5 m² and the packing height is 3.5 m. It is required to calculate the following:

a. The required number of theoretical stages, using the Kremser equation.
b. The number of transfer units, N_{OG}.
c. The height equivalent of a theoretical plate, HETP.
d. The height of a transfer unit, H_{OG}.
e. The capacity coefficient, $K_Y a$.

REFERENCES

Bolles, W. L. and J. R. Fair, *I. Chem. Eng. Symp.* Series, No. 56, 1979.

Eckert, J. S., *Chem. Eng.*, 82(8), 70, 1975.

Seader, J. D., and E. J. Henley, *Separation Process Principles*, New York, John Wiley and Sons, 1998.

Vital, T. J., S. S. Grossel, and P. I. Olsen, *Hydrocarbon Processing*, December 1984, p. 75.

16 Control and Optimization of Separation Processes

As in most processes, a distillation column and other separation processes must be maintained at operating conditions that result in products meeting certain specifications. To achieve this objective on a continuous basis the process is equipped with an automatic control system. Various disturbances can occur during the operation of the process, such as variations in ambient conditions or in the feed flow rate or composition. This can move the process away from design steady-state conditions, causing the products to be off-specification. The automatic controller counters the disturbances by adjusting the operating conditions such as to maintain the process variables at acceptable values.

Quite often, the desired performance of a process is defined in terms of constraints, and within these constraints variations in the operating conditions are possible. Under these circumstances it is desirable not only to satisfy the performance specifications, but also to achieve this objective at an optimum level, economic or otherwise. Optimization could be another task of the overall control system.

In a basic process control action a manipulated variable is adjusted by the controller such as to influence a controlled variable to stay at a given set point. The control action in multistage columns can be a challenge for a number of reasons, some of which are briefly discussed here.

Properties to be controlled may not be measurable online fast enough to allow for a timely action by the manipulated variable. Such properties may have to be inferred from other measured properties. A column product purity or composition, for example, could be inferred from measured column temperatures on a number of trays. The required property is related to the measurements by inferential property correlations whose parameters must be determined. In the composition-temperature example, the correlation parameters are evaluated from measured temperatures and laboratory composition analysis, and are updated every time laboratory analyses become available.

Inverse responses can occur in multistage columns due to different dynamics having opposite effects. The result is that the initial response of a column variable to a sudden change in the manipulated variable is in the opposite direction to the ultimate response. An example is an initial rise in the reboiler liquid level resulting from a sudden increase in the reboiler steam rate, followed by a gradual decrease in the liquid level. The initial rise is caused by excessive liquid overflow from the trays above the reboiler resulting from excessive frothing following the sudden increase in the steam rate. As time progresses, the higher steam rate increases the vapor boilup and lowers the liquid level.

The nature of the dynamics in a multistage column is influenced by the fact that the column is a series of interconnected dynamic stages. As a result, the lag between an input disturbance and the response at different points in the column is particularly significant. These conditions of column dynamics should be given special consideration in the design of a control system.

Another general characteristic of multistage columns is that they usually require controlling several interrelated variables using a number of interrelated manipulated variables. This is a multiple-input, multiple-output (MIMO) problem as compared with the basic control action in a single-input, single-output (SISO) problem. In a multiple variable process each manipulated variable may affect more than one controlled variable due to process interactions. The multiple variable control problems in multistage columns may be handled either by multiple control loops or by a multivariable controller. In multiple control loops each manipulated variable is associated with one controlled variable. A multivariable control strategy may involve a dynamic predictive feature as well as a process optimization step that would maximize an objective function by manipulating additional variables.

16.1 MULTILOOP CONTROLLERS

In the multiloop controller strategy each manipulated variable controls one variable in a feedback proportional integral derivative (PID) control loop. Taking a single-feed, two-product distillation column with a total condenser and a reboiler as an example, a basic list of possible controlled variables includes the distillate and bottoms compositions, the liquid levels in the reflux accumulator and the column bottom, and the column pressure. The main manipulated variables are the reflux, distillate, and bottoms flow rates and the condenser and reboiler heat duties.

The obvious question is how to pair the manipulated and controlled variables. One manipulated variable can potentially affect more than one controlled variable, resulting in interactions among the control loops. The concern is that adjusting a manipulated variable to move one controlled variable in the desired direction could cause another variable to move away from its set point. It is clear that a proper design of the control loops is necessary to prevent such a scenario. Toward this goal, certain strategies should be followed as described below.

16.1.1 PAIRING THE MANIPULATED AND CONTROLLED VARIABLES

The following method, based on steady-state gains, is recommended for pairing the manipulated and controlled variables (Bristol, 1966). The gains may be obtained from steady-state plant data or from a steady-state process simulator. The steady-state gain K_{ij} is defined as

$$K_{ij} = \frac{dC_i}{dM_j} \qquad (16.1)$$

where dC_i is the change in controlled variable i caused by dM_j, the change in manipulated variable j. Holding constant all the manipulated variables except M_j defines the open-loop, steady-state gain:

$$K_{ij}^O = \left(\frac{\partial C_i}{\partial M_j} \right)_M \tag{16.2}$$

The closed-loop gain K_{ij}^C is the gain in the $i\text{--}j$ loop when all the other loops are closed, that is, their controlled variables are held at their set points:

$$K_{ij}^C = \left(\frac{\partial C_i}{\partial M_j} \right)_C \tag{16.3}$$

The relative gains are defined as the ratios of the open-loop to closed-loop gains:

$$\mu_{ij} = \frac{K_{ij}^O}{K_{ij}^C} = \frac{(\partial C_i/\partial M_j)_M}{(\partial C_i/\partial M_j)_C} \tag{16.4}$$

A process with n controlled variables and n manipulated variables is characterized by an n by n matrix of relative gains, the relative gain matrix.

The pairing of variables to minimize interaction between the loops is done based on the following rule: Pair those controlled and manipulated variables whose relative gains are positive and closest to 1.0. This rule is consistent with the fact that if $\mu_{ij} = 1$, $K_{ij}^C = K_{ij}^O$. That is, opening or closing other loops has no effect on loop $i\text{--}j$, and thus no interactions exist with the other loops. The farther away μ_{ij} is from 1.0, the greater the difference between K_{ij}^O and K_{ij}^C: Closing or opening other loops affects loop $i\text{--}j$, which means interaction does exist with other loops.

The relative gains may be evaluated from open-loop gains obtained from a steady-state simulator, and from derived closed-loop gains. To determine the open-loop gains from the simulator, the manipulated variables are perturbed one at a time, and the variation in each controlled variable is recorded. If a manipulated variable M_j is perturbed by ΔM_j, resulting in a variation ΔC_i in controlled variable C_j, the gain is approximated as

$$K_{ij}^O = \frac{\Delta C_i}{\Delta M_j} \tag{16.5}$$

The closed-loop gains can be derived mathematically from the open-loop gains. For a two by two process, the variations ΔC_1 and ΔC_2 resulting from changes in both manipulated variables are given by the equations

$$\Delta C_1 = K_{11}^O \Delta M_1 + K_{12}^O \Delta M_2 \tag{16.6}$$

$$\Delta C_2 = K_{21}^O \Delta M_1 + K_{22}^O \Delta M_2 \tag{16.7}$$

The closed-loop gains are approximated as follows:

$$K_{11}^C = \left(\frac{\Delta C_1}{\Delta M_1} \right)_{C2} \qquad K_{12}^C = \left(\frac{\Delta C_1}{\Delta M_2} \right)_{C2}$$

$$K_{21}^C = \left(\frac{\Delta C_2}{\Delta M_1} \right)_{C1} \qquad K_{22}^C = \left(\frac{\Delta C_2}{\Delta M_2} \right)_{C1}$$

Gains K_{11}^C and K_{12}^C are evaluated at constant C_2, that is, with $\Delta C_2 = 0$. Similarly, K_{21}^C and K_{22}^C are evaluated with $\Delta C_1 = 0$. Thus, for instance, to evaluate K_{11}^C set ΔC_2 to zero in Equation 16.7 and eliminate ΔM_2 from Equation 16.6:

$$\Delta M_2 = -\Delta M_1 (K_{21}^O / K_{22}^O)$$

$$\Delta C_1 = \Delta M_1 \left(\frac{K_{11}^O K_{22}^O - K_{12}^O K_{21}^O}{K_{22}^O} \right)$$

This derivation was carried out at constant C_2; hence,

$$K_{11}^C = \left(\frac{\Delta C_1}{\Delta M_1} \right)_{C2} = \frac{K_{11}^O K_{22}^O - K_{12}^O K_{21}^O}{K_{22}^O} \tag{16.8}$$

The other closed-loop gains may be obtained in a similar manner, and the relative gains follow from their definition by Equation 16.4:

$$\mu_{11} = \frac{K_{11}^O K_{22}^O}{K_{11}^O K_{22}^O - K_{12}^O K_{21}^O} \tag{16.9a}$$

$$\mu_{12} = \frac{K_{12}^O K_{21}^O}{K_{12}^O K_{21}^O - K_{11}^O K_{22}^O} \tag{16.9b}$$

$$\mu_{21} = \frac{K_{12}^O K_{21}^O}{K_{12}^O K_{21}^O - K_{11}^O K_{22}^O} \tag{16.9c}$$

$$\mu_{22} = \frac{K_{11}^O K_{22}^O}{K_{11}^O K_{22}^O - K_{12}^O K_{21}^O} \tag{16.9d}$$

A quick inspection of these expressions shows that the terms in each column or row of the relative gain matrix add up to unity. This is true for matrices of any order n:

$$\sum_i \mu_{ij} = 1 \quad j = 1, \ldots, n$$

$$\sum_j \mu_{ij} = 1 \quad i = 1, \ldots, n$$

Thus, there are only $(n - 1)^2$ independent terms in an n by n matrix, and only that many terms need to be evaluated.

The procedure described above for evaluating relative gains for a two by two process may be extended to an n by n process using matrix algebra. Equations 16.6 and 16.7 are written in matrix notation as

$$\Delta C = K^O \Delta M \tag{16.10}$$

The relative gain matrix is calculated as

$$\mu = K^O[(K^O)^{-1}]^T \tag{16.11}$$

Equation 16.21 states that the terms of the relative gain matrix are evaluated by multiplying each term of the steady-state open-loop gain matrix by the corresponding term of the transpose of the inverse of that matrix.

Following are two examples (16.1 and 16.2) of a distillation column that demonstrate the effect of applying different pairing strategies. In both examples the control loops for the column pressure and the liquid levels in the condenser accumulator and the column bottom are determined independently based on practical considerations. Thus, the column pressure is controlled by various techniques that may involve the condenser coolant rate, and the liquid levels are controlled by the product flow rates. What remains to be decided is how to pair the distillate and bottoms compositions with the reflux rate and the reboiler heat duty. The same distillation column is used in both examples, having a total condenser and a reboiler, one feed and two products. The column is designed to separate a benzene–toluene mixture into benzene and toluene products with specified purities.

EXAMPLE 16.1: CONTROL OF DISTILLATION PRODUCT PURITIES, CASE I

A 100 kmol/h stream containing 60% mole benzene and 40% mole toluene is sent to a 12-stage distillation column on the sixth stage from the top. The column pressure is 100 kPa, with a total condenser and a reboiler. The distillate is the benzene product with a specification of 6.0% mole toluene, and the bottom is the toluene product with a specification of 6.0% mole benzene. These specifications will be met by manipulating the reflux rate and the reboiler heat duty. It is required to determine the best pairing between the manipulated and controlled variables.

SOLUTION

The open-loop gains are determined from steady-state simulation. The responses of the controlled to manipulated variables are expected, in general, to be nonlinear, and therefore the gains are best determined by perturbing each manipulated variable in both directions around the design conditions. For the specified compositions, the design reflux rate is 99.91 kmol/h and the design reboiler duty is 3.491×10^6 kJ/h. Each variable is perturbed by $\pm 0.25\%$ while the other is held constant. The results are tabulated below.

Manipulated Variable		Controlled Variables	
1	2	1	2
Reflux Rate	Reboiler Duty	Toluene in dist.	Benzene in btms
(kmol/h)	(kJ/h)	(mole%)	(mole%)
100.16	3.491×10^6	5.8265	6.1805
99.66	3.491×10^6	6.1748	5.8309
99.91	3.500×10^6	6.1780	5.7699
99.91	3.482×10^6	5.8297	6.2385

Equation 16.5 is used to calculate the open-loop gains:

$$K_{11}^O = \frac{5.8265 - 6.1748}{100.16 - 99.66} = -0.6966$$

$$K_{12}^O = \frac{6.1780 - 5.8297}{(3.500 - 3.482) \times 10^6} = 19 \times 10^{-6}$$

$$K_{21}^O = \frac{6.1805 - 5.8309}{100.16 - 99.66} = 0.6992$$

$$K_{22}^O = \frac{5.7699 - 6.2385}{(3.500 - 3.482) \times 10^6} = -26 \times 10^{-6}$$

The relative gains are calculated by Equation 16.9:

$$\mu_{11} = \mu_{22} = \frac{(-0.6966)(-26 \times 10^{-6})}{(-0.6966)(-26 \times 10^{-6}) - (19 \times 10^{-6})(0.6992)} = 3.75$$

$$\mu_{12} = \mu_{21} = \frac{(19 \times 10^{-6})(0.6992)}{(19 \times 10^{-6})(0.6992) - (-0.6966)(-26 \times 10^{-6})} = -2.75$$

As noted earlier, the sum of the relative gains on each row or each column is 1.0, and only one relative gain needs to be calculated for this two by two problem $(3.75 - 2.75 = 1)$.

Following the Bristol rule, the reflux rate should control the distillate composition, and the reboiler duty the bottoms composition. This pairing is the only one with a positive relative gain. It would be the better choice, since the alternative would cause negative interactions. The effectiveness of the recommended pairing should still be checked, since the relative gain of 3.75 is considerably distanced from 1.0, indicating possible controller interaction.

The interaction between the controller loops is evaluated by calculating the change in a controlled variable that results from a change in a manipulated variable. In the preferred pairing, the reflux rate, M_1, controls the distillate composition C_1. The change in C_1 caused by a change in M_1 is calculated by Equation 16.5. If $\Delta M_1 = 0.1$ kmol/h, then

$$\Delta C_1 = K_{11}^O \Delta M_1 = (-0.6966)(0.1) = -0.06966\% \quad \text{mole toluene}$$

The change in the bottoms composition, C_2, caused by the change in the reflux rate is

$$\Delta C_2 = K_{21}^O \Delta M_1 = (0.6992)(0.1) = 0.06992\% \quad \text{mole benzene}$$

The reboiler duty, M_2, will move to reverse the change in C_2:

$$\Delta M_2 = -\Delta C_2/K_{22}^O = -0.06992/(-26 \times 10^{-6}) = 2.689 \times 10^3 \text{ kJ/h}$$

The change in the distillate composition resulting from changes in both manipulated variables is calculated as

$$\Delta C_1 = K_{11}^O \Delta M_1 + K_{12}^O \Delta M_2 = (-0.6966)(0.1) + (19 \times 10^{-6})(2.689 \times 10^3)$$
$$= -0.06966 + 0.0511 = -0.01856\% \text{ mole toluene.}$$

It is observed that some interaction takes place. As the reboiler duty tries to counter the effect of the reflux rate on the bottoms composition, it also affects the distillate composition. However, the reflux rate is still reasonably effective in controlling this composition, and this is indicated by the fact that a net change in the distillate composition in the correct direction results from the changes in both manipulated variables.

EXAMPLE 16.2: CONTROL OF DISTILLATION PRODUCT PURITIES, CASE II

The same distillation column of Example 16.1 is operated such that the benzene product contains 2.0% mole toluene and the toluene product contains 2.0% mole benzene. The manipulated variables are the reflux rate and the reboiler heat duty. At design conditions the reflux rate is 208.0 kmol/h and the reboiler duty is 6.7915×10^6 kJ/h. Evaluate the control-loop options.

SOLUTION

Steady-state simulation is used as in Example 16.1 to determine the open-loop gains. The simulation results are as follows:

Manipulated Variables		Controlled Variables	
1	**2**	**1**	**2**
Reflux Rate	**Reboiler Duty**	**Toluene is dist.**	**Benzene in btms**
(kmol/h)	**(kJ/h)**	**(mole%)**	**(mole%)**
208.52	6.7915×10^6	1.6528	2.3798
207.48	6.7915×10^6	2.3241	1.7608
208.0	6.8085×10^6	2.3392	1.7416
208.0	6.7745×10^6	1.6417	2.4100

Equation 16.5 is used to calculate the open-loop gains:

$$K_{11}^O = \frac{1.6528 - 2.3241}{208.52 - 207.48} = -0.6455$$

$$K_{12}^O = \frac{2.3392 - 1.6417}{(6.8085 - 6.7745) \times 10^6} = 20.5147 \times 10^{-6}$$

$$K_{21}^O = \frac{2.3798 - 1.7608}{208.52 - 207.48} = 0.5952$$

$$K_{22}^O = \frac{1.7416 - 2.4100}{(6.8085 - 6.7745) \times 10^6} = -19.6588 \times 10^{-6}$$

The relative gains are calculated by Equation 16.9:

$$\mu_{11} = \mu_{22} = \frac{(-0.6455)(-19.6588)}{(-0.6455)(-19.6588) - (20.5147)(0.5952)} = 26.47$$

$$\mu_{12} = \mu_{21} = \frac{(20.5147)(0.5952)}{(20.5147)(0.5952) - (-0.6455)(-19.6588)} = -25.47$$

In this column operation neither of the pairing options is promising: one pair of relative gains is negative and the other is too high. The controller interaction is evaluated as in Example 16.1. For a reflux rate change, $\Delta M_1 = 0.1$ kmol/h, using Equation 16.5,

$$\Delta C_1 = K_{11}^O \Delta M_1 = (-0.6455)(0.1) = -0.06455\% \quad \text{mole toluene}$$

The resulting change in the bottoms composition,

$$\Delta C_2 = K_{21}^O \Delta M_1 = (0.5952)(0.1) = 0.05952\% \quad \text{mole benzene}$$

The reboiler duty moves to reverse the change in C_2,

$$\Delta M_2 = -\Delta C_2 / K_{22}^O = -0.05952/(-19.6588 \times 10^{-6}) = 3.028 \times 10^3 \text{ kJ/h}$$

The change in C_1 resulting from changes in both M_1 and M_2,

$$\Delta C_1 = K_{11}^O \Delta M_1 + K_{12}^O \Delta M_2 = (-0.6455)(0.1) + (20.5147 \times 10^{-6})(3.028 \times 10^3)$$
$$= -0.06455 + 0.06212 = -0.00243\% \text{ mole toluene}$$

A strong interaction between the controllers is evident, with the move by the reboiler duty to hold down the bottoms composition almost completely canceling out the effect of the reflux rate on the distillate composition.

It is clear from Example 16.2 that correct pairing of the variables does not always adequately reduce the interaction between the control loops. In such cases consideration should be given to decoupling techniques, where additional controllers are used to eliminate interactions between the loops (Seborg et al., 1989; Smith and Corripio, 1997). Alternatively, a radically different approach, predictive control, should be considered.

16.2 DYNAMIC PREDICTIVE MULTIVARIABLE CONTROL

As noted in the introduction to this chapter and in Section 16.1, several factors may compromise the effectiveness and stability of conventional control techniques for multiple variable processes, and specifically for multistage separations. A successful, comprehensive control strategy for such processes must take into account these factors, most of which are interrelated.

Effective decoupling of interacting control loops may be hampered by different response times for each of the interacting loops. Decouplers are intended to cancel out interactions by implementing certain adjustments in each control loop. Predictors are also used with decouplers to forecast the dynamic responses. These techniques require considerable periodic tuning due to changes in feed flow rates, feed compositions, and other external conditions, and their success is limited.

Process dynamics is another important factor that must be considered. In a distillation column, for instance, the time elapsed between changing the reflux rate and observing a change in a product composition could be measured in hours. With this response time, and in the absence of dynamic prediction capability, the controller will start taking action hours after a disturbance occurs, and it would take even longer for the correction to take effect. Linear predictions are commonly used to forecast trends of process variables; but many processes, particularly multistage separations, are often highly nonlinear. Substantial improvement can be achieved with a nonlinear model.

Another contributor to the lag between a disturbance and controller action is associated with product analyzers response time. Inferential property models that correlate product properties to readily measurable column variables can cut that response time (Smith, 2002).

16.2.1 MODEL-BASED CONTROL AND OPTIMIZATION

The growing power of computers and the availability of efficient computation algorithms have led to the development of integrated systems that address the issues discussed above. Several commercial model-based packages exist for process control and online optimization using a variety of methods (STAR, NLC, 2002; RMPC; Friedman, 1977). The basic concepts of a typical system are discussed below.

A controller/optimizer strategy may include some or all of the following tasks:

1. Generate steady-state data from snap shot plant measurements which generally may not represent steady-state conditions.
2. Use the steady-state data to adjust the feed composition and to calculate a complete and consistent set of steady-state process variables.
3. If extra degrees of freedom are available, optimize the process to determine values for the manipulated variables that maximize an objective function such as the profit.
4. Use a multivariable controller (MVC) to determine the action and timing of each manipulated variable based on information from a dynamic model of the process.

Tasks 1 and 4 receive dynamic data of the process from a dynamic simulation model. This model predicts dynamic responses of the process variables to step changes of the manipulated variables.

In Task 1 the manipulated variables are fixed at snap shot values, and the dynamic responses forecast steady-state process variables corresponding to current manipulated variable values.

Task 2 is a data reconciliation step based on a rigorous steady-state simulator. Discrepancies between plant measurements and simulator predictions are minimized by adjusting the feed composition and other available free variables. Plant measurements may include inferred properties, that is, properties that are not measured directly but inferred from other measurements. Task 3 performs online optimization if extra degrees of freedom exist. Tasks 2 and 3 may be executed in one step.

The MVC, Task 4, adjusts the manipulated variables to satisfy the specifications of the controlled variables. It accomplishes its objective using dynamic response forecasts predicted by the dynamic simulator. The entire computational cycle is repeated every time interval, typically a few minutes, and each cycle is based on data from the previous cycle. In each cycle the controller outputs are computed such as to minimize the difference between the predicted dynamic responses and the desired targets. The MVC action is updated every cycle based on new predicted dynamic responses and new targets.

NOMENCLATURE

C_i Controlled variable
K_{ij} Steady-state gain
M_j Manipulated variable
n Number of variables
μ_{ij} Relative gain

SUBSCRIPTS

i Variable designation
j Variable designation

SUPERSCRIPTS

O Open loop
C Closed loop

PROBLEMS

16.1. A liquid stream at a flow rate w_i is fed to the top of a vertical cylindrical container with cross-sectional area A. At the bottom of the container the liquid flows out of the container in a pipe through a valve at a flow rate $w_e = (\Delta P/k)^{0.5}$, where ΔP is the pressure drop across the valve, and k is the valve constant. The container cross-sectional area, $A = 9$ m^2, and the liquid density, $\rho = 1000$ kg/m^3.

    a. With $w_i = 5$ kg/s, steady state is reached when the liquid level in the container is $h = 2$ m. Find the value of the valve constant k.

    b. Write a differential equation representing the liquid level in an unsteady-state condition.

    c. If the liquid level at time $t = 0$ is 5 m, and the inlet flow rate $w_i = 5$ kg/s, find the liquid level at 1600 s. Use numerical integration.

    d. At $h = 2$ m the steady-state flow rate is $w_e = w_i = 5$ kg/s. If the level set point is changed from 2 m to 2.1 m, what should be the adjusted flow rate of w_i?

16.2. Two streams, v_1 and v_2, are sent to the top of a stirred tank with an outlet stream v. Stream v_1 is pure water and stream v_2 is a salt solution at concentration $c_2 = 100$ kg/m³. At steady state, $v_1 = 0.6$ m³/min, $v_2 = 0.2$ m³/min, and the volume of solution in the tank is $V = 2$ m³. The specific volumes of the liquids are assumed constant and do not depend on the salt concentration.

    a. Find the steady-state concentration c in the outlet stream v.

    b. Flow rates v_1 and v_2 are used as the manipulated variables to meet the specifications of the controlled variables V and c. If each of the manipulated variables is perturbed one at a time by +1% from steady-state conditions, find the resulting values of the controlled variables and the relative gains.

    c. Based on the relative gains obtained in (b), how would you pair the manipulated variables with the controlled variables?

    d. Calculate the change in V resulting from a change in v_1 with v_2 held constant, and the change in c resulting from a change in v_2 with v_1 held constant. Additionally, find the change in V resulting from a change in v_2 with v_1 held constant, and the change in c resulting from a change in v_1 with v_2 held constant.

    e. Calculate the change in each one of the controlled variables resulting from changes in both manipulated variables.

REFERENCES

Bristol, E. H., On a new measure of interaction for multivariable control, *IEEE Trans. Auto. Control*, AC-11, 133, 1966.

Friedman, Y. Z., Advanced control and on-line optimization, *AIChE Houston Chapter CAST Meeting*, Houston, February 1977.

RMPC, Honeywell, Inc., Phoenix, Arizona.

Seborg, D. E., T. F. Edgar, and D. A. Mellichamp, *Process Dynamics and Control*, New York, Wiley, 1989.

Smith, C. A., *Automated Continuous Process Control*, Wiley, New York, 2002.

Smith, C. A. and A. B. Corripio, *Principles and Practice of Automatic Process Control*, 2nd ed, New York, Wiley, 1997.

STAR, NLC, *Dot Products*, PAS, Inc., Houston, Texas, 2002.

17 Batch Distillation

The processes discussed so far in this book are all treated as continuous operations. Although dynamics were considered, for the most part those processes were considered as steady state, where the properties or state variables at any given point in the process are considered time invariant. An important separation process that differs in this respect is batch distillation, a time-dependent process.

A batch distillation apparatus consists of a reboiler attached to the bottom of a trayed or packed column with a condenser at the top. A batch of feedstock is charged to the reboiler, and then the mixture in the reboiler is gradually vaporized. The vapor flows up the column, is condensed in the condenser, and part of the condensate is returned to the column as reflux, flowing countercurrent to the vapor. The rest of the condensate is collected over a period of time as a single distillate product or a series of distillate fractions. Generally, there are no side feeds or side products and the entire column acts as a rectifier.

The same principles of continuous fractionation apply in the batch rectifier, except in batch distillation the compositions and column conditions change continuously with time.

The initial distillate cut is the lightest and, as the distillation progresses, the liquid remaining in the reboiler becomes continuously richer in the heavier components, and subsequent distillate cuts become increasingly heavier. The residue remaining in the reboiler after the last distillate cut is the heaviest cut. A multicomponent feed mixture may be separated in one batch distillation column into a number of products with specified purities. Given the required number of trays and reflux ratio, a batch distillation column could, in principle, separate a normal feed mixture (one that is not reactive or azeotrope forming) into its pure constituents.

Batch distillation is commonly used for separating and purifying materials of high value that are generally handled in relatively small quantities, such as pharmaceuticals and specialty chemicals. A continuous process usually requires long periods of time to reach steady state, during which large quantities of the products are off specification. The batch still also adds versatility to the plant since the same unit may be used to process different feeds and generate multiple products.

Column hydraulics calculations for the design and operation of batch distillation rectifiers, trayed or packed, are similar to those of continuous columns (Chapters 14 and 15). Since the same column may be used for different mixtures at different operating conditions and since the operating conditions change during a given distillation task, it is important to check the column hydraulics for all the expected operating conditions. The reboiler and condenser must be designed to handle the expected rates of vaporization and condensation.

The economical design and operation of batch distillation systems must take into account many factors, most of which are time-dependent variables. The mathematical complexity of the process provides a strong incentive for using computer simulation to solve the problem.

17.1　PRINCIPLES OF BATCH DISTILLATION

The batch distillation system schematic shown in Figure 17.1 consists of a reboiler, a rectifying column, a condenser, and a series of receivers. The plant is piped so that each product may be directed to any one of the receivers.

A batch distillation task is started up by boiling the liquid in the reboiler and vaporizing it at a rate determined by the heat transfer rate. The vapor generated is sent through the column and into the condenser, where it is totally condensed. Initially, the entire condensate is refluxed back to the column with no product drawn. This is a condition of total reflux, similar to the total reflux operation of a continuous column. Under these conditions the batch column operates at steady state, that is, the column operating parameters at any given point do not change with time. At total reflux, the maximum separation is achieved with the given number of trays or packing height.

Depending on the tray geometry (diameter, weir height, downcomer design, etc.) in trayed columns or on the type of packing in packed columns, a certain amount of liquid is retained within the column as liquid holdup. An additional amount of liquid holdup is associated with the condenser assembly. The remaining space within the distillation system contains the vapor holdup. In continuous columns, holdup is usually not a factor since it does not affect the separation as long as the column operates at steady state.

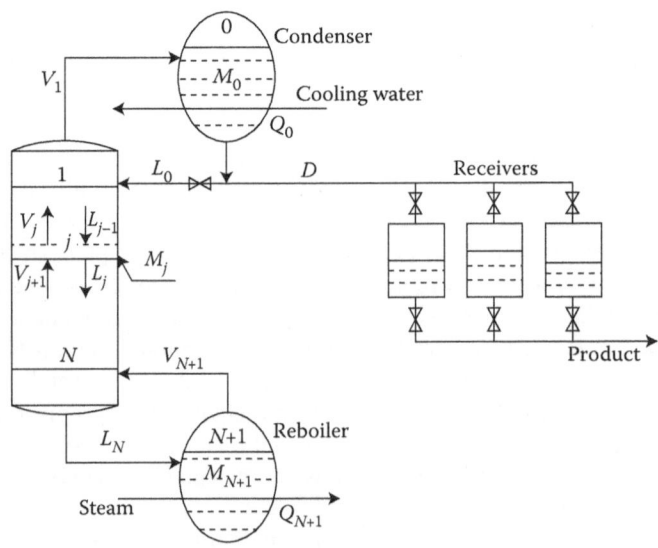

FIGURE 17.1 Batch distillation system schematic.

17.1.1 Effect of Holdup

As distillate starts to be drawn from the condenser following total reflux operation, the compositions throughout the column start changing and unsteady-state conditions ensue. It is generally accepted that the effect of the vapor holdup on the batch column performance may be neglected. The liquid holdup affects the column performance in different ways and to varying degrees. The effects become more noticeable with larger holdups relative to the amount of liquid in the reboiler.

One effect is that the holdup lowers the original concentration of the light components in the feed charge. Distillate production starts with a lower light component concentration in the reboiler, making the separation more difficult. Another effect, which seems to counter the first, is associated with the lag in composition changes in the column resulting from the holdup. The delayed drop of the light components' concentration in the holdup can enhance the separation.

These observations are too qualitative, but it is important to note that holdup is a factor that must be taken into consideration when evaluating batch processes, especially when holdups reach 5–10% or higher relative to the liquid in the reboiler. The more rigorous mathematical solution methods incorporate liquid holdup and quantify its effects.

17.1.2 Operating Strategies

As the distillate is withdrawn from the condenser, its composition changes continuously, becoming progressively heavy. The composition of a cut is an integrated average of the instantaneous distillate composition over the time period of collecting that cut. Subsequent cuts are made by redirecting the distillate to different receivers each time the required composition of the current cut has been attained. It often happens that, in order to meet certain purity cuts, other intermediate, off-specification cuts must be produced. The intermediate products are collected and usually blended with raw feed for recycling.

In batch distillation the column conditions vary throughout the distillation cycle. Therefore, it is possible, in principle, to conduct the distillation in an infinite number of ways by arbitrarily changing the column conditions with time. Any two of the variables: reflux ratio, distillate rate, boilup rate, condenser duty, and reboiler duty could be controlled independently during the distillation cycle (Section 17.1.3). Operating strategies are sought that would meet performance objectives such as product purity specifications. Additionally, the operation may be driven by optimization criteria, such as minimizing the distillation time or maximizing profit. A discussion of some of the possible operating strategies follows.

17.1.2.1 Constant Reflux

Operating the column at a constant reflux rate results in a distillate with a continuously changing composition. The composition profile over time is determined by the reflux rate and the distillate rate. The lightest component starts at its highest value, followed by concentration peaks of components with descending volatility.

The sharpness of the separation is determined by the reflux ratio. Since the reflux rate is limited by hydraulics considerations (Chapters 14 and 15), higher reflux ratios would require lowering the distillate rate, thus lengthening the distillation cycle. The separation sharpness must therefore be weighed against the time required to complete the distillation.

17.1.2.2   Constant Distillate Composition

A constant-composition distillate may be produced by progressively increasing the reflux rate. When it reaches the maximum allowable from a hydraulic or economic standpoint, the distillate is redirected to a new receiver, and the reflux rate is lowered to a new starting point. The procedure is repeated for subsequent cuts, some of which may be off specification. Here again, lower distillate rates generally result in better separation at the expense of longer distillation times.

17.1.2.3   Cycling Operation

The column is initially operated at total reflux until equilibrium is reached. The condensate is then totally diverted to a receiver for a short period of time, during which the column operates at zero reflux. Before the composition of the liquid in the receiver goes off specification, the condensate is diverted back to the column, and the unit is operated again at total reflux. The cycle is repeated for all the remaining cuts.

17.1.2.3.1   Optimized Operation

Given an objective function to be maximized or minimized, it is possible, in principle, to compute an optimum profile of reflux ratio and distillate rate for the entire distillation cycle. The principles of such approaches are introduced in Section 17.2.4.

17.1.3   Conceptual Control and Degrees of Freedom

Reference again is made to Figure 17.1 for a degrees of freedom analysis and a study of possible control strategies. The description rule (Section 5.2.1) is applied here to an existing system. The variables that can be set by external independent means include the steam rate to the reboiler, the cooling water rate to the condenser, and the reflux and product valve positions. At a given time, only one product receiver is active, and the valves of all the other receivers are closed.

At total reflux the reboiler duty is balanced by the condenser duty. If, at this point, a distillate product is drawn at a certain rate without changing the reflux rate, both reboiler and condenser duties must be increased to handle the higher rate of vaporization and condensation. In general, any set of reboiler and condenser duties determines the reflux and distillate rates. The reflux and product valve positions must be controlled to maintain reboiler and condenser liquid levels. The boilup rate is also determined from the other variables by material balance. Thus, of the five variables—reboiler duty, condenser duty, boilup rate, reflux rate, and product rate—two are independent and the system has two degrees of freedom.

The overall control strategy depends on the reflux policy selected and can vary from simple manual control to elaborate systems designed to control the reflux based on some optimization scheme. Aside from the overall control strategy, the primary controllers most commonly used include the following: the reboiler steam rate controls the pressure drop across the column, the reflux rate controls the temperature at some point in the column, the condenser cooling water rate controls the condensate temperature, and the product rate controls the liquid level in the accumulator.

17.2 SOLUTION METHODS

The complexity of batch distillation calculations is due to the unsteady-state, dynamic nature of the process. Most of the variables describing the column are constantly changing with time. Hence, an accurate mathematical model must include differential as well as algebraic equations. With the ever-increasing computing power available, rigorous simulation using reasonable computing time is now possible for the design and performance evaluation of batch distillation operations. As is always the case, the use of simulators is most effective when based on a qualitative understanding of the problem, aided by graphical or shortcut preliminary calculations. These methods may themselves be adequate for small applications where the economic penalty of overdesign is insignificant.

17.2.1 GRAPHICAL AND SHORTCUT METHODS: BINARY SYSTEMS

Binary batch distillation can be represented graphically on a Y–X diagram using the McCabe–Thiele method as in continuous binary distillation. Since the compositions throughout the system change continuously with time, one plot represents one instant in the distillation cycle. In addition to the McCabe–Thiele assumptions in continuous distillation, the graphical method for batch distillation also assumes negligible column and condenser holdups compared to the amount of liquid in the reboiler. With low percentage holdups (0–10% of the liquid in the reboiler), the accumulation terms in the mass balance equations (discussed in subsequent sections) can be neglected, and instantaneous pseudo-steady-state conditions can be assumed.

At a given point in time, t^1, the amount of liquid in the reboiler is M_{N+1}^1, and its composition in terms of the mole fraction of the lighter component is X_{N+1}^1. The distillate rate and composition at time t^1 are D^1 mol/h and X_D^1, respectively. The internal liquid and vapor flows at the same instant are L^1 and V^1 mol/h, assumed constant from tray to tray. A material balance at t^1 around an envelope that includes the condenser and a column section from the condenser to any tray gives the following operating line equation if the holdups are neglected:

$$Y^1 = (L^1/V^1)X^1 + (D^1/V^1)X_D^1 \tag{17.1}$$

The operating line equation at time t^2, when the reboiler liquid amount and composition are M_{N+1}^2 and X_{N+1}^2, is given by

$$Y^2 = (L^2/V^2)X^2 + (D^2/V^2)X_D^2 \tag{17.2}$$

A material balance around the condenser relates the L/V ratio to the reflux ratio at any instant:

$$\frac{L}{V} = \frac{R}{R+1} \tag{17.3}$$

Instantaneous operating lines may be constructed based on given operating conditions. Figure 17.2 shows two operating lines drawn for a constant L/V ratio (or constant reflux ratio) at two different points in time. From the intersection point of the vertical line at X_D^1 and the 45° diagonal, an operating line is drawn with slope L/V. Next, the theoretical stages (three in this case) are stepped off between the operating line and the equilibrium curve. The equilibrium reboiler composition is X_{N+1}^1. At a later time the distillate composition X_D^2 is less than X_D^1 since this is the mole fraction of the lighter component, which is decreasing as the distillation progresses. The operating line is drawn parallel to the first one with the same slope, L/V, and the three stages are stepped off, giving a reboiler composition of X_{N+1}^2.

For an operation at constant distillate composition X_D, the reflux ratio is constantly changing. The operating lines at different times all pass through the X_D intersection with the diagonal, each with a different slope. In order to maintain a constant distillate composition, the reflux ratio, and hence the operating line slope, must increase with time. Figure 17.3 shows two operating lines at two different points in time, giving reboiler compositions X_{N+1}^1 and X_{N+1}^2. As time progresses, the concentration of the lighter component in the reboiler decreases, that is, $X_{N+1}^2 < X_{N+1}^1$.

The graphical procedure is applicable to any binary batch distillation process and is not limited to operations at constant reflux ratio or constant distillate composition.

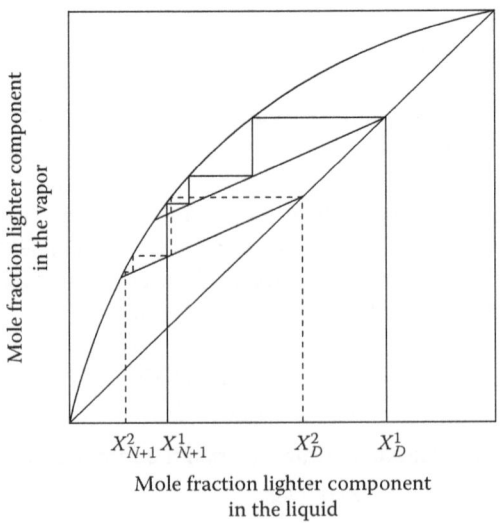

FIGURE 17.2   Graphical method for constant reflux ratio.

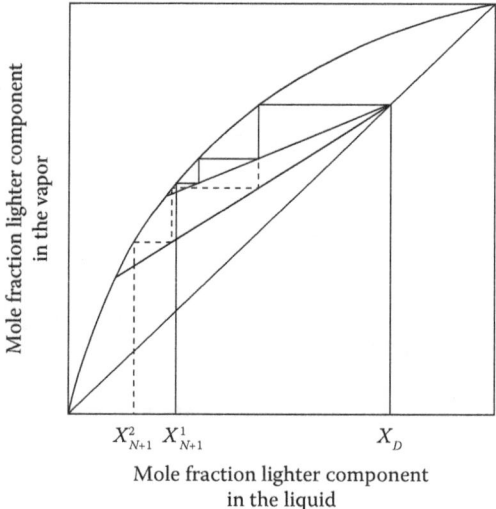

FIGURE 17.3   Graphical method for constant distillate composition.

The result for a given number of stages is a tabular relationship between the distillate composition, the reboiler composition, and the L/V ratio:

$$X_D = f(X_{N+1}, L/V) \qquad (17.4)$$

Other variables may be calculated based on this relationship. Since the holdup is neglected, the rate of change in the reboiler liquid equals the distillate rate. By total material balance,

$$-\frac{dM_{N+1}}{dt} = D \qquad (17.5)$$

and by material balance on the lighter component,

$$-\frac{d(M_{N+1}X_{N+1})}{dt} = DX_D \qquad (17.6)$$

By expanding the derivative in Equation 17.6, combining with Equation 17.5 and rearranging, the following is obtained:

$$\frac{dM_{N+1}}{M_{N+1}} = \frac{dX_{N+1}}{X_D - X_{N+1}} \qquad (17.7)$$

Integrating this equation from initial conditions ($t = 0$) to current conditions gives (Smoker et al., 1940)

$$\ln\left(\frac{M_{N+1}}{M_{N+1}^0}\right) = \int_{X_{N+1}^0}^{X_{N+1}} \frac{dX_{N+1}}{X_D - X_{N+1}} \tag{17.8}$$

The right-hand side is evaluated by graphical integration using the Equation 17.4. If this integral is denoted by I, then

$$M_{N+1} = M_{N+1}^0 e^{-I} \tag{17.9}$$

The amount of distillate cut, C, and its composition are calculated by total and lighter component mass balances:

$$C = M_{N+1}^0 - M_{N+1}$$

$$CX_C = M_{N+1}^0 X_{N+1}^0 - M_{N+1} X_{N+1}$$

where X_C is the cumulative mole fraction of the lighter component in C. The two mass balance equations are combined to give

$$X_C = \frac{M_{N+1}^0 X_{N+1}^0 - M_{N+1} X_{N+1}}{M_{N+1}^0 - M_{N+1}} \tag{17.10}$$

The time required to distill an amount C at constant reflux and distillate rates is calculated as

$$t = \frac{C}{D} = \frac{M_{N+1}^0 - M_{N+1}}{D} \tag{17.11}$$

The distillate rate may be expressed in terms of the boilup rate and reflux ratio:

$$D = \frac{V}{R+1} \tag{17.12}$$

Substituting Equations 17.9 and 17.12 in Equation 17.11 gives

$$t = (R+1)\frac{M_{N+1}^0}{V}(1 - e^{-I}) \tag{17.13}$$

In general, the distillate rate may vary, and the distillation time is written in differential form:

$$dt = \frac{dC}{D} = -\frac{dM_{N+1}}{D}$$

This is combined with Equation 17.7 to give

$$dt = -\frac{M_{N+1}dX_{N+1}}{D(X_D - X_{N+1})} \tag{17.14}$$

The distillation time is obtained by graphical or numerical integration of this equation. If the process is at constant distillate composition, the instantaneous and cumulative distillate compositions are identical:

$$X_D = X_C$$

Thus, Equation 17.10 may be rearranged to give

$$M_{N+1} = \frac{M_{N+1}^0(X_D - X_{N+1}^0)}{X_D - X_{N+1}} \tag{17.15}$$

The distillate rate is given in terms of the boilup rate and the L/V ratio:

$$D = V - L = V(1 - L/V) \tag{17.16}$$

Equations 17.15 and 17.16 are combined with Equation 17.14 to give

$$dt = -\frac{M_{N+1}^0(X_D - X_{N+1}^0)dX_{N+1}}{V(1 - L/V)(X_D - X_{N+1})^2} \tag{17.17}$$

The time required to change the reboiler composition from X_{N+1}^0 to X_{N+1} is obtained by integration:

$$t = \int_{X_{N+1}}^{X_{N+1}^0} \frac{M_{N+1}^0(X_D - X_{N+1}^0)}{V(1 - L/V)(X_D - X_{N+1})^2} \, dX_{N+1}$$

The integration is carried out graphically using data as expressed by Equation 17.4.

17.2.1.1   Differential Distillation

A simple application for batch distillation is what is known as *differential distillation*. It consists of a boiler containing a load of liquid. The liquid is progressively vaporized, with the vapor flowing to a condenser, producing a liquid distillate. There is no reflux, so the only stage with vapor–liquid equilibrium is the boiler.

The model assumes perfectly mixed liquid in the boiler. As the distillation process progresses, the liquid composition changes with time, as does the instantaneous composition of the vapor, being at equilibrium with the current boiler liquid composition. The vapor is condensed to a liquid distillate, and with no liquid accumulation, the instantaneous vapor and liquid distillate are of the same composition.

With a single equilibrium stage and no reflux, the separation power in differential distillation is obviously limited. It is the equivalent of a batch flash operation. Consequently, practical applications would include the separation of wide-boiling mixtures, with low expectations on the purity of the products.

For a single stage, Equation 17.8 may be applied where X_D is replaced with Y, the mole fraction of one of the components in the equilibrium vapor distillate:

$$\ln\left(\frac{M}{M^0}\right) = \int_{X^0}^{X} \frac{dX}{Y - X} \tag{17.8a}$$

The integral may be evaluated graphically, with X varying in small increments from initial conditions, X_0, to the final state, X. The corresponding Y values represent the equilibrium vapor composition. The amount of distillate and its cumulative composition are calculated by equations equivalent to Equations 17.9 and 17.10.

Analytical solution to Equation 17.8a is possible for a binary system with constant relative volatility. The relative volatility is expressed as

$$\alpha = \frac{Y}{X} \frac{1 - X}{1 - Y}$$

This is rearranged for an expression explicit in Y:

$$Y = \frac{\alpha X}{1 + (\alpha - 1)X}$$

Substitution in Equation 17.8a gives

$$\ln\left(\frac{M}{M^0}\right) = \int_{X^0}^{X} \left[\frac{1}{(\alpha - 1)X(1 - X)} + \frac{1}{1 - X}\right] dX$$

Upon integration and simplification the following equation is obtained:

$$\ln\left(\frac{M^0}{M}\right) = \frac{1}{\alpha - 1}\left[\ln\left(\frac{X^0}{X}\right) - \alpha \ln\left(\frac{1 - X^0}{1 - X}\right)\right]$$

This equation, relating current liquid holdup and composition to initial liquid holdup and composition, may also be written in the form

$$\frac{M}{M^0} = \left(\frac{X}{X^0}\right)^{\frac{1}{\alpha - 1}} \left(\frac{1 - X^0}{1 - X}\right)^{\frac{\alpha}{\alpha - 1}} \tag{17.9a}$$

EXAMPLE 17.1: BENZENE–TOLUENE DIFFERENTIAL BATCH DISTILLATION

A batch still contains 100 kmol of a benzene–toluene solution at 85°C and 100 kPa. The mixture is distilled with no reflux in the column, in a differential batch distillation process. Determine the equilibrium liquid and vapor compositions at 5°C increments up to 105°C. It is also required to find at each time interval the amount of solution in the still and the total distillate composition.

The K-value of benzene at 1 atm is given by the following equation with the temperature in degrees Celsius: $K_B = 0.0921\exp(0.0296T)$. A constant relative volatility of benzene to toluene is given as 2.40.

SOLUTION

Given, $\alpha_{BT} = K_B/K_T = 2.40 = (Y_B/X_B)(1 - X_B)/(1 - Y_B) = K_B(1 - X_B)/(1 - K_B X_B)$

This equation is rearranged to calculate X_B at each given temperature:

$$X_B = (K_B - \alpha_{BT})/[K_B(1 - \alpha_{BT})]$$

This is followed by calculating Y_B, X_T, and Y_T. At each temperature, M, the amount of solution in the still, is calculated from Equation 17.9a. M^0 is the starting value of M, at 85°C. The total distillate composition is calculated by material balance:

$$M^0 X^0 = MX + (M^0 - M)X_C$$

$$X_C = (M^0 X^0 - MX)/(M^0 - M)$$

The results are tabulated as follows:

T (°C)	K_B	K_T	X_B	X_T	Y_B	Y_T	M (kmol)	$X_{C,B}$	$X_{C,T}$
85	1.140	0.475	0.789	0.211	0.900	0.100	100.000	0.900	0.100
90	1.322	0.551	0.582	0.418	0.770	0.230	24.914	0.858	0.142
95	1.533	0.639	0.404	0.596	0.619	0.381	10.426	0.834	0.166
100	1.777	0.741	0.250	0.750	0.445	0.555	4.994	0.818	0.182
105	2.061	0.859	0.118	0.882	0.242	0.758	2.202	0.804	0.196

In a rectifying process the column has reflux, and evaluating its performance must consider the reflux ratio as well as the other factors. Example 17.2 uses the McCabe–Thiele approach to solve such a column at certain defined conditions.

EXAMPLE 17.2: BATCH DISTILLATION OF A BINARY SYSTEM BASED ON THE MCCABE–THIELE METHOD

A mixture of n-butane and n-pentane is loaded into a batch distillation reboiler to produce a distillate at constant rate and composition. The column pressure is constant at 500 kPa, the initial charge is 100 kmol, and the distillate rate is 10 kmol/h. The column has three theoretical trays plus the reboiler, giving it four theoretical stages. The initial charge composition, the required distillate composition, and the component relative volatilities, assumed constant, are as follows:

	Initial Charge Composition	Distillate Composition	Relative Volatility
	$X^0_{N+1,i}$	$X_{D,i}$	$\alpha_{i,2}$
1. n-Butane	0.6	0.9	2.315
2. n-Pentane	0.4	0.1	1.0

It is required to find the amount and composition of the residue, the total distillate amount, and the L/V ratio as they change with time. Tray holdups may be neglected.

SOLUTION

With constant distillate rate and composition, the amount and composition of the residue at any point in time can be determined by material balance. The number of stages is also fixed, so that leaves the reflux ratio (or L/V ratio) as the only variable to be adjusted to satisfy the phase equilibrium conditions.

The McCabe–Thiele method uses a graphical approach, where the equilibrium data, the operating lines, and the stepping of stages are done graphically. In this example, where the relative volatility is assumed constant, a numerical approach may be implemented.

The relative volatility of a binary system at equilibrium is defined as

$$\alpha_{12} = (Y_1/X_1)(1 - X_1)/(1 - Y_1)$$

With the given constant value of the relative volatility, the equilibrium curve is expressed as

$$Y_1 = 2.315X_1/(1 + 1.315X_1) \quad \text{or} \quad X_1 = Y_1/[Y_1 - 2.315(Y_1 - 1)]$$

The operating line relates liquid and vapor compositions between stages:

$$Y_{2,1} = (L/V)X_{1,1} + b$$

The distillate composition is to be kept constant at $X_{D1} = 0.9 \ (= Y_{D1})$. Therefore, the operating line must always pass through this point, so that

$$b = 0.9 - 0.9(L/V)$$

Substituting in the operating line equation,

$$Y_{2,1} = (L/V)(X_{1,1} - 0.9) + 0.9$$

At the initial state, $t = 0$, the operating line passes through $X_{4,1} = 0.6$, the initial residue composition. The calculations alternate between the equilibrium equation and the operating line equation:

Equilibrium,

$$X_{1,1} = Y_{D,1}/[Y_{D,1} - 2.315(Y_{D,1} - 1)] = 0.9/[0.9 - 2.315(0.9 - 1)] = 0.7954$$

Operating,

$$Y_{2,1} = (L/V)(0.7954 - 0.9) + 0.9$$

Equilibrium,

$$X_{2,1} = Y_{2,1}/[Y_{2,1} - 2.315(Y_{2,1} - 1)]$$

Operating,

$$Y_{3,1} = (L/V)(X_{2,1} - 0.9) + 0.9$$

Equilibrium,

$$X_{3,1} = Y_{3,1}/[Y_{3,1} - 2.315(Y_{3,1} - 1)]$$

Operating,

$$Y_{4,1} = (L/V)(X_{3,1} - 0.9) + 0.9$$

Equilibrium,

$$X_{4,1} = 0.6 = Y_{4,1}/[Y_{4,1} - 2.315(Y_{4,1} - 1)]$$

These equations may be solved by trial and error, by assuming a value for L/V, then solving them from the top down or from the bottom up. The L/V value is updated until all the equations are satisfied. (An instant solution may be obtained by a function such as "Goal Seek" in The Excel spreadsheet program).

The residue amount and composition are calculated by material balance at subsequent points in time. Following each time interval the calculations proceed as above for the L/V ratio and stage compositions at each successive residue condition.

At a distillate rate of V kmol/h, and at time t min, the amount distilled is

$$C^t = Vt/60 \text{ kmol}$$

Amount remaining in the reboiler,

$$M_4^t = M_4^0 - C^t$$

Residual composition at interval end, by component material balance,

$$M_4^0 X_{4,i}^0 = M_4^t X_{4,i}^t + C^t X_{D,i}$$

To illustrate, at $t = 30$ min,

$$C = (10)(30)/60 = 5 \text{ kmol}$$

$$M_4 = 100 - 5 = 95 \text{ kmol}$$

$$(100)(0.6) = 95X_{4,1} + (5)(0.9)$$

$$X_{4,1} = 0.5842$$

The results are summarized below:

Steps	Start	1	2	3
Elapsed time, t (min)	0	30	90	150
Initial charge, M_4 (kmol)	100	95	85	75
Amount distilled, C (kmol)	0	5	15	25
Operating line slope, L/V	0.4838	0.5088	0.5656	0.6342
Stage 1, X_1	0.795	0.795	0.795	0.795
Y_1	0.849	0.847	0.841	0.834
Stage 2, X_2	0.709	0.705	0.695	0.684
Y_2	0.808	0.801	0.784	0.763
Stage 3, X_3	0.645	0.634	0.611	0.582
Y_3	0.776	0.765	0.736	0.698
Reboiler, X_4	0.600	0.584	0.547	0.500

As time goes by, and the distillation continues, the residue and light key concentration in it go down. In order to maintain the required concentration in the distillate, the reflux or L/V ratio has to increase with time. Depending on the column conditions, the L/V ratio may at some point reach 1 (total reflux), at which point further distillation will bring down the light key component concentration in the distillate.

17.2.2   SHORTCUT METHODS: MULTI-COMPONENT DISTILLATION

Multi-component batch distillation computation time can be excessive with rigorous methods, especially for problems that require multiple runs, such as optimization, and for online applications. It is desirable to have shortcut alternatives for such situations.

The graphical-based shortcut methods for binary batch distillation may be applied to multicomponent distillation only when the separation is between two key components to produce one distillate product and the residue. In this case the calculations may be approximated by lumping the other components with either of the key components and treating the system as a pseudo-binary.

Truly multicomponent solutions based on continuous distillation shortcut methods have been proposed for batch distillation. The Fenske, Underwood, and Gilliland equations or correlations are commonly used in conjunction with each other to solve continuous distillation problems as described in Section 12.3. Diwekar and Madhavan (1991) describe how these techniques may be modified for the design of batch distillation columns for variable and constant reflux cases.

The unsteady-state operation is approximated by a pseudo-steady-state model using a series of time steps, during each of which steady-state conditions are assumed. At the end of each time step the bottoms product becomes the feed for the next time step.

For a variable reflux case, the procedure initially assumes constant product (distillate) compositions X_{Di} for all the components. The minimum reflux ratio, R_{min}, required to produce a distillate with composition X_{Di} from a feed with composition $X_{N+1,i}^0$ (which is the reboiler initial composition) is calculated by the Underwood method (Equations 12.29 and 12.30). The ratio of refluxes is

$$r = \frac{R}{R_{min}}$$

where the starting reflux ratio, R, is given. The minimum number of trays is calculated by the Fenske equation from the distillate composition X_{Di} and the bottoms composition, assumed initially the same as $X_{N+1,i}^0$. Equation 12.14 may be used for this purpose, rewritten in terms of the ratio of light key to heavy key mole fractions and relative volatilities:

$$N_{min} = \frac{\ln(X_{Dl}X_{N+1,h}^0/X_{N+1,l}^0 X_{Dh})}{\ln(\alpha_l/\alpha_h)}$$

The Gilliland correlation is used to determine the ratio N/N_{min} corresponding to r. This determines the number of trays N, which is held constant in subsequent time steps in which the reflux ratio becomes the calculated variable.

During a small time step, a small amount of liquid in the reboiler is vaporized. As a result, the composition of the liquid remaining in the reboiler changes. The amount vaporized and the change in the reboiler composition during the time step is determined by the rate of heat transfer in the reboiler. The computations proceed by incrementing the composition of a reference component in the reboiler, which is equivalent to incrementing the time. By material balance, the new reboiler composition for any component i is given in terms of the new composition of the reference component, $X_{N+1,r}$ (Diwekar and Madhavan, 1991):

$$X_{N+1,i} = X_{Di} - \left[\frac{X_{Di} - X_{N+1,i}^0}{X_{Dr} - X_{N+1,r}^0}(X_{Dr} - X_{N+1,r})\right]$$

The Fenske–Underwood–Gilliland methods are again applied to the distillate composition, X_{Di} (assumed constant), the current reboiler composition, $X_{N+1,i}$, and the number of trays, N, to determine r and hence the reflux ratio, R. The procedure is repeated by further incrementing the reference component composition for each time step until a target composition $X_{N+1,r}^F$ is reached.

This simplified procedure is based on the assumption that the concentrations of all the components in the distillate are constant if the reference component concentration in the distillate is constant. If the concentrations of the nonreference components change, the authors (Diwekar and Madhavan, 1991) recalculate the concentrations based on the Hengstebeck–Geddes equation (Hengstebeck, 1946; Geddes, 1958). The authors also describe a procedure for the constant reflux case, where all the component compositions in the distillate change as the distillation progresses.

EXAMPLE 17.3: BATCH DISTILLATION BY THE SHORTCUT METHOD

A process is defined similar to Example 17.2, with the same components, initial charge and composition, the same constant distillate rate and required composition, and the same column pressure. In this case the reflux ratio is maintained at twice the minimum value. The column has six actual trays plus the reboiler, the equivalent of seven actual trays. The overall tray efficiency is 65%, and the relative volatility of butane to pentane is assumed constant at 2.315.

It is required to find the amount and composition of the residue, the total distillate amount, and the reflux rate as a function of time. At what point will it not be possible to meet the required distillate composition? Tray holdups may be neglected.

SOLUTION

Whereas Example 17.2 was solved based on the McCabe–Thiele method, the way this example is defined favors the shortcut, Fenske–Underwood–Gilliland approach. This method may be used for multicomponent mixtures, but it is not a method selection criterion in this example, which is a binary system.

The process is approximated as a number or relatively small steps, each of which corresponds to a defined time interval. In each step the shortcut distillation model is applied based on the initial reboiler charge for that step and the desired constant distillate composition. The resulting product is used as the reboiler charge for the next step. The computations are continued until the available number of stages cannot produce the required distillate composition.

In Example 17.2, the number of theoretical stages is fixed but the reflux ratio (or L/V) could be varied to a point, to satisfy the specifications. In this example the reflux ratio is set by the problem statement, and the required number of stages is to be determined.

INITIAL STATE

Given the initial reboiler and the distillate compositions, calculate the minimum reflux ratio and the minimum number of stages by the Fenske and Underwood methods. With a reflux ratio that is twice the minimum, calculate the number of theoretical stages by the Gilliland method. The outcome is the initial reboiler charge and composition for the next step.

Calculate θ by Equation 12.30

$$(2.315 \times 0.6)/(2.315 - \theta) + (1.0 \times 0.4)/(1.0 - \theta) = 0 \quad \rightarrow \theta = 1.294$$

Calculate R_m by Equation 12.29:

$$R_m = (2.315 \times 0.9)/(2.315 - 1.294) + (1.0 \times 0.1)/(1.0 - 1.294) - 1 = 0.7005$$

$$M_N^0 = 100 \text{ kmol}$$

$$R = 2R_m = 2 \times 0.7005 = 1.401$$

Calculate N_m by Equation 12.17a (or the equivalent equation in Section 17.2.2):

$$N_m = \log[(0.9/0.6)(0.4/0.1)]/\log(2.315/1.0) = 2.1345$$

Calculate N by the Gilliland correlation,

$$N = 4.11$$

The actual number of stages,

$$N_{act} = N/\eta = 4.11/0.65 = 6.3$$

$$\text{Initial charge, } M_7^0 = 100 \text{ kmol}$$

$$\text{Distillate rate, } V = 10 \text{ kmol/h}$$

$$\text{Interval time, } t = 30 \text{ min}$$

$$\text{Amount distilled, } C = Vt/60 = (10)(30)/60 = 5 \text{ kmol}$$

$$\text{Amount remaining in the reboiler, } M_7^1 = M_7^0 - C^1 = 100 - 5 = 95 \text{ kmol}$$

Residual composition at interval end, by component material balance,

$$M_7^0 X_{7,i}^0 = M_{7,i}^1 X_{7,i}^1 + C^1 X_{D,i}$$

$$(100)(0.6) = 95\, X_{7,1}^1 + (5)(0.9)$$

$$X_{7,1}^1 = 0.584, \; X_{7,2}^1 = 0.416$$

The calculations are repeated for additional steps until the required number of stages for meeting the required distillate composition exceeds the available number of stages. This is when the distillation process should stop. The results for the next steps are listed below:

Steps	Start	1	2	3
Elapsed time, t (min)	0	30	90	150
Initial charge, M_7 (kmol)	100	95	85	75
Amount distilled, C (kmol)	0	5	15	25
Residual $X_{7,1}$	0.6	0.584	0.547	0.50
Residual $X_{7,2}$	0.4	0.416	0.453	0.50
Minimum reflux ratio, R_m	0.7005	0.748	0.862	1.017
Reflux ratio, R	1.401	1.496	1.725	2.034
Minimum stages, N_m	2.135	2.212	2.393	2.618
Theoretical stages, N	4.11	4.19	4.38	4.61
Actual stages, N_{act}	6.04 (6)	6.44 (7)	6.74 (7)	7.09 (>7)

The results indicate that after 25 kmol are distilled, the distillate cannot be maintained at the required composition with the existing number of stages.

17.2.3   RIGOROUS METHODS

The rigorous solution of batch distillation columns carries an extra dimension of complexity over continuous steady-state distillation because it is inherently a transient operation. The basic assumption of steady-state operation in the continuous column model obviously does not apply for batch distillation. The only possible steady-state operation in batch distillation is at total reflux, which is commonly used as the initial condition for the dynamic solution of the column.

As in continuous columns, the rigorous solution of batch columns is obtained by solving the energy and material balance and phase equilibrium equations. While holdup terms are not considered in continuous operations, they must be included in batch distillation. The column schematic of Figure 17.1 is the problem model with nomenclature similar to continuous columns. The condenser is designated as stage 0, the reboiler as stage $N + 1$, and any of the N column trays (or equivalent packing stages) is designated as stage j. The liquid molar holdup on stage j is designated as M_j.

In addition to the basic continuous column model assumptions of equilibrium stages and adiabatic operation, dynamics-related assumptions are made for the batch model. Distefano (1968) assumed constant volume of liquid holdup, negligible vapor holdup, and negligible fluid dynamic lag. Although different solution strategies may be employed, the fundamental model equations are the same.

Condenser total material balance:

$$\frac{dM_0}{dt} = V_1 - (L_0 + D) \tag{17.18}$$

Condenser component material balances:

$$\frac{d(M_0 X_{0i})}{dt} = V_1 Y_{1i} - (L_0 + D) X_{0i} \tag{17.19}$$

for $i = 1, ..., C$.

Condenser energy balance:

$$\frac{d(M_0 u_0)}{dt} = V_1 H_1 - (L_0 + D) h_0 - Q_0 \tag{17.20}$$

where u_0 is the molar internal energy of the liquid in the condenser.

Total material balance on stage j ($1 \leq j \leq N$):

$$\frac{dM_j}{dt} = V_{j+1} + L_{j-1} - V_j - L_j \tag{17.21}$$

Component material balances on stage j $(1 \leq j \leq N)$:

$$\frac{d(M_j X_{ji})}{dt} = V_{j+1} Y_{j+1,i} + L_{j-1} X_{j-1,i} - V_j Y_{ji} - L_j X_{ji} \qquad (17.22)$$

for $i = 1, \ldots, C$.

Energy balance on stage j $(1 \leq j \leq N)$:

$$\frac{d(M_j u_j)}{dt} = V_{j+1} H_{j+1} + L_{j-1} h_{j-1} - V_j H_j - L_j h_j \qquad (17.23)$$

Reboiler total material balance:

$$\frac{dM_{N+1}}{dt} = L_N - V_{N+1} \qquad (17.24)$$

Reboiler component material balances:

$$\frac{d(M_{N+1} X_{N+1,i})}{dt} = L_N X_{Ni} - V_{N+1} Y_{N+1,i} \qquad (17.25)$$

for $i = 1, \ldots, C$.

Reboiler energy balance:

$$\frac{d(M_{N+1} u_{N+1})}{dt} = L_N h_N - V_{N+1} H_{N+1} + Q_{N+1} \qquad (17.26)$$

Phase equilibrium relations:

$$Y_{ji} = K_{ji} X_{ji} \qquad (17.27)$$

for $i = 1, \ldots, C; j = 1, \ldots, N+1$.

Mole fraction summation equations:

$$\sum_i X_{ji} = 1 \qquad (17.28a)$$

for $j = 0, 1, \ldots, N+1$, and

$$\sum_i Y_{ji} = 1 \qquad (17.28b)$$

for $j = 1, \ldots, N+1$.

Equations 17.18 through 17.28 encompass a set of $2NC + 4N + 3C + 7$ equations to be solved for the time-dependent, unknown variables. These variables are listed as follows:

X_{ji}	$(N+2)C$
Y_{ji}	$(N+1)C$
L_j	$N+1$
V_j	$N+1$
D	1
T_j	$N+2$
M_j	$N+2$
Q_0	1
Q_{N+1}	1
Total	$2NC + 4N + 3C + 9$

Thus, in order to define the column operation uniquely, two specifications are required, as already concluded using the description rule (Section 17.1.3). These could be the reflux rate and distillate rate, L_0 and D. Note that a subcooled condenser is assumed so that no phase equilibrium equation is written for stage 0 and no Y_{0i} variables exist. The column pressure profile is assumed fixed or determined independently from hydraulics calculations and is not included in the column variables. Also, the enthalpies and phase equilibrium coefficients are, in general, functions of the temperature, pressure, and composition (Chapter 1) and are therefore not considered as additional unknown variables.

Equations 17.18 through 17.28 may be solved using different strategies, including simultaneous or iterative methods. Distefano (1968) describes a computational procedure that solves the equations one at a time. Boston et al. (1981) describe a more flexible and efficient method that can handle a variety of specifications and column configurations, including multiple feeds, side draws, and side heaters. The method uses advanced numerical procedures to handle nonlinearity and stability problems. The object here is not to describe the details of the methods, which may be found in the references. The Distefano method is outlined here to illustrate briefly some of the potential numerical computational problems.

The column is initially solved at total reflux, constant overhead vapor rate, and steady-state conditions. At time $t = 0$ transient conditions develop as distillate draw-off begins. Numerical integration of the differential equations is performed for finite amounts of distillate draw-off, using initial values of the variables from the steady-state solution.

Equations 17.18 through 17.26 are first manipulated and rearranged to express composition derivatives, liquid and vapor flow rates, and heat duties explicitly:

$$\frac{dX_{0i}}{dt} = \frac{V_1}{M_0}(Y_{1i} - X_{0i}) \tag{17.29}$$

$$\frac{dX_{ji}}{dt} = \frac{V_{j+1}}{M_j}(Y_{j+1,i} - X_{ji}) + \frac{L_{j-1}}{M_j}(X_{j-1,i} - X_{ji}) - \frac{V_j}{M_j}(Y_{ji} - X_{ji}) \tag{17.30}$$

$$\frac{dX_{N+1,i}}{dt} = -\frac{V_{N+1}}{M_{N+1}}(Y_{N+1,i} - X_{N+1,i}) + \frac{L_N}{M_{N+1}}(X_{Ni} - X_{N+1,i}) \qquad (17.31)$$

$$V_1 = L_0 + D + \frac{dM_0}{dt} \qquad (17.32)$$

$$L_j = V_{j+1} + L_{j-1} - V_j - \frac{dM_j}{dt} \qquad (17.33)$$

$$V_{j+1} = V_j \frac{H_j - h_j}{H_{j+1} - h_j} - L_{j-1}\frac{h_{j-1} - h_j}{H_{j+1} - h_j} + \frac{M_j}{H_{j+1} - h_j}\frac{dh_j}{dt} \qquad (17.34)$$

$$Q_0 = V_1(H_1 - h_0) - M_0\frac{du_0}{dt} \qquad (17.35)$$

$$Q_{N+1} = V_{N+1}(H_{N+1} - h_{N+1}) - L_N(h_N - h_{N+1}) + M_{N+1}\frac{du_{N+1}}{dt} \qquad (17.36)$$

Starting from total reflux conditions, the distillate rate is incremented from zero to D, thereby lowering the reflux rate to $L_0 - D$. Since negligible fluid dynamic lags are assumed, all the liquid rates are instantly lowered to $L_j - D$. The vapor rates are maintained at V_j. These flow rates and the steady-state total reflux mole fractions are used to calculate the mole fraction derivatives by Equations 17.29 through 17.31. The molar holdups in these equations are calculated from the assumed constant volume holdups multiplied by calculated molar densities.

The mole fractions at the end of the time increment (equivalent to the distillate increment) are calculated by numerical integration. The magnitude of the time increment is determined by stability and truncation considerations. The batch distillation model contains tray holdups with time constants much smaller than the reboiler time constant. These conditions ("stiff systems") can cause computational instability unless very small time increments are used. The penalty is excessive computing time and the likelihood of incurring truncation errors. Distefano (1968) provides values for the maximum time increment size consistent with stability for a number of integration schemes. The same time increment is used to determine the incremental distillate rate for the first step.

The calculated mole fractions on a given stage generally may not sum up to unity due to inaccuracies, truncation errors, and so on, and must therefore be normalized. With the normalized compositions, a bubble point calculation is performed on each stage to determine the temperatures T_j, and the vapor mole fractions Y_{ji}. This is equivalent to solving Equations 17.27 and 17.28. (The solution of similar equations is discussed in Chapter 2.)

Based on current temperatures and compositions, the liquid molar densities and the liquid and vapor enthalpies are now calculated. The molar holdups are calculated, and the holdup and enthalpy derivatives with respect to time are computed numerically from current and latest values of these variables.

Next, Equations 17.32 through 17.34 are used to update the vapor and liquid flow rates, and Equations 17.35 and 17.36 are used to calculate the condenser and reboiler duties. The reboiler molar holdup is calculated by over all material balance:

$$M_{N+1} = M_{N+1}^0 - \sum_{j=0}^{N} M_j - \int_0^t D\,dt \qquad (17.37)$$

The computational cycle is repeated until the required amount of distillate has been produced.

17.2.4   OPTIMIZATION

Optimizing batch distillation operations can have a significant economic impact especially when the separation of high-value chemicals is involved. The control variable for optimizing a batch distillation product is the reflux ratio policy, that is, the variation of the reflux ratio with time. As mentioned in Section 17.1.2, two reflux ratio policies are commonly in use: a constant reflux ratio or a variable reflux ratio that produces a constant-composition distillate. The alternative is to optimize the reflux ratio policy to achieve certain objectives, such as maximizing the amount of product with a specified composition in a given length of time or minimizing the distillation time required to produce a given amount of product with a specified composition.

The optimum operation of a multiproduct column is one that maximizes an overall objective function, such as the annual profit. The optimization decision variables are the duration and reflux ratio policy of each product.

The problem of extensive computing time typical of batch distillation is compounded with the superimposition of an optimization routine. For this reason batch distillation optimization algorithms are usually built around shortcut batch distillation methods. Possible approaches to formulating the optimization problem are presented, although a detailed mathematical discussion is outside the scope of this book.

With negligible liquid holdup, negligible pressure drop, constant vapor and liquid flow rates from tray to tray, time-invariant vapor flow rate, and ideal thermodynamics, the batch distillation problem for product k may be stated as follows (Farhat et al., 1990):

$$f_k\left[t, M_{N+1}(t), X_{ji}(t), T_j(t), M'_{N+1}(t), X'_{ji}(t), T'_j(t), R(t), v\right] = 0 \qquad (17.38)$$

where $j = 1, \ldots, N$, $i = 1, \ldots, C$, $t_{k-1} \leq t \leq t_k$, $R(t)$ is the reflux ratio function, and v is a vector of time-invariant parameters: the initial compositions, operating pressure, number of trays, and vapor flow rate. The prime indicates derivatives with respect to time. Since

$$D = \frac{V}{R(t) + 1}$$

the amount of product k collected between times t_{k-1} and t_k is given by

$$C_k = \int_{t_{k-1}}^{t_k} \frac{V}{R(t) + 1} dt \qquad (17.39)$$

The principal component in product k is also designated as k, and its amount in the product is given as

$$C_{kk} = \int_{t_{k-1}}^{t_k} \frac{VX_{Dk}}{R(t) + 1} dt \qquad (17.40)$$

The purity of component k in product k is, therefore,

$$X_{Ck} = \frac{C_{kk}}{C_k} \qquad (17.41)$$

The optimization problem for a given product is to maximize C_k subject to the constraints

$$X_{Ck} = X_{Ck}^*$$

and

$$f_k = 0$$

for a given time interval t_{k-1} to t_k and a specified purity X_{Ck}^*. This optimization problem may be formulated using a linear programming approach and solved by sequential quadratic programming.

NOMENCLATURE

C Amount of distillate (moles)
C Number of components
D Distillate molar rate
h Liquid molar enthalpy
H Vapor molar enthalpy
I Integral in Equation 17.7
K Vapor–liquid equilibrium coefficient
L Liquid molar flow rate in the column
M Liquid molar holdup
N Number of stages in the column
N_{min} Minimum number of stages
Q Heat duty
R Reflux ratio

R_{min} Minimum reflux ratio
r R/R_{min}
t Time
u Liquid molar internal energy
V Vapor molar flow rate in the column
v Time-invariant parameters
X Mole fraction in the liquid
Y Mole fraction in the vapor
α Relative volatility

SUBSCRIPTS

C Condenser or accumulator
D Distillate
h Heavy key component
i Component designation
j Stage designation, 0 for the condenser
k Product number
l Light key component
r Reference component

SUPERSCRIPTS

0,1,2   Points in time
*   Specified value
F Final conditions

PROBLEMS

17.1. In a binary batch distillation process with no reflux (differential distillation), constant relative volatility is assumed throughout the process, $\alpha = 2$. If the initial liquid composition $X^0 = 0.4$, what is the initial distillate composition? What is the composition of the liquid remaining in the boiler when 50% of the original liquid has been distilled? When 99% has been distilled?

17.2. In a single-stage batch distillation with no reflux (differential distillation) at 100 kPa, the still is charged with 100 kmol of an equimolar binary mixture of hexane and heptane. The K-values are given by $\ln K_i = A_i - B_i/T$, where T is in degrees Kelvin.

		A_i	B_i
1.	Hexane	11.08	3789
2.	Heptane	11.29	4192

Calculate the following quantities when the boiler temperature is 360 K: (a) the instantaneous vapor composition; (b) the cumulative distillate composition; and (c) the amount and composition of the liquid remaining in the boiler. Neglect the holdup between the reboiler and the accumulator.

17.3. A three-stage batch distillation column is charged with a 100 kmol mixture containing 60% mole component 1 and 40% mole component 2. The column pressure is maintained at 100 kPa, and the distillate rate is 20 kmol/h. It is desired to produce two distillation cuts. The first cut will be produced by continuously adjusting the reflux ratio to maintain the distillate composition at 90% mole component 1. Production of the second cut starts when the L/V ratio is 0.80. The L/V ratio will be fixed at this value until the second cut cumulative composition is 75% mole component 1. Determine the amount of each cut. Assume negligible tray holdups and use vapor–liquid equilibrium data from Problem 6.1.

17.4. A hydrocarbon mixture is to be separated by batch distillation at 520 kPa to produce a constant-composition distillate, the key components being propane and n-butane. The starting charge composition, the relative volatilities, and the desired distillate composition are as follows:

	$X^0_{N+1,i}$	α_{i4}	X_{Di}
Ethane	0.12	13.304	0.200
Propane	0.48	5.553	0.793
n-Butane	0.25	2.315	0.007
n-Pentane	0.15	1.000	0.000

If the distillation were to be started at twice the minimum reflux ratio, determine the required number of stages. If the initial charge is 100 kmol and the distillate rate is 10 kmol/h, calculate the reflux rate, the amounts of distillate and residue, and the residue composition as a function of time. Irrespective of tray hydraulics and reboiler and condenser capacity constraints, when should the distillation be stopped? Assume negligible tray holdups and use shortcut methods.

17.5. An initial charge of a 100 kmol mixture of 40 mol% benzene (1) and 60 mol% toluene (2) is distilled by differential batch distillation with no reflux. The process is stopped when 70% of the benzene is distilled. Find the amount and average composition of the distillate and the amount and composition of the residue. The relative volatility of benzene to toluene is $\alpha_{1,2} = 2.5$.

17.6. The charge to a differential batch still (no reflux) is 100 kmol of an equimolar mixture of components 1 and 2, with a relative volatility $\alpha_{1,2}$, assumed constant at 2.0. Determine the amounts of distillate and residue and the residue composition when the distillate contains 60 mol% of component 1.

REFERENCES

Boston, J. F., H. I. Britt, S. Jirapongphan, and V. B. Shah, *Foundations of Computer-Aided Chemical Process Design, 2*, Edited by R. S. Mah and W. D. Seider, New York, American Institute of Chemical Engineers, 1981, 203.

Distefano, G. P., *AIChE J.*, 14, 190, 1968.

Diwekar, U. M. and K. P. Madhavan, *Ind. Eng. Chem. Res.*, 30, 713, 1991.

Farhat, S., M. Czernicki, L. Pibouleau, and S. Domenech, *AIChE J.*, 36, 1349, 1990.

Geddes, R. L., *AIChE J.*, 4, 389, 1958.

Hengstebeck, R. J., *Trans. Am. Inst. Chem. Eng.*, 42, 309, 1946.

Smoker, E. H. and A. Rose, *Trans. Am. Inst. Chem. Eng.*, 36, 285, 1940.

18 Membrane Separation Operations

In contrast to operations discussed in all previous chapters where separation is the result of different phases in contact having different compositions, membrane separation is achieved by a membrane, or a barrier that is preferentially permeable to different species. The outcome is a process where fluids (gas or liquid) of different compositions flow on each side of the membrane.

Depending on the membrane structure and the separation mechanism, several membrane separation processes are in use for different applications.

The term *membrane* in general refers to a barrier through which molecules can traverse by one mechanism or another. Membranes used in separation processes may be broadly classified as porous or nonporous.

For porous membranes to bring about separation, their pore size must be comparable to the molecular size of the species to be separated. Separation takes place due to different species having different molecular weights.

With nonporous membranes the species to be separated dissolve in the membrane and diffuse through it. Separation results from different species having different diffusivities through the membrane.

18.1 GENERAL MEMBRANE SEPARATION PROCESS

A basic separation process may be represented schematically by a unit as shown in Figure 18.1. The membrane separator consists of two regions, one on each side of the membrane. The feed, containing the mixture to be separated, is introduced on one side of the membrane. The permeate, containing higher concentrations of the species that preferentially permeate through the membrane, flows on the other side. The remainder of the feed, the residue, exits on the same side of the membrane as the feed.

In general, a molecule of component i in the fluid (gas or liquid) traversing the membrane encounters resistance through the membrane as well as in the fluid on both sides. The resistance in the fluid is in the films adjacent to the membrane on each side. If G_i is the general driving force property of component i (partial pressure for gases, concentration for liquids) its value changes along a path from the bulk phase on one side of the membrane to the bulk phase on the other side. As shown in Figure 18.2, G_{1i} and G_{2i} are the values of the property in the bulk phase on each side, G_{1if} and G_{2if} are the values in each fluid film in contact with the membrane, and G_{1im} and G_{2im} are the values on each side of the membrane surface.

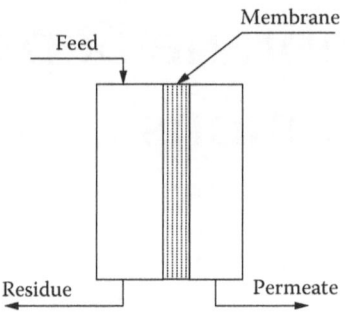

FIGURE 18.1   General membrane separation process.

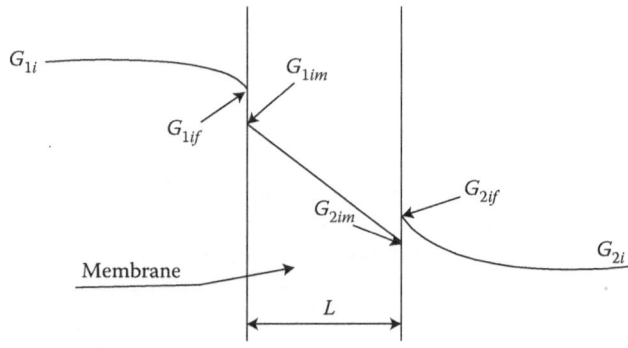

FIGURE 18.2   Permeation mechanism.

The property values on the membrane surface and in the film at the interface are considered at equilibrium, related to each other by equilibrium distribution coefficients, assumed equal on both sides of the membrane:

$$K_i = \frac{G_{1im}}{G_{1if}} = \frac{G_{2im}}{G_{2if}} \tag{18.1}$$

The flux N_i of component i across the membrane is the flow rate of that component per unit of membrane area. The molecules must also flow through the fluid films on both sides of the membrane. (In certain cases of gas permeation the resistances in the films may be neglected.) At steady state the flux is the same through the films and through the membrane. The flux through the films is controlled by mass transfer:

$$N_i = k_{1i}(G_{1i} - G_{1if}) = k_{2i}(G_{2if} - G_{2i}) \tag{18.2}$$

where k_{1i} and k_{2i} are mass transfer coefficients. The flux of component i through the membrane is a function of the driving force gradient and the diffusivity of the component in the membrane:

$$N_i = \frac{D_i}{L}(G_{1im} - G_{2im}) \tag{18.3}$$

where D_i is the permeation diffusivity and L is the membrane thickness. In terms of the driving force property values in the film at the interface,

$$N_i = \frac{D_i K_i}{L}(G_{1if} - G_{2if}) \tag{18.4}$$

It is common practice to define the permeance, p_{Mi}, and the permeability, P_{Mi} as follows:

$$p_{Mi} = \frac{D_i K_i}{L} \tag{18.5}$$

$$P_{Mi} = D_i K_i = p_{Mi} L \tag{18.6}$$

With these definitions, Equation 18.4 may be written as

$$N_i = p_{Mi}(G_{1if} - G_{2if}) \tag{18.7}$$

or

$$N_i = \frac{P_{Mi}}{L}(G_{1if} - G_{2if}) \tag{18.8}$$

The permeance (or permeability) and the permeation diffusivity are properties relating to component i and the particular membrane type.

The general driving force across the films and the membrane is proportional to the flux. The proportionality constants are the resistances to flow. Equations 18.2 and 18.7 may be written in terms of resistances:

$$G_{1i} - G_{1if} = (1/k_{1i})N_i$$

$$G_{2if} - G_{2i} = (1/k_{2i})N_i$$

$$G_{1if} - G_{2if} = (1/p_{Mi})N_i$$

where $(1/k_{1i})$, $(1/k_{2i})$, and $(1/p_{Mi})$ are the resistances in the films and in the membrane. The resistances are additive and the flux may be expressed in terms of the combined resistance between G_{1i} and G_{2i}:

$$N_i = \frac{G_{1i} - G_{2i}}{(1/k_{1i}) + (1/p_{Mi}) + (1/k_{2i})} \tag{18.9}$$

In liquid permeation the driving force is usually expressed in terms of the molar concentration of the permeating component, c_i. Equation 18.1 is written as

$$K_i = \frac{c_{1im}}{c_{1if}} = \frac{c_{2im}}{c_{2if}} \tag{18.10}$$

The flux is given as

$$N_i = \frac{D_i}{L}(c_{1im} - c_{2im}) = \frac{D_i K_i}{L}(c_{1if} - c_{2if}) \tag{18.11}$$

$$= p_{Mi}(c_{1if} - c_{2if}) = \frac{P_{Mi}}{L}(c_{1if} - c_{2if}) \tag{18.12}$$

$$= k_{1i}(c_{1i} - c_{1if}) = k_{2i}(c_{2if} - c_{2i}) \tag{18.13}$$

$$= \frac{c_{1i} - c_{2i}}{(1/k_{1i}) + (1/p_{Mi}) + (1/k_{2i})} \tag{18.14}$$

A consistent set of units may have the flux in kmol/m².s, the concentrations in kmol/m³, the diffusivity in m²/s, the membrane thickness in m, the permeance in m/s, the permeability in m²/s, and the mass transfer coefficients in m/s.

18.1.1   POSSIBLE CONSISTENT SETS OF UNITS

N_i (Flux)	$= p_{Mi}$ (Permeance)	$\times \Delta c_{if}$ (Concentration)
kmol/(m²·s)	m/s	kmol/m³
N_i (Flux)	$= p_{Mi}$ (Permeance)	$\times \Delta p_{if}$ (Partial pressure)
kmol/(m²·s)	kmol/(s.m²·cm Hg)	cm Hg
cm³(STP)/(cm²·s)	cm³(STP)/(s·cm²·cm Hg)	cm Hg
kmol/(m²·h)	kmol/(h.m²·kPa)	kPa

STP = Standard temperature (0°C) and pressure (1 atm).
1 kmol gas = 22.4 m³ (STP)

The diffusivity D_i in liquid permeation of component i in a given membrane is related to d_i, the diffusivity of solute i in bulk solution, and to the membrane characteristics. These include the porosity or void fraction of the membrane, ε, where permeation takes place. Another membrane characteristic is the tortuosity τ of the pores, which represents their deviation from being straight pores. It is the ratio of the actual length of the pore to its length if it were straight along the diffusion path. A parameter that relates the size of the diffusing molecule relative to the pore diameter is the restrictive factor K_r (Beck and Schultz, 1970) expressed in terms of d_m and d_p, the molecular and pore diameters:

$$K_r = \left[1 - \frac{d_m}{d_p}\right]^4$$

Taking into account all these parameters, the diffusivity in liquid permeation is expressed as

$$D_i = \frac{\varepsilon d_i}{\tau} K_{ri} \tag{18.14a}$$

EXAMPLE 18.1: LIQUID PERMEATION

Water containing urea at a bulk phase concentration of $c_1 = 0.008$ kmol/m³ is flowing on one side of a membrane. The bulk phase concentration of urea on the other side of the membrane is $c_2 = 0.002$ kmol/m³. The diffusivity of urea in the membrane is $D = 2.5 \times 10^{-11}$ m²/s and the membrane thickness is $L = 4 \times 10^{-6}$ m. The equilibrium distribution coefficient on either side of the membrane is $K = 1.4$. The mass transfer coefficients are $k_1 = 3 \times 10^{-2}$ m/s and $k_2 = 2 \times 10^{-5}$ m/s. Calculate the flux of urea through the membrane, the permeance and permeability, and the concentrations of urea at the film interface with the membrane and on the membrane surface on both sides.

SOLUTION

The permeance and permeability are calculated from Equations 18.5 and 18.6:

$$p_M = \frac{DK}{L} = \frac{(2.5 \times 10^{-11})(1.4)}{4 \times 10^{-6}} = 8.75 \times 10^{-6} \text{ m/s}$$

$$P_M = DK = (2.5 \times 10^{-11})(1.4) = 3.5 \times 10^{-11} \text{ m}^2/\text{s}$$

The flux is calculated from Equation 18.14:

$$N = \frac{0.008 - 0.002}{1/(3 \times 10^{-2}) + 1/(8.75 \times 10^{-6}) + 1/(2 \times 10^{-5})} = 3.65 \times 10^{-8} \text{ kmol/m}^2.\text{s}$$

The concentrations at the film interface are calculated from Equation 18.13:

$$c_{1f} = c_1 - N/k_1 = 0.008 - (3.65 \times 10^{-8}/3 \times 10^{-2}) \approx 0.008 \text{ kmol/m}^3$$

$$c_{2f} = c_2 + N/k_2 = 0.002 + (3.65 \times 10^{-8}/2 \times 10^{-5}) = 0.003825 \text{ kmol/m}^3$$

The concentrations on the membrane surfaces are calculated from Equation 18.10:

$$c_{1m} = c_{1f} K = (0.008)(1.4) = 0.0112 \text{ kmol/m}^3$$

$$c_{2m} = c_{2f} K = (0.003825)(1.4) = 0.005355 \text{ kmol/m}^3$$

Check the flux using Equation 18.11:

$$N = \frac{2.5 \times 10^{-11}}{4 \times 10^{-6}}(0.0112 - 0.005355) = 3.65 \times 10^{-8} \, kmol/m^2.s$$

$$= \frac{2.5 \times 10^{-11} \times 1.4}{4 \times 10^{-6}}(0.008 - 0.003825) = 3.65 \times 10^{-8} \, kmol/m^2.s$$

In gas permeation the driving force across the membrane may be expressed in terms of concentrations, but is usually defined in terms of partial pressures. The partial pressures of component i are p_{1i} and p_{2i} in the bulk gas phase, and p_{1if} and p_{2if} in the gas phase at the gas–membrane interface. The concentration in the membrane at the interface is related to the partial pressure in the gas at the interface by Henry's law, the equivalent to Equation 18.1 for gas permeation:

$$H_i = \frac{c_{1im}}{p_{1if}} = \frac{c_{2im}}{p_{2if}} \tag{18.15}$$

where H_i represents Henry's constant. The flux is given as

$$N_i = \frac{D_i}{L}(c_{1im} - c_{2im}) = \frac{D_i H_i}{L}(p_{1if} - p_{2if}) \tag{18.16}$$

$$= P_{Mi}(p_{1if} - p_{2if}) = \frac{P_{Mi}}{L}(p_{1if} - p_{2if}) \tag{18.17}$$

$$= k_{1i}(p_{1i} - p_{1if}) = k_{2i}(p_{2if} - p_{2i}) \tag{18.18}$$

$$= \frac{p_{1i} - p_{2i}}{(1/k_{1i}) + (1/p_{Mi}) + (1/k_{2i})} \tag{18.19}$$

Possible consistent units could be the same as above for the flux, the diffusivity, the thickness, and the concentration, with the pressure in cm Hg (cm mercury), Henry's constant in kmol/m³.cm Hg, the permeability in kmol/s.m.cm Hg, and the mass transfer coefficients and the permeance in kmol/s.m².cm Hg. An alternative set would have the flux in cm³ (STP, standard temperature and pressure, 0°C, 1 atm)/cm².s, Henry's constant in cm³ (STP)/cm³.cm Hg, the thickness in cm, the pressure in cm Hg, the permeability in cm³ (STP)/s.cm.cm Hg, and the mass transfer coefficients and permeance in cm³ (STP)/s.cm².cm Hg. For converting molar flow rates to volumetric flow rates, the volume of 1 kmol gas is 22.4 m³ (STP).

The diffusivity in gas permeation may include ordinary molecular diffusion and/or what is known as *Knudsen* diffusion. Ordinary diffusion involves mainly collisions between the gas molecules. In the Knudsen diffusion the molecules are flowing in tubes, the membrane pores. If the pore diameter is smaller than the mean free path of the gas molecules, the collisions in the diffusion process are mainly between the gas molecules and the pore wall. Assuming ideal gas behavior, the Knudsen diffusivity is expressed as

$$d_{Ki} \ (\text{cm}^2/\text{s}) = 4850 \ d_p \ (\text{cm}) \ [T(\text{K})/M_i]^{0.5}$$

where M_i is the molecular weight of the diffusing component. Taking into account both the molecular and Knudsen diffusivities, as well as the membrane void fraction and the pore tortuosity, the diffusivity in gas permeation is expressed as

$$D_i = \frac{\varepsilon}{\tau} \left[\frac{1}{(1/d_i) + (1/d_{Ki})} \right] \quad\quad (18.19a)$$

EXAMPLE 18.2: GAS PERMEATION

A membrane 2×10^{-5} cm thick is to be used in one of the steps in a process to produce oxygen (1) and nitrogen (2) from air. It is required to estimate the fluxes of oxygen and nitrogen through the membrane based on the following data:

Diffusivities in the membrane (cm²/s)	$D_1 = 5 \times 10^{-7}$	$D_2 = 3 \times 10^{-7}$
Henry's constants (cm³ (STP)/cm³.cm Hg)	$H_1 = 6 \times 10^{-4}$	$H_2 = 3 \times 10^{-4}$
Partial pressures (cm Hg) Oxygen:	$p_{11} = 160$	$p_{21} = 24$
Nitrogen:	$p_{12} = 600$	$p_{22} = 56$

Calculate the permeances, permeabilities, fluxes, and the concentrations of oxygen and nitrogen on the membrane surfaces. The film resistances may be neglected.

SOLUTION

The permeances and permeabilities are calculated from equations equivalent to 18.5 and 18.6, written in terms of Henry's constants:

$$P_{M1} = D_1 H_1 = (5 \times 10^{-7})(6 \times 10^{-4}) = 3 \times 10^{-10} \ \text{cm}^3(\text{STP})/\text{cm.s.cm Hg}$$

$$P_{M2} = D_2 H_2 = (3 \times 10^{-7})(3 \times 10^{-4}) = 9 \times 10^{-11} \ \text{cm}^3(\text{STP})/\text{cm.s.cm Hg}$$

$$p_{M1} = \frac{P_{M1}}{L} = \frac{3 \times 10^{-10}}{2 \times 10^{-5}} = 1.5 \times 10^{-5} \ \text{cm}^3(\text{STP})/\text{cm}^2.\text{s.cm Hg}$$

$$p_{M2} = \frac{P_{M2}}{L} = \frac{9 \times 10^{-11}}{2 \times 10^{-5}} = 4.5 \times 10^{-6} \ \text{cm}^3(\text{STP})/\text{cm}^2.\text{s.cm Hg}$$

The fluxes are calculated from Equation 18.17, with the assumption that $p_{1if} \approx p_{1i}$ and $p_{2if} \approx p_{2i}$, since the film resistances are neglected:

$$N_1 = p_{M1}(p_{11} - p_{21}) = (1.5 \times 10^{-5})(160 - 24) = 2.04 \times 10^{-3} \ \text{cm}^3(\text{STP})/\text{cm}^2.\text{s}$$

$$N_2 = p_{M2}(p_{21} - p_{22}) = (4.5 \times 10^{-6})(600 - 56) = 2.45 \times 10^{-3} \ \text{cm}^3(\text{STP})/\text{cm}^2.\text{s}$$

The concentrations on the membrane surfaces are calculated from Equation 18.15:

$$c_{11m} = H_1 p_{11} = (6 \times 10^{-4})(160) = 9.60 \times 10^{-2} \, cm^3(STP)/cm^3$$
$$= (9.60 \times 10^{-2})/22.4 = 4.29 \times 10^{-3} \, kmol/m^3$$

$$c_{12m} = H_2 p_{12} = (3 \times 10^{-4})(600) = 1.80 \times 10^{-1} \, cm^3(STP)/cm^3$$
$$= (1.80 \times 10^{-1})/22.4 = 8.04 \times 10^{-3} \, kmol/m^3$$

$$c_{21m} = H_1 p_{21} = (6 \times 10^{-4})(24) = 1.44 \times 10^{-2} \, cm^3(STP)/cm^3$$
$$= (1.44 \times 10^{-2})/22.4 = 6.43 \times 10^{-4} \, kmol/m^3$$

$$c_{22m} = H_2 p_{22} = (3 \times 10^{-4})(56) = 1.68 \times 10^{-2} \, cm^3(STP)/cm^3$$
$$= (1.68 \times 10^{-2})/22.4 = 7.50 \times 10^{-4} \, kmol/m^3$$

The fluxes may also be calculated in $kmol/m^2.s$ by Equation 18.16:

$$N_1 = \frac{5 \times 10^{-7}}{(2 \times 10^{-5})(100)}(4.29 \times 10^{-3} - 6.43 \times 10^{-4}) = 9.12 \times 10^{-7} \, kmol/m^2.s$$

$$N_2 = \frac{3 \times 10^{-7}}{(2 \times 10^{-5})(100)}(8.04 \times 10^{-3} - 7.50 \times 10^{-4}) = 1.09 \times 10^{-6} \, kmol/m^2.s$$

18.2   PERFORMANCE OF MEMBRANE SEPARATORS

Membrane separators are built with a variety of membrane types assembled in modules that can take any of several forms such as spiral-wound, tubular, plate-and-frame, and so on (Hsieh et al., 1988).

The performance of a membrane separator may be predicted on the basis of permeation fluxes and material balances. The formulation of these relationships for a particular module depends on the flow patterns of the fluid on both sides of the membrane.

The equations derived in the previous section represent permeation rates and flux, and Examples 18.1 and 18.2 are applications to a specific model. The model assumes the fluid on each side of the membrane to have a constant composition parallel to the membrane. The compositions normal to the membrane are also assumed constant, with the possible exception of composition gradients in the films adjacent to the membrane. The bulk phase compositions on both sides of the membrane had to be given since no material balances were considered. The flow pattern implied in this model is that of perfect mixing (if the film resistances next to the membrane are neglected). Other flow patterns include cross flow, countercurrent flow, and cocurrent flow.

18.2.1   PERFECT MIXING MODEL

This model is represented in Figure 18.3, where the feed with molar flow rate F_F enters the separator on the residue side of the membrane. It separates into a

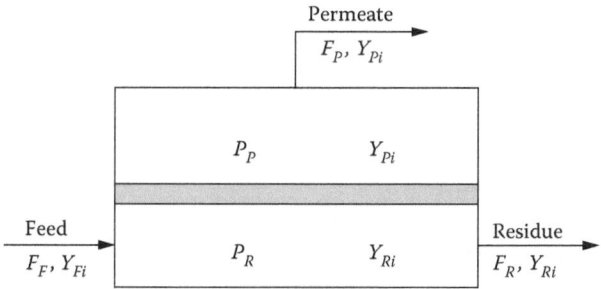

FIGURE 18.3 Perfect mixing membrane separator.

permeate with molar flow rate F_P and a residue with molar flow rate F_R. As implied by the model definition, the feed entering the separator instantly mixes with the fluid on the residue side of the membrane to form a mixture of uniform composition that is the residue. The permeating fluid forms a permeate of uniform composition on the permeate side. The film resistances are neglected. The compositions of the permeate and residue products are identical to the compositions inside the separator on the permeate side and the residue side. Since perfect mixing is likely to occur primarily in gases, the perfect mixing model is analyzed specifically for gas permeation.

By overall material balance on the separator at steady state,

$$F_F = F_P + F_R \tag{18.20}$$

A material balance on component i is as follows:

$$F_F Y_{Fi} = F_P Y_{Pi} + F_R Y_{Ri} \tag{18.21}$$

where Y_{Fi}, Y_{Pi}, and Y_{Ri} are the mole fractions of component i in the feed, the permeate, and the residue. Equation 18.21 is combined with Equation 18.20 to eliminate F_R, and the resulting equation is divided through by F_F to give

$$Y_{Fi} = \theta Y_{Pi} + (1 - \theta) Y_{Ri} \tag{18.22}$$

where θ is the fraction of feed permeated:

$$\theta = F_P / F_F \tag{18.23}$$

The flow rate of component i in the permeate equals the product of the flux of that component times the membrane area A:

$$F_P Y_{Pi} = F_F \theta Y_{Pi} = A N_i$$

This equation is combined with Equation 18.17:

$$F_F \theta Y_{Pi} = A p_{Mi}(p_{Ri} - p_{Pi})$$

The partial pressures at the film interface are replaced by the bulk partial pressures because the film resistances are neglected. Also, subscripts 1 and 2 are replaced by R and P to indicate the residue side and the permeate side. The partial pressures are now replaced by the total pressure and mole fractions of the products:

$$F_F \theta Y_{Pi} = A p_{Mi}(P_R Y_{Ri} - P_P Y_{Pi}) \tag{18.24}$$

where P_R and P_P are the total pressures on the residue and permeate sides. Another relationship that must be satisfied is the mole fraction sum equality to unity:

$$\sum_i Y_{Ri} = \sum_i Y_{Pi} = 1.0 \tag{18.25}$$

In a multicomponent mixture with C components $i = 1,\ldots,C$, Equations 18.22, 18.24, and 18.25 comprise $2C + 1$ equations. For a given separation operation F_F, Y_{Fi}, p_{Mi}, P_R, and P_P are considered fixed or determined by conditions external to the model. The remaining variables are θ, Y_{Pi}, Y_{Ri}, and A, totaling $2C + 2$ variables. Hence, one variable must be specified in order to define the process. This could be the fraction permeated, the membrane area, or one of the components' mole fraction in the permeate or residue. For existing equipment the membrane area is known and the products flow rates and compositions can be calculated. In a design situation the membrane area can be calculated to satisfy a performance specification such as the fraction permeated or a component mole fraction in one of the products. In this discussion the permeances are assumed to be known, which implies given permeabilities and membrane thickness.

Equations 18.22, 18.24, and 18.25, together with the required specification, may be solved simultaneously using a method such as Newton–Raphson's (Sections 2.3.2 and 4.2.3). Alternatively, the equations may be manipulated and solved iteratively. Equation 18.24 is rearranged as

$$Y_{Pi} = \frac{P_R A p_{Mi}}{P_P A p_{Mi} + F_F \theta} Y_{Ri} \tag{18.26}$$

And from Equation 18.22,

$$Y_{Ri} = \frac{Y_{Fi} - \theta Y_{Pi}}{1 - \theta} \tag{18.27}$$

These two equations are combined by eliminating Y_{Ri} to give the following:

$$Y_{Pi} = \frac{P_R A p_{Mi} Y_{Fi}}{(1 - \theta)(P_P A p_{Mi} + F_F \theta) + P_R A p_{Mi} \theta} \tag{18.28}$$

Equation 18.24 is again rearranged to calculate the membrane area:

$$A = \frac{F_F \theta Y_{Pi}}{p_{Mi}(P_R Y_{Ri} - P_P Y_{Pi})} \tag{18.29}$$

For an existing separator where the membrane area is known, the performance of the separator is determined by an iterative computational procedure as follows:

1. Assume a value for θ.
2. Calculate Y_{Pi} for $i = 1,\ldots,C$ by Equation 18.28.
3. Check if Equation 18.25 is satisfied. If $\sum Y_{Pi} \neq 1$ repeat the calculations beginning at step 1 with a new value for θ until $\sum Y_{Pi} = 1$.
4. Calculate Y_{Ri} from Equation 18.27, F_P from Equation 18.23, and F_R from Equation 18.20.

EXAMPLE 18.3: MEMBRANE SEPARATOR RATING

The air separation process introduced in Example 18.2 is to be carried out in a perfect mixing membrane separator. The feed air flow rate is $F_F = 1 \times 10^6$ cm³ (STP)/s with mole fraction oxygen $Y_{F1} = 0.21$ and mole fraction nitrogen $Y_{F2} = 0.79$. The membrane area $A = 1 \times 10^8$ cm². The permeances from Example 18.2 will be used: $p_{M1} = 1.5 \times 10^{-5}$, $p_{M2} = 4.5 \times 10^{-6}$ cm³ (STP)/cm².s.cm Hg. The pressures on the residue side and on the permeate side are $P_R = 760$ cm Hg and $P_P = 80$ cm Hg, respectively. Calculate the flow rates and compositions of the permeate and residue streams.

SOLUTION

Following the algorithm described above, a value for θ is initially assumed at 0.5. From Equation 18.28,

$$Y_{P1} = \frac{760 \times 1 \times 10^8 \times 1.5 \times 10^{-5} \times 0.21}{(1-0.5)(80 \times 1 \times 10^8 \times 1.5 \times 10^{-5} + 1 \times 10^6 \times 0.5) + 760 \times 1 \times 10^8 \times 1.5 \times 10^{-5} \times 0.5}$$
$$= 0.2720$$

$$Y_{P2} = \frac{760 \times 1 \times 10^8 \times 4.5 \times 10^{-6} \times 0.79}{(1-0.5)(80 \times 1 \times 10^8 \times 4.5 \times 10^{-6} + 1 \times 10^6 \times 0.5) + 760 \times 1 \times 10^8 \times 4.5 \times 10^{-6} \times 0.5}$$
$$= 0.6154$$

$$Y_{P1} + Y_{P2} = 0.2720 + 0.6154 = 0.8874$$

The iterations to satisfy Equation 18.25 converge at $\theta = 0.3927$. The other variables are calculated directly from the appropriate equations. The results are summarized below:

$$F_F = 1 \times 10^6, \ F_P = 3.927 \times 10^5, \ F_R = 6.073 \times 10^5 \ \text{cm}^3 \ \text{(STP)/s}$$

	Y_{Fi}	Y_{Pi}	Y_{Ri}
Oxygen	0.21	0.3154	0.1418
Nitrogen	0.79	0.6846	0.8582

In a design situation, if the permeate fraction or a component mole fraction in one of the products is specified, the following iterative procedure may be used to calculate the membrane area and the products:

1. a. If θ is specified, assume a value for Y_{Pk}, where k is any component in the feed that is distributed in the permeate and the residue. The assumed value should be greater than or less than Y_{Fk}, depending on whether k is expected to be mostly permeated or rejected. Calculate Y_{Rk} from Equation 18.27.
   b. If Y_{Pk} or Y_{Rk} is specified, assume a value for θ. Calculate Y_{Rk} or Y_{Pk} from Equation 18.27.
2. Calculate the membrane area from Equation 18.29.
3. Calculate Y_{Pi} for $i = 1,...,C$ by Equation 18.28.
4. Check if Equation 18.25 is satisfied. If $\sum Y_{Pi} \neq 1$ repeat the calculations beginning at step 1(a) or 1(b) with an updated value of the iteration variable until $\sum Y_{Pi} = 1$.
5. Calculate Y_{Ri} from Equation 18.27, F_P from Equation 18.23, and F_R from Equation 18.20.

EXAMPLE 18.4: MEMBRANE SEPARATOR DESIGN AND RATING

A membrane separator using 1 mil thickness low-density polyethylene membrane is to be designed for concentrating hydrogen in a hydrogen–methane–carbon monoxide gas mixture. The separator performance may be approximated by a perfect mixing model. The feed flow rate is $1.0 \times 10^4 \ \text{cm}^3$ (STP)/s, and its composition and component permeabilities in polyethylene membrane are given below:

	Y_{Fi}	P_{Mi} (cm³ (STP)/cm.s.cm Hg)
1. Hydrogen	0.27	9.9×10^{-10}
2. Methane	0.54	2.7×10^{-10}
3. Carbon monoxide	0.19	1.3×10^{-10}

a. *Design.* If the pressure is 320 cm Hg on the residue side and 50 cm Hg on the permeate side, calculate the membrane area required to achieve a specified mole fraction hydrogen in the permeate as $Y_{P1} = 0.50$.

b. *Rating*. For a membrane separator area of 3.00×10^7 cm^2 find the products rates and compositions using the same property data as above. Do the calculations first using the perfectly mixed model. Next, consider a membrane model that performs as a cross-flow model discussed in the next section. Use two stages for the separator, each having an area of 1.50×10^7. Compare the results of the perfectly mixed model and the two-stage approximation to the cross-flow model.

Solution

The permeances are first calculated from the permeabilities and the membrane thickness:

$$p_{M1} = 9.9 \times 10^{-10}/2.54 \times 10^{-3} = 3.8976 \times 10^{-7} \text{ cm}^3(\text{STP})/\text{cm}^2.\text{s.cmHg}$$

$$p_{M2} = 2.7 \times 10^{-10}/2.54 \times 10^{-3} = 1.0630 \times 10^{-7} \text{ cm}^3(\text{STP})/\text{cm}^2.\text{s.cmHg}$$

$$p_{M3} = 1.3 \times 10^{-10}/2.54 \times 10^{-3} = 0.5118 \times 10^{-7} \text{ cm}^3(\text{STP})/\text{cm}^2.\text{s.cmHg}$$

a. Perfectly Mixed Model

The iterations are started by assuming a fraction permeated. Assume $\theta = 0.25$ and calculate Y_{R1} from Equation 18.27:

$$Y_{R1} = \frac{0.27 - (0.25)(0.50)}{1 - 0.25} = 0.1933$$

Calculate the membrane area from Equation 18.29:

$$A = \frac{1.0 \times 10^4 \times 0.25 \times 0.5}{3.8976 \times 10^{-7}(320 \times 0.1933 - 50 \times 0.5)} = 8.7017 \times 10^7 \text{ cm}^2$$

Calculate Y_{Pi} from Equation 18.28:

$$Y_{P1} = \frac{320 \times 8.7017 \times 10^7 \times 3.8976 \times 10^{-7} \times 0.27}{(1 - 0.25)(50 \times 8.7017 \times 10^7 \times 3.8976 \times 10^{-7} + 1.0 \times 10^4 \times 0.25) + 320 \times 8.7017 \times 10^7 \times 3.8976 \times 10^{-7} \times 0.25} = 0.50$$

$$Y_{P1} = \frac{320 \times 8.7017 \times 10^7 \times 1.0630 \times 10^{-7} \times 0.54}{(1 - 0.25)(50 \times 8.7017 \times 10^7 \times 1.0630 \times 10^{-7} + 1.0 \times 10^4 \times 0.25) + 320 \times 8.7017 \times 10^7 \times 1.0630 \times 10^{-7} \times 0.25} = 0.5396$$

$$Y_{P1} = \frac{320 \times 8.7017 \times 10^7 \times 0.5118 \times 10^{-7} \times 0.19}{(1 - 0.25)(50 \times 8.7017 \times 10^7 \times 0.5118 \times 10^{-7} + 1.0 \times 10^4 \times 0.25) + 320 \times 8.7017 \times 10^7 \times 0.5118 \times 10^{-7} \times 0.25} = 0.1129$$

$$\sum Y_{Pi} = 1.1525$$

The iterative calculations are restarted with a new value for θ until $\Sigma Y_{Pi} = 1.0$. The results are as follows:

$$\theta = 0.1236$$

$$A = 3.1078 \times 10^7 \text{ cm}^2$$

	Y_{Pi}	Y_{Ri}
Hydrogen	0.50	0.2375
Methane	0.4203	0.5569
Carbon monoxide	0.0797	0.2056

b. Two-Stage Approximation to a Cross-Flow Model
Perfect Mixing Model

The feed stream is sent to the residue side of the separator with a membrane area of $3.0 \times 10^7 \text{ cm}^2$. A value is determined for the fraction permeated, θ, that satisfies the summation equations. The product rates and compositions are calculated, giving the following results:

$$\theta = 0.117, F_F = 10,000 \text{ cm}^3(\text{STP})/\text{s}, A = 3.0 \times 10^7 \text{ cm}^2$$

	Y_{Fi}	F_{Pi} (cm³(STP)/s)	Y_{Pi}	F_{Ri} (cm³(STP))	Y_{Ri}
H_2	0.27	593.2	0.5070	2106.0	0.2385
CH_4	0.54	482.5	0.4124	4916.5	0.5568
CO	0.19	94.3	0.0806	1807.5	0.2047
Total	1.00	1170.0	1.0000	8830.0	1.0000

Cross-Flow Approximated as a Two-Stage Model

The feed is sent to the residue side of the first stage with a membrane area of $1.5 \times 10^7 \text{ cm}^2$. The value of θ is determined to satisfy the summation equations, and the product rates and compositions are calculated. The residue from the first stage is sent to the residue side of the second stage which also has a membrane area of $1.5 \times 10^7 \text{ cm}^2$. The residue from the second stage is the residue product of the cross-flow model, and the combined stream of the permeates from the two stages constitutes the permeate product of the model. The results are tabulated as follows:

Stage 1, $\theta = 0.0603, F_F = 10,000 \text{ cm}^3(\text{STP})/\text{s}, A = 1.5 \times 10^7 \text{ cm}^2$

	Y_{Fi}	F_{Pi} (cm³(STP)/s)	Y_{Pi}	F_{Ri} (cm³(STP))	Y_{Ri}
H_2	0.27	318.5	0.5282	2381.2	0.2534
CH_4	0.54	240.2	0.3983	5159.9	0.5491
CO	0.19	44.3	0.0735	1855.9	0.1975
Total	1.00	603.0	1.0000	9397.0	1.0000

Stage 2, $\theta = 0.0622$, $F_f = 9397.0$ cm³(STP)/s, $A = 1.5 \times 10^7$ cm²

	Y_{Fi}	F_{Pi} (cm³(STP)/s)	Y_{Pi}	F_{Ri} (cm³(STP))	Y_{Ri}
H₂	0.2534	294.8	0.5043	2086.8	0.2368
CH₄	0.5491	242.9	0.4156	4918.3	0.5581
CO	0.1975	46.8	0.0801	1807.4	0.2051
Total	1.0000	584.5	1.0000	8812.5	1.0000

The permeate product of the two-stage cross-flow model is obtained by combining the permeate streams from both stages. The cross-flow residue is the same as stage 2 residue:

	Permeate		Residue	
	F_{Pi} (cm³(STP)/s)	Y_{Pi}	F_{Ri} (cm³(STP)/s)	Y_{ri}
H₂	613.3	0.5165	2086.8	0.2368
CH₄	483.1	0.4068	4918.3	0.5581
CO	91.1	0.0767	1807.4	0.2051
Total	1187.5	1.0000	8812.5	1.0000

The purpose of this example process is to produce a permeate stream with a higher concentration of hydrogen than in the feed, and methane and carbon monoxide with lower concentrations than in the feed. Comparing the perfect mixing model to the two-stage cross-flow approximation model, it can be seen that the hydrogen concentration went up from 0.27 to 0.5070 in the perfect mixing model, and to 0.5165 in the cross-flow model. For methane and carbon monoxide the concentrations went down from 0.54 and 0.19 to 0.4124 and 0.0806 in the perfect mixing model and to 0.4068 and 0.0767 in the cross-flow model. Although the improvement may seem marginal, it does show a trend that cross flow improves the separation between hydrogen on one side, and methane and carbon monoxide on the other. Increasing the number of stages would bring the system closer to the cross-flow model.

18.2.2 CROSS-FLOW MODEL

In this model the flow on the residue side is parallel to the membrane and on the permeate side it is normal to the membrane and away from it as shown in Figure 18.4. This flow pattern is typical in membrane separators where the velocity is high on the high-pressure residue side. On the permeate side the pressure is low and the permeate is forced to flow away from the membrane.

If the film resistances are neglected on both sides of the membrane, the fluid composition on each side may be assumed constant across a plane perpendicular to the membrane. On the residue side the composition changes continuously in the direction of flow (parallel to the membrane) as in plug flow. On the permeate side the composition also changes continuously in the direction parallel to the membrane due to the changing composition on the residue side.

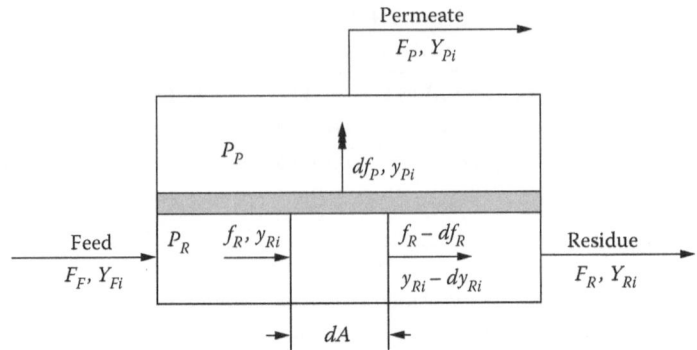

FIGURE 18.4 Cross-flow membrane separator.

Referring to Figure 18.4, a total material balance is written around a differential volume defined by membrane area dA:

$$df_P = df_R \qquad (18.30)$$

Molar flow rates f are the variable rates inside the separator, with subscripts P and R designating, respectively, the permeate side and residue side of the membrane. A material balance around the differential volume is written for component i. The component differential permeate rate equals the differential change in the component residue rate:

$$y_{Pi}df_P = d(f_R y_{Ri}) = f_R dy_{Ri} + y_{Ri}df_R$$

Mole fractions y represent the variable compositions inside the separator. The above equation is combined with Equation 18.30 to eliminate df_R, and then rearranged:

$$f_R dy_{Ri} = (y_{Pi} - y_{Ri})df_P \qquad (18.31)$$

The differential fraction permeated is defined as

$$d\theta = \frac{df_P}{f_R} \qquad (18.32)$$

Equation 18.31 may be written in terms of $d\theta$:

$$dy_{Ri} = (y_{Pi} - y_{Ri})d\theta \qquad (18.33)$$

The permeation rate through dA is written as

$$y_{Pi}df_P = p_{Mi}(P_R y_{Ri} - P_P y_{Pi})dA \qquad (18.34)$$

where P_R and P_P are the total pressures on each side of the membrane and p_{Mi} is the permeance of component i. The permeate rate and composition are calculated by integration:

$$F_P = \int_0^A df_P \tag{18.35}$$

$$F_P Y_{Pi} = \int_0^A y_{Pi} df_P \tag{18.36}$$

The above differential equations along with the summation Equation 18.25 may be solved numerically.

The cross-flow model may also be approximated by a series of stages each of which is represented by a perfect mixing model as depicted in Figure 18.5. The accuracy of the model increases with the number of stages used. This is the equivalent to a numerical solution of the equations above. Each stage receives feed f_F with composition y_{Fi} from the upstream stage (or F_F and Y_{Fi} if it is the first stage). The permeate is Δf_P, composition y_{Pi}, and the residue is f_R, composition y_{Ri}. Stream f_R (y_{Ri}) becomes the feed, f_F (y_{Fi}), to the next stage. The residue from the final stage is F_R, composition Y_{Ri}. By a total material balance on each stage,

$$f_F = \Delta f_P + f_R \tag{18.37}$$

Component i material balance on the stage,

$$f_F y_{Fi} = \Delta f_P \cdot y_{Pi} + f_R y_{Ri} \tag{18.38}$$

The fraction permeated on a stage,

$$\Delta\theta = \frac{\Delta f_P}{f_F} \tag{18.39}$$

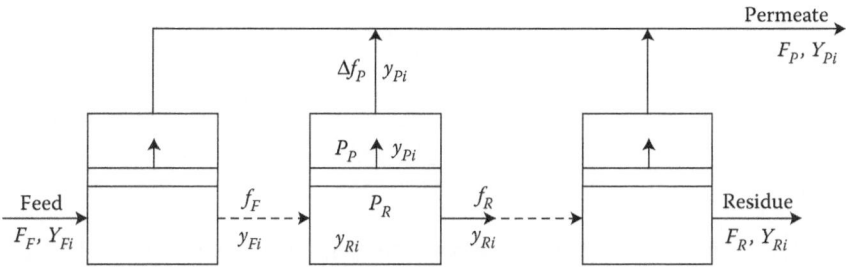

FIGURE 18.5 Cross-flow representation by multiple stages.

Equations 18.37, 18.38, and 18.39 are combined to eliminate f_F, f_R, and Δf_P:

$$y_{Fi} = \Delta\theta.y_{Pi} + (1 - \Delta\theta)y_{Ri} \tag{18.40}$$

The permeation rate in a stage is expressed as

$$y_{Pi}\Delta f_P = p_{Mi}(P_R y_{Ri} - P_P y_{Pi})\Delta A$$

where ΔA is the stage membrane area. Substituting for Δf_P from Equation 18.39,

$$y_{Pi}f_F\Delta\theta = p_{Mi}(P_R y_{Ri} - P_P y_{Pi})\Delta A$$

If y_{Ri} is eliminated between this equation and Equation 18.40, the following expression is obtained for the permeate composition in a stage:

$$y_{Pi} = \frac{P_R \Delta A p_{Mi} y_{Fi}}{(1 - \Delta\theta)(P_P \Delta A p_{Mi} + f_F\Delta\theta) + P_R \Delta A p_{Mi}\Delta\theta} \tag{18.41}$$

For rating an existing membrane separator with a given membrane area, Equation 18.41 is solved starting with the external feed and the first stage. The solution of each stage follows the methods described for the perfect mixing model. The membrane area for each stage, ΔA, is obtained by dividing the total membrane area A by the number of stages. The number of stages is chosen in accordance with the required level of accuracy.

1. Given F_F, Y_{Fi} (or f_F, y_{Fi} calculated in the upstream stage), ΔA, p_{Mi}, P_R, P_P.
2. Assume a value for $\Delta\theta$.
3. Calculate y_{Pi} from Equation 18.41.
4. If $\sum y_{Pi} = 1$ calculate y_{Ri} (Equation 18.40) and f_R (Equation 18.37) and go to the next stage. Otherwise restart at step 2 with an updated value for $\Delta\theta$.

The total permeate flow rate and composition is calculated by summation over all the stages:

$$F_P = \sum_{stages} \Delta f_P \tag{18.42}$$

$$Y_{Pi}F_P = \sum_{stages} y_{Pi}\Delta f_P \tag{18.43}$$

In a design situation a value is chosen for ΔA. This area may be obtained by dividing an estimated total membrane area by a provisional number of stages, based on the required accuracy. The estimated total area may be based on a perfect mixing model. The computations are then carried out as described above. Stages are added progressively until the performance specifications of the separator are met.

The procedures described above can be applied to multicomponent systems and are not limited to binary systems. Although the multistage approach is only an approximation to the rigorous cross-flow model, it is a useful tool for preliminary studies. It could show the enrichment and depletion trends of the components in the mixture and their dependence on the number of stages.

The accuracy of the method can be improved by increasing the number of stages while keeping the total permeation area unchanged. The second part of Example 18.4 uses two stages and shows the effect of multiple stages compared to the perfectly mixed model.

18.2.3 COUNTERCURRENT AND COCURRENT FLOW MODELS

In these models parallel flow to the membrane is assumed on both sides of it. If the film resistances are neglected the composition is considered constant on each side of the membrane across a plane perpendicular to it, and the streams on both sides are in plug flow.

The models are represented in Figure 18.6. The following derivations apply to both the countercurrent and cocurrent flow models. A total material balance around the differential volume surrounding membrane area dA on each side of it gives

$$df_R = df_P = \sum_i N_i dA \qquad (18.44)$$

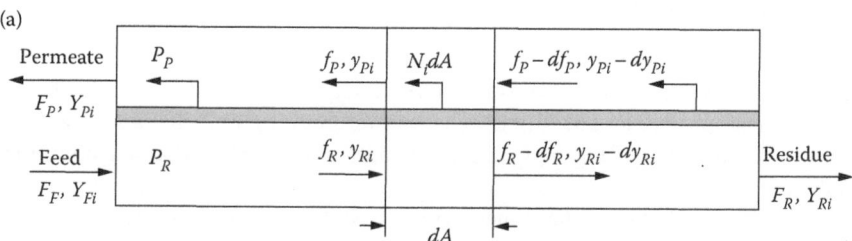

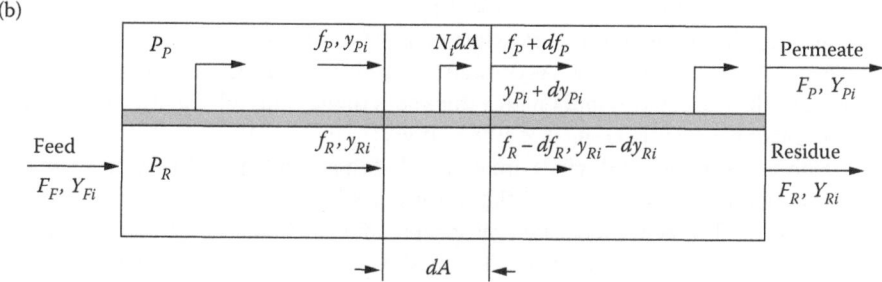

FIGURE 18.6 (a) Countercurrent flow and (b) cocurrent flow membrane separators.

A material balance on component i around the differential volume on the residue side gives

$$y_{Ri}f_R = N_i dA + (f_R - df_R)(y_{Ri} - dy_{Ri})$$

Or,

$$N_i dA = f_R dy_{Ri} + y_{Ri} df_R \tag{18.45}$$

A similar equation may be derived for the permeate side:

$$N_i dA = f_P dy_{Pi} + y_{Pi} df_P \tag{18.46}$$

The flow through the membrane is determined by the flux equation:

$$N_i = p_{Mi}(P_R y_{Ri} - P_P y_{Pi}) \tag{18.47}$$

The above differential and algebraic equations may be solved numerically to calculate the separation in these flow pattern models.

The countercurrent and cocurrent flow models may also be approximated by a series of perfect mixing blocks as described for the cross-flow model. The countercurrent flow model would require an additional iterative loop to converge the recycle created by the counterflowing permeate stream.

18.3 APPLICATIONS

Common practical applications of permeation separation phenomena include gas permeation, dialysis, and reverse osmosis. A variety of equipment modules for carrying out these processes are in use. The membrane type and the way it is incorporated in the device define the different separator assemblies that are commonly available.

Dense phase polymer membranes with no supportive substructure, so-called symmetric membranes, must have a minimum thickness of about 1 mil (1×10^{-3} in. or 2.54×10^{-3} cm) to ensure mechanical integrity and freedom from imperfections such as holes in the membrane. Possible disadvantages of these membranes are low fluxes and limited selectivity. Asymmetric membranes overcome these limitations, being very thin dense polymers with a thickness in the order of 1×10^{-7} m, supported on a porous layer that is about 1×10^{-4} m thick.

Membrane separation devices are assembled in a number of forms. In a flat sheet form the membrane is laid over a flat porous support. A unit would include a large number of the flat sheets separated by spacers and stacked together. In another configuration the flat sheet may be spiral-wound with spacers around a perforated tube. Other arrangements involve tubular membranes or hollow fiber membranes assembled in bundles. In the tubular module the membrane is wrapped around a tubular porous support. In hollow fiber units the membrane is a hollow polymer fiber, and is supported by a porous cylinder surrounding it.

18.3.1 Gas Permeation

Gas permeation is a process where gas components with different permeabilities are separated using a membrane device. Permeation selectivity usually depends on the relative molecular weights of the components to be separated. However, since the permeability is a specific property of both the component and the membrane, selectivity may depend on factors other than the molecular weight. An example is the separation of oxygen and nitrogen. Even though oxygen has a higher molecular weight than nitrogen, its permeability in polyethylmethacrylate is about five times that of nitrogen.

Other gas permeation applications include separation of hydrogen from methane, hydrogen from carbon monoxide, and removal of components such as carbon dioxide, helium, moisture, and organic solvents from gas streams. Gas permeation for such operations may provide a more economical and more practical alternative than conventional separation processes such as cryogenic distillation, absorption, or adsorption.

The driving force in gas permeation may be expressed in terms of the difference between a component partial pressure on the residue side and the permeate side of the membrane. The feed is introduced to the separator at a high pressure, while the permeate side is controlled at a low pressure. Examples 18.3 and 18.4 use the perfect mixing model for the performance evaluation and the design of two gas permeation processes.

18.3.2 Dialysis

Dialysis is liquid permeation where solutes with different permeabilities and film resistances are transported at different fluxes through the membrane, resulting in separation of the solutes. The driving force for each solute is its concentration gradient across the membrane, and the flux is given by Equations 18.11 through 18.14.

The feed stream containing the solvent and the solutes enters the separator on one side of the membrane and leaves as the dialysate (equivalent to the residue, Figure 18.7). On the other side of the membrane a stream is fed that is pure solvent (or a solvent with much lower solute concentrations than the feed). This stream exits as diffusate (equivalent to the permeate). The pressures on each side of the membrane need not be different since the driving force is the concentration gradient rather than the pressure gradient. However, if the pressures are equal, the solvent

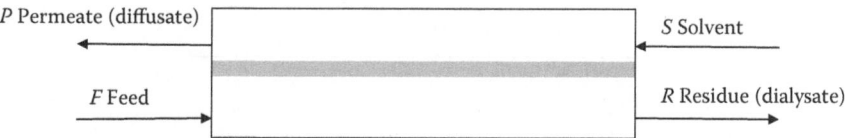

FIGURE 18.7   General dialysis model.

may diffuse from the diffusate side to the dialysate side by osmosis (see Section 18.3.3). This solvent transport would tend to lower the driving force of the solutes. To counter this effect the pressure may have to be raised on the dialysate side.

Dialysis has many applications, perhaps the most important of which is blood purification with the artificial kidney, where urea and other impurities are removed from the blood by dialysis. Applications in the chemical process industry include the removal of metal ions from acid solutions, and the purification of many other products such as polymers and pharmaceuticals.

In many applications dialysis is carried out in countercurrent flow, flat sheet equipment, where laminar flow is assumed. Countercurrent flow calculations similar to those discussed in Section 18.2 could be applied to countercurrent dialysis. Alternatively, a simplified model may be used where the driving force is expressed as a log mean average of the concentration differences, as shown in Example 18.5.

EXAMPLE 18.5: SEPARATION BY DIALYSIS

An aqueous solution of H_2SO_4 (1) and $CuSO_4$ (2) is fed to a dialyzer to remove some of the $CuSO_4$ from the acid. The feed flow rate is 0.30 m³/h and the concentrations of the solutes in the feed are 0.2 kmol/m³ H_2SO_4 and 0.1 kmol/m³ $CuSO_4$. The solvent is pure water with a flow rate of 0.32 m³/h, and is sent to the dialyzer in countercurrent flow to the feed. The permeances and mass transfer coefficients (assumed equal on both sides of the membrane) are given below for the solutes:

	p_{Mi}, m/s	k_i, m/s
1 H_2SO_4	1.8×10^{-5}	3.7×10^{-5}
2 $CuSO_4$	7.0×10^{-6}	1.4×10^{-5}

It is required to lower the concentration of $CuSO_4$ in the diffusate to 0.075 kmol/m³. Calculate the required membrane area and the compositions of the diffusate and dialysate. It may be assumed that no water permeation occurs.

Solution

The solutions are sufficiently dilute to justify the assumption of constant volumetric flow rates on each side of the membrane:

Feed rate = dialysate rate = F = 0.30 m³/h
Solvent rate = diffusate rate = S = 0.32 m³/h

The concentrations are as follows, in kmol/m³:

H_2SO_4 in the feed	$c_{F1} = 0.2$
$CuSO_4$ in the feed	$c_{F2} = 0.1$
H_2SO_4 in the solvent	$c_{S1} = 0.0$
$CuSO_4$ in the solvent	$c_{S2} = 0.0$

H_2SO_4 in the diffusate	c_{P1}
$CuSO_4$ in the diffusate	$c_{P2} = 0.075$
H_2SO_4 in the dialysate	c_{R1}
$CuSO_4$ in the dialysate	c_{R2}

The overall mass transfer coefficients are calculated from the permeances and the mass transfer coefficients on each side of the membrane:

$$\frac{1}{K_1} = \frac{1}{k_{11}} + \frac{1}{p_{M1}} + \frac{1}{k_{21}} = \frac{1}{3.7 \times 10^{-5}} + \frac{1}{1.8 \times 10^{-5}} + \frac{1}{3.7 \times 10^{-5}} = 109609 \text{ s/m}$$

$$K_1 = 9.12 \times 10^{-6}$$

$$\frac{1}{K_2} = \frac{1}{k_{12}} + \frac{1}{p_{M2}} + \frac{1}{k_{22}} = \frac{1}{1.4 \times 10^{-5}} + \frac{1}{7.0 \times 10^{-6}} + \frac{1}{1.4 \times 10^{-5}} = 285715 \text{ s/m}$$

$$K_2 = 3.50 \times 10^{-6}$$

Overall material balance for $CuSO_4$:

$$Fc_{F2} + Sc_{S2} = Fc_{R2} + Sc_{P2}$$

$$(0.30)(0.1) + 0.0 = 0.30c_{R2} + (0.32)(0.075)$$

$$c_{R2} = 0.02 \text{ kmol/m}^3$$

Material balance for $CuSO_4$ on the diffusate side:

Flow rate of $CuSO_4$ through the membrane (base on Equation 18.14)
= Flow rate of $CuSO_4$ in the diffusate

$$A\Delta c_{LM2}K_2 = Sc_{P2}$$

where Δc_{LM2} is the log mean average of the concentration differences:

$$\Delta c_{LM2} = \frac{(c_{F2} - c_{P2}) - (c_{R2} - c_{S2})}{\ln \dfrac{c_{F2} - c_{P2}}{c_{R2} - c_{S2}}} = \frac{(0.1 - 0.075) - (0.02 - 0.0)}{\ln \dfrac{0.1 - 0.075}{0.02 - 0.0}} = 0.0224 \text{ kmol/m}^3$$

The membrane area is calculated from the equation above:

$$A = \frac{Sc_{P2}}{\Delta c_{LM2}K_2} = \frac{(0.32)(0.075)}{(0.0224)(3.50 \times 10^{-6})(3600)} = 85 \text{ m}^2$$

Overall material balance for H_2SO_4:

$$Fc_{F1} + Sc_{S1} = Fc_{R1} + Sc_{P1}$$

$$(0.30)(0.2) + 0.0 = 0.30c_{R1} + 0.32c_{P1}$$

$$c_{R1} = 0.2 - 1.0667c_{P1}$$

Material balance for H_2SO_4 on diffusate side:

Flow rate of H_2SO_4 through the membrane (based on Equation 18.14)
= Flow rate of H_2SO_4 in the diffusate

$$A\Delta c_{LM1}K_1 = Sc_{P1}$$

where Δc_{LM1} is the log mean average of the concentration differences:

$$\Delta c_{LM1} = \frac{(c_{F1} - c_{P1}) - (c_{R1} - c_{S1})}{\ln\dfrac{c_{F1} - c_{P1}}{c_{R1} - c_{S1}}} = \frac{0.2 - c_{P1} - c_{R1}}{\ln\dfrac{0.2 - c_{P1}}{c_{R1}}}$$

Substitute for Δc_{LM1} in the equation above:

$$c_{P1} = \frac{A\Delta c_{LM1}K_1}{S} = \frac{(85)(3600)(9.12 \times 10^{-6})}{0.32}\frac{(0.2 - c_{P1} - c_{R1})}{\ln\dfrac{0.2 - c_{P1}}{c_{R1}}} = \frac{8.721(0.2 - c_{P1} - c_{R1})}{\ln\dfrac{0.2 - c_{P1}}{c_{R1}}}$$

Substitute for c_{R1} from the H_2SO_4 overall material balance equation:

$$c_{P1} = \frac{(8.721)(0.0667)c_{P1}}{\ln\dfrac{0.2 - c_{P1}}{0.2 - 1.0667c_{P1}}}$$

Solving for c_{P1} then c_{R1},

$$c_{P1} = 0.1737 \text{ kmol/m}^3$$
$$c_{R1} = 0.2 - 1.0667c_{P1} = 0.0147 \text{ kmol/m}^3$$

The complete results are tabulated as follows:

	Feed	Solvent	Dialysate	Diffusate
H_2SO_4, kmol/m³	0.20	0.0	0.0147	0.1737
$CuSO_4$, kmol/m³	0.10	0.0	0.0200	0.0750
H_2SO_4, kmol/h	0.06	0.0	0.0044	0.0556
$CuSO_4$, kmol/h	0.03	0.0	0.0060	0.0240
Total, m³/h	0.30	0.32	0.30	0.32

18.3.3   REVERSE OSMOSIS

Osmosis is the transport of a solvent through a dense membrane from the dilute solution side of the membrane to the more concentrated side, noting that the concentration of the solvent is higher on the dilute side. This transport can take place if the membrane is permeable to the solvent but impermeable to the solutes, and if the pressure on both sides of the membrane is the same. The result is a mixing process rather than a separation process since the solutes concentration is raised on the dilute side while it is lowered on the concentrated side. If the pressure on the concentrated side is raised relative to the dilute side, a condition of equilibrium will be reached at a certain pressure difference at which no net solvent transport through the membrane takes place.

If the liquid on one side of the membrane is a pure solvent and on the other side is a solution at a given concentration, the pressure difference required to result in zero net solvent transport is the osmotic pressure of that solution at its existing concentration and temperature. Thus, the osmotic pressure is defined as

$$\pi_1 = P_1 - P_0 \tag{18.48}$$

where P_1 is the pressure on the solution side and P_0 is the pressure on the pure solvent side of the membrane when no net solvent transport exists.

If P_1 is raised above $\pi_1 + P_0$ the solvent will transport from the solution side to the solvent side by the process known as reverse osmosis. In general the dilute solution may not be a pure solvent but a solution with its own osmotic pressure π_2 (Figure 18.8). The osmotic pressures are defined in terms of equilibrium pressures as

$$\pi_1 = P_1 - P_0$$

$$\pi_2 = P_2 - P_0$$

where P_0 is the pressure of a pure solvent. Thus, at equilibrium,

$$P_1 - \pi_1 = P_2 - \pi_2$$

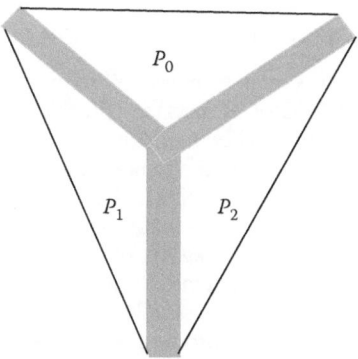

0–Pure solvent, 1–Solution to be purified, 2–Low concentration solution.
At equilibrium (no permeation), $P_0 = P_1 - \pi_1 = P_2 - \pi_2$.

FIGURE 18.8   Osmotic pressures.

In order for reverse osmosis to take place, P_1 must be greater than $P_2 + \pi_1 - \pi_2$, and the driving force is

$$DF = (P_1 - P_2) - (\pi_1 - \pi_2)$$

Subscripts 1 and 2 refer to the feed (or residue) side and the permeate side, respectively. If mass transfer resistances are neglected, the solvent flux is given by

$$N_S = \frac{P_{MS}}{L}\left[(P_1 - P_2) - (\pi_1 - \pi_2)\right] = p_{MS}\left[(P_1 - P_2) - (\pi_1 - \pi_2)\right] \quad (18.49)$$

where $P_{MS}, p_{MS},$ and L are the solvent permeability and permeance, and the membrane thickness. Equation 18.49 is a special application of the general principle expressed in Equation 18.17. The solute flux that could occur concurrently with reverse osmosis may be calculated by any one of Equations 18.11 through 18.14.

The osmotic pressure of a solution increases as its absolute temperature and/or the total molar concentration of the solutes are raised. For dilute solutions the osmotic pressure may be estimated by an equation similar to the ideal gas equation:

$$\pi = RTc \quad (18.50)$$

where R is the universal gas constant, T is the absolute temperature, and c is the total molar concentration of the solutes.

Like other membrane separation processes, reverse osmosis has the advantage of affecting the separation without the need for the creation of a second phase as in distillation, absorption, and so on. Thus, no thermal energy or separating agents are required, which makes the process economically attractive. A major application for reverse osmosis is the desalination of seawater, as well as other water purification operations. Another category of applications includes separating or purifying thermally unstable products as in food processing.

EXAMPLE 18.6: SEAWATER DESALINATION BY REVERSE OSMOSIS

It is required to design a reverse osmosis unit to process 2500 m³/h of seawater at 25°C containing 3.5 wt% dissolved salts, and produce purified water with 0.05 wt% dissolved salts. The pressure will be maintained at 135 atm on the residue side and 3.5 atm on the permeate side, and the temperature on both sides at 25°C. The dissolved salts may be assumed to be NaCl. With the proposed membrane, the salt permeance is 8.0×10^{-5} m/h and the water permeance is 0.085 kg/m².h.atm. The density of the feed seawater is 1020 kg/m³, of the permeate, 997.5 kg/m³, and of the residue (with an estimated salt content of 5 wt%), 1035 kg/m³. Assuming a perfect mixing model and neglecting the mass transfer resistances, determine the required membrane area and calculate the product flow rates and compositions.

SOLUTION

Overall NaCl material balance,

$$F_F x_F = F_R x_R + (F_F - F_R)x_P$$

Stream mass flow rates are represented by F and mass fractions of salt by x. Subscripts F, R, and P stand for feed, residue, and permeate.

$$F_F = (2500 \text{ m}^3/\text{h})(1020 \text{ kg/m}^3) = 2{,}550{,}000 \text{ kg/h}$$

$$2{,}550{,}000 \times 0.035 = F_R x_R + (2{,}550{,}000 - F_R)(0.0005)$$

$$F_R(x_R - 0.0005) = 87975$$

Guess $x_R = 0.05$

$$F_R = 87{,}975/(0.05 - 0.0005) = 1{,}777{,}273 \text{ kg/h}$$

$$F_P = 2{,}550{,}000 - 1{,}777{,}273 = 772{,}727 \text{ kg/h}$$

Salt concentration on the residue side,

$$c_R = \frac{x_R \rho_R}{M} = \frac{0.05 \times 1035}{58.5} = 0.8846 \text{ kmol/m}^3$$

where ρ_R is the salt solution density and M is the salt molecular weight. Salt concentration on the permeate side,

$$c_P = \frac{x_P \rho_P}{M} = \frac{0.0005 \times 997.5}{58.5} = 0.008526 \text{ kmol/m}^3$$

Osmotic pressures (Equation 18.50, with two ions per NaCl molecule),

$$\pi_R = 0.082057 \times 298.15 \times 0.8846 \times 2 = 43.28 \text{ atm}$$

$$\pi_P = 0.082057 \times 298.15 \times 0.008526 \times 2 = 0.417 \text{ atm}$$

The membrane area based on water transport is calculated by equating the water flow rate in the permeate product to the rate of water permeation from Equation 18.49:

$$F_P(1 - x_P) = A \rho_{MW} \left[(P_R - P_P) - (\pi_R - \pi_P) \right]$$

$$A = \frac{(772{,}727)(1 - 0.0005)}{(0.085)(135 - 3.5 - 43.28 + 0.417)} = 102512 \text{ m}^2$$

The membrane area may also be calculated based on the salt transport by equating the salt flow rate in the permeate product to the rate of salt permeation from Equation 18.14:

$$F_P x_P = A \rho_{MS}(c_R - c_P)M$$

$$A = \frac{(772{,}727)(0.0005)}{(8.0 \times 10^{-5})(0.8846 - 0.008526)(58.5)} = 94234 \text{ m}^2$$

Since the calculated membrane areas based on water transport and salt transport do not match, a new guess for x_R should be used. A solution by trial and error is found at $x_R = 0.04723$. The residue solution density is assumed to remain at 1035 kg/m³.

$$F_R = 87,975/(0.04723 - 0.0005) = 1,882,624 \text{ kg/h} = 1819.0 \text{ m}^3/\text{h}$$

$$F_P = 2,550,000 - 1,882,624 = 667,376 \text{ kg/h} = 669.0 \text{ m}^3/\text{h}$$

$$c_R = \frac{0.04723 \times 1035}{58.5} = 0.8356 \text{ kmol/m}^3$$

$$c_P = \frac{0.0005 \times 997.5}{58.5} = 0.008526 \text{ kmol/m}^3$$

Osmotic pressures,

$$\pi_R = 0.082057 \times 298.15 \times 0.8356 \times 2 = 40.89 \text{ atm}$$

$$\pi_P = 0.082057 \times 298.15 \times 0.008526 \times 2 = 0.417 \text{ atm}$$

Membrane area based on water transport,

$$A = \frac{(667,376)(1 - 0.0005)}{(0.085)(135 - 3.5 - 40.89 + 0.417)} = 86211 \text{ m}^2$$

Membrane area based on salt transport,

$$A = \frac{(667,376)(0.0005)}{(8.0 \times 10^{-5})(0.8356 - 0.008526)(58.5)} = 86209 \text{ m}^2$$

NOMENCLATURE

A Membrane area
C Number of components
c Concentration
d Molecular diffusivity
d_K Knudsen diffusivity
d_m Molecular diameter
d_p Pore diameter
D Permeation diffusivity
F Stream flow rate
f Local flow rate inside the membrane separator
G General driving force property
H Henry's constant
K Equilibrium distribution coefficient
K_r Restrictive factor
k Mass transfer coefficient

L	Membrane thickness
M	Molecular weight
N	Flux
P	Total pressure
P_M	Permeability
p	Partial pressure
p_M	Permeance
T	Temperature
x	Mass fraction
Y	Mole fraction
y	Local mole fraction inside the membrane separator
ε	Porosity
θ	Fraction of feed permeated
π	Osmotic pressure
ρ	Density
τ	Tortuosity

SUBSCRIPTS

1,2	Membrane sides
F	Feed
f	Film interface
i	Component
k	Component
m	Membrane surface
P	Permeate
R	Residue
S	Solvent

PROBLEMS

18.1. It is required to design a membrane separator for the separation of a gas mixture of nitrogen and methane. The stream flow rate and composition, and the component permeances in the proposed membrane are given below. The pressures on the residue side and permeate side are 5475 kPa and 100 kPa, respectively. Assuming a perfect-mixing model for the separator, determine the membrane area required to bring the nitrogen concentration in the permeate to 30 mol%. Calculate the products' flow rates and compositions.

Feed stream:

	kmol/h	p_{Mi} (kmol/m².h.kPa)
1. Nitrogen	175	6.00×10^{-6}
2. Methane	825	1.20×10^{-6}
	1000	

18.2.   For the process described in Problem 18.1, calculate the products' flow rates and compositions if the membrane area is 45,000 m^2.

18.3.   Acetone may be recovered from air by gas permeation using a membrane that is highly selective to acetone. The air–acetone feed to the membrane separator, at a rate of 700,000 cm^3(STP)/s, has the following composition and component permeabilities in cm^3(STP)/cm.s.cm Hg:

	Y_{Fi}	P_{Mi}
1. Acetone	0.005	2×10^{-6}
2. Air	0.995	4×10^{-10}

The film resistances on both sides of the membrane may be neglected. The pressures in the separator are 90 cmHg on the residue side and 5 cmHg on the permeate side. The membrane thickness is 2×10^{-4} cm and its area is 5×10^8 cm^2.

a. Assuming a perfect-mixing model, calculate the permeate and residue rates and compositions.

b. In a first approximation to a cross-flow model, calculate the products' rates and compositions using two perfect-mixing stages similar to the arrangement shown in Figure 18.5. Assume each stage has half the total membrane area.

18.4.   A 280,000 cm^3(STP)/s gas stream containing 60% mole propylene (1) and 40% mole propane (2) is to be processed in a membrane separator to raise the concentration of propylene in the permeate. The pressure is 1550 cm Hg on the residue side and 76 cm Hg on the permeate side. The permeabilities of propylene and propane are 9.0×10^{-10} and 2.8×10^{-10} cm^3(STP)/(cm.s.cmHg), respectively, and the membrane thickness is 0.1×10^{-4} cm.

a. Calculate the membrane area required to raise the concentration of propylene in the permeate to 75% mole, and determine the flow rates and compositions of the products. Assume a perfect-mixing model separator.

b. For the membrane area calculated in (a), determine the flow rates and compositions of the products if the flow pattern in the separator is cross-flow. Use a series of five perfect-mixing models, each having one fifth the membrane area, as an approximate representation of the cross-flow model.

18.5.   In Example 18.6, a reverse osmosis unit was designed to desalinate seawater based on provided data. It is required to rate an existing reverse osmosis unit, to determine its product rates and compositions. Data such as permeances, pressures, solution properties given in Example 18.6 remain unchanged. The unit membrane area is 86,000 m^2 and the seawater flow rate is 2,550,000 kg/h.

18.6. A membrane separator is to be used for separating nitrogen from methane as specified below:

	Feed, Y_{Fi}	Permeate, Y_{Pi}	Permeance $(cm^3(STP)/cm^2.s.cm\ Hg)$
Nitrogen (1)	0.2	0.327	5×10^{-6}
Methane (2)	0.8		1×10^{-6}
Flow (cm³(STP)/s)	6.0×10^6		
Pressure (kPa)	5475	100	

The residue pressure may be assumed equal to the feed pressure. In order to estimate the required membrane surface area, calculations are to be made at assumed values of θ, the fraction of feed permeated, between 0.4 and 0.5. Based on these calculations what is the recommended range of membrane surface area and what are the expected ranges of product flows and compositions (1 kPa = 0.75 cm Hg).

18.7. Nitrogen is to be separated from methane in a perfect-mixing membrane separator as specified below. Calculate the required membrane area, the methane permeance, and the product rates.

Component	Permeance, p_{Mi} $(cm^3(STP)/cm^2.s.cm\ Hg)$	Feed, Y_{Fi}	Permeate, Y_{Pi}	Residue, Y_{Ri}
1. Nitrogen	5.5×10^{-6}	0.30	0.40	0.10
2. Methane		0.70	0.60	0.90
Flow rate (cm³(STP)/s)		1.0×10^6		

The pressure on the residue side is 4000 cm Hg and on the permeate side 76 cm Hg.

18.8. In an air separation process using a perfect-mixing membrane separator, the feed air flow rate is $n_F = 1 \times 10^6$ cm³ (STP)/s, with mole fraction oxygen $y_{F1} = 0.21$ and mole fraction nitrogen $y_{F2} = 0.79$. The membrane permeances for oxygen and nitrogen are, respectively, 1.5×10^{-5} and 4.5×10^{-6} cm³ (STP)/cm².s.cm Hg. Film resistances may be neglected. The pressures on the residue side and the permeate side are $P_R = 760$ cm Hg and $P_P = 80$ cm Hg. The permeate rate is 390,840 cm³ (STP)/s and contains 32 mol% oxygen. Find the mole fractions of oxygen and nitrogen on both sides of the membrane, and calculate the membrane area based on oxygen permeation and on nitrogen permeation. Do the areas match?

18.9. A membrane separator is used to enrich the hydrogen in a hydrogen–methane mixture defined below:

	F_{Fi} (kmol/h)	p_{Mi} (kmol/h.m².kPa)
1. H_2	200	2.44×10^{-4}
2. CH_4	25	3.94×10^{-5}

The pressure is 3500 kPa on the residue side and 140 kPa on the permeate side. The composition on the permeate side is specified at 95 mol%. Assuming a perfect-mixing model, calculate the required membrane area. Use a fraction of feed permeated, $\theta = 0.15$. Calculate the product rates and compositions, and check if θ needs to be modified.

18.10 The composition of an aqueous solution of HCl and NaCl is to be altered to raise the concentration of HCl relative to that of NaCl, using a dialyzer. The membrane is microporous, 100 m^2 in area. The feed components permeances and mass transfer coefficients ($k_{1i} = k_{2i}$) are given below:

	p_{Mi}, m/h	k_i, m/h
1. HCl	0.030	0.52
2. NaCl	0.009	0.47

The feed solution flow rate is 0.5 m^3/h, with 0.24 kmol/m^3 HCl and 0.08 kmol/m^3 NaCl. The solvent is pure water, also at 0.5 m^3/h. It is required to find the flow rates and compositions of the diffusate and dialysate. Water permeation may be neglected.

18.11 A desalination plant uses reverse osmosis to lower seawater salt concentration from 3.6 wt% to 600 ppm. The seawater feed rate is 5,000 m^3/h, at 20°C and 14,000 kPa, which is also the residue side pressure. The permeate side is at 20°C and 350 kPa. For a membrane area of 200,000 m^2, find the flow rates and compositions of the product streams, assuming a perfect mixing model. The water permeance is 0.00082 kg/m^2-h-kPa, and that of the salt is 8.5×10^{-5} m/h. The densities are $\rho_F = 1025$ kg/m^3, $\rho_P = 1000$ kg/m^3, $\rho_R = 1050$ kg/m^3, and mass fraction of salt in the residue, $x_R = 0.055$ (estimated).

REFERENCES

Beck, R.E. and J.S. Schultz, *Science*, 170, 1302–1305, 1970.
Hsieh, H.P., R.R. Bhave, and H.L. Fleming, *J. Membrane Sci.*, 39, 221–241, 1988.

19 Fluid–Solid Operations

The unit operations considered so far in all past chapters involved mass transfer between vapor and liquid phases. Other processes exist where solids interact with liquids and/or gases. Examples of such processes include adsorption, ion exchange, and chromatography.

19.1 FLUID–SOLID INTERACTION MODELS

The solid phase in fluid–solid interaction processes is the *sorbent*, which is used for carrying out the above operations. The sorbent used for adsorption is the adsorbent, and for ion exchange it is the ion exchanger. In the chromatographic process the sorbents are stationary phases of various compositions.

In adsorption, certain gas or liquid components are attracted ("adsorbed") selectively to the adsorbent, thereby resulting in separation between the fluid components. Adsorbed components constitute the *adsorbate*. An example application is the removal of organic compounds from water, using an adsorbent that is selective for organics.

Ion exchange is a chemical reaction between ions in solution and ions in the solid. Typically, the ion in solution replaces an atom in an insoluble solid, the ion exchanger, and that solid atom is turned into an ion in solution. An example of such a process is the replacement of Ca^{2+} ions in solution with Na^+ ions. The sodium ions exist in combination with an organic polymer, R, to form the ion exchanger Na_2R. The reaction between Ca^{2+} ions and the sorbent is representative of a water-softener process,

$$Na_2R + Ca^{2+} \leftrightarrow CaR + 2Na^+$$

The ion exchanger can be regenerated by interaction with a sodium chloride solution to reverse the reaction above.

In chromatography, the sorbents' different selectivities for different components allow its use for separating components in a gas or liquid as in distillation.

The necessary feature for all sorbents is their selectivity, their distinct level of interaction with any given fluid component relative to the others. This property enables the sorbent to accomplish the required separation or purification. Another necessary sorbent property is the ratio of its active surface area relative to its mass. This allows reasonable unit operations equipment size for high-volume processing. The sorbents should also be possible to be regenerated for extended use (Reynolds, 1982; Perry and Green, 1997).

19.1.1 Adsorbents

These are solid structures in various forms with the primary property of selectivity with given vapor or liquid systems. Another important feature of the adsorbent is

631

the active surface area per unit adsorbent mass, A_s/m. A large surface area relative to mass requires a porous or microporous structure for the adsorbent. The active surface area is the combined surface area of all the pores (Cusack and Karr, 1991).

The A_s/m ratio may be estimated based on some inherent properties of the adsorbent: its density ρ, porosity ε, and the shape and size of the pores. The porosity is pores volume fraction of the total adsorbent particle volume, $\varepsilon = v_p/V$. If the pores are assumed to be cylindrical, with diameter d and length L, then the pore surface area per unit pore volume is

$$A_s/v_p = \pi dL/(\pi d^2 L/4) = 4/d$$

Substituting $v_p = V\varepsilon$,

$$A_s/V = 4\varepsilon/d$$

For adsorbent particle of mass m and density ρ, the pore surface area per unit adsorbent mass is

$$S_p = A_s/m = 4\varepsilon/\rho d \tag{19.1}$$

EXAMPLE 19.1: POROUS ADSORBENT PROPERTIES

An activated carbon adsorbent has a pore diameter of 2×10^{-7} cm, a porosity of 0.5, and a density of 0.8 g/cm³. If the area occupied on the pore surface per molecule of adsorbed water vapor is 1×10^{-15} cm²/molecule, find the adsorbent pore surface area/gram adsorbent, and the amount of water adsorbed per gram adsorbent.

SOLUTION

From Equation 19.1, the pore surface area per gram adsorbent,

$$S_p = [4(0.5)]/[(0.8 \text{ g/cm}^3)(2 \times 10^{-7} \text{ cm})] = 1.25 \times 10^7 \text{ cm}^2/\text{g}$$

Number of adsorbed water molecules per gram adsorbent,

$$= (1.25 \times 10^7 \text{ cm}^2/\text{g})/(1 \times 10^{-15} \text{ cm}^2/\text{molecule})$$
$$= 1.25 \times 10^{22} \text{ molecules/g adsorbent}$$

Mass of adsorbed water per gram adsorbent,

$$= [(1.25 \times 10^{22} \text{ molecules/g adsorbent})(18.02 \text{ g water/mol})]/[6.023$$
$$\times 10^{23} \text{ molecules/mol}] = 0.374 \text{ g water/g adsorbent}$$

19.1.2 ION EXCHANGERS

Most ion exchangers are solid synthetic polymers with ionic functional groups. The ion exchangers are placed in a solvent to carry out the ion exchange. Resins can be converted to cation-exchangers or anion-exchangers by treatment with a strong acid or a strong base. A cation in solution can be exchanged with H^+ in the solid exchanger:

$$Na^+ + RH \leftrightarrow RNa + H^+$$

Similarly, an anion in solution can be exchanged with OH^- in the solid exchanger:

$$Cl^- + ROH \leftrightarrow RCl + OH^-$$

The ion exchange capacity of an exchanger is expressed as equivalents per kg of dry resin, or equivalents per liter of wet resin. Thus, one mole of an ion exchanger resin of 100 molecular weight with one mobile charge having an ionic valence of one would have a capacity of

$$(1 \text{ eq/mol})(1 \text{ mol/100 g})(1000 \text{ g/kg}) = 10 \text{ eq/kg}$$

This value corresponds to a mobile ion with ionic valence of 1, such as H^+. The capacity of an exchanger with one mobile ion but with ionic valence of 2, such as Ca^{2+}, would be 20 eq/kg.

Consider the following sulfonation reaction of a resin to produce an ion exchanger:

$$\underset{\text{(solid resin)}}{RH} \quad + \quad \underset{\text{(solution)}}{H_2SO_4} \quad \leftrightarrow \quad \underset{\text{(solid exchanger)}}{RSO_3^-} \quad + \quad \underset{\text{(solution)}}{H^+} \quad + \quad H_2O$$

The capacity is expressed based on a valence of 1 (H^+) per kg of ion exchanger (RSO_3^-).

EXAMPLE 19.2: ION-EXCHANGE CAPACITY

An ion-exchange resin is made by styrene sulfonation. What is the maximum ion-exchange capacity of the resin?

SOLUTION
The sulfonation reaction is written as

$$C_6H_5CH = CH_2 + H_2SO_4 \leftrightarrow C_6H_5CH = CHSO_3^- + H^+ + H_2O$$

Since in this reaction each mole of styrene gives one mobile charge, the number of moles styrene defines the number of mobile charges (each mole produces one mobile charge), which is also equal to the number of moles ion-exchanger resin. Using 100 g styrene as a point of reference, the number of moles styrene is

$$100/\text{styrene MW} = 100/104.14 = 0.96$$

The ion-exchange resin mass,

$$100 \text{ g (MW of ion-exchange resin/MW of styrene)} = 100(183.21/104.14)$$
$$= 175.93 \text{ g}$$

Relating the number of equivalents to the ion-exchange resin mass,

The capacity of the resin = Number of mobile charges
$$\times \text{ Mobile ion valence/Mass of resin (kg)}$$
$$= [(0.96 \text{ mol} \times 1 \text{ Mobile charge/mol})$$
$$\times (1 \text{ Ionic valence/Mobile charge)]/Mass of resin (kg)}$$
$$= 0.96 \text{ eq} \times 1 \times 1 \times 1000/175.93 = 5.46 \text{ eq/kg}$$

19.1.3 Chromatographic Processes

Various sorbents may be required for these processes, as any such process may include several steps, each of which may operate with a different sorbent. Besides, the fluid to be processed could be a gas or a liquid, which expands the selection of suitable sorbents.

The fundamentals of chromatographic separation involves a *mobile phase* (gas or liquid) which consists of a carrier fluid (gas or liquid) into which the components to be separated are injected. The carrier should be inert relative to the sorbent, that is, it will not interact with it.

The mobile phase flows through a packing where it contacts the *stationary phase* which contains the sorbent as a solid or liquid supported on a solid. In ion-exchange chromatographic processes the stationary phase would be a synthetic ion exchanger.

The mobile phase—stationary phase contact process is followed by an *elution* process where an *eluent* is fed and flows through the length of the stationary phase structure. Different adsorbed species have different affinities to the sorbent, and are therefore eluted at different rates, thereby bringing about their separation.

19.2 PHASE EQUILIBRIUM

As in liquid–liquid or vapor–liquid equilibria, when a liquid or vapor is in contact with a sorbent, equilibrium is established at the solid surface between the compositions of a solute in the two phases. This is expressed in terms of the concentration of the solute in the sorbent as a function of its concentration in the fluid phase. Whereas phase equilibrium in vapor–liquid or liquid–liquid systems can be estimated based on the thermodynamic condition of equality of component fugacities in the phases, no valid theory exists for predicting solid–fluid systems. Equilibrium concentrations for these systems must be based on experimental data.

19.2.1 Isotherms

Equilibrium data depend on the system temperature, therefore an expression for these data sets applies to the temperature at which the data were obtained, hence the name *isotherm* for these expressions. Isothermal equilibrium data may be fitted to a mathematical equation that defines the adsorbate composition in the adsorbant at a given solute composition in the fluid, and vice versa. Any one of a number of mathematical functions may be used for fitting a given set of data. These functions are empirical, based on experimental data. The selection of one of them for a given physical application depends on how well that function represents the data over a reasonable range of compositions (Treybal, 1980).

The general form of the isothermal functions can be expressed as

$$q = f(c)$$

where q is the composition of the of the adsorbate in the adsorbent, kg adsorbate/kg adsorbent, and c the composition of the adsorbate in the fluid, kg/m^3. The following functions are in common use; the one that gives the best fit for the data should be selected.

The *linear* isotherm, which is empirical, takes the form

$$q = Kc \tag{19.2}$$

where K is a constant derived from the experimental data. Using the dimensions as defined above, the K dimensions are (m^3 fluid)/(kg adsorbent). The use of this isotherm is generally limited to dilute solutions.

The *Freundlich* isotherm is also empirical, and is expressed as

$$q = Kc^n \tag{19.3}$$

The constants K and n are evaluated based on the experimental data. The exponent n is dimensionless and the K dimensions are consistent with the equation and units defined above:

$$[K] = [(\text{kg adsorbate})^{1-n} \, (\text{m}^{3n})]/(\text{kg adsorbent})$$

This isotherm provides a good fit for experimental data over reasonably extended compositions.

The *Langmuir* isotherm is the only one that is developed on the basis of theoretical considerations. It takes the form

$$q = q_0 c/(K + c) \tag{19.4}$$

Using consistent units, q_0 is in kg adsorbate/kg solid and K is in kg adsorbate/m^3. Although the equation itself is based on theory, its constants must still be determined to match experimental data. This equation has been found to fit certain processes, but as is the case of the other isotherms, it has to be tested for any given application.

19.2.1.1 Gas Adsorption

The isotherm equations introduced above apply to fluids in general, (liquid or gas), where the fluid composition was defined as kg/m^3. For gas adsorption it may be more convenient to express the gas composition in terms of partial pressure of the adsorbate in the gas. If the gas phase is assumed to follow the ideal gas model, then the partial pressure of the adsorbate in the gas is expressed as

$$p = nRT/V = mRT/MV$$

where m is the adsorbate mass and M its molar mass, V is the gas volume, T its absolute temperature, and R is the universal gas constant. If m is in kg, V in m³, T in Kelvin, then p in kPa is given as

$$p = 8.3145 \ cT/M$$

$$c = pM/8.3145T$$

where c is in kg adsorbate/m³ gas. The linear isotherm defines the adsorbate composition in kg adsorbate/kg adsorbent as

$$q = Kc = KpM/8.3145T$$

For a given adsorbate at a given temperature this relation may be abbreviated as

$$q = kp \tag{19.2a}$$

Typically, this equation applies to low concentrations of the adsorbate, where k may be predicted or derived from low concentration data. If k is to be determined by a least squares fit of q versus p data, and also ensure the function passes through the origin ($p = q = 0$), then k can be determined as

$$k = (\Sigma p_i q_i)/(\Sigma p_i^2)$$

The Freundlich isotherm,

$$q = kp^n \tag{19.3a}$$

This equation may be converted to a linear form by taking the logarithm on both sides:

$$\log q = \log k + n \log p \tag{19.3b}$$

The Langmuir isotherm is defined as

$$q = q_0 p/(K' + p) \tag{19.4a}$$

A linear representation of this equation can be obtained by rearranging it:

$$p/q = (K' + p)/q_0 \tag{19.4b}$$

EXAMPLE 19.3: COMPARING EQUILIBRIUM ISOTHERMS

The following equilibrium data were obtained for the batch adsorption of phenol from a water solution, using activated carbon as the adsorbent. Among the linear, Freundlich, and Langmuir isotherms, it is required to determine the one that best fits the data.

c, kg phenol/m³ solution	0.32	0.12	0.04	0.006	0.001
q, kg phenol/kg adsorbent	0.15	0.12	0.10	0.06	0.04

SOLUTION

Linear: The constant K in the linear isotherm expression was determined by fitting that equation to the data. This was accomplished by minimizing the sum of the errors squared, $\Sigma[q(data)-q(calculated)]^2$. The resulting equation is

$$q = 0.564\ c$$

q (calculated)	0.180	0.068	0.023	0.003	0.001

The sum of errors squared is 0.0142.

Freundlich: The Freundlich isotherm was fitted to the data by expressing the equation in its linearized form,

$$\log q = \log K + \log c$$

The following equation was obtained by minimizing the sum of errors squared based on the linearized equation. The resulting isotherm is

$$q = 0.19888\ c^{0.2312}$$

q (calculated)	0.153	0.122	0.094	0.061	0.040

The sum of errors squared is 0.00005.

Langmuir: This isotherm can be linearized by expressing it as

$$c/q = (K + c)/q_0$$

Minimizing the sum of errors squared led to the equation

$$c/q = 6.5194\ c + 0.0966$$

Or,

$$q = 0.1534\ c/(0.0148 + c)$$

c (calculated)	0.146	0.137	0.112	0.044	0.010

The sum of errors squared is 0.0016.

It can be seen that the Freundlich isotherm gave the best fit, followed by the Langmuir, and finally the linear.

19.2.2 ION-EXCHANGE EQUILIBRIUM

A chemical equation is at equilibrium when the forward and reverse rates of the reaction are equal. Consider a reaction such as

$$aA + bB \leftrightarrow cC + dD$$

The forward and reverse rates are written as

$$\text{Forward} = k[A]^a[B]^b$$

$$\text{Reverse} = k'[C]^c[D]^d$$

where [A], [B], [C], and [D] are the concentrations of each specie, A, B, C, and D, and a, b, c, and d are the numbers of molecules of each specie taking part in the reaction. The coefficients k and k' are proportionality constants. At equilibrium,

$$k[A]^a[B]^b = k'[C]^c[D]^d$$

and the equilibrium constant is

$$K = k/k' = \{[C]^c[D]^d\}/\{[a]^a[B]^b\}$$

Applying the equilibrium condition to an ion-exchange reaction such as

$$Na^+ + HR \leftrightarrow NaR + H^+$$

At equilibrium,

$$K = \{[NaR][H^+]\}/\{[Na^+][HR]\}$$

For ion exchangers, the concentrations may be expressed in terms of q, the equivalents/L of wet resin, and c, the equivalents/L of solution (see Example 19.2). Applied to the Na^+, H^+ reaction, the equilibrium K takes the form

$$K = (q_{NaR} \cdot c_{H+})/(q_{HR} \cdot c_{Na+})$$

For a general reaction where a pair of counter ions A and B, one replacing the other in an ion exchanger R, the equation for cations is written as

$$A^{n+} + nBR \leftrightarrow AR_n + nB^+$$

Similarly for anions,

$$A^{n-} + nBR \leftrightarrow AR_n + nB^-$$

The equilibrium constant for either of these reactions is expressed as

$$K_{AB} = (q_{ARn} c_B^n)/(q_{BR}^n c_{An}) \tag{19.5}$$

The equilibrium constant can be estimated based on relative molar *selectivity* coefficients. These are coefficients K_A, K_B, ..., of ions A, B, ..., interacting with an ion-exchange resin, R, for which data are available. The equilibrium constant for a pair of ions A and B is estimated as

$$K_{AB} = K_A/K_B \tag{19.6}$$

During the ion exchange equal equivalents are exchanged, so the total concentration in solution and in the resin remains constant. The total in solution is designated as C equivalents/L:

$$C = c_A + c_B = \text{constant} \tag{19.7}$$

The total concentration in the resin is the resin capacity, Q equivalents/L:

$$Q = q_A + q_B = \text{constant} \tag{19.8}$$

EXAMPLE 19.4: CALCIUM/ACID ION EXCHANGE

It is required to remove calcium from a water solution. Initially, the calcium solution is at 0.03 mol/L (0.06 eq/L). The calcium cations will be removed by ion exchange with a strong acid cation. The equilibrium constant K_{AB} may be estimated based on relative selectivity constants, K_A (Ca^{2+}) = 5.15 and K_B (H^+) = 1.26. The total concentrations in the phases are $C = 0.08$ and $Q = 2.0$ eq/L of wet bed. If at equilibrium the concentration of Ca^{2+} in solution is 0.03 mol/L, find the equilibrium concentration of Ca^{2+} in the resin.

SOLUTION

The ion-exchange equation written for the current system takes the form

$$Ca^{2+} + 2HR \leftrightarrow CaR_2 + 2H^+$$

From Equation 19.6,

$$K_{AB} = 5.15/1.26 = 4.087$$

The equivalent of Equation 19.5 applied to this system,

$$4.087 = [(q_{CaR2})(c_{H+})^2]/[(q_{HR})^2(c_{Ca}^{2+})] \tag{19.9}$$

From Equation 19.7,

$$0.08 = c_{H+} + 2c_{Ca}^{2+} \tag{19.10}$$

From Equation 19.8,

$$2.00 = q_{HR} + 2q_{CaR2} \tag{19.11}$$

The last three equations have three unknowns: q_{CaR2}, c_{H+}, q_{CaR2}. The fourth variable is defined, $c_{Ca}^{2+} = 2 \times 0.03 = 0.06$ eq/L. From Equation (19.10),

$$c_{H+} = 0.08 - 2(0.03) = 0.02$$

Equation (19.11) rearranged,

$$q_{CaR2} = 2.00/2 - q_{HR}/2 = 1.00 - 0.5 \, q_{HR}$$

Substituting in Equation (19.9),

$$4.087 = [(1 - 0.5 q_{HR})(0.02)^2]/[(q_{HR})^2(0.03)]$$

Rearranging,

$$(q_{HR})^2 + 0.00163 q_{HR} - 0.00326 = 0 \quad \text{Solution, } q_{HR} = 0.05629$$

$$q_{CaR2} = 1 - (0.5)(0.05629) = 0.9719$$

Equivalent of Ca^{2+} in resin at equilibrium $= 2 \times (0.9719) = 1.9437$. This should be compared with the original value of 2.0 eq/L.

19.3   APPLICATIONS

The different solid–fluid interaction models and the various types and sizes of the sorbent media particles, as well as the purpose of a sorption operation, separation, purification, and so on, led to the development of a variety of processes, along with the required equipment types. These processes are associated with the main interaction models discussed earlier: adsorption, ion exchange, and chromatography.

An adsorption process involves essentially three steps: bringing into mutual contact the adsorbent and the fluid to be treated (liquid or gas), allowing adequate contact time for the adsorption to take place, and separating the fluid from the adsorbent. An additional related step, the adsorbent recovery, may also be included in the process. Depending on the particular adsorption application, any of the following procedures may be used:

19.3.1   SINGLE-STAGE BATCH EQUILIBRIUM

This process is generally used when a small amount of a fluid, the solution, is to be treated, such as in the case of pharmaceuticals. The adsorbent and the solution are mixed thoroughly in a container until equilibrium is reached between the phases.

If the solution volume, assumed fixed, is S m^3 and the adsorbent mass, also constant, is M kg, then by adsorbate material balance the amount of adsorbate leaving the solution equals what is adsorbed:

$$S(c - c_e) = M(q_e - q) \tag{19.12}$$

where c is the initial concentration of the adsorbate in the solution, kg/m^3, and q is the initial adsorbate concentration in the adsorbent, kg/kg. Subscript e indicates equilibrium conditions. Given the initial concentrations, Equation 19.12 along with an isotherm equation can be solved for the final, equilibrium concentrations (Collins, 1967).

EXAMPLE 19.5: BATCH EQUILIBRIUM ADSORPTION

Activated granular carbon is used to treat water contaminated with phenol. The granular carbon is mixed with the water at 25°C and stirred thoroughly until equilibrium is reached. The equilibrium at 25°C can be represented by the Freundlich isotherm,

$$q_e = 0.2c_e^{0.23}$$

In one such operation 1 m³ of water containing 0.3 kg/m³ phenol was treated with 1.5 kg activated carbon with no residual contaminants. Determine the final phenol concentration in the water and in the adsorbent.

SOLUTION

Initial conditions: $S = 1.0$ m³, $M = 1.5$ kg, $c = 0.3$ kg/m³, $q = 0$.

Equation 19.9 is solved with the Freundlich equation to determine c_e and q_e. For the given values, Equation 19.9 is written as

$$1.0(0.3 - c_e) = 1.5(q_e - 0)$$

Or,

$$q_e = 0.2 - 0.667c_e$$

Combining this equation with the Freundlich equation gives

$$c_e^{0.23} + 3.335c_e - 1 = 0$$

Solving this equation by trial and error or using a solver gives

$$c_e = 0.11685 \text{ kg/m}^3$$

$$q_e = 0.1221 \text{ kg/kg}$$

While the Example 19.5 provides the final equilibrium results, in practice it may be required to determine the minimum amount of adsorbent required to achieve a given purity of the solution. Equilibrium is again assumed, as indicated in Example 19.6.

EXAMPLE 19.6: MINIMUM ADSORBENT

For the system described in Example 19.5, find the minimum amount of adsorbent that would lower the phenol concentration in the liquid to $c = 0.1$ kg/m³.

SOLUTION

At equilibrium, $c_e = c = 0.1$ kg/m³
From the Freundlich equation,

$$q_e = 0.2 \, c_e^{0.23} = 0.1178$$

From the mass balance equation,

$$1.0(0.3 - 0.1) = M(0.1178 - 0)$$

$$M = 0.2/0.1178 = 1.7 \text{ kg}$$

19.3.2 Nonequilibrium Processes

The rate at which a solid–fluid system can reach equilibrium is significant in the design of systems for these applications. Compared to vapor–liquid or liquid–liquid systems approaching phase equilibrium, solid-related equilibria require additional considerations.

In the batch equilibrium process, where agitation and contact time can be controlled to enhance the transfer of solutes from the solution to the adsorbent, phase equilibrium can be reached or approached. Under these conditions the end results depend on equilibrium rather than kinetics.

In continuous or semicontinuous operations the fluid flows through or in contact with the adsorbent. The approach to equilibrium is hindered by the brief contact time at a given interfacial point. As a result, reaching equilibrium is primarily controlled by mass transfer, a process that includes solute transfer through the fluid as well as the solid. Solute molecules in the fluid must transfer through a fluid film surrounding the solid adsorbent particles, move through pores in the particles, then diffuse through the particle structure. Theoretical methods for the prediction of mass transfer rates have been published. However, these methods do not always predict accurate results, and experimental data should be available to verify or supplement the theoretical predictions. What follows is a case where experimental data can be used for designing adsorption columns.

19.3.3 Fixed-Bed Adsorption Columns

A fixed-bed adsorption column is a vertical column packed with granular adsorption particles. A liquid stream to be purified is fed above the packing and flows down through the granules. If the feed stream is a gas or vapor, it may be sent to the bottom of the column and flows upward through the packing.

One advantage of this type of columns is the observation that with proper design, the solute could be completely adsorbed, and the entire sorbent can approach saturation. These conditions require a certain ratio of the amount of fluid treated to the amount of sorbent in the fixed bed, and this ratio is determined by system isotherms. A performance model of the column requires that the following assumptions be met or approximated:

1. Negligible mass transfer resistance.
2. The adsorbent is initially adsorbate-free.
3. The fluid is in plug flow with no axial dispersion.

Assuming phase equilibrium conditions, a problem solution strategy may be used where the length of the column is expressed as a series of low-height adsorbent-packed sections treated as equilibrium units and stacked on top of each other in the column, each containing M kg adsorbent. The fluid feed enters the top section (or bottom, if gas or vapor) at a flow rate F m³/s and the sections are solved in sequence. The calculations are done for a residence time of t s in each section. The volume of fluid in each packed section is

$$S = Ft \text{ m}^3$$

As in the single-stage analysis, and assuming a Freundlich isotherm, Equations 19.9 and 19.3 are used on each section i (i inlet, $i + 1$ outlet)

$$c_{i+1} = c_i - (M/S)(q_{i+1} - q_i)$$

$$q_{i+1} = K(c_{i+1})^n$$

For the top or bottom section, the inlet is the fluid feed stream.

In order to complete the model, additional calculations at subsequent time intervals should be carried out. The calculations defined above reflect initial conditions at time interval #1. This should be followed by calculations for time intervals #2, #3, ... Justification for this procedure is the fact that the process is unsteady state. The analysis at each time interval shows the change in phase conditions along the column height, at a given time. Subsequent time interval calculations represent time progression, where the column performance can be observed as a function of elapsed time. This allows monitoring the column conditions stage-wise and time-wise.

During time interval #1 the fluid stream is sent to packing section 1. The exit fluid is sent to section 2, ... until the other end of the column is reached. For known values of c_0, q_0 (the starting values for section 1), the above equations are used to calculate c_1, q_1, then c_2, q_2, and so on.

During time interval #2, (same residence time as t) c_1 and q_1 are calculated based on the external fluid stream and the initial conditions c_1 and q_1 values from the previous time interval.

Additional time intervals are calculated and concentrations are determined for all the column sections. As more time intervals and/or more column sections are added, the adsorbent in the top sections and down the column becomes progressively closer to saturation with the solute, and the solute concentrations in the solution begin to rise. The effectiveness of the column starts dropping at that point, and this should allow the optimum column length to be determined.

Available experimental data can be used in preliminary calculations to fine tune parameters K, n, and t, as well as the number of sections and time intervals. Moreover, equilibrium isotherms other than the Freundlich may be tested. The tuned parameters can then be used for calculations of the actual industrial operation. For example, for the same fluid and adsorbent materials and conditions used in the preliminary calculations, and for the actual fluid stream flow rate, a column diameter is determined for the actual operation that will result in the same fluid velocity as in the preliminary calculations. Additional calculations at increments of adsorbent packing and elapsed time can be carried out as described earlier in this section. The object is to find a packing height and running time that would minimize the solute concentration in the product fluid.

REFERENCES

Collins, J., *Chem. Eng. Progr. Symp. Series*, 63(74), 31, 1967.

Cusack, R. W. and A. E. Karr, *Chem. Eng.*, 98(4), 112, 1991.

Perry, R. H. and D. Green, *Perry's Chemical Engineers' Handbook*, 7th edition, New York, McGraw-Hill Book Company, 1997.

Reynolds, T. D., *Unit Operations and Processes in Environmental Engineering*, Boston, PWS Publishers, 1982.

Treybal, R. E., *Mass Transfer Operations*, 3rd edition, New York, McGraw-Hill Book Company, 1980

Index